Electromagnetics Engineering Handbook

Analysis and Design of Electrical and Electronic Devices and Systems

About the Authors

Prof. Paul R.P. Hoole was born in Jaffna, Sri Lanka in 1958. After having his basic schooling in Jaffna, he earned all his degrees, first degree to postgraduate, in the United Kingdom. He holds an MSc degree in Electrical Engineering with a Mark of Distinction from the University of London and an MSc degree in Plasma Science from University of Oxford. His doctorate, the DPhil. degree, is from the University of Oxford. In his engineering career he has spent time in Singapore, Papua New Guinea, USA, Sri Lanka and Malaysia. His most recent published research has been in the areas of electromagnetic wireless communication/navigation antennas, electromagnetic brain technology and science, climate electromagnetics and lightning-aircraft electrodynamics. Beyond the time he devotes to engineering teaching and research, Prof. Hoole also spends time studying and teaching the Bible applied to contemporary times in seminaries and churches. He is married to a medical doctor, Chrishanthy, and they have three children: Esther, Ezekiel and Elisabeth.

Dr K. Pirapaharan, BSc. Eng. Hons (Peradeniya), M.Eng (Kinki, Japan), PhD (Kinki, Japan), is Senior Lecturer at School of Engineering, Taylor's University in Malaysia. His PhD degree is in the area of microwave and millimeter waves. From 2001 to 2003, he was a postdoctoral research associate at the Centre for Computational Electromagnetics, University of Illinois at Urbana Champaign, USA. From 2004 to 2011, he was a Senior Lecturer in the Electrical and Information Engineering Department at University of Ruhuna, Sri Lanka where he was the head of the department from 2005 to 2008. Dr Pirapaharan's research interests include wave propagation in inhomogeneous media, adaptive antenna techniques and computational electromagnetics. He is a member of the IEEE and IET.

Prof. S. Ratnajeevan H. Hoole, BSc. Eng. Hons Cey., MSc with Mark of Distinction London, PhD. Carnegie Mellon, is Professor of Electrical and Computer Engineering at Michigan State University in the US. For his accomplishments in electromagnetic product synthesis the University of London awarded him its higher doctorate, the DSc. (Eng.) degree, in 1993, and the IEEE elevated him to the grade of Fellow in 1995. Prof. Hoole has been Vice Chancellor of University of Jaffna in Sri Lanka, and as Member of the University Grants Commission there, was responsible with six others for the regulation of the administration of all 15 Sri Lankan universities and their admissions and funding. He has contributed widely to the learned literature on Tamil studies and been a regular columnist in newspapers. Prof. Hoole has been trained in Human Rights Research and Teaching at The René Cassin International Institute of Human Rights, Strasbourg, France, and has pioneered teaching human rights in the engineering curriculum.

Electromagnetics Engineering Handbook

Analysis and Design of Electrical and Electronic Devices and Systems

P.R.P. Hoole
University of Malaya, Malaysia

K. Pirapaharan
Taylor's University, Malaysia

S.R.H. Hoole
Michigan State University, USA

WIT*PRESS* Southampton, Boston

P.R.P. Hoole
University of Malaya, Malaysia

K. Pirapaharan
Taylor's University, Malaysia

S.R.H. Hoole
Michigan State University, USA

Published by

WIT Press
Ashurst Lodge, Ashurst, Southampton, SO40 7AA, UK
Tel: 44 (0) 238 029 3223; Fax: 44 (0) 238 029 2853
E-Mail: witpress@witpress.com
http://www.witpress.com

For USA, Canada and Mexico

WIT Press
25 Bridge Street, Billerica, MA 01821, USA
Tel: 978 667 5841; Fax: 978 667 7582
E-Mail: infousa@witpress.com
http://www.witpress.com

British Library Cataloguing-in-Publication Data

A Catalogue record for this book is available
from the British Library

ISBN: 978-1-84564-798-8
eISBN: 978-1-84564-799-5

Library of Congress Catalog Card Number: 2013934186

The texts of the papers in this volume were set individually by the authors or under their supervision.

Printed in Great Britain by Lightning Source UK.

To our children

Mahilmany, Elilini, Anbini and Yovahn (SRHH)

Vicknesh (KP)

Esther, Ezekiel and Elisabeth (PRPH)

and to all our students

Great are the works of the LORD; they are studied by all who delight in them.
Full of splendor and majesty is His work. (Psalm 111: verses 2, 3a)

Contents

Introduction

Electromagnetic fields, both static and dynamic, form the foundational basis of all electrical and electronic engineering devices and systems. The book aims to present an in-depth, simple and comprehensive reference book on electromagnetics engineering. The audience which the book seeks to address includes undergraduate students, university teachers, and design and consultant engineers and researchers. In much of electrical and electronics engineering – including analogue and digital telecommunications engineering, biomedical monitoring and diagnostic equipment, power systems engineering, and sensor technology, getting back to the fundamental principles that govern these technologies, namely electromagnetic fields and waves – has become crucial for future, customer-friendly technology and systems.

To give a few examples, consider first wireless, mobile technology: radio frequency identifiers, microwave safety and security systems, computing and communications. Whereas probability mathematics helped the field to handle additive and multiplicative noise, for instance, the need now to have better electromagnetic models of signals and electromagnetic image/signal processing has become pertinent to design and implement multi-input, multi-output systems, and handle fast-moving mobile transceivers (transmitter/receivers), and to reduce the computational burden (both memory wise and in power consumption), and increase the computational speed and accuracy of the technical software that takes a center stage in such systems. Another example is in the field of medical diagnosis, where monitoring and treatment tools depend on electromagnetics engineering. The future systems that will depend on accurate manipulation and use of electromagnetic fields and waves include noninvasive hyperthermia treatment, permanent burning of cancer cells, and localization of the brain signals and their changes for better diagnosis and early treatment of neurological illnesses. A third example is the use of electromagnetic sensors and safety systems for vehicle collision avoidance and for the early warning of impending (man-made/natural) disasters, disaster management and intelligent decision-making equipment. In order to understand, design and implement new technology, in-depth knowledge and application of electromagnetic fields and waves are important. The following portions of the book (Chapters 9 and 10) are written in research paper style since they are still being extensively researched into.

The book aims to be simple so that an undergraduate student studying electromagnetics engineering for the first time will find this book easy reading to understand the essentials of the largely invisible, but powerful, electromagnetic fields, governed by the four elegant Maxwell's equations. Moreover, this book also helps to apply that knowledge through analytical and computational solutions of these frequency and material dependent electric and magnetic fields. As electrical and electronic engineering grows and subdivides into many specialities, this book aims to inform the reader the basic principles that govern all these specialized systems, and on how to apply that knowledge to understand and design both devices and systems that may operate at vastly different frequencies, in different media (e.g., semiconductor materials, magnetic materials, biological tissues, outer space and seawater). It is also aimed at very different functions depending on the area of application. For example, at very low-power frequencies, electromagnetic fields perform vastly different functions from device to device, such as in power transformers, current transformers, infrared sensors, synchronous generators, superconducting devices, electric motors and electric powered transport systems. Our hope is that this handbook will be of much help to students, engineers, innovators and researchers working in very different disciplines.

Having taught and researched and done consultation work on electromagnetic fields and waves applied to engineering problems ranging from direct current to daylight devices and systems, we seek to consider the subject in depth both in its mathematical formulation and analysis, to come to a firm understanding of electromagnetic engineering, as well as to gain the intuition and knowledge needed to integrate the basic principles for conceiving, designing, implementing and operating highly complex electrical and electronic devices and systems. An in-depth and comprehensive view of devices and systems is presented through the function and interaction of the two physical quantities of electric field intensity (**E**) and magnetic flux density (**B**), and their mathematical companions electric flux density (**D**) and magnetic field intensity (**H**). The material presented in this book will help undergraduates, engineers and research scientists understand electromagnetic fields and their applications, and address present and future electrical and electronic devices and systems using the foundational principles of electromagnetics.

We have made every possible effort to present a book without errors. But inadvertently they may have crept in. We have also tried to cover the broad range of engineering practice in which electromagnetic engineering plays a key role. However, if the reader should detect any errors, or identify any important area of application that has not been addressed in this handbook, the authors will be grateful to get feedback, so that we could correct and expand the handbook if a revised, future edition of the book is done. The personal e-mail of the first author may be used toward this end: prphoole@gmail.com.

1

Vector analysis

Vector analysis is a mathematical tool with which electromagnetic concepts are most conveniently expressed and best comprehended. Vector analysis may be divided into two general categories such as vector algebra and vector calculus. Vector algebra concerns algebraic operations on vectors, while vector calculus concerns differential and integral operations involving vector functions. Electromagnetic fields are functions of positions in space as well as in time. Various vector calculus operations allow us to deliberately state the fundamental laws governing these electromagnetic fields and to perform calculations involving these field quantities.

1.1 Vectors and scalars

Quantities such as time, mass, temperature, charge, electric potential and energy are scalars, while velocity, force, displacement, electric field intensity and flux density are vectors. A vector quantity contains two items of information: a direction and magnitude. On the other hand, a scalar quantity possesses no directional component.

Thus a vector $\mathbf{A}$ has both magnitude and direction. The magnitude of $\mathbf{A}$ is a scalar written as $|\mathbf{A}|$. A unit vector $\mathbf{u}$ along $\mathbf{A}$ is defined as a vector whose magnitude is unity. That is,

$$\mathbf{u} = \frac{\mathbf{A}}{|\mathbf{A}|} \tag{1.1}$$

1.2 Orthogonal coordinate system

If we are to utilize vector concepts fully in understanding electromagnetic field behavior, we must introduce a coordinate system. Coordinate systems enable us to locate a point in three-dimensional space as the intersection of three surfaces. The coordinate variables in the generalized coordinate systems that we will discuss are denoted as n_1, n_2 and n_3. For reasons that will soon become apparent, it is desirable to choose the respective surfaces of the coordinate system to be orthogonal to each other at every point in the space. This is referred to as an orthogonal coordinate system.

In order to define a vector $\mathbf{A}$ in an orthogonal coordinate system, unit vectors $\mathbf{u}_{n1}$, $\mathbf{u}_{n2}$ and $\mathbf{u}_{n3}$ are introduced. These unit vectors are of unit length and are defined orthogonal to the appropriate surface. Thus we may decompose **A** into components that are tangential to these unit vectors:

$$\mathbf{A} = A_{n1}\mathbf{u}_{n1} + A_{n2}\mathbf{u}_{n2} + A_{n3}\mathbf{u}_{n3} \tag{1.2}$$

Quite often we are interested in differential path lengths, surface areas and volumes in these coordinate systems. Some of these coordinates may not be lengths. Thus, we need a conversion coefficient to relate a differential change in a coordinate variable, $\mathrm{d}n_i$, to a differential change in length along that coordinate surface, $\mathrm{d}l_i$. These conversion factors are denoted for the coordinate surface as h_i. Thus,

$$\mathrm{d}l_i = h_i\,\mathrm{d}n_i \tag{1.3}$$

A vector representing a differential change in path in some arbitrary direction becomes

$$\begin{aligned}\mathrm{d}\mathbf{l} &= \mathrm{d}l_1\,\mathbf{u}_{n1} + \mathrm{d}l_2\,\mathbf{u}_{n2} + \mathrm{d}l_3\,\mathbf{u}_{n3}\\ &= h_1\,\mathrm{d}n_1\,\mathbf{u}_{n1} + h_2\,\mathrm{d}n_2\,\mathbf{u}_{n2} + h_3\,\mathrm{d}n_3\,\mathbf{u}_{n3}\end{aligned} \tag{1.4}$$

If the surfaces are incremented by the differential amounts $\mathrm{d}n_1$, $\mathrm{d}n_2$ and $\mathrm{d}n_3$ in directions $\mathbf{u}_{n1}$, $\mathbf{u}_{n2}$ and $\mathbf{u}_{n3}$, the differential volume is formed:

$$\begin{aligned}\mathrm{d}v &= \mathrm{d}l_1\,\mathrm{d}l_2\,\mathrm{d}l_3\\ &= h_1\,h_2\,h_3\,\mathrm{d}n_1\,\mathrm{d}n_2\,\mathrm{d}n_3\end{aligned} \tag{1.5}$$

Similarly, the differential surface areas are

$$\begin{aligned}\mathrm{d}s_1 &= \mathrm{d}l_2\,\mathrm{d}l_3\\ &= h_2\,h_3\,\mathrm{d}n_2\,\mathrm{d}n_3\end{aligned} \tag{1.6a}$$

$$\begin{aligned}\mathrm{d}s_2 &= \mathrm{d}l_1\,\mathrm{d}l_3\\ &= h_1\,h_3\,\mathrm{d}n_1\,\mathrm{d}n_3\end{aligned} \tag{1.6b}$$

$$\begin{aligned}\mathrm{d}s_3 &= \mathrm{d}l_1\,\mathrm{d}l_2\\ &= h_1\,h_2\,\mathrm{d}n_1\,\mathrm{d}n_2\end{aligned} \tag{1.6c}$$

However, the major interest in differential surface areas involves the normal to the surfaces. Looking forward to this requirement, we state

$$\mathrm{d}\mathbf{s}_1 = h_2\,h_3\,\mathrm{d}n_2\,\mathrm{d}n_3\,\mathbf{u}_{n1} \tag{1.7a}$$

$$\mathrm{d}\mathbf{s}_2 = h_1\,h_3\,\mathrm{d}n_1\,\mathrm{d}n_3\,\mathbf{u}_{n2} \tag{1.7b}$$

$$\mathrm{d}\mathbf{s}_3 = h_1\,h_2\,\mathrm{d}n_1\,\mathrm{d}n_2\,\mathbf{u}_{n3} \tag{1.7c}$$

1.3 The Cartesian coordinate system

The Cartesian coordinate system is possibly the most common orthogonal coordinate system. This is a right-handed coordinate system in the sequence of $x \to y \to z$. The unit vectors $\mathbf{u}_x$, $\mathbf{u}_y$ and $\mathbf{u}_z$ are orthogonal to appropriate planes and are pointed in the direction of increasing coordinate value. A vector in this coordinate system can be written as (Fig. 1.1)

$$\mathbf{A} = A_x\mathbf{u}_x + A_y\mathbf{u}_y + A_z\mathbf{u}_z \tag{1.8}$$

The conversion factors h_x, h_y and h_z are all unity since each coordinate variable represents a length:

$$h_x = h_y = h_z = 1 \tag{1.9}$$

Therefore, the differential arc length is

$$\mathrm{d}\mathbf{l} = \mathrm{d}x\,\mathbf{u}_x + \mathrm{d}y\,\mathbf{u}_y + \mathrm{d}z\,\mathbf{u}_z \tag{1.10}$$

The differential surface elements turn into

$$\mathrm{d}\mathbf{s}_x = \mathrm{d}y\,\mathrm{d}z\,\mathbf{u}_x \tag{1.11a}$$

$$\mathrm{d}\mathbf{s}_y = \mathrm{d}x\,\mathrm{d}z\,\mathbf{u}_y \tag{1.11b}$$

$$\mathrm{d}\mathbf{s}_z = \mathrm{d}x\,\mathrm{d}y\,\mathbf{u}_z \tag{1.11c}$$

and the differential volume element is

$$\mathrm{d}v = \mathrm{d}x\,\mathrm{d}y\,\mathrm{d}z \tag{1.12}$$

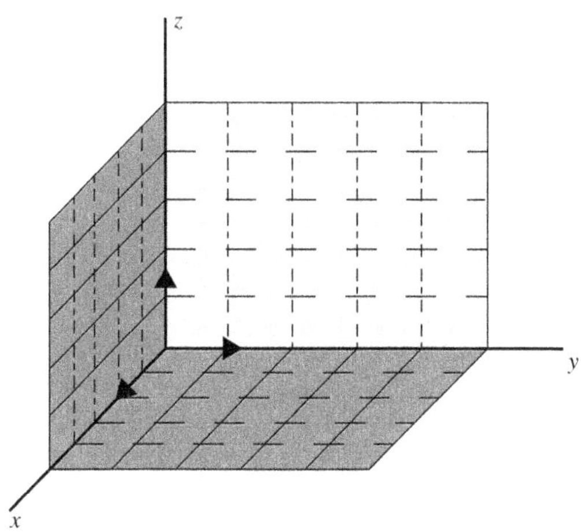

Figure 1.1: Illustration of the Cartesian coordinate system.

1.4 The cylindrical coordinate system

The cylindrical coordinate system is very convenient whenever we are dealing with problems having cylindrical symmetry. The surfaces used in this coordinate system consist of a plane of constant height z, a cylindrical surface of radius r with the z axis as its axis and a half plane perpendicular to the xy plane at an angle ϕ with respect to the xz plane. This is also a right-handed coordinate system in the sequence of $r \rightarrow \phi \rightarrow z$. Unit vectors $\mathbf{u}_r$, $\mathbf{u}_\phi$ and $\mathbf{u}_z$ are orthogonal at every point. However, the directions of the unit vectors $\mathbf{u}_r$ and $\mathbf{u}_\phi$ vary point to point in contrast with the Cartesian coordinate system (Fig. 1.2).

A vector in this coordinate system can be written as

$$\mathbf{A} = A_r \mathbf{u}_r + A_\phi \mathbf{u}_\phi + A_z \mathbf{u}_z \tag{1.13}$$

The conversion factors h_r and h_z are unity since these two coordinate variables represent lengths. However, the ϕ coordinate is an angle. A differential change in length in the ϕ direction on this cylinder of radius r is given by $r\,\mathrm{d}\phi$. Therefore,

$$h_r = h_z = 1 \quad h_\phi = r \tag{1.14}$$

Hence, the differential arc length is

$$\mathrm{d}\mathbf{l} = \mathrm{d}r\,\mathbf{u}_r + r\,\mathrm{d}\phi\,\mathbf{u}_\phi + \mathrm{d}z\,\mathbf{u}_z \tag{1.15}$$

The differential surface elements turn into

$$\mathrm{d}\mathbf{s}_r = r\,\mathrm{d}\phi\,\mathrm{d}z\,\mathbf{u}_r \tag{1.16a}$$

$$\mathrm{d}\mathbf{s}_\phi = \mathrm{d}r\,\mathrm{d}z\,\mathbf{u}_\phi \tag{1.16b}$$

$$\mathrm{d}\mathbf{s}_z = r\,\mathrm{d}r\,\mathrm{d}\phi\,\mathbf{u}_z \tag{1.16c}$$

and the differential volume element is

$$\mathrm{d}v = r\,\mathrm{d}r\,\mathrm{d}\phi\,\mathrm{d}z \tag{1.17}$$

A vector represented in the Cartesian coordinate system can be transformed to an equivalent one in the cylindrical coordinate system:

$$A_r = A_x \cos\phi + A_y \sin\phi \tag{1.18a}$$

$$A_\phi = -A_x \sin\phi + A_y \cos\phi \tag{1.18b}$$

$$A_z = A_z \tag{1.18c}$$

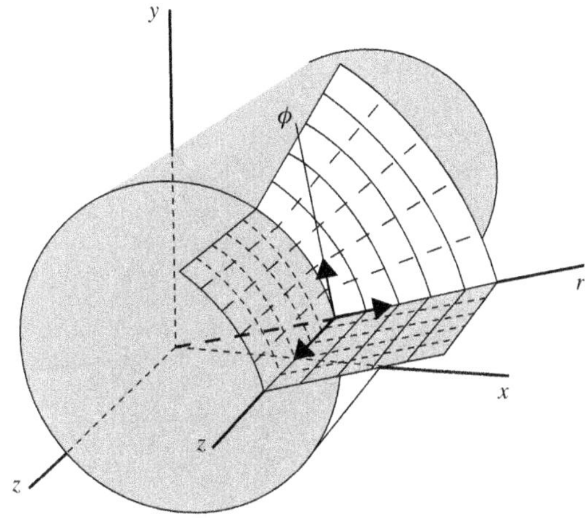

Figure 1.2: Illustration of the cylindrical coordinate system.

and vice versa

$$A_x = A_r \cos\phi - A_\phi \sin\phi \tag{1.19a}$$

$$A_y = A_r \sin\phi + A_\phi \cos\phi \tag{1.19b}$$

$$A_z = A_z \tag{1.19c}$$

Also, we convert the x, y and z variables in each of the components to the r, ϕ and z variables using

$$r = \sqrt{x^2 + y^2} \tag{1.20a}$$

$$\phi = \tan^{-1}\left(\frac{y}{x}\right) \tag{1.20b}$$

$$z = z \tag{1.20c}$$

and vice versa

$$x = r\cos\phi \tag{1.21a}$$

$$y = r\sin\phi \tag{1.21b}$$

$$z = z \tag{1.21c}$$

1.5 The spherical coordinate system

The spherical coordinate system is the most appropriate when dealing with problems having a degree of spherical symmetry. The surfaces used to define a spherical

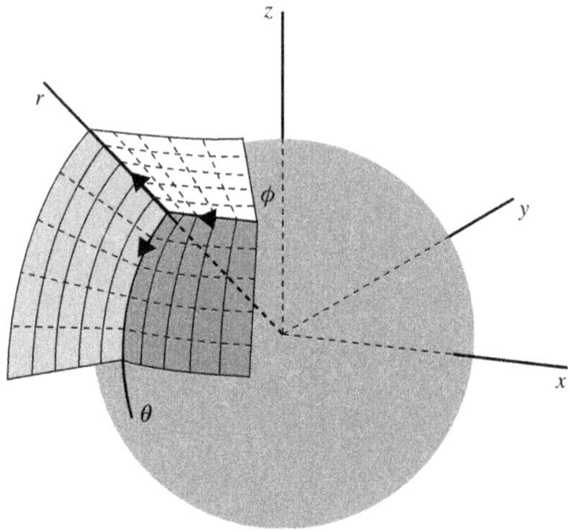

Figure 1.3: Illustration of the spherical coordinate system.

coordinate system consist of a sphere of radius r centered at the origin, a right circular cone with its apex at the origin with its axis as the z axis and having a half angle θ, and a half plane perpendicular to the xy plane and making an angle ϕ with the xz plane. Unit vectors $\mathbf{u}_r$, $\mathbf{u}_\theta$ and $\mathbf{u}_\phi$ are orthogonal at every point. However, the directions of the unit vectors $\mathbf{u}_r$, $\mathbf{u}_\theta$ and $\mathbf{u}_\phi$ vary point to point in contrast to the rectangular coordinate system (Fig. 1.3).

A vector in this coordinate system can be written as

$$\mathbf{A} = A_r\mathbf{u}_r + A_\theta\mathbf{u}_\theta + A_\phi\mathbf{u}_\phi \tag{1.22}$$

The conversion factors h_r is unity since the coordinate variable represents lengths. However, the θ and ϕ coordinates are angles. A differential change in lengths involves the coordinates. Therefore,

$$h_r = 1 \quad h_\theta = r \quad h_\phi = r\sin\theta \tag{1.23}$$

Hence, the differential arc length is

$$\mathbf{dl} = \mathrm{d}r\,\mathbf{u}_r + r\,\mathrm{d}\theta\,\mathbf{u}_\theta + r\sin\theta\,\mathrm{d}\phi\,\mathbf{u}_\phi \tag{1.24}$$

The differential surface elements turn into

$$\mathbf{ds}_r = r^2\sin\theta\,\mathrm{d}\theta\,\mathrm{d}\phi\,\mathbf{u}_r \tag{1.25a}$$

$$\mathbf{ds}_\theta = r\sin\theta\,\mathrm{d}r\,\mathrm{d}\phi\,\mathbf{u}_\theta \tag{1.25b}$$

$$\mathbf{ds}_z = r\,\mathrm{d}r\,\mathrm{d}\theta\,\mathbf{u}_\phi \tag{1.25c}$$

and the differential volume element is

$$dv = r^2 \sin\theta \, dr \, d\theta \, d\phi \tag{1.26}$$

A vector represented in the Cartesian coordinate system can be transformed to an equivalent one in the spherical coordinate system:

$$A_r = A_x \sin\theta \cos\phi + A_y \sin\theta \sin\phi + A_z \cos\theta \tag{1.27a}$$

$$A_\theta = A_x \cos\theta \cos\phi + A_y \cos\theta \sin\phi - A_z \sin\theta \tag{1.27b}$$

$$A_\phi = -A_x \sin\phi + A_y \cos\phi \tag{1.27c}$$

and vice versa

$$A_x = A_r \sin\theta \cos\phi + A_\theta \cos\theta \cos\phi - A_\phi \sin\phi \tag{1.28a}$$

$$A_y = A_r \sin\theta \sin\phi + A_\theta \cos\theta \sin\phi - A_\phi \cos\phi \tag{1.28b}$$

$$A_z = A_r \cos\theta - A_\theta \sin\theta \tag{1.28c}$$

Also, we convert the x, y and z variables in each of the components to the r, θ and ϕ variables using

$$r = \sqrt{x^2 + y^2 + z^2} \tag{1.29a}$$

$$\theta = \tan^{-1}\left(\frac{\sqrt{x^2 + y^2}}{z}\right) \tag{1.29b}$$

$$\phi = \tan^{-1}\left(\frac{y}{x}\right) \tag{1.29c}$$

and vice versa

$$x = r \sin\theta \cos\phi \tag{1.30a}$$

$$y = r \sin\theta \sin\phi \tag{1.30b}$$

$$z = r \cos\theta \tag{1.30c}$$

1.6 Products of vectors

The product of two vectors takes two forms: the dot product and the cross product. Thus two definitions of vector products lie in their appearance in the equations governing electromagnetic fields.

1.6.1 The scalar or dot product

The dot product of two vectors **A** and **B** is the product of the magnitudes of **A** and **B** and the cosine of the smallest angle between them. The outcome of the dot product

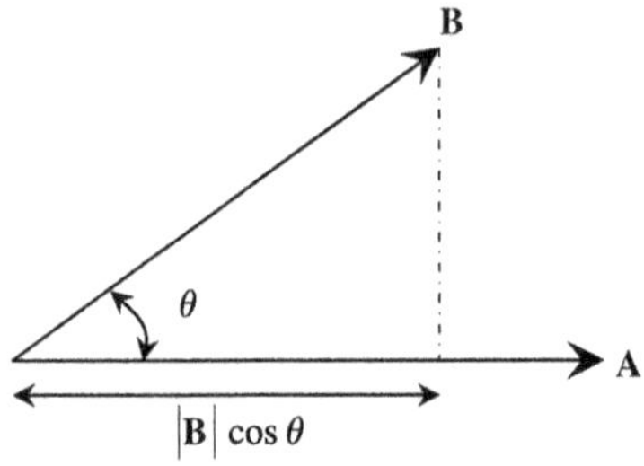

Figure 1.4: Schematic illustration of the dot product.

of vectors is a scalar. Thus this product rule is called the *scalar product* (Fig. 1.4).

$$\mathbf{A} \cdot \mathbf{B} = |\mathbf{A}||\mathbf{B}|\cos\theta \tag{1.31}$$

Hence, $\mathbf{A} \cdot \mathbf{A} = |\mathbf{A}|^2$

$$|\mathbf{A}| = \sqrt{\mathbf{A} \cdot \mathbf{A}} \tag{1.32}$$

The angle between two vectors can therefore be given as

$$\theta = \cos^{-1}\left(\frac{\mathbf{A} \cdot \mathbf{B}}{|\mathbf{A}||\mathbf{B}|}\right) \tag{1.33}$$

The dot product can also be interpreted as the product of the magnitude of **A** and the magnitude of the projection of **B** onto **A** or vice versa.

For orthogonal coordinate systems

$$\mathbf{u}_{n1} \cdot \mathbf{u}_{n1} = \mathbf{u}_{n2} \cdot \mathbf{u}_{n2} = \mathbf{u}_{n3} \cdot \mathbf{u}_{n3} = 1 \tag{1.34}$$

$$\mathbf{u}_{n1} \cdot \mathbf{u}_{n2} = \mathbf{u}_{n2} \cdot \mathbf{u}_{n3} = \mathbf{u}_{n3} \cdot \mathbf{u}_{n1} = 0 \tag{1.35}$$

Thus, we obtain the dot product in the Cartesian coordinate system as

$$\mathbf{A} \cdot \mathbf{B} = A_x B_x + A_y B_y + A_z B_z \tag{1.36a}$$

Similarly, in cylindrical and spherical coordinate systems, we respectively obtain

$$\mathbf{A} \cdot \mathbf{B} = A_r B_r + A_\phi B_\phi + A_z B_z \tag{1.36b}$$

$$\mathbf{A} \cdot \mathbf{B} = A_r B_r + A_\theta B_\theta + A_\phi B_\phi \tag{1.36c}$$

1.6.2 The vector or cross product

The cross product is classified as a vector product since the cross product of two vectors yields another vector as the result. The cross product of two vectors **A** and

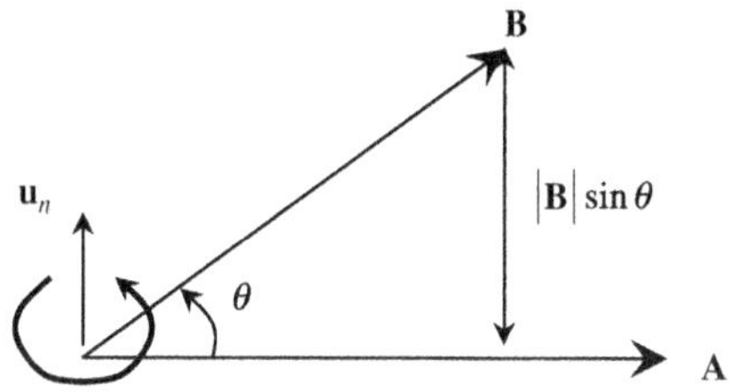

Figure 1.5: Schematic illustration of the cross product.

B is defined as

$$\mathbf{A} \times \mathbf{B} = |\mathbf{A}||\mathbf{B}|\sin\theta\, \mathbf{u}_n \tag{1.37}$$

where $\mathbf{u}_n$ is a unit vector normal to the plane that contains **A** and **B** as shown in Fig. 1.5. The direction of $\mathbf{u}_n$ is the direction that a right-hand screw would advance if turned in the direction from **A** to **B**. Since the order of multiplication is important in determining the direction of the resultant vector, the directions of resultant vectors are opposite to each other when the order of the product is exchanged. Hence,

$$\mathbf{A} \times \mathbf{B} = -\mathbf{B} \times \mathbf{A} \tag{1.38}$$

The angle θ between two vectors **A** and **B** could be obtained from the cross product as

$$\theta = \sin^{-1}\left(\frac{|\mathbf{A} \times \mathbf{B}|}{|\mathbf{A}||\mathbf{B}|}\right) \tag{1.39}$$

The cross product can be thought of as projections. The magnitude of $\mathbf{A} \times \mathbf{B}$ is the product of magnitude of **A** and the length of the projection of **B** onto a line perpendicular to **A** or vice versa.

In orthogonal coordinate systems, the cross products of the unit vector yield

$$\mathbf{u}_{n1} \times \mathbf{u}_{n1} = \mathbf{u}_{n2} \times \mathbf{u}_{n2} = \mathbf{u}_{n3} \times \mathbf{u}_{n3} = 0 \tag{1.40}$$

$$\begin{aligned} \mathbf{u}_{n1} \times \mathbf{u}_{n2} &= \mathbf{u}_{n3} \\ \mathbf{u}_{n2} \times \mathbf{u}_{n3} &= \mathbf{u}_{n1} \\ \mathbf{u}_{n3} \times \mathbf{u}_{n1} &= \mathbf{u}_{n2} \end{aligned} \tag{1.41}$$

With the results in (1.40) and (1.41), we can obtain the cross product of two vectors in any orthogonal coordinate system as

$$\mathbf{A} \times \mathbf{B} = (A_{n2}B_{n3} - A_{n3}B_{n2})\mathbf{u}_{n1} + (A_{n3}B_{n1} - A_{n1}B_{n3})\mathbf{u}_{n2} + (A_{n1}B_{n2} - A_{n2}B_{n1})\mathbf{u}_{n3} \tag{1.42}$$

Alternatively, the component of the cross product can be given by using the following determinant rule

$$\mathbf{A} \times \mathbf{B} = \begin{vmatrix} \mathbf{u}_{n1} & \mathbf{u}_{n2} & \mathbf{u}_{n3} \\ A_{n1} & A_{n2} & A_{n3} \\ B_{n1} & B_{n2} & B_{n3} \end{vmatrix} \tag{1.43}$$

The evaluation of the cross product when the two vectors are expressed in Cartesian, cylindrical and spherical coordinate systems is identical to that in the general orthogonal coordinate system. Thus the respective cross product results of the Cartesian, cylindrical and spherical coordinate systems are given

$$\mathbf{A} \times \mathbf{B} = (A_y B_z - A_z B_y)\mathbf{u}_x + (A_z B_x - A_x B_z)\mathbf{u}_y + (A_x B_y - A_y B_x)\mathbf{u}_z \tag{1.44a}$$

$$\mathbf{A} \times \mathbf{B} = (A_\phi B_z - A_z B_\phi)\mathbf{u}_r + (A_z B_r - A_r B_z)\mathbf{u}_\phi + (A_r B_\phi - A_\phi B_r)\mathbf{u}_z \tag{1.44b}$$

$$\mathbf{A} \times \mathbf{B} = (A_\theta B_\phi - A_\phi B_\theta)\mathbf{u}_r + (A_\phi B_r - A_r B_\phi)\mathbf{u}_\theta + (A_r B_\theta - A_\theta B_r)\mathbf{u}_\phi \tag{1.44c}$$

1.7 Fields

There are two types of fields we will deal with in electromagnetic field theory; these are scalar fields and vector fields.

1.7.1 Scalar field

The essential concept of a scalar field is that the quantity of interest is a scalar quantity defined over a continuum since it has no direction associated with its amplitude of values that depends on the location of the point of interest and the time of observation.

1.7.2 Vector field

The basic difference between a scalar field and a vector field is that the quantity of interest in a vector field is assumed to have a directional property as well as a magnitude at the location of the point of interest and the time of observation.

1.8 The gradient of the scalar field

The rate of change of the scalar field with respect to its orthogonal coordinate distances at a point is defined as the gradient of a scalar field. Since the rate of change is different in each direction, the gradient is a vector. There are numerous examples of scalar electromagnetic fields for which we would also like to determine the gradient properties. These properties are determined from gradient operators.

Consider a scalar field $f(n_1, n_2, n_3)$ defined in a generalized orthogonal coordinate system. The total differential change of $f(n_1, n_2, n_3)$ is

$$\begin{aligned} df &= \frac{\partial f}{\partial n_1} dn_1 + \frac{\partial f}{\partial n_2} dn_2 + \frac{\partial f}{\partial n_3} dn_3 \\ &= \frac{\partial f}{\partial l_1} dl_1 + \frac{\partial f}{\partial l_2} dl_2 + \frac{\partial f}{\partial l_3} dl_3 \end{aligned} \tag{1.45}$$

where $dl_i = h_i \, dn_i$.

Further df can be written as

$$\begin{aligned} df &= \left(\frac{\partial f}{\partial l_1} \mathbf{u}_{n1} + \frac{\partial f}{\partial l_2} \mathbf{u}_{n2} + \frac{\partial f}{\partial l_3} \mathbf{u}_{n3} \right) \cdot (dl_1 \mathbf{u}_{n1} + dl_2 \mathbf{u}_{n2} + dl_3 \mathbf{u}_{n3}) \\ &= \left(\frac{\partial f}{\partial l_1} \mathbf{u}_{n1} + \frac{\partial f}{\partial l_2} \mathbf{u}_{n2} + \frac{\partial f}{\partial l_3} \mathbf{u}_{n3} \right) \cdot \mathbf{dl} \end{aligned} \tag{1.46}$$

The vector component $\left(\frac{\partial f}{\partial l_1} \mathbf{u}_{n1} + \frac{\partial f}{\partial l_2} \mathbf{u}_{n2} + \frac{\partial f}{\partial l_3} \mathbf{u}_{n3} \right)$ is defined as the gradient of $f(n_1, n_2, n_3)$ and will be given by the symbol

$$\mathbf{grad}\, f = \left(\frac{\partial f}{\partial l_1} \mathbf{u}_{n1} + \frac{\partial f}{\partial l_2} \mathbf{u}_{n2} + \frac{\partial f}{\partial l_3} \mathbf{u}_{n3} \right) \tag{1.47}$$

Thus we may write

$$df = \mathbf{grad}\, f \cdot \mathbf{dl} \tag{1.48}$$

As a shortened notation, we will denote

$$\mathbf{grad}\, f = \nabla f \tag{1.49}$$

Substituting the conversion factors $dl_i = h_i \, dn_i$ into (1.47), we obtain

$$\mathbf{grad}\, f(n_1, n_2, n_3) = \nabla f(n_1, n_2, n_3) = \left(\frac{1}{h_1} \frac{\partial f}{\partial n_1} \mathbf{u}_{n1} + \frac{1}{h_2} \frac{\partial f}{\partial n_2} \mathbf{u}_{n2} + \frac{1}{h_3} \frac{\partial f}{\partial n_3} \mathbf{u}_{n3} \right) \tag{1.50}$$

The evaluation of the gradient operation for the Cartesian, cylindrical and spherical coordinate systems is derivable from the operation for the general orthogonal coordinate system. Thus the respective gradient operations for the Cartesian, cylindrical and spherical coordinate systems are given by

$$\nabla f(x, y, z) = \left(\frac{\partial f}{\partial x}\mathbf{u}_x + \frac{\partial f}{\partial y}\mathbf{u}_y + \frac{\partial f}{\partial z}\mathbf{u}_z \right) \tag{1.51a}$$

$$\nabla f(r, \phi, z) = \left(\frac{\partial f}{\partial r}\mathbf{u}_r + \frac{1}{r}\frac{\partial f}{\partial \phi}\mathbf{u}_\phi + \frac{\partial f}{\partial z}\mathbf{u}_z \right) \tag{1.51b}$$

$$\nabla f(r, \theta, \phi) = \left(\frac{\partial f}{\partial r}\mathbf{u}_r + \frac{1}{r}\frac{\partial f}{\partial \theta}\mathbf{u}_\theta + \frac{1}{r\sin\theta}\frac{\partial f}{\partial \phi}\mathbf{u}_\phi \right) \tag{1.51c}$$

1.9 The line integral of a vector field

The evaluation of the line integral of a vector field **F** in an orthogonal coordinate system is straightforward. In the generalized orthogonal system,

$$\begin{aligned} \int_a^b \mathbf{F}(n_1, n_2, n_3) \cdot d\mathbf{l} &= \int_a^b (F_{n1}\mathbf{u}_{n1} + F_{n2}\mathbf{u}_{n2} + F_{n3}\mathbf{u}_{n3}) \cdot (\mathrm{d}l_1\mathbf{u}_{n1} + \mathrm{d}l_2\mathbf{u}_{n2} + \mathrm{d}l_3\mathbf{u}_{n3}) \\ &= \int_{l_1} F_{n1}\,\mathrm{d}l_1 + \int_{l_2} F_{n2}\,\mathrm{d}l_2 + \int_{l_3} F_{n3}\,\mathrm{d}l_3 \end{aligned} \tag{1.52}$$

Substituting the conversion factors, $\mathrm{d}l_i = h_i\,\mathrm{d}n_i$, we obtain

$$\int_a^b \mathbf{F}(n_1, n_2, n_3) \cdot \mathrm{d}\mathbf{l} = \int_{n_1} h_1 F_{n1}\,\mathrm{d}n_1 + \int_{n_2} h_2 F_{n2}\,\mathrm{d}n_2 + \int_{n_3} h_3 F_{n3}\,\mathrm{d}n_3 \tag{1.53}$$

Each of the above integrals requires integration with respect to the individual coordinates. Thus, the components F_{n1}, F_{n2} and F_{n3} must be converted to only one coordinate variable: n_1, n_1 and n_1, respectively. These conversions will be made using the function(s) of the integral path. In Cartesian, cylindrical and spherical coordinate systems, these become

$$\int_a^b \mathbf{F}(x, y, z) \cdot \mathrm{d}\mathbf{l} = \int_{x_a}^{x_b} F_x\,\mathrm{d}x + \int_{y_a}^{y_b} F_y\,\mathrm{d}y + \int_{z_a}^{z_b} F_z\,\mathrm{d}z \tag{1.54a}$$

$$\int_a^b \mathbf{F}(r, \phi, z) \cdot d\mathbf{l} = \int_{r_a}^{r_b} F_r\,\mathrm{d}r + \int_{\phi_a}^{\phi_b} rF_\phi\,\mathrm{d}\phi + \int_{z_a}^{z_b} F_z\,\mathrm{d}z \tag{1.54b}$$

$$\int_a^b \mathbf{F}(r, \theta, \phi) \cdot d\mathbf{l} = \int_{r_a}^{r_b} F_r\,\mathrm{d}r + \int_{\theta_a}^{\theta_b} rF_\theta\,\mathrm{d}\theta + \int_{\phi_a}^{\phi_b} r\sin\theta\, F_\phi\,\mathrm{d}\phi \tag{1.54c}$$

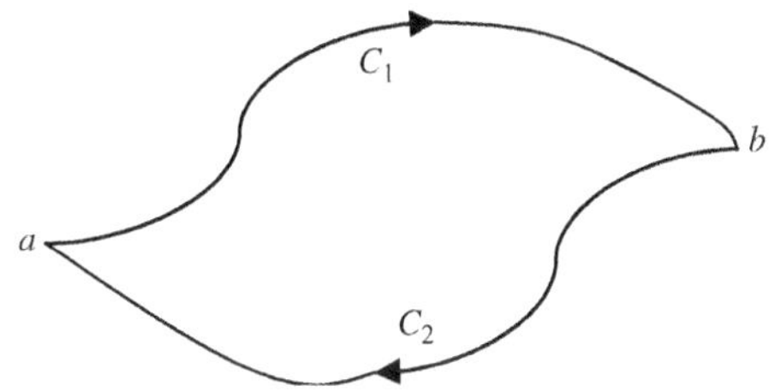

Figure 1.6: Line integral along a contour.

If we integrate the differential change of a scalar f along a contour, we obtain

$$\int_a^b \mathrm{d}f = \int_a^b \nabla f \cdot \mathrm{d}\mathbf{l} = f_b - f_a \tag{1.55}$$

Since the integral of $\mathrm{d}f$ depends only on the value of f at the end points of the contour, the result in (1.55) is independent of the path taken from a to b. If we choose a contour C_1 from a to b and a return along contour C_2 from b to a (Fig. 1.6), we find that

$$\int_{C_1} \mathrm{d}f + \int_{C_2} \mathrm{d}f = \oint \mathrm{d}f = \oint \nabla f \cdot \mathrm{d}\mathbf{l} = 0 \tag{1.56}$$

We thus find that the line integral of the gradient of a scalar field around a closed contour is identically zero. More importantly, we observe that if a vector field $\mathbf{F}$ can be written as the gradient of some scalar field of f such as

$$\mathbf{F} = \nabla f \tag{1.57}$$

then (1.56) shows that

$$\oint \nabla f \cdot \mathrm{d}\mathbf{l} = \oint \mathbf{F} \cdot \mathrm{d}\mathbf{l} = 0 \tag{1.58}$$

regardless of the contour path. Such a vector field is said to be conservative. Hence we find that if a vector field $\mathbf{F}$ can be written as the gradient of the scalar field f, then the line integral of $\mathbf{F}$ between two points is independent of the path taken between these two points. The line integral of $\mathbf{F} = \nabla f$ depends only on the value of f at the end points of the path:

$$\int_{a,C_1}^{b} \mathbf{F} \cdot \mathrm{d}\mathbf{l} = \int_{a,C_2}^{b} \mathbf{F} \cdot \mathrm{d}\mathbf{l} \quad \text{if } \mathbf{F} = \nabla f \tag{1.59}$$

1.10 Divergence of a vector field

It is very useful in the study of electromagnetic fields to determine the net outflow of the flux of a vector field through a given surface. Consider a vector field in region of space shown in Fig. 1.7. The flux of the vector field **F** through a surface S is defined as

$$\Phi = \int_S |\mathbf{F}| \cos\theta \, \mathrm{d}s = \int_S \mathbf{F} \cdot \mathbf{ds} \qquad (1.60)$$

where ds is a differential surface area at a point on the surface and θ is the angle between the vector **F** and the unit normal $\mathbf{u}_n$ to the surface at that point. The vector surface area is defined as, see Fig. 1.7,

$$\mathbf{ds} = \mathrm{d}s \, \mathbf{u}_n \qquad (1.61)$$

The component of **F** that is tangential to the surface contributes nothing to the net flux flow of the vector through the surface. Also there are two possibilities for the direction of **ds** for an open surface. Both are normal to the surface and opposite to each other as there are two sides to the surface. This ambiguity is resolved by considering a closed surface and defining **ds** as pointing out of the volume enclosed by surface, if the surface S is closed. Then the net outward flux Φ of the vector from the surface can be given as follows:

$$\Phi = \oint_S \mathbf{F} \cdot \mathbf{ds} \qquad (1.62)$$

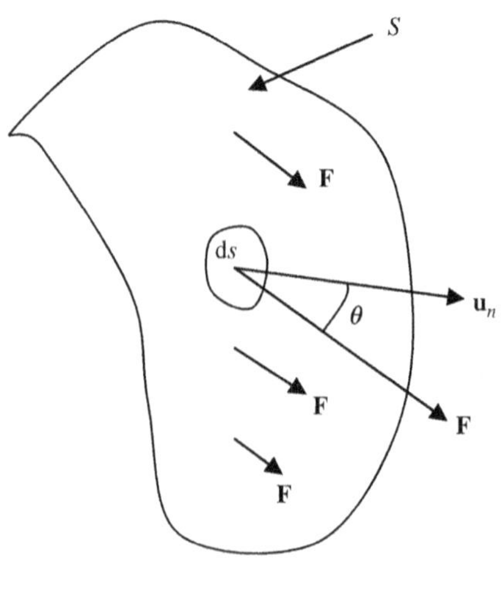

Figure 1.7: Flux of vector field through a surface.

Substituting $F_{n1}\mathbf{u}_{n1} + F_{n2}\mathbf{u}_{n2} + F_{n3}\mathbf{u}_{n3}$ and $\mathbf{ds} = \mathrm{d}l_2\,\mathrm{d}l_3\,\mathbf{u}_{n1} + \mathrm{d}l_3\,\mathrm{d}l_1\,\mathbf{u}_{n2} + \mathrm{d}l_1\,\mathrm{d}l_2\mathbf{u}_{n3}$ for $\mathbf{F}$ in (1.62), we obtain

$$\Phi = \oint_S F_{n1}\,\mathrm{d}l_2\,\mathrm{d}l_3 + F_{n2}\,\mathrm{d}l_3\,\mathrm{d}l_1 + F_{n3}\,\mathrm{d}l_1\,\mathrm{d}l_2 \tag{1.63}$$

Substituting for $\mathrm{d}l_i = h_i\,\mathrm{d}n_i$ in (1.63),

$$\Phi = \oint_S F_{n1}h_2h_3\,\mathrm{d}n_2\,\mathrm{d}n_3 + F_{n2}h_3h_1\,\mathrm{d}n_3\,\mathrm{d}n_1 + F_{n3}h_1h_2\,\mathrm{d}n_1\,\mathrm{d}n_2 \tag{1.64}$$

For a closed boundary of the surface, (1.64) can be further modified as

$$\begin{aligned}\Phi &= \iiint \frac{\partial(F_{n1}h_2h_3)}{\partial n_1}\mathrm{d}n_1\,\mathrm{d}n_2\,\mathrm{d}n_3 + \frac{\partial(F_{n2}h_3h_1)}{\partial n_2}\,\mathrm{d}n_2\,\mathrm{d}n_3\,\mathrm{d}n_1 \\ &\quad + \frac{\partial(F_{n3}h_1h_2)}{\partial n_3}\mathrm{d}n_3\,\mathrm{d}n_1\,\mathrm{d}n_2 \\ &= \iiint \frac{1}{h_1h_2h_3}\left[\frac{\partial(F_{n1}h_2h_3)}{\partial n_1} + \frac{\partial(F_{n2}h_3h_1)}{\partial n_2} + \frac{\partial(F_{n3}h_1h_2)}{\partial n_3}\right]h_1h_2h_3\,\mathrm{d}n_1\,\mathrm{d}n_2\,\mathrm{d}n_3\end{aligned} \tag{1.65}$$

Since $h_1h_2h_3\,\mathrm{d}n_1\,\mathrm{d}n_2\,\mathrm{d}n_3 = \mathrm{d}l_1\,\mathrm{d}l_2\,\mathrm{d}l_3 = \mathrm{d}v$, the expression for the net flux leaving the volume enclosed by the closed surface S is obtained from (1.65) as follows:

$$\Phi = \oint_S \mathbf{F}\cdot\mathbf{ds} = \int_v \frac{1}{h_1h_2h_3}\left[\frac{\partial(F_{n1}h_2h_3)}{\partial n_1} + \frac{\partial(F_{n2}h_3h_1)}{\partial n_2} + \frac{\partial(F_{n3}h_1h_2)}{\partial n_3}\right]\mathrm{d}v \tag{1.66}$$

Hence the net flux leaving from volume δv can be given as

$$\lim_{\delta v\to 0}\oint_S \mathbf{F}\cdot\mathbf{ds} = \lim_{\delta v\to 0}\frac{1}{h_1h_2h_3}\left[\frac{\partial(F_{n1}h_2h_3)}{\partial n_1} + \frac{\partial(F_{n2}h_3h_1)}{\partial n_2} + \frac{\partial(F_{n3}h_1h_2)}{\partial n_3}\right]\delta v \tag{1.67}$$

Providing that $\operatorname{div}\mathbf{F} = \dfrac{1}{h_1h_2h_3}\left[\dfrac{\partial(F_{n1}h_2h_3)}{\partial n_1} + \dfrac{\partial(F_{n2}h_3h_1)}{\partial n_2} + \dfrac{\partial(F_{n3}h_1h_2)}{\partial n_3}\right]$ for the generalized orthogonal coordinate system, we can obtain the definition for div$\mathbf{F}$ from (1.67) as.

$$\operatorname{div}\mathbf{F} = \lim_{\delta v\to 0}\frac{\oint_S \mathbf{F}\cdot\mathbf{ds}}{\delta v} \tag{1.68}$$

The divergence operator is commonly written as

$$\operatorname{div}\mathbf{F} = \nabla\cdot\mathbf{F} = \frac{1}{h_1h_2h_3}\left[\frac{\partial(F_{n1}h_2h_3)}{\partial n_1} + \frac{\partial(F_{n2}h_3h_1)}{\partial n_2} + \frac{\partial(F_{n3}h_1h_2)}{\partial n_3}\right] \tag{1.69}$$

For the Cartesian, cylindrical and spherical coordinates, the respective divergence operators are

$$\nabla \cdot \mathbf{F}(x,y,z) = \frac{\partial F_x}{\partial x} + \frac{\partial F_y}{\partial y} + \frac{\partial F_z}{\partial z} \tag{1.70a}$$

$$\nabla \cdot \mathbf{F}(r,\phi,z) = \frac{1}{r}\frac{\partial (rF_r)}{\partial r} + \frac{1}{r}\frac{\partial F_\phi}{\partial \phi} + \frac{\partial F_z}{\partial z} \tag{1.70b}$$

$$\nabla \cdot \mathbf{F}(r,\theta,\phi) = \frac{1}{r^2}\frac{\partial (r^2 F_r)}{\partial r} + \frac{1}{r\sin\theta}\frac{\partial (F_\theta \sin\theta)}{\partial \theta} + \frac{1}{r\sin\theta}\frac{\partial F_\phi}{\partial \phi} \tag{1.70c}$$

1.11 The divergence theorem

The net flux of a vector field **F** out of a closed surface involves the component of **F** normal to the surface over the surface. An alternative expression for determining the net flux can be given by the divergence theorem:

$$\oint_S \mathbf{F} \cdot \mathrm{d}\mathbf{s} = \int_v \nabla \cdot \mathbf{F}\, \mathrm{d}v \tag{1.71}$$

The divergence theorem states that the integral of the normal component of a vector field over a closed surface yields the same results as the integral of the divergence of the vector field through the volume enclosed by the surface.

1.12 Curl of a vector field

The net motion of a vector field **F** about some closed contour C is the line integral of **F** along C:

$$m = \oint_C \mathbf{F} \cdot \mathrm{d}\mathbf{l} \tag{1.72}$$

Consider a contour C on the surface $\mathbf{s}_3$ of the generalized orthogonal coordinate system. We can write the differential arc length $\mathrm{d}\mathbf{l} = \mathrm{d}l_1\,\mathbf{u}_{n1} + \mathrm{d}l_2\,\mathbf{u}_{n2} = h_1\,\mathrm{d}n_1\,\mathbf{u}_{n1} + h_2\,\mathrm{d}n_2\,\mathbf{u}_{n2}$ from (1.4). Substituting for the vector field and differential arc length in (1.72), we obtain the net motion of the vector field in the $\mathbf{s}_3$ surface as

$$m_3 = \oint_C F_{n1}\,\mathrm{d}l_1 + F_{n2}\,\mathrm{d}l_2 = \oint_C F_{n1}h_1\,\mathrm{d}n_1 + F_{n2}h_2\,\mathrm{d}n_2 \tag{1.73}$$

For the closed boundary of the contour, (1.73) can be further modified as

$$\begin{aligned} m_3 &= \iint -\frac{\partial (F_{n1}h_1)}{\partial n_2}\mathrm{d}n_2\,\mathrm{d}n_1 + \frac{\partial (F_{n2}h_2)}{\partial n_1}\mathrm{d}n_1\,\mathrm{d}n_2 \\ &= \iint \frac{1}{h_1 h_2}\left\{\frac{\partial (F_{n2}h_2)}{\partial n_1} - \frac{\partial (F_{n1}h_1)}{\partial n_2}\right\}\mathrm{d}l_1\,\mathrm{d}l_2 \end{aligned} \tag{1.74}$$

Equation (1.74) can be written in terms of surface integration as follows:

$$m_3 = \int_{S_3} \frac{1}{h_1 h_2}\left\{\frac{\partial(F_{n2}h_2)}{\partial n_1} - \frac{\partial(F_{n1}h_1)}{\partial n_2}\right\} ds_3 \tag{1.75}$$

Generalizing that the contour path by stating that C is run through all three surfaces $\mathbf{s}_1$, $\mathbf{s}_2$ and $\mathbf{s}_3$, the net motion can be given as

$$\begin{aligned} m = m_1 + m_2 + m_3 = & \int_{S_1} \frac{1}{h_2 h_3}\left\{\frac{\partial(F_{n3}h_3)}{\partial n_2} - \frac{\partial(F_{n2}h_2)}{\partial n_3}\right\} ds_1 \\ & + \int_{S_2} \frac{1}{h_3 h_1}\left\{\frac{\partial(F_{n1}h_1)}{\partial n_3} - \frac{\partial(F_{n3}h_3)}{\partial n_1}\right\} ds_2 \\ & + \int_{S_3} \frac{1}{h_1 h_2}\left\{\frac{\partial(F_{n2}h_2)}{\partial n_1} - \frac{\partial(F_{n1}h_1)}{\partial n_2}\right\} ds_3 \end{aligned} \tag{1.76}$$

Equation (1.76) can be further written as vector expressions as given in:

$$\oint_C \mathbf{F} \cdot \mathbf{dl} = \int_S \begin{bmatrix} \dfrac{1}{h_2 h_3}\left\{\dfrac{\partial(F_{n3}h_3)}{\partial n_2} - \dfrac{\partial(F_{n2}h_2)}{\partial n_3}\right\}\mathbf{u}_{n1} \\ +\dfrac{1}{h_3 h_1}\left\{\dfrac{\partial(F_{n1}h_1)}{\partial n_3} - \dfrac{\partial(F_{n3}h_3)}{\partial n_1}\right\}\mathbf{u}_{n2} \\ +\dfrac{1}{h_1 h_2}\left\{\dfrac{\partial(F_{n2}h_2)}{\partial n_1} - \dfrac{\partial(F_{n1}h_1)}{\partial n_2}\right\}\mathbf{u}_{n3} \end{bmatrix} \cdot \mathbf{ds} \tag{1.77}$$

The expression in the parenthesis of (1.77) is defined as either curl $\mathbf{F}$ or $\nabla \times \mathbf{F}$. Thus (1.77) can alternatively be written as

$$\oint_C \mathbf{F} \cdot \mathbf{dl} = \int_S \text{curl } \mathbf{F} \cdot \mathbf{ds} = \int_S \nabla \times \mathbf{F} \cdot \mathbf{ds} \tag{1.78}$$

Observe that, curl $\mathbf{F}$ or $\nabla \times \mathbf{F}$ can be easily remembered from the following expression:

$$\text{curl } \mathbf{F} = \nabla \times \mathbf{F} = \frac{1}{h_1 h_2 h_3}\begin{vmatrix} h_1\mathbf{u}_{n1} & h_2\mathbf{u}_{n2} & h_3\mathbf{u}_{n3} \\ \dfrac{\partial}{\partial n_1} & \dfrac{\partial}{\partial n_2} & \dfrac{\partial}{\partial n_3} \\ h_1 F_{n1} & h_2 F_{n2} & h_3 F_{n3} \end{vmatrix} \tag{1.79}$$

For the Cartesian, cylindrical and spherical coordinates, the respective curl operators are given by

$$\nabla \times \mathbf{F}(x,y,z) = \left\{ \frac{\partial F_z}{\partial y} - \frac{\partial F_y}{\partial z} \right\} \mathbf{u}_x + \left\{ \frac{\partial F_x}{\partial z} - \frac{\partial F_z}{\partial x} \right\} \mathbf{u}_y + \left\{ \frac{\partial F_y}{\partial x} - \frac{\partial F_x}{\partial y} \right\} \mathbf{u}_z \tag{1.80a}$$

$$\nabla \times \mathbf{F}(r,\phi,z) = \left\{ \frac{1}{r}\frac{\partial F_z}{\partial \phi} - \frac{\partial F_\phi}{\partial z} \right\} \mathbf{u}_r + \left\{ \frac{\partial F_r}{\partial z} - \frac{\partial F_z}{\partial r} \right\} \mathbf{u}_\phi + \frac{1}{r}\left\{ \frac{\partial (rF_\phi)}{\partial r} - \frac{\partial F_r}{\partial \phi} \right\} \mathbf{u}_z \tag{1.80b}$$

$$\nabla \times \mathbf{F}(r,\theta,\phi) = \frac{1}{r\sin\theta}\left\{ \frac{\partial (F_\phi \sin\theta)}{\partial \theta} - \frac{\partial F_\theta}{\partial \phi} \right\} \mathbf{u}_r + \frac{1}{r}\left\{ \frac{1}{\sin\theta}\frac{\partial F_r}{\partial \phi} - \frac{\partial (rF_\phi)}{\partial r} \right\} \mathbf{u}_\theta + \frac{1}{r}\left\{ \frac{\partial (rF_\theta)}{\partial r} - \frac{\partial F_r}{\partial \theta} \right\} \mathbf{u}_\phi \tag{1.80c}$$

1.13 Stokes' theorem

While deriving the expression for the net motion in the previous section, we have proved Stokes' theorem which is expressed as

$$\oint_C \mathbf{F} \cdot \mathbf{dl} == \int_S (\nabla \times \mathbf{F}) \cdot \mathbf{ds}$$

In words, Stokes' theorem states that the net flux of $(\nabla \times \mathbf{F})$ over any open surface S is equal to the line integral of $\mathbf{F}$ along the closed contour C that bounds S.

Since S is an open surface, we must be careful to define the direction of **ds** that is normal to the surface. This is defined by the right-hand rule. If we arch and place the four fingers of our right hand in the direction of the contour C, then the direction in which the thumb will point is the direction of **ds**.

1.14 Vector identities

On numerous occasions we need to call on vector identities to simplify expressions. Here some of the more useful vector identities are listed. In the following expressions, **A**, **B** and **C** are vector quantities, while ϕ and ψ are scalar quantities.

1. $\nabla \times \nabla\phi = 0$
2. $\nabla \cdot \nabla \times \mathbf{A} = 0$
3. $\nabla(\phi\psi) = \phi(\nabla\psi) + \psi(\nabla\phi)$
4. $\nabla \cdot (\phi\mathbf{A}) = \mathbf{A} \cdot (\nabla\phi) + \phi(\nabla \cdot \mathbf{A})$

5. $\nabla \times (\phi \mathbf{A}) = (\nabla \phi) \times \mathbf{A} + \phi (\nabla \times \mathbf{A})$

6. $\mathbf{A} \cdot (\mathbf{B} \times \mathbf{C}) = \mathbf{B} \cdot (\mathbf{C} \times \mathbf{A}) = \mathbf{C} \cdot (\mathbf{A} \times \mathbf{B})$

7. $\mathbf{A} \times \mathbf{B} \times \mathbf{C} = \mathbf{B}(\mathbf{A} \cdot \mathbf{C}) - \mathbf{C}(\mathbf{A} \cdot \mathbf{B})$

For further reading

[1] Balanis, C.A., *Advanced Engineering Electromagnetics*, 2nd Edition, USA: John Wiley, 1989.

[2] Clayton, R.P., Keith W.W., and Syed A.N., *Introduction to Electromagnetic Fields,* McGraw-Hill, 1997.

[3] Griffiths, D.J., *Introduction to Electrodynamics*, USA: Prentice Hall, 1998.

[4] Hoole, S.R.H. Hoole and Hoole, P.R.P., *A Modern Short Course in Engineering Electromagnetics*, USA: Oxford University Press, 1996

[5] Sadiku, M.N.O., *Principals of Electromagnetics*, UK: Oxford University Press, 2009.

2

Static fields and Maxwell's equations

2.1 Introduction

The set of Maxwell's four equations and the Lorentz force law are five elegant equations of natural laws that underlie the entire discipline of electrical, electronic and communication engineering. The basic theory that governs electrical or electronic device performance, as well as electromagnetic field based signal or image processing – as in magnetic resonance imaging (MRI), satellite images of earth, fast, reliable telecommunication systems, collision avoidance radar – is the electromagnetic field theory defined by Maxwell's four equations. Whether it is a low-frequency device such as an induction motor (operating at 50 or 60 Hz) or a high-frequency device like a 10 GHz traveling wave tube (TWT) amplifier used in microwave satellites, or a low-frequency (50 or 60 Hz) system such as the electric power system or a high-frequency (e.g. 2 GHz) systems such as wireless, mobile communication system, all these electric devices and systems are characterized and controlled by Maxwell's equations. In order to get exact equations that define the operating characteristic of a passive device such as an antenna, which is a powerful electromagnetic sensor when coupled to signal processing software, or the equations that govern an active device such as an electric induction motor or electronic transistor, we must solve Maxwell's equations specifically applied to the geometry and material properties of the device.

Similarly when analyzing and designing telecommunications, electric power, biomedical, remote sensing, safety or security systems – solutions of Maxwell's equations must be sought either in their original Maxwell's form or the electromagnetic field action approximated to electric circuit systems. In this book, we shall be looking at the fundamental action of Maxwell's equations in a variety of devices and systems. Maxwell's equations may well be applied to all macroscopic electromagnetic devices and systems. Moreover, Maxwell's equations will yield the wave equation for electromagnetic waves, where in free space, electromagnetic waves, over almost the entire frequency spectrum (e.g. 1 kHz to 10^{12} Hz) will travel at the velocity of light ($c = 3 \times 10^8$ m/s). These waves, which may be used to carry energy or information, are extensivley used in developing high-impact, technical devices and systems.

In this chapter, we shall first summarize the general properties of Maxwell's equations and then eleaborate on the important principles and applications of electrostatics and magnetostatics. Electrostatics, for instance, has become a major field of technical applications both in high-power electrical engineering (new electrostatic power storage) and in mechanical or radio frequency (RF) microelectromechanical system (MEMS) and nano electromechanical system (NEMS), both as actuators and sensors.

2.2 Basic electromagnetic theory: Maxwell's equations

The electromagnetic force exerted on an electric charge q (C) moving at velocity $\mathbf{u}$ (m/s) in an electric field $\mathbf{E}$ and a magnetic field $\mathbf{B}$ is given by

$$\mathbf{F} = q\mathbf{E} + q\mathbf{u} \times \mathbf{B} \tag{2.1}$$

Equation (2.1) is known as the Lorentz force equation. The force exerted by the electric field is in the direction of the electric field and will be present whether the electric charge is stationary or moving. The force exerted by the magnetic field is perpendicular to the magnetic field, and it is absent if the electric charge is stationary. The magnetic force may change the direction of the motion of a charged particle, but it expends no energy in moving the electric charge. The electric force alone changes the kinetic energy and velocity of the electric charge.

The Lorentz force defines the physical effects of the unseen electric and magnetic fields. We observe that it is the electric field intensity $\mathbf{E}$ and the magnetic flux density $\mathbf{B}$ that are the physical parameters of the electromagnetic field evidently existing. The electric field intensity $\mathbf{E}$ (in V/m) and magnetic flux density $\mathbf{B}$ (in T or Webers/m^2) are the real, although invisible (except in the visible light frequency spectrum), physical quantities that produce flow of electric currents in conductors, or movement of current carrying conductors immersed in a magnetic field, and the transfer of electric energy and electronic information over copper or aluminum conductors and free space. Both magnetic field intensity $\mathbf{H}$ and electric flux density $\mathbf{D}$ are fictitious quantities that are mathematically defined to make the field definitions complete. We have $\mathbf{H} = \mathbf{B}/\mu$ and $\mathbf{D} = \boldsymbol{\varepsilon}\mathbf{E}$ and $\mathbf{J} = \sigma\mathbf{E}$, where material properties μ, ε and σ are, respectively, the permeability, permittivity and electric conductivity of the material in which the fields exist. The conduction current $\mathbf{J}$ ($=\sigma\mathbf{E}$) will not exist in regions of zero conductivity, that is, in dielectric medium such as free space. However, $\partial\mathbf{D}/\partial t = \partial\varepsilon\mathbf{E}/\partial t = \mathbf{J}_D$, called the displacement current density, may flow in dielectric materials too. It is due to this displacement current that we can have wireless communications. Along the antenna conductor itself we have conduction currents, but once the signals are launched out from the antenna, the conduction current becomes zero. The signal energy and information are now carried by displacement currents in free space. The displacement currents flow along the closed electric field lines of each wave packet in free space.

According to Lenz's law, the induced load voltage V_L (volts) in a coil cut by time-changing magnetic field will be in such a direction as to produce a current that opposes the voltage-inducing magnetic flux change. The two quantities **E** and **B** are mutually related and defined by the four Maxwell's equations given below:

Faraday's law, $$\nabla \times \mathbf{E} = -\frac{\partial \mathbf{B}}{\partial t} \tag{2.2}$$

Ampere's law, $$\nabla \times (\mu^{-1}\mathbf{B}) = \sigma \mathbf{E} + \frac{\partial(\varepsilon \mathbf{E})}{\partial t} \tag{2.3}$$

Gauss's law, $$\nabla \cdot (\varepsilon \mathbf{E}) = \rho \tag{2.4}$$

And Gauss's law for magnetism, $$\nabla \cdot \mathbf{B} = 0 \tag{2.5}$$

where σ is the conductivity of the medium, ε is the permittivity of the medium, μ is the permeability of the medium and ρ is the volume electric charge density (C/m^3).

As already indicated, $\mathbf{E}$ = vector electric field intensity in V/m, $\mathbf{B}$ = vector magnetic flux density in Webers/m^2, $\mathbf{D}$ = vector electric flux density in C/m^2, $\mathbf{H}$ = vector magnetic field intensity in A/m. The Maxwell equations are linear since they contain **E** and **B** and their derivatives to the first power only.

We shall define two mathematical quantities, the electric flux density **D** and magnetic field intensity **H** (A/m) as counterparts to the physical quantities **B** and **E**respectively, as given by the constitutive equations:

$$\mathbf{D} = \varepsilon \mathbf{E} \tag{2.6}$$

$$\mathbf{B} = \mu \mathbf{H} \tag{2.7}$$

Thus the four Maxwell's equations may be rewritten as follows:

$$\nabla \times \mathbf{E} = -\frac{\mu \partial \mathbf{H}}{\partial t} \tag{2.8}$$

$$\nabla \times \mathbf{H} = \mathbf{J} + \frac{\partial \mathbf{D}}{\partial t} = \mathbf{J} + \mathbf{J_d} \tag{2.9}$$

$$\nabla \cdot \mathbf{D} = \rho \tag{2.10}$$

$$\nabla \cdot \mu \mathbf{H} = 0 \tag{2.11}$$

where we have defined vector conduction current density $\mathbf{J} = \sigma \mathbf{E}$ A/m^2, and $\mathbf{J_d} = \frac{\partial \mathbf{D}}{\partial t}$ A/m^2 is the displacement current. In conductors, the conduction current

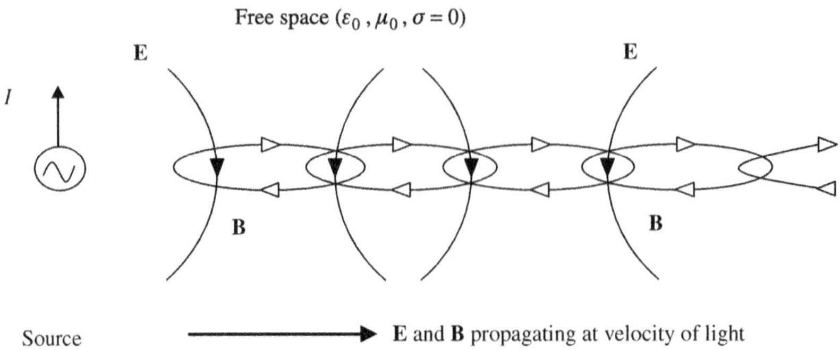

Figure 2.1: Propagation of electromagnetic waves.

dominates, and the displacement current is almost zero. However, in dielectrics (insulators) such as air, the displacement current $\mathbf{J_d}$ may exist especially at high frequency, whereas the conduction current is zero (unless there is electric break-down of the dielectric, forming free space charges). It is because of the displacement current that we have electromagnetic energy (or information) flowing through free space. Looking at (2.9) we note that when one of the two terms of the right-hand side is not zero, that the nonzero current component will produce **H** (i.e., $\mathbf{B}/\mu$). Now looking at (2.2), when **B** exists at any point in space, it will produce **E**. What happens in free space at high frequencies is shown in Fig. 2.1.

An electrical source such as a telecommunication signal generator delivers high-frequency (e.g. 1000 MHz = 10^9 Hz) current I to an antenna. At the antenna, conduction current density J ($= I/A$, where A is the cross-sectional area over which the current flows in the antenna) becomes displacement current $\mathbf{J_d}$ flowing along the electric field lines that are produced in the free space adjacent to the antenna. This electric field is produced by the electric charges Q that flow with I ($I = \frac{\mathrm{d}Q}{\mathrm{d}t} = \mathbf{J} \cdot \mathbf{A}$). These time-varying high-frequency electric fields, according to Maxwell equation (2.3), produce a time-varying magnetic flux density **B** on the left-hand side of (2.3). This time varying quantity **B**, according to (2.2), produces the time varying quantity **E** on the left-hand side of (2.2). The curl operators in (2.8) and (2.9) indicate that the field produced on the left-hand side of each equation is produced ahead of the field producing it on the right-hand side of each equation. Thus by this back-and-forth mutual production of **E** and **B** in free space, the electromagnetic fields **E** and **B** propagate along free space. Any other antenna (receiver) placed in its path will capture a part of that analogue or digital communication information carrying **E** and **B** at distances very far from the transmitter. For instance, communication signals transmitted out from a geostationary satellite antenna will be captured by an earth station antenna at a distance of about 3600 km away from the satellite antenna.

The integral forms of Maxwell's equations are

Faraday's law,
$$\oint_l \mathbf{E} \cdot d\ell = -\int_s \frac{\partial \mathbf{B}}{\partial t} \cdot \mathbf{ds} \quad (2.12)$$

Ampere's law,
$$\oint \mathbf{H} \cdot d\ell = \int_s \left(\mathbf{J} + \frac{\partial \mathbf{D}}{\partial t}\right) \cdot \mathbf{ds} \quad (2.13)$$

Gauss's law,
$$\oint \mathbf{D} \cdot \mathbf{ds} = Q \quad (2.14)$$

And Gauss's law for magnetism
$$\oint_s \mathbf{B} \cdot \mathbf{ds} = 0 \quad (2.15)$$

the latter based on the observation that isolated magnetic charges do not exist. In the integral form, the first two Maxwell's equations may be rewritten using the total flux ψ and current I:

Faraday's law

$$\oint \mathbf{E} \cdot \mathbf{dl} = -\frac{d\varphi}{dt} \quad (2.16)$$

Ampere's law

$$\oint_c \mathbf{H} \cdot \mathbf{dl} = I + \int_s \frac{\partial \mathbf{D}}{\partial t} \cdot \mathbf{ds} \quad (2.17)$$

Some important material property values, which we shall frequently be referring to, are listed below:

Permittivity of free space, $\varepsilon_0 = 8.854 \times 10^{-12}$ F/m

Permeability of free space, $\mu_0 = 4\pi \times 10^{-7}$ H/m

Boltzmann's constant, $k = 1.38 \times 10^{-23}$ J/K

Electric charge of an electron, $e = -1.602 \times 10^{-19}$ C

Speed of electromagnetic waves in free space, $c = 3 \times 10^8$ m/s

Impedance of free space, Z_0 or $\eta_0 = 376\,\Omega$

In electromagnetic engineering, we often deal with several powers to the ten, and the following notations are used: Exa (10^{18}), peta (10^{15}), tera (10^{12}), giga (10^9), mega (10^6), kilo (10^3), centi (10^{-2}), milli (10^{-3}), micro (10^{-6}), nano (10^{-9}), pico (10^{-12}), femto (10^{-15}) and atto (10^{-18}). For example, 1 micrometer means 10^{-6} m.

2.3 Vector potential functions

Often it is difficult to solve Maxwell's equations directly for **E** and **B** fields. Therefore, an indirect method has been developed to solve for static and dynamic electromagnetic fields. In this indirect method, we shall solve for the electric potential ϕ and vector potential **A** which are defined as follows.

Consider the Maxwell's equation (2.2). Let us define the vector potential (a mathematical fiction) **A** as satisfying $\mathbf{B} = \nabla \times \mathbf{A}$. Thus

$$\nabla \times \mathbf{E} = -\frac{\partial}{\partial t}(\nabla \times \mathbf{A}) \tag{2.18}$$

which may be rewritten as

$$\nabla \times \left(\mathbf{E} + \frac{\partial \mathbf{A}}{\partial t}\right) = 0 \tag{2.19}$$

Let

$$\mathbf{E}' = \mathbf{E} + \frac{\partial \mathbf{A}}{\partial t} \tag{2.20}$$

resulting in

$$\nabla \times \mathbf{E}' = 0 \tag{2.21}$$

where $\mathbf{E}'$ is a conservative vector. A conservative quantity may be expressed us the gradient of a scalar, leading us to the definition of scalar potential (voltage in electric current language) as

$$\mathbf{E}' = -\nabla\phi \tag{2.22}$$

In electric circuit theory, the scalar potential Φ is often referred to as a potential drop or voltage. Therefore, electric field **E** may be expressed as

$$\mathbf{E} = \mathbf{E}' - \frac{\partial \mathbf{A}}{\partial t} \tag{2.23}$$

$$= -\nabla\phi - \frac{\partial \mathbf{A}}{\partial t} \tag{2.24}$$

Thus, if we solve for ϕ and **A** from Maxwell's equations, we can obtain the electric field **E** from (2.24) and magnetic flux density **B** from

$$\mathbf{B} = \nabla \times \mathbf{A} \tag{2.25}$$

When there is no time variation,

$$\frac{\partial}{\partial t}(\mathbf{A}) = 0 \tag{2.26}$$

and (2.24) reduces to

$$\mathbf{E} = -\nabla\phi \tag{2.27}$$

which is the case for electrostatic fields.

2.4 Boundary conditions

What happens to the electric field and the magnetic field when they encounter a material boundary? For instance, in wireless telecommunications, what happens to a satellite transmitted microwave signal traveling in the dielectric medium of air, when it hits the metal roof of a motor vehicle, the surface of the sea or the glass windows of a high rise building? What happens to the low-frequency (e.g. 50 Hz) magnetic field in the air-gap of an electric machine, when it contacts the magnetic material of the machine stator? How do we design to reduce the risk of electric field caused breakdown in a power cable, by using multidielectric insulation, that is dielectric materials of different permittivity? To answer each one of these three questions – whether it be the microwave frequency electromagnetic fields at the dielectric–metal interface, or the magnetic field at the air–magnetic material interface or the electrostatic field at the dielectric interface of a multi-dielectric cable insulation – we need to understand how the electric and magnetic fields behave at material interfaces.

2.4.1 General boundary conditions

Consider the material boundary shown in Fig. 2.2. The material properties (electrical permittivity, magnetic permeability and electrical conductivity) of the upper and lower media are $(\varepsilon_1, \mu_1, \sigma_1)$ and $(\varepsilon_2, \mu_2, \sigma_2)$ respectively. The boundary conditions answer the question: How are the electric fields and magnetic fields that are tangential and normal to the boundary, and at the boundary, related to each other just above and below the boundary?

The electric and magnetic fields in medium 1 of Fig. 2.2 are $(\mathbf{E}_1, \mathbf{B}_1)$ and in medium 2 are $(\mathbf{E}_2, \mathbf{B}_2)$. The surface charge density at the interface is ρ_s C/m^2 and the current density is $\mathbf{J}_s$ A/m^2. The normal unit vectors in medium 1 and 2 are $\mathbf{n}_1$ and $\mathbf{n}_2$, respectively. Each of both normal unit vectors, as shown in Fig. 2.2, point from the interface out into their respective medium. The tangential $(\mathbf{E}_{1t}, \mathbf{E}_{2t})$ and normal $(\mathbf{E}_{1n}, \mathbf{E}_{2n})$ components of the fields are illustrated for the electric fields $\mathbf{E}_1$ and $\mathbf{E}_2$ shown in Fig. 2.2.

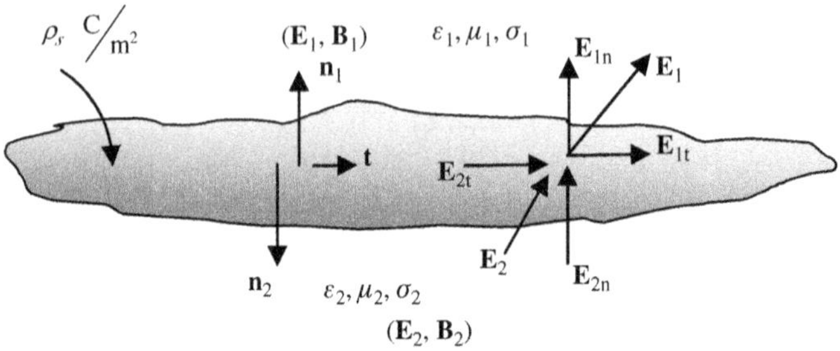

Figure 2.2: Defining boundary conditions.

The tangential boundary conditions are

$$\mathbf{n}_2 \times (\mathbf{E}_1 - \mathbf{E}_2) = 0 \tag{2.28}$$

$$\mathbf{n}_2 \times (\mathbf{H}_1 - \mathbf{H}_2) = \mathbf{J}_s \tag{2.29}$$

where $\mathbf{H}_1 = \mathbf{B}_1/\mu_1$ and $\mathbf{H}_2 = \mathbf{B}_2/\mu_2$, and $\mathbf{J}_s$ is a surface current density in A/m. The normal boundary conditions are

$$\mathbf{n}_2 \cdot (\mathbf{D}_1 - \mathbf{D}_2) = \rho_s \tag{2.30}$$

$$\mathbf{n}_2 \cdot (\mathbf{B}_1 - \mathbf{B}_2) = 0 \tag{2.31}$$

where $\mathbf{D}_1 = \varepsilon_1 \mathbf{E}_1$, $\mathbf{D}_2 = \varepsilon_2 \mathbf{E}_2$ and ρ_s is the surface charge density in C/m^2.

2.4.2 Dielectric–dielectric boundary conditions

For dielectric–dielectric boundary conditions, we have $\sigma_1 = \sigma_2 = 0$, and the surface conduction current density $\mathbf{J}s = 0$. In this case, the boundary conditions simplify to

$$\mathbf{E}_{1t} = \mathbf{E}_{2t} \quad \text{and} \quad \mathbf{H}_{1t} = \mathbf{H}_{2t} \tag{2.32}$$

and

$$\mathbf{D}_{1n} - \mathbf{D}_{2n} = \rho_s \quad \text{and} \quad \mathbf{B}_{1n} = \mathbf{B}_{2n} \tag{2.33}$$

where the subscripts t and n signifying tangential and normal components.

2.4.3 Dielectric–conductor boundary conditions

If medium 1 is a dielectric (i.e. $\sigma_1 = 0$) and medium 2 is a conductor ($\sigma_2 \neq 0$), we have

$$\mathbf{E}_{1t} = \mathbf{E}_{2t} = 0, \quad \mathbf{H}_{1t} = \mathbf{J}_s, \quad \mathbf{H}_{2t} = 0 \tag{2.34}$$

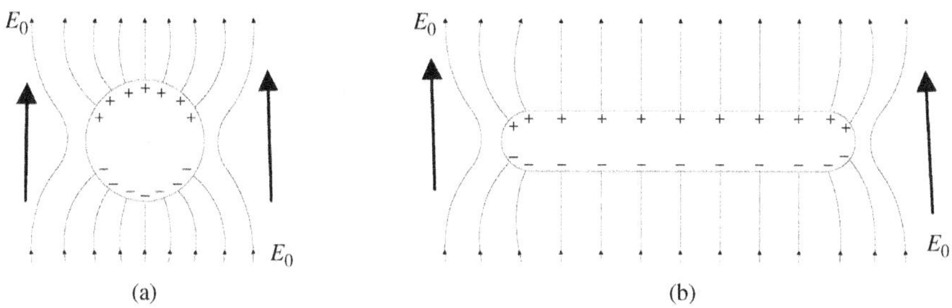

Figure 2.3: Electric fields at conductor boundaries. Upward electric field induces positive electric charges on the top, and negative electric charges at the bottom of the conducting surface immersed in the electric field.

$$\mathbf{D}_{1\mathrm{n}} = \rho_\mathrm{s} \text{ (i.e. } \varepsilon_1 \mathbf{E}_{1\mathrm{n}} = \rho_\mathrm{s}), \quad \mathbf{D}_{2\mathrm{n}} = 0, \quad \mathbf{B}_{1\mathrm{n}} = \mathbf{B}_{2\mathrm{n}} = 0 \qquad (2.35)$$

The fact that the tangential electric fields are zero at the surface of the conductor indicates that the electric field at the surface of a conductor is normal to the conductor surface. In Fig. 2.3(a) is shown a conducting sphere placed in an electric field, and in Fig. 2.3(b) a conducting cylinder placed in an electric field. The presence of the metallic sphere or the cylindrical conductor makes the electric field lines to bend to satisfy the condition that electric lines are always normal to a conductor boundary.

The external electric field E_0 causes positive electric charges to accumulate at the top, and negative electric charges to accumulate at the lower parts of the conductors. This is also in keeping with the fact that electric fields originate from positive electric charges and terminate at negative electric charges. Since the electric field within a conductor is zero – otherwise we would have infinite current by virtue of $\mathbf{J} = \sigma \mathbf{E}$ as $\sigma \to \infty$ – (2.33) requires a surface charge ρ_s to form so that D_n outside is matched by the ρ_s.

2.5 Continuity equation

Let us consider the continuity equation $\nabla \cdot \mathbf{J} = -\frac{\partial \rho}{\partial t}$ under two special conditions.

(i) First, consider the situation where the electric charge density ρ does not change with time, that is

$$\frac{\partial \rho}{\partial t} = 0 \qquad (2.36)$$

as when direct current (DC) flows through a conductor or at the node (or junction) of a network of interconnected conductors. In this case

$$\nabla \cdot \mathbf{J} = 0 \qquad (2.37)$$

which on application of Gauss's theorem yields

$$\int_v \nabla \cdot \mathbf{J}\, \mathrm{d}v = \oint_s \mathbf{J} \cdot \mathbf{ds} = 0 \tag{2.38}$$

This surface integration tells that, whether in a single conductor or at the junction of a set of conductors, the total electric current coming in is equal to the total current going out. This law, in electric network theory, is called Kirchhoff's current law.

(ii) Second, consider an electric conductor which is subjected to an electric field along its axial direction. Although the electric charges inside and along the surface of the conductor were loosely and randomly moving in the conductor, the volume charges density was zero before the application of the electric field from an external source. But the moment that the electric field is applied, say at one end of the conductor by connecting a battery on to it, at the point of application the free electrons are moved along the electric field lines (not randomly any longer) to form a collection, or a volume density of electric charges (ρ), governed by the continuity equation

$$\nabla \cdot \mathbf{J} = -\frac{\partial \rho}{\partial t} \tag{2.39}$$

Using $\mathbf{J} = \sigma \mathbf{E}$ and $\nabla \cdot \mathbf{E} = \rho/\varepsilon$, we have

$$\nabla \cdot \mathbf{J} = \sigma \nabla \cdot E = \sigma \cdot \frac{\rho}{\varepsilon} = -\frac{\partial \rho}{\partial t} \tag{2.40}$$

or

$$\frac{\partial \rho}{\partial t} + \left(\frac{\sigma}{\varepsilon}\right) \rho = 0 \tag{2.41}$$

Let $\rho = \rho_0$ at $t = 0$, then the solution of (2.41) becomes

$$\rho(t) = \rho_0 \mathrm{e}^{-(\sigma/\varepsilon)t} = \rho_0 \mathrm{e}^{-t/T_\mathrm{r}} \tag{2.42}$$

where $T_\mathrm{r} = \varepsilon/\sigma$ is called the relaxation time constant. Thus what happens is that at the instant that the electric field is applied, there is a collection of electric charge ρ_0 C/m^2 that arises at that point, which will decay to $\mathrm{e}^{-1} = 1/\mathrm{e} \cong 37\%$ of ρ_0 within a time interval of $t = T_\mathrm{r}$ s. But the current flow has commenced and will continue along the conductor. How is that? The building up of the electric charge ρ_0 at, say point $z = 0$ m of the conductor, will produce its own electric field (which will also decay) which will drive the randomly moving electrons in front of it, say at $z = 1$ μm. Thus there is an electric current that flows along the conductor, away from $z = 0$. That is, say, every 1 μm by 1 μm, there will be rapid movement of electron clusters forming a continuous flow of electric current along the conductor.

For a conducting material, such as copper, $\varepsilon = \varepsilon_0 = 8.854 \times 10^{-12}$ F/m; $\sigma = 6 \times 10^7$ S/m, the relaxation time $T = \varepsilon/\sigma \cong 2 \times 10^{-19}$ s, which is very short indeed. However, for an insulator such as mica, $\sigma = 10^{-15}$ S/m, and $\varepsilon = \varepsilon_r\varepsilon_0 = 6\varepsilon_0$, yielding a relaxation time constant of about 15 hours! This means that an application of an electric field to an insulating dielectric material will cause an electric charge build-up, but there is no movement of electric charge from say $z = 0$ to $z = 1\ \mu$m and onward. Thus there is no electric current flow in an insulator such as mica, unless the applied electric field is large enough to cause electric breakdown of the insulator. For electric breakdown of an insulator to take place, the electric field should be large enough to tear out the electrons bound to the nucle, setting the electrons free to move.

2.6 Electrostatics

2.6.1 Electric and magnetic fields between two parallel plates connected to a current source

Consider the two parallel plates shown in Fig. 2.4(a), which are separated by a distance d and connected to a time-changing current source causing a conduction current I_c to flow through the plates. The conduction current flows through the two conducting (e.g. copper) wires connected to the parallel plates, and between the plates it flows as the displacement current I_d which is equal in magnitude to the conduction current. We want to determine the electric and magnetic fields inside this parallel plate capacitor arrangement.

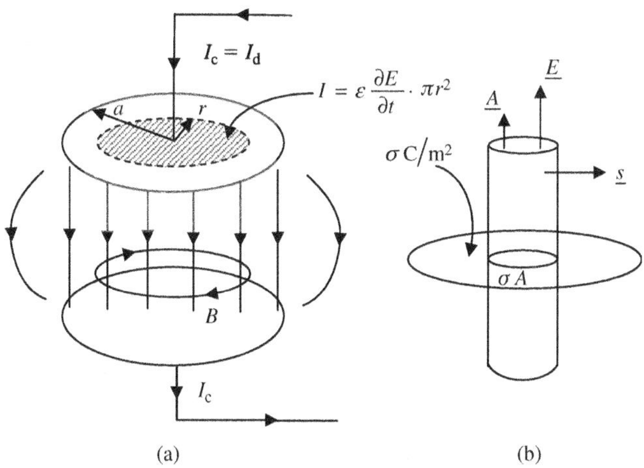

Figure 2.4: Electric fields from two parallel plates and a single charged plate. (a) Two parallel plates and the electric and magnetic fields formed between the plates. (b) A single infinitely large plate and the electric field in the upper area of the plate, neglecting fringing at the edge of the plate.

Before we determine the fields of the parallel plates of Fig. 2.4(a), consider first a single plate as shown in Fig. 2.4(b). The surface electric charge density σ C/m^2 on the plate is related to the total charge Q C on the plate by the equation

$$\sigma = \frac{Q}{\pi a^2} \tag{2.43}$$

where a is the radius of the plate.

For a single infinite plane with an electric charge density σ C/m^2 on it, as shown in Fig 2.4(b), according to Gauss's theorem

$$\int_s \varepsilon E \cdot \mathrm{d}s = \int_r \rho \, \mathrm{d}v = \int \sigma \, \mathrm{d}s \tag{2.44}$$

Thus, for electric fields on both sides of the single plate shown in Fig 2.4(b), we have

$$\varepsilon EA + \varepsilon EA = \sigma A \tag{2.45}$$

or

$$E = \frac{\sigma}{2\varepsilon} \tag{2.46}$$

For the two-plate capacitor arrangement shown in Fig. 2.4(a), the electric field between the two plates is a summation of the electric field produced by each plate, giving us $E = \sigma/\varepsilon$ between the two plates.

Applying Ampere's law

$$\nabla \times \mathbf{H} = \varepsilon \frac{\partial E}{\partial t} \quad \text{or} \quad \oint_c \mathbf{B} \cdot \mathrm{d}\mathbf{l} = \mu_0 I \tag{2.47}$$

to a ring of radius r, with displacement current $\varepsilon \, \partial \mathbf{E}/\partial t$ flowing in the middle (Fig. 2.4(a)), yields

$$B_\phi 2\pi r = \mu_0 \pi r^2 \varepsilon \frac{\partial \mathbf{E}}{\partial t} \tag{2.48}$$

Therefore,

$$H_\phi = \mu_0^{-1} B_\phi = \frac{1}{2} r \varepsilon \frac{\partial \mathbf{E}}{\partial t} \tag{2.49}$$

The conduction current between the plates is $J = 0$; the energy between the two plates is associated with the displacement current $J_\mathrm{d} = \varepsilon \, \partial E/\partial t$. The energy contained in the electric field, in Joules, and the surface integral of the Poynting vector $\mathbf{E} \times \mathbf{H}$ are

$$-\frac{\partial}{\partial t}\int_{v} \frac{1}{2}\varepsilon \mathbf{E}^2 \, \mathrm{d}v = -\frac{\partial}{\partial t}\int \frac{\varepsilon \mathbf{E}^2}{2}\pi a^2 \, \mathrm{d}z \tag{2.50}$$

$$= -\frac{\varepsilon \pi a^2}{2} 2 \, \mathrm{d}\mathbf{E}\frac{\mathrm{d}\mathbf{E}}{\mathrm{d}t} = -\pi a^2 \, \mathrm{d}\varepsilon \mathbf{E}\frac{\mathrm{d}\mathbf{E}}{\mathrm{d}t}$$

$$\int_{s} (\mathbf{E} \times \mathbf{H})\mathrm{d}s = -[\mathbf{EH}]_{r=a} \cdot 2\pi a d \tag{2.51}$$

$$= -\pi a^2 \, \mathrm{d}\varepsilon \mathbf{E}\frac{\mathrm{d}\mathbf{E}}{\mathrm{d}t}$$

is the term that describes the propagation of energy from the upper plate to the lower plate of the two-plate system, and it is equal to the energy contained in the electric field as shown in (2.50). The energy contained in the magnetic field can also be shown to be the same. What takes place is that energy propagates from the upper plate to the lower plate by the electric field and the magnetic field exchanging the energy from one to the other, carrying the energy from the upper to the lower plate. The energy (or information) transfer by electromagnetic fields shall be more fully discussed in Section 2.11.2.

2.6.2 Electric fields in insulator materials

2.6.2.1 Electric field and the dielectric atom

In a conductor or a semiconductor, the electrons are not tightly bound to the nucleus at room temperature and may thus freely move around when subject to an electric field of the order of 1 mV/m. In dielectric or insulator materials (e.g. mica or cross-linked polyethylene, XLPE), the electrons in the inner and the outer shells are strongly bound to the nucleus by electric forces and the electrons may be assumed to spread themselves equally around the nucleus as shown in Fig. 2.5(a). When subject to an electric field, the electron cloud and the nucleus may reorient themselves as shown in Fig. 2.5(b), forming an atomic dipole.

2.6.2.2 The electric dipole

Now this dipole shown in Fig. 2.5(b) will produce its own electric field, called the polarization electric field **P**. Consider an isolated electric dipole shown in Fig. 2.6(a).

In order to determine the electric field that the dipole produces at a distance r from the middle of the dipole, consider an isolated charge q and an observation point $O(r, \theta, \phi)$, as shown in Fig. 2.6(a). Gauss's law states that

$$\nabla \cdot \mathbf{D} = \rho \tag{2.52}$$

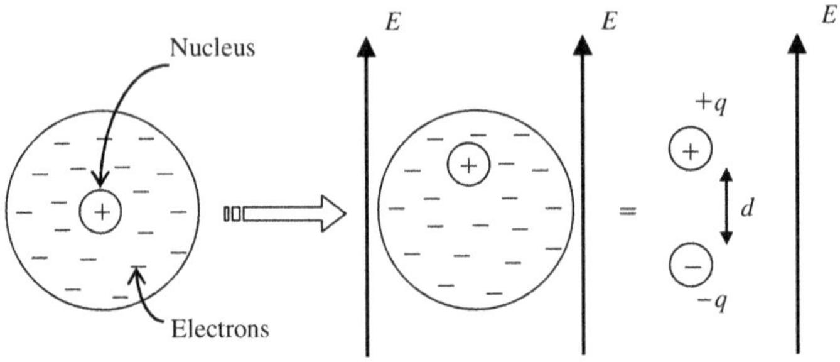

Figure 2.5: Electric field and a dielectric atom: (a) dielectric atom; (b) dielectric atom subject to an external field E.

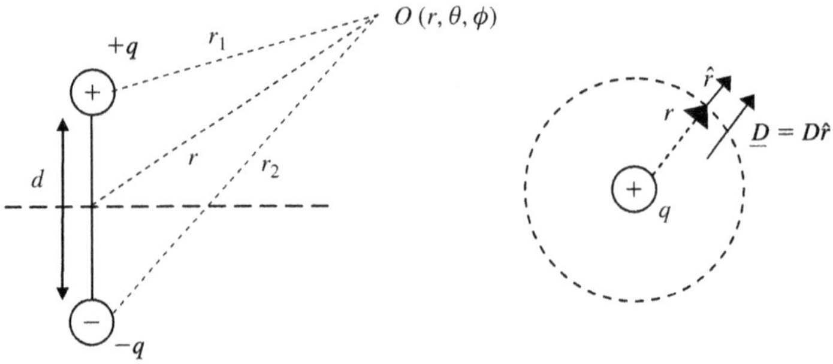

(a) (b)

Figure 2.6: An electric dipole.

Taking a volume integral of (2.52) at distance r (Fig. 2.6(b)), we have

$$\int_v \nabla \cdot \mathbf{D}\, \mathrm{d}v = \int_v \rho\, \mathrm{d}v = q \tag{2.53}$$

where q is the total charge enclosed by the spherical surface at radial distance r. Using the divergence theorem

$$\int_v \nabla \cdot \mathbf{D} = \oint_s \mathbf{D} \cdot \mathrm{d}s \tag{2.54}$$

yields

$$\oint_s \mathbf{D} \cdot \mathbf{ds} = q \tag{2.55}$$

which implies that integrating the electric flux density over a closed surface around an electric charge is equal to the electric charge. For a spherical surface, and constant **D** at its surface, we get that

$$\mathbf{D} \cdot 4\pi r^2 = q \tag{2.56}$$

Thus

$$\mathbf{E} = \frac{\mathbf{D}}{\varepsilon} = \frac{q}{4\pi \varepsilon r^2} u_\mathrm{r} \ \mathrm{V/m} \tag{2.57}$$

where u_r is the unit vector shown in Fig. 2.6(b).

The scalar potential (or voltage) at $O(r, Q, \phi)$ is obtained from

$$\mathbf{E} = -\nabla\phi \quad \text{or} \quad \phi = -\int_{\alpha}^{r} \mathbf{E}(r) \cdot \mathrm{d}\mathbf{r} \tag{2.58}$$

That is

$$\phi = -\int_{\alpha}^{r} \mathbf{E}(r) u_{\hat{r}} \cdot \mathrm{d}r \, u_{\hat{r}} = \frac{q}{4\pi \varepsilon r} \ \mathrm{V} \tag{2.59}$$

For the dipole shown in Fig.2.6(a), using the principle of superposition, the total electric potential at the observation point is given by

$$\phi = \frac{q}{4\pi \varepsilon r_1} - \frac{q}{4\pi \varepsilon r_2} = \frac{q}{4\pi \varepsilon}\left(\frac{r_2 - r_1}{r_1 r_2}\right) \tag{2.60}$$

For $d << r_1, \ r_2$, we have

$$r_2 - r_1 \cong \mathrm{d}\cos\theta \quad \text{and} \quad r_1 r_2 \cong r^2 \tag{2.61}$$

Therefore,

$$\phi = \frac{qd\cos\theta}{4\pi \varepsilon r^2} \tag{2.62}$$

In vector form

$$qd\cos\theta = q\mathbf{d} \cdot \mathbf{u}_r = \mathbf{p} \cdot \mathbf{u}_r \tag{2.63}$$

where $\mathbf{p} = q\mathbf{d}$ is the dipole moment. We may rewrite the scalar potential as

$$\phi = \frac{\mathbf{p} \cdot \mathbf{u}_r}{4\pi \varepsilon_0 r^2} \tag{2.64}$$

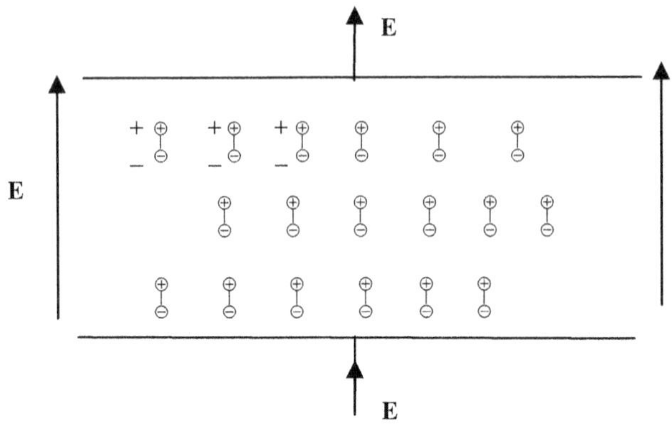

Figure 2.7: Polarization of a dielectric material.

Hence the polarization electric field due to a single dipole is given by

$$\mathbf{P} = \mathbf{E}_0 = -\nabla\varphi = -\left(\mathbf{u}_r\frac{\partial\varphi}{\partial r} + \mathbf{u}_\theta\frac{1}{r}\frac{\partial\varphi}{\partial\theta} + \mathbf{u}_\varphi\frac{1}{r\sin\theta}\frac{\partial\varphi}{\partial\varphi}\right)$$
$$= \frac{qd}{4\pi\varepsilon r^3}(2\cos\theta\,\mathbf{u}_r + \sin\theta\,\mathbf{u}_\theta)\ \mathrm{V/m} \qquad (2.65)$$

This is the electric field produced by a dipole, for distances $r >> d$. Close to the dipole, computation of **E** may require numerical methods to solve the Laplace's equation $\nabla^2\phi = 0$.

2.6.2.3 Interaction of an electric field with dielectric material

Inside the dielectric material which is subject to an external electric field **E**, we have the situation shown in Fig. 2.7.

The atoms of the dielectric material become polarized, producing at any given point inside the material a net polarization field **p**. In Fig. 2.7, we have shown all the atomic dipoles aligned in the same direction as the external field **E**. Such is the case for isotropic dielectric materials. In anisotropic materials, atomic dipoles are allowed to align themselves only along certain (e.g. crystal) axes. We assume isotropic materials where the atomic dipoles are parallel to **E**, and homogeneous such that material quantities such as conductivity (σ), permittivity ($\varepsilon_0\varepsilon_r = \varepsilon$) and permeability ($\mu_0\mu_r = \mu$) are constant at every point in the material. Therefore, inside the dielectric material subject to an external electric field **E**, the total electric flux density

$$\mathbf{D} = \varepsilon_0\mathbf{E} + \mathbf{P}$$
$$= \varepsilon_0\mathbf{E} + \varepsilon_0 X_e\mathbf{E}$$
$$= \varepsilon_0(1 + X_e)\mathbf{E} = \varepsilon\mathbf{E} \qquad (2.66)$$

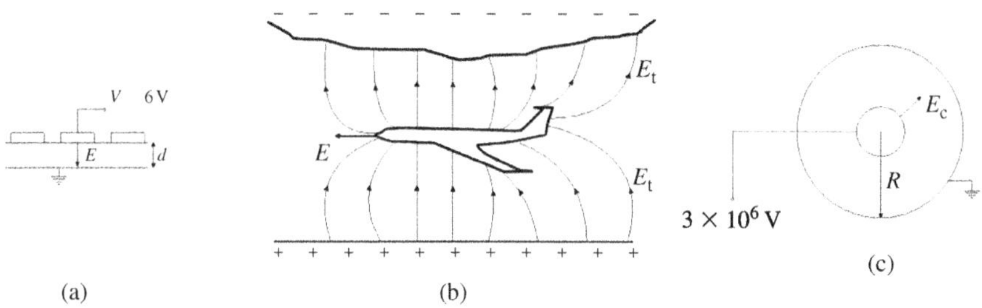

Figure 2.8: Electric field (a) in a semiconductor device (b) around an aircraft beneath a thundercloud, and (c) in a coaxial cable.

where χ_e is the electric susceptibility of the material, $\varepsilon = \varepsilon_0(1 + \chi_e) = \varepsilon_0\varepsilon_r$ is the permitivitty of the material and $\varepsilon_r = (1 + \chi_e)$ is the relative permittivity of the material.

For certain dielectric (insulator) media such as free space (or air) X_e is very small (e.g. $X_e = 0.0006$) and we may assume that $\varepsilon = \varepsilon_0$, as for vacuum. The permittivity of vacuum is $\varepsilon_0 = 8.854 \times 10^{-12}$ F/m.

If the externally applied electric field is increased to a critical value, the force (eE) that it exerts on each electron of charge e may be large enough to wrench it away from the nucleus and the atom. As such freed electrons increase, the dielectric material becomes more like a conductor through which the bunched up, freed electrons move along the electric field lines giving rise to an electric current. This process is called electric breakdown. The critical electric field at which such breakdown occurs is called the dielectric strength, or the breakdown electric field, of the material. The dielectric strength for air ($\varepsilon_r = 1.0006$), petroleum oil ($\varepsilon_r = 2.1$), polystyrene ($\varepsilon_r = 2.6$), glass ($\varepsilon_r \cong 10$) and mica ($\varepsilon_r = 6$) are 3×10^6 V/m, 12×10^6 V/m, 20×10^6 V/m, 40×10^6 V/m and 200×10^6 V/m, respectively.

Prevention of electrostatic field breakdown is a field of major interest in the design of microelectronic devices, power system high-voltage equipment and aircraft electronic systems, to name but three specialist areas.

In Fig. 2.8(a) is shown a microelectronic semiconductor device, where the distance d is of the order of micro (10^{-6}) to nano (10^{-9}) meters. How do we design it in such a way as to make it as small as possible, while at the same time prevent the electric field E from exceeding the dielectric strength of the material between the live conductor and the ground plane? In future electronic devices the operating voltages are expected to reduce below 2 V, allowing nanometer, miniaturized circuits become common. However, with such small nanometer distances between electrodes, even the slightest increase in voltages due to interference or lightning sourced high voltages, the device may break down due to electrostatic discharge (ESD). This is the case for about 30% of failures in electronic circuits in the present 6 V devices and circuits. These ESD caused 30% of circuit failures may very well increase with nanometer electronic devices, unless solutions are found to reduce over voltages and device failures due to electromagnetic interference (EMI) problems.

In Fig. 2.8(b) is shown an aircraft flying under a thundercloud. How do we prevent the intense electric field at the random (or nose) of the aircraft from exceeding the dielectric strength of air?

In Fig. 2.8(c) is shown a single phase electric cable at a voltage of 3×10^6 V. How do we design the cable such that its geometry and the dielectric material used will permit a smaller cable (i.e. if R minimized) in which the electric field at the conductor, E_c, does not exceed the dielectric strength of the insulating material?

When designing, analyzing and testing of electrostatic devices, in order to ensure high reliability, we seek to minimize the occurrence of ESD under normal operating conditions as well as under severe electric stress due to near breakdown electric fields produced by transient voltages in the circuit and induced electric fields due to interference.

2.6.3 Volume electric space charge

Consider now a volume electric charge density $\rho\,\mathrm{C/m^3}$ spread over a volume V as shown in Fig. 2.9.

If the electric charge density is made up by electrostatic, then the electrostatic potential $\phi(r)$ at a distance r from the origin is given by

$$\phi(r) = \frac{1}{4\pi\varepsilon} \int_{v'} \frac{\rho(r_s)}{r'} \mathrm{d}v' \tag{2.67}$$

and magnetic vector potential $\mathbf{A} = 0$ when there is no magnetic source present.

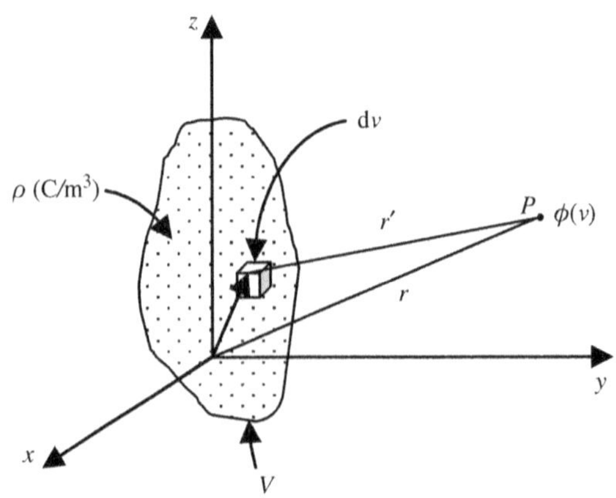

Figure 2.9: Electric field produced by volume electric charge.

If, however, the electric charge is time varying, then the effect of the electric charge $\rho \cdot dv$ at r_s, at instant of time $t = t_0$ will be felt at distance r at time instant $t = t_0 + r'/u$, where u is the velocity at which the electromagnetic wave travels in the medium. In free space $u = c$, the velocity of light. Therefore, for time-varying electric charge in a volume v', the electrostatic (or scalar) potential at observation point P is given by

$$\phi(r,t) = \frac{1}{4\pi\varepsilon} \int\limits_{v'} \frac{\rho\left(r_s, t - \frac{r'}{u}\right)}{r'} dv' \text{ V} \tag{2.68}$$

and the vector potential

$$\mathbf{A}(r,t) = \frac{\mu}{4\pi} \int\limits_{v'} \frac{\mathbf{J}\left(r_s, t - \frac{r'}{u}\right)}{r'} dv' \text{ (Webers/m)} \tag{2.69}$$

Because of the time retardation $t - r'/u$ due to the finite velocity of propagation u, the quantities $\phi(r,t)$ and $\mathbf{A}(r,t)$ are called retarded scalar potential and time retarded vector potential respectively.

2.6.4 The Poisson equation

Using the definition for the electrostatic potential,

$$\mathbf{E} = -\nabla\phi \tag{2.70}$$

the Gauss law

$$\nabla \cdot \mathbf{D} = \rho \tag{2.71}$$

becomes the Poisson equation

$$-\varepsilon\nabla^2\phi = \rho \tag{2.72}$$

In the case of stationary electric charges, Maxwell's equations that govern the electric fields are given by

$$\nabla \times \mathbf{E} = 0 \tag{2.73}$$

$$\nabla \cdot \varepsilon\mathbf{E} = \rho \tag{2.74}$$

which are totally independent of magnetic fields. Defining $\mathbf{E} = -\nabla\phi$, the scalar electrostatic potential is defined by the Poisson equation

$$\nabla^2\phi = -\frac{\rho}{\varepsilon} \tag{2.75}$$

In this case the electric field is decoupled from the magnetic field. There is a vast number of electric devices that make use of electrostatic fields, including the following: electrocardiogram (measuring the blood pumping pattern of the heart), electrostatic ink-jet printers, solid-state control devices, MEMS, electrostatic pollution control, noncontact voltmeters and electrostatic photography.

The integral solution of the scalar potential ϕ, given in (2.75), may be expressed as

$$\phi(r) = \frac{1}{4\pi\varepsilon}\int_{v'} \frac{\rho}{r'}\mathrm{d}v' = \frac{1}{4\pi\varepsilon}\int_z\int_y\int_x \frac{\rho}{r'}\mathrm{d}x\,\mathrm{d}y\,\mathrm{d}z \tag{2.76}$$

for volume electric charge density ρ C/m^3,

$$\phi(r) = \frac{1}{4\pi\varepsilon}\int_{s'} \frac{\rho_s}{r'}\mathrm{d}s' = \frac{1}{4\pi\varepsilon}\int_y\int_x \frac{\rho_s}{r'}\mathrm{d}x\,\mathrm{d}y \tag{2.77}$$

for surface electric charge density ρ_s C/m^3,

$$\phi(r) = \frac{1}{4\pi\varepsilon}\int_{\ell} \frac{\rho_L}{r'}\mathrm{d}\ell = \frac{1}{4\pi\varepsilon}\int_x \frac{\rho_L}{r'}\mathrm{d}x \tag{2.78}$$

for line charge distribution of ρ_L C/m. Note that in all three cases, the integration is over the sources contained in the volume v'.

In the case of an electric charge-free region, where the Poisson's equation reduces to Laplace's equation

$$\nabla^2\phi = 0 \tag{2.79}$$

and the electric charges that appear in the integral solutions will be due to ρ_s or ρ_L on conductor surfaces, or the volume charge density ρ in a cloud of electric charges, with the region of interest being outside the region occupied by the electric charges.

In a conductor material such as copper, where the electrons are loosely bound to the nucleus, an electric field will cause the electrons to move along the electric field lines due to the electrostatic force

$$\mathbf{F_e} = q\mathbf{E} \tag{2.80}$$

where q is the electric charge experiencing the force. For electrons, $q = e$, the electric charge of an electron, $e = 1.6 \times 10^{-19}$ C. The electrostatic force F_e causes the electrons to drift at a drift velocity of

$$\mathbf{u_e} = -\mu_e\mathbf{E} \tag{2.81}$$

where μ_e is the electron mobility and the negative sign indicates that electrons drift against the direction of the electric fields, since they possess negative electric charge.

When solving for the scalar potential ϕ, we need to specify ϕ at some points (e.g. at conductor surfaces with known potentials, including earthed conductor potential $\phi = 0$); specification of ϕ along some surface points is known as the Dirchlet boundary condition. At some other points, the normal derivative of $\phi(\partial\phi/\partial n)$ may be specified; this is known as the Neumann boundary condition.

For the solution of the Poisson equation to be unique, we need to specify ϕ at least at one point (i.e. the Dirchlet condition applies at that one point) and at all the other boundary points, the Neumann condition or the Dirchlet condition must apply.

2.6.5 Electric field in a coaxial cable with space electric charges

In this section we shall consider the analytical solutions of the scalar potentia, in the absence of spatial electric charges (resulting in Laplace's equation) and in the presence of spatial electric charges (resulting in Poisson's equation) in the solution regions. It will serve to illustrate the influence of electric charges on electric fields, as well as the distinct roles of Laplace's equation and Poisson's equation in the analysis and design of electrostatic devices and systems.

Consider the two long concentric cylindrical electrodes shown in Fig. 2.10. The source voltage applied between the inner conductor and the outer sheath is V_s. We want to determine the electric field at radius r from the centerline. Consider

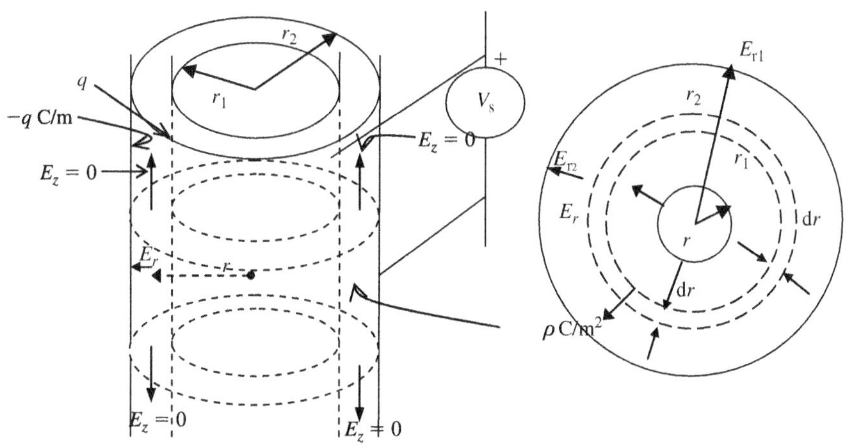

Figure 2.10: Electric fields inside a coaxial cable center.

first a uniform electric charge density of ρ C/m^3 filling the space between the two conductors.

Case 1 ρ = 0

If $\rho = 0$, we have Laplace's equation $\nabla^2\phi = 0$ applying to the space between the two cylindrical conductors, and for radial symmetry, Laplace's equation becomes

$$\frac{1}{r}\frac{\partial}{\partial r}\left(\frac{r\,\partial\phi}{\partial r}\right) = 0 \tag{2.82}$$

The electric field at radial distance r is $E_r = -\frac{\partial\phi}{\partial r}$. Consider a length L of the concentric cylindrical conductor. For this case the flux crossing the cylindrical surface of circumference $2\pi r$ and length L cuts a surface area of $2\pi rL$. Since the electric field is radial (E_r), no electric field will cross the two end surfaces closing the cylindrical surface being considered ($E_z = 0$). Applying Gauss's law stated in (2.14) over the closed surface, we have

$$\varepsilon E 2\pi rL = qL \tag{2.83}$$

yielding

$$\mathbf{E} = \frac{q}{2\pi\varepsilon r}\mathbf{u}_r \tag{2.84}$$

Since

$$E = -\nabla\phi \tag{2.85}$$

Using (2.84) in the right-hand side of (2.85), we have

$$\begin{aligned}\frac{q}{2\pi\varepsilon r}\mathbf{u}_r &= -\left(\mathbf{u}_r\frac{\partial\phi}{\partial r} + \mathbf{u}_\theta\frac{1}{r}\frac{\partial\phi}{\partial\theta} + \mathbf{u}_z\frac{\partial\phi}{\partial z}\right)\\ &= -\mathbf{u}_r\frac{\mathrm{d}\phi}{\mathrm{d}r}\end{aligned} \tag{2.86}$$

Rearranging (2.86), we get

$$\mathrm{d}\phi = -\frac{q}{2\pi\varepsilon}\frac{\mathrm{d}r}{r} \tag{2.87}$$

which on integrating, to get the scalar potential at r, yields

$$\int_{V_S}^{\phi_r}\mathrm{d}\phi = -\frac{q}{2\pi\varepsilon}\int_{r_1}^{r}\frac{\mathrm{d}r}{r} \tag{2.88}$$

Hence the electrostatic potential in between the two concentric cylindrical conductors with no space charges in the inner region between them is given by

$$\phi_r - V_s = \frac{q}{2\pi\varepsilon}[\ln r - \ln r_1]$$
$$= \frac{q}{2\pi\varepsilon}\ln\left(\frac{r}{r_1}\right) \quad (2.89)$$

Or, rearranging (2.89), we have

$$\phi_r = V_s + \frac{q}{2\pi\varepsilon}[\ln r - \ln r_1] \quad (2.90)$$

Note that $E_r \propto 1/r$ and $\phi_r \propto \ln(r)$.

Case 2 ρ ≠ 0

With the presence of space charges between the two cylinders, Poisson's equation to be solved is $\nabla^2\phi = -\rho/\varepsilon$, which for the axial field problem yields may be written in terms of $\nabla \cdot \mathbf{E} = \rho/\varepsilon$, after bringing ε out of the divergence operator assuming a homogeneous medium (e.g. air) in the space between the two conductors. In cylindrical coordinates,

$$\nabla \cdot E = \frac{1}{r}\frac{\partial}{\partial r}(rE_r) + \frac{1}{r}\frac{\partial E_\theta}{\partial r} + \frac{\partial E_z}{\partial z} = \frac{\rho}{\varepsilon} \quad (2.91)$$

or

$$\frac{1}{r}\frac{\partial(rE_r)}{\partial r} = \frac{\rho}{\varepsilon} \quad (2.92)$$

since $E_\theta = E_z = 0$. The electric space charge density $\rho = \rho_p - \rho_e \approx \rho_p$ assuming that the positive ion space charge density ρ_p is very much greater than the electron space charge density ρ_e ($\rho_p >> \rho_e$). The current density (in A/m^2) $J = \rho_p u_p$ where u_p is the drift velocity of the ions in m/s. Hence the space current in A/m (current per unit length) is given by

$$I_p = 2\pi r j = 2\pi r \rho_p u_p \quad (2.93)$$
$$= 2\pi r \rho_p \mu_m E_r \quad (2.94)$$

where, we have used drift velocity $u_p = \mu_m E_r$, where μ_m is the ion mobility, and $\rho = \rho_p = \frac{I_P}{2\pi\rho_p\mu_m r E_r}$ in $(2\pi\rho_p\mu_m r E_r)$ of (2.94), and integrating (2.92) we get

$$(rE_r)^2 - (r_1E_1)^2 = \frac{I_p}{2\pi\varepsilon\mu_m}(r^2 - r_0^2) \quad (2.95)$$

For $r >> r_1$, and $rE_r >> r_1E_1$ then rE_r reduces to

$$r^2E_r^2 = \frac{I_\mathrm{p}}{2\pi\varepsilon\mu_\mathrm{m}}r^2 \tag{2.96}$$

Hence the radial electric field at r is

$$E_r = \sqrt{\frac{I_\mathrm{p}}{2\pi\mu_\mathrm{p}\varepsilon}}\ \mathrm{V/m} \tag{2.97}$$

Comparing this result to the solution (2.84) in the absense of space charge, we note that with ions filling the inter-cylindrical space, we note that $E_r \propto \frac{1}{r}$ for $\rho = 0$, we now find the E_r for distances $r >> r_1$ is independent of r and is only a function of the space ion current I_P. The ion current I_P distorts the electric fields. In Fig. 2.11 is shown the electric field distribution $E_r(r)$ as a function of r, for $\rho = 0$ and $\rho \neq 0$.

We shall now work out ion current I_P as a function of supply voltage V_S. At the outer sheath of the concentric cylinders, $r = r_2$ and $E_r = E_{r2}$. For $r_1 << r_2$, we get

$$(E_{r2}r_2)^2 - (E_{r1}r_1)^2 = \frac{I_\mathrm{p}}{2\pi\varepsilon\mu_\mathrm{m}}r_2^2 \tag{2.98}$$

with $\rho \neq 0$ or $I_\mathrm{p} \neq 0$. We note the E_r is constant, between $r = r_1$ and $r = r_2$ (see Fig. 2.11). Thus we may take $E_{r2} \approx \frac{V_\mathrm{s}}{r_2}$. Therefore, for $r = r_1$, $E = E_{r1}$.

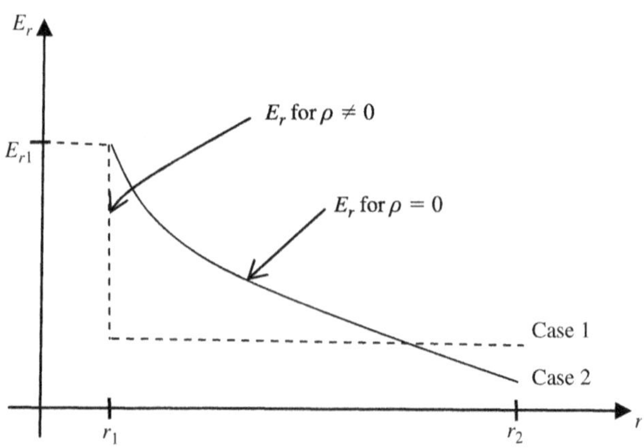

Figure 2.11: Electric field distribution inside a coaxial cable with and without free space charges.

Then $(E_{r2}r_2)^2 - (E_{r1}r_1)^2 = \frac{I_p}{2\pi\varepsilon\mu_m}r_2^2$ becomes

$$(V_s)^2 - (V_1)^2 = \frac{I_p}{2\pi\varepsilon\mu_m}r_2^2 \tag{2.99}$$

or

$$I_p = \frac{2\pi\varepsilon\mu_m}{r_2^2}[(V_s)^2 - (V_1)^2] \tag{2.100}$$

where

$$V_1 = E_{r1}r_1 \ln\left(\frac{r_2}{r_1}\right) \tag{2.101}$$

Experimented investigations give a slightly different dependence of I_p on V_s: $V_s(V_s - V_1)$ instead of $(V_s^2 - V_1^2)$, due to greater accumulation of electric space charges closer to the inner conductor as patches and other simplifications in our analytical solution given in (2.101). On a cylindrical surface, the onset of corona ionization is at 70 kV/cm, whereas for parallel plate electrodes it is 30 kV/cm. Therefore $E_1 = 70$ kV/m for ionization start inside the coaxial conductor system. For $r_1 = 0.1$ cm, $r_2 = 10$ cm, $\mu_m = 2\,\text{cm}^2$ V/S, we get $V_1 = 7 \times 10^4 \times 0.1 \times \ln(10^2) = 32$ kV. For length $L = 100$ cm, $V_s = 40$ kV, we get $I_p = 10\,\mu$A.

2.6.6 Charge simulation method for computation of electrostatic fields

For the electrode system shown in Fig. 2.12, we seek to determine the capacitance of the cylindrical conductor. The charge simulation method is used to determine the charge to potential ratio (q/V), where V is the potential of the wire with respect to earth. The geometry of the conducting wire is shown in Fig. 2.13. The wire is at a height h above the earthed plane, where h is taken to be approximately equal to the radius a of the wire; the effect of the earthed plane will be tackled by replacing the

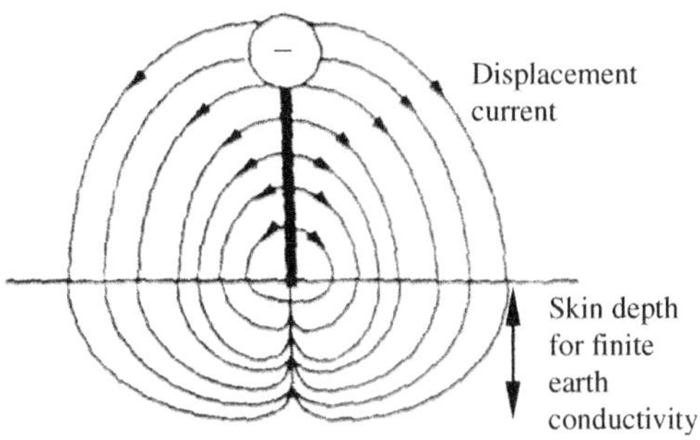

Figure 2.12: An electrically charged sphere connected to a vertical, conducting wire held over perfectly conducting earth.

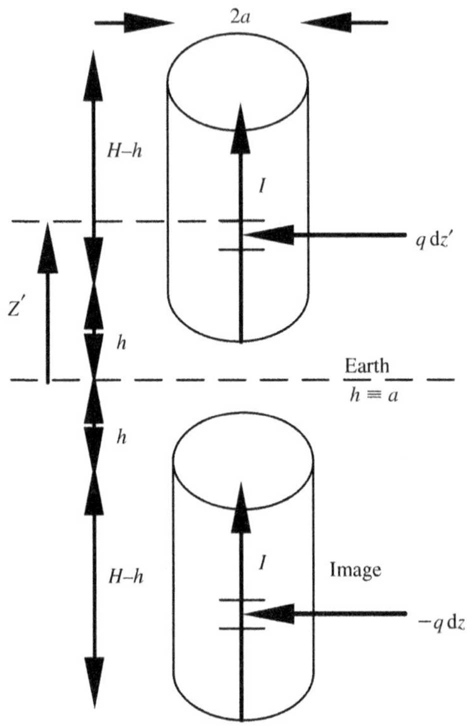

Figure 2.13: Definition of the electrode parameters.

earthed plate by the mirror image of the wire as shown in Fig. 2.13. The length of the wire is H. If the electric charge on the wire is $+q$ C/m, then the electric charge on the mirror image is $-q$ C/m.

For the electrode system shown in Fig. 2.13 ignoring first the presence of the image, the potential at point p due to a line charge of q C/m placed along the centerline of the conductor of length l is given by

$$V_{\mathrm{p}} = \frac{q}{4\pi\varepsilon_0}\int_0^l \left(\frac{1}{\sqrt{(z'-z)^2+a^2}}\right)\mathrm{d}z' \tag{2.102}$$

$$= \frac{q}{4\pi\varepsilon_0}(\log_{\mathrm{e}} z_1 - \log_{\mathrm{e}} z_2) \tag{2.103}$$

where

$$z_1 = \log_{\mathrm{e}}\left[z_{\mathrm{A}} + \sqrt{z_{\mathrm{A}}^2+1}\right] \tag{2.104}$$

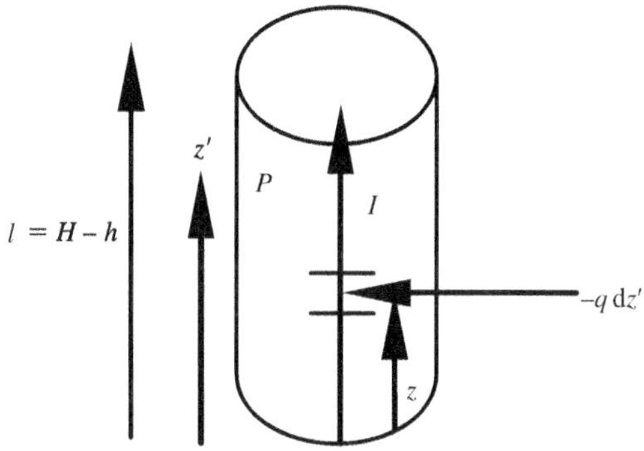

Figure 2.14: Application of the charge simulation method.

$$z_{\mathrm{A}} = \frac{1 - z'}{a} \tag{2.105}$$

$$z_2 = \log_e\left[z_{\mathrm{B}} + \sqrt{z_{\mathrm{B}}^2 + 1}\right] \tag{2.106}$$

$$z_{\mathrm{B}} = -z'/a \tag{2.107}$$

The image may be placed to replace the earth (which is assumed to be a perfect conductor) and its contribution added to the potential due to the real conductor by superposition. The effect of the earth is to reduce the potential on the conductor, particularly close to the earth. For the lower segments, we compute the potential at three points and take the average. The capacitance (in F/m) is determined from q/V_{p}. The lowest point at the capacitance calculated is at a point which is h ($= a$ m) above the earth. It was found that except for points very close to the earth, $1/(LC)^{0.5} \approx c$, where c is the velocity of light. For $a = 4$ mm, very close to the earth, $C = 24.5$ pF/m, $L = 3.3$ μH/m, whereas further above the ground, $C = 4.3$ pF/m, $L = 3.3$ μH/m.

In order to ensure that the electric charge structure we used to determine the capacitance does not give any significant discrepancy when applied to a vertical conductor held above the ground (a vertical wire antenna held above ground and a lightning return stroke channel are two possible examples), we examine a charge distribution structure which is close to the expected distribution of the electric charge. Consider a drop in charge from the tip (q_2 C/m) to the spherical electrode end of the channel (q_1 C/m) in a step like fashion which is linearized to give the electric charge at height z (zero at ground) to be

$$q = q_2 + \frac{z}{l}(q_1 - q_2) \tag{2.108}$$

where l is the length of the channel. Now the potential at point P on the conductor channel at height z' is given by

$$V_{\text{p}} = \frac{1}{4\pi\varepsilon_0} \times \left[\frac{z'}{l}q_1 + \left(1 - \frac{z'}{l}q_2\right)\right](\log_e z_1 - \log_e z_2) + \frac{1}{4\pi\varepsilon_0}(q_1 - q_2)\frac{a}{l}\left[\sqrt{z_{\text{A}}^2 + 1}\sqrt{z_{\text{B}}^2 + 1}\right] \quad (2.109)$$

The q_1/q_2 ratio will not significantly influence the final results, in which case we could set q C/m as uniform and equal over the entire length of the wire. Then the capacitance per unit length along the wire is q/V_{p}. Let

$$T_1 = \frac{z'}{l}(\log_e z_1 - \log_e z_2) + I_1) \quad (2.110)$$

$$T_2 = \left(1 - \frac{z'}{l}\right)(\log_e z_1 - \log_e z_2) + I_2) \quad (2.111)$$

$$T_3 = \frac{a}{l}\left[\sqrt{z_{\text{A}}^2 + 1} - \sqrt{z_{\text{B}}^2 + 1}\right] + I_3 \quad (2.112)$$

where the ground plane has been replaced by the image of the source, and I_1, I_2 and I_3 indicate the contributions of the images carrying opposite polarity of charge.

The capacitance of the upper spherical conductor depends on the radius of the sphere. The capacitance of the sphere may be calculated using the model of an isolated charged sphere from $C_c = 4\pi\varepsilon_0 r_1$ F, where r_1 is the radius of the spherical conductor. When the sphere is close to the earth, the capacitance is pC_c, where $p = (1 - k + k^2), k = r_1/2h$ and h is the height of the center of the sphere above the earth. For most cases $p = 1$.

2.6.7 Microelectromechanical systems

2.6.7.1 MEMS actuators and sensors

MEMS are used as transducers which convert energy from one form to another. They are used as both sensors and actuators. As actuators, they perform mechanical movement. Electric and magnetic fields produced by a stator are used to produce both linear and rotational movements of the rotor. Electrostatic fields are widely used to produce MEMS actuators. As sensors, they obtain information. As mechanical sensors, MEMS are used to measure velocity, acceleration, position, pressure and force. As magnetic sensors, they are used to measure flux density, flux and polarization. As electric field sensors, they are used to measure electric field strength, wavelength, phase and electric field polarization. MEMS chemical sensors are used to measure composition, reaction rate and concentration. As thermal sensors, they are used to measure heat energy, heat flow and temperature. The output or input of MEMS will be voltage, current or electric charge.

The electrical properties of various materials are exploited to construct MEMS (or NEMS). These properties are as follows:

(a) Magnetostriction effect. A ferromagnetic material is squeezed and deformed when a magnetic field penetrates the material. When the field is switched off, the deformed material returns to its former shape and size. When the frequency is around that of sound, sound is produced as in an earphone.

(b) Photoelectric effects. p–n semiconductor junctions yield electron-hole pairs when extremely high-frequency electromagnetic fields, such as visible light, impinge on the semiconductor.

(c) Piezoelectric effects. Materials called electrets, such as quartz crystal and zirconate titanate ceramics, develop an electric charge separation across them when mechanical pressure is exerted on them at frequencies less than 100 MHz. Strain gauges, small earphones or microphones make use of electrets. Normally, for small pressures, the charge separation is proportional to the pressure exerted.

(d) Piezoresistive effect. The resistance of certain materials, such as those used in strain gauges, change when an external force is applied to them.

(e) Thermoelectric effect. When two wires made up of different materials (e.g. NiCr–NiAl) are connected at the ends and their other two ends are heated up to different temperatures, an electric current is observed to flow through the two conductors which together form a closed electric path. The so-called Seebeck's voltage across the junctions is produced by Petlier's junction voltage and Thompson's internal voltage across both conductors. Temperature measurements from 250°C up to 1500°C can be measured. This effect being reversible, an external current can produce cooling at the cold junction.

(f) Thermoresistive effect. The electric resistance of thermistors (e.g. sintered mixtures of metallic oxides) exhibit large changes in resistance (e.g. from 100 Ω to 100 kΩ) for temperature changes in the range of −50°C to 300°C.

MEMS technology is a relatively new development that has potential to make a great change in the way that electronics is done. In future fourth generation (4G) wireless, mobile telecommunication systems, for instance, there will be increasing use of array antennas in both the base stations (BS) and mobile stations (MS) such as the handheld telephone. In array antennas, the use of MEMS as RF switches, phase shifters and other devices is set to have a great impact on telecommunication electronic circuits. At its basic implementation point, MEMS are used as miniature devices that use mechanical movements to produce a short circuit or open circuit in a transmission line. In phased array antennas, a MEMS switch can be used at any frequency, since it can handle DC to 120 GHz. As phase shifters in array antennas, they are used to switch lengths of transmission lines to produce various phase shifts.

When compared to its electronic semiconductor component, competitors such as PIN diodes and FET transistors, which are currently used for switching, MEMS has low-power consumption (less than 0.05 mW compared 50 mW for a PIN diode and 0.1 mW for FETs), very high isolation (>20 dB), low loss (insertion loss of 0.1 dB, compared to about 1 dB for PIN diode and 1.0 dB for FETs), and excellent intermodulation performance. They can be built on glass and low-cost silicon substrates. Future research work will seek to improve its switching speed (100 μs, compared to 50 ns for PIN diodes and 50 ns for FETs) and power handling capability (less than 1 W compared to 10 W for the semiconductor devices).

A major hindrance to power handling capacity in MEMS is the ESD that may occur at higher voltages and currents. In addition to internal voltages related to power flow, for antennas that are mounted in outdoor applications, thunder storms induced voltages in the antenna electronic system, could also lead to another source of ESD threat to MEMS.

The use of MEMS (and NEMS, nano-EMS) is expected to increase in producing smart communication systems with an intelligent combination of MEMS with solid-state circuits. They are attractive alternatives to semiconductor devices from the fabrication point of view as well; the MEMS fabrication process using standard foundry silicon processes is low cost, compared to other RF and microwave solid-state integrated electronic fabrication processes.

The electric field based MEMS are set to become a major, billion dollar industry within the semiconductor industry. However, for future breakthroughs to replace semiconductor devices by MEMS and NEMS, a major issue to be dealt with is designing for ESD so as to improve reliability and power handling capacity, and to reduce switching times. To improve the power handling capacity and reliability of MEMS/NEMS, it is crucial to handle ESD so that MEMS based systems are reliable even in electronically threatening environments (such as in thunder storms and when handling high-power radar signals) and are able to handle power closer to the 10 W handled by semiconductor devices.

2.6.7.2 A MEMS actuator

Consider a cantilever beam consisting of two $a \times b$ parallel metal plates, riveted along a side of length b each plate. One of the two plates is fixed so that it cannot move, whereas the other (e.g. upper) plate is allowed to bend at the side that is riveted and not mobile. If a voltage V is applied across the two plates, and the separation between the two plates is c before the voltage is applied and the free end of the upper plate bends by a distance z toward the lower plate, the deformation force on the upper

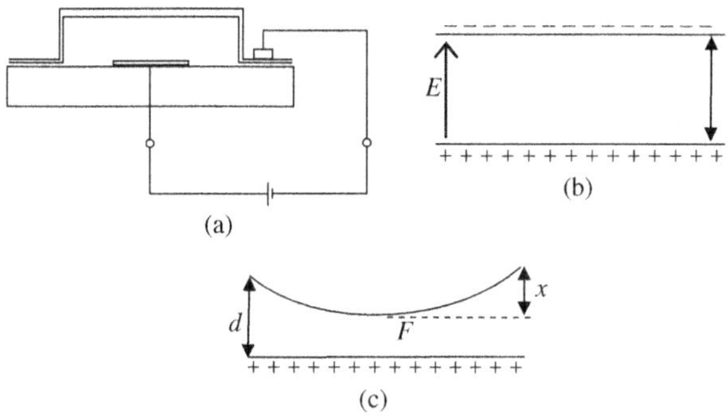

Figure 2.15: MEMS device.

plate is $F_D(z) = K_0(c - z)$, where the constant K_0 (e.g. 3000 nN/μm) is the beam's modulus of elasticity. Let us work out an expression for the displacement distance c in terms of the applied voltage V. When no voltage is applied, the capacitance of the two-plate arrangement is given by $C_0 = \varepsilon(ab)/c$, where the area of each plate is given by ab. The capacitance $C(z)$ of the two-plate arrangement when the free side of the upper beam has bent by z toward the lower plate is $C(z) = (2c)/(2c - z)C_0$. The energy stored is given by $W(z) = (C(z)V^2)/2$. Hence the force is given by $F_E(z) = \mathrm{d}W(z)/\mathrm{d}z = -(C_0V^2)/[4c(1 - z/(2c))^2]$. We may obtain a relationship between the bending distance z and the applied voltage V by equating the deformation and electrostatic forces, yielding $K_0(c - z) = -(C_0V^2)/[4c(1 - z/(2c))^2]$.

2.6.8 Minitiarized sensors

2.6.8.1 Microphone

Consider the MEMS device shown in Fig. 2.15(a), where the top conductor is a movable diaphragm. When electric charges build up on the small, lower conductor and the upper diaphragm (Fig. 2.15(b)), the electrostatic force will cause the diaphragm to be displaced as shown in Fig. 2.15(c).

The energy stored in the charged electrode system of Fig. 2.15(b) is

$$w = \frac{1}{2}CV^2 \tag{2.113}$$

where C is the capacitance in terms of the area A, the distance d is the maximum distance between the lower electrode and the charged diaphragm, and the sag distance x is given by

$$C = \frac{\varepsilon_0 A}{d} \rightarrow \frac{\varepsilon_0 A}{(d - x)} \tag{2.114}$$

where the change in the distance before the electric charge is applied (d) and after the electric charge appears ($d - x$) is indicated in (2.114). The electrostatic force

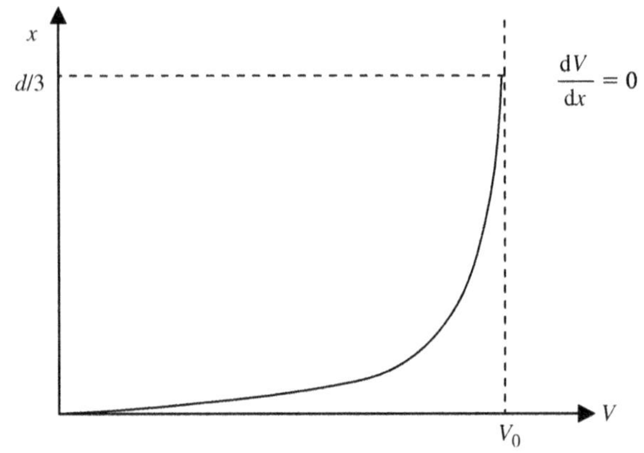

Figure 2.16: The relationship between applied voltage (V) and the distance of deflection (x).

is found by virtual work.

$$F_e = \frac{\mathrm{d}}{\mathrm{d}x}\left(\frac{1}{2}CV^2\right) = \frac{\varepsilon_0 A V^2}{(d-x)^2} \tag{2.115}$$

and the mechanical force is given by

$$F_{\mathrm{m}} = -kx \tag{2.116}$$

Using (2.115) and (2.116), when

$$F_{\mathrm{m}} = F_{\mathrm{e}} \tag{2.117}$$

we get

$$kx = \frac{\varepsilon_0 A V^2}{(d-x)^2} \tag{2.118}$$

In Fig. 2.16, using (2.118), we have sketched the deflection x of the upper diaphragm to close the valve when the voltage V is increased.

To find the value of x at which when

$$\frac{\mathrm{d}V}{\mathrm{d}x} = 0, \tag{2.119}$$

using (2.118), we get

$$V = \sqrt{\frac{kx}{\varepsilon_0 A}}(d-x) \tag{2.120}$$

From (2.120),

$$\frac{\mathrm{d}V}{\mathrm{d}x} = 0, \quad \text{when } x = \frac{d}{3} \tag{2.121}$$

2.6.8.2 Resistance based sensor

Another MEMS device uses serpentine wire resistors to monitor the air pressure of a motor car tyre, or sag in a building. A serpentine resistor and its alternative, which is the piezoelectric wire, are made up of a very thin conductor which easily bends when the surface to which it is attached bends or sags. When it is placed on a tyre, when the tyre pressure is too high, it will bend out and its resistance will increase since the length of the conductor will increase, and the cross-sectional area of the wire will decrease as the wire is stretched out. Under rest conditions, the resistance is given by

$$R = \rho\frac{L}{A} \tag{2.122}$$

When the wire is stretched out by a tyre with excessive air inside, or a building wall that is sagging, the resistance is

$$R_n = \rho\frac{(L + \mathrm{d}L)}{(A - \mathrm{d}A)} \tag{2.123}$$

where the length of the conductor has increased by $\mathrm{d}L$ and the cross-sectional area has decreased by $\mathrm{d}A$. Therefore, the resistance of the wire increases and may be detected by a simple Wheastone bridge. In such a manner, it is possible to measure the pressure of a tyre or the sag of a concrete or iron structure.

2.6.8.3 Electric field measurements

Consider a rotating plate with a total area A with holes in it, that is cutting an electric field that is to be measured. Underneath the rotating plate is a stationary plate, as in the case of a two-plate capacitor. The surface electric charge σ_s C/m^2 induced on the plate is given by (see Section 1.6.1)

$$\varepsilon E = \sigma_s \tag{2.124}$$

or

$$E = \frac{\sigma_s}{\varepsilon} \tag{2.125}$$

Therefore, the current that will flow between the two plates is given by

$$I = \frac{\mathrm{d}Q}{\mathrm{d}t} = \frac{\mathrm{d}(\sigma_s A)}{\mathrm{d}t} = \varepsilon E\frac{\mathrm{d}A}{\mathrm{d}t} \tag{2.126}$$

where the time-varying area is given by

$$A = \frac{A_0}{2}(1 + \sin \omega t) \tag{2.127}$$

where A_0 is the steady-state area of each plate and ω is the radial speed of the top plate. Therefore, the peak current to the sensing electrode is given by

$$I_p = \frac{1}{2}\omega \varepsilon E A_0 \tag{2.128}$$

If the peak current is measured, then the unknown electric field being measured may be obtained from (2.128).

2.6.9 Semiconductors

In semiconductor materials such as silicon (Si) or germanium (Ge), each atom possesses four electrons in the outer orbit (called valence electrons), as shown in Fig. 2.17(a) and (b)). A strong crystal lattice requires each atom to share the four electrons with a neighboring atom to possess the eight electrons needed in the outer orbit to have a stable lattice as shown in Fig. 2.17(c). These intrinsic semiconductor materials may be doped with electron donors such as phosphorus which has five valence electrons to form an n-type extrinsic semiconductor (Fig. 2.17(d)). Alternatively they may be doped with boron, where each atom has only three valence (outer shell) electrons, thus forming a p-type extrinsic semiconductor with an excess of positive holes (Fig. 2.17(e)).

Semiconductors such as silicon are very good insulators at −273°C. At room temperature, some of the outer shell electrons (eight of them, each atom sharing one electron with four neighboring silicon atoms) become freed to move around (Fig. 2.17(c)). When doped the silicon crystal lattice has more free electrons (n-type) or holes (p-type) to conduct electromagnetic energy and information.

The basic field equations that characterize the behaviour of semiconductors will be defined in the following:

(i) Poisson's equation

$$\frac{\partial E}{\partial x} = \frac{\rho - \rho_0}{\varepsilon} \tag{2.129}$$

where E is the electric field and $\rho - \rho_0$ C/m^3 is the net electric charge density.

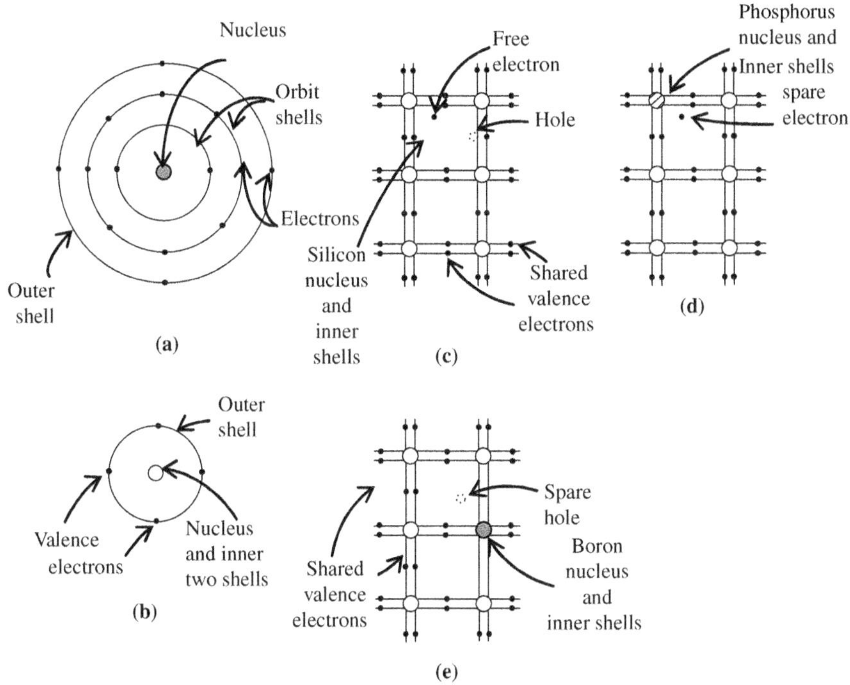

Figure 2.17: Semiconductor lattice structure: (a) silicon atom, (b) simplified model of silicon atom, (c) crystal lattice of silicon, (d) crystal lattice of n-type silicon (doped with phosphorus) and (e) crystal lattice of p-type silicon (doped with boron).

(ii) The current density equations

$$J_{\mathrm{e}} = e\mu_{\mathrm{e}} N_{\mathrm{e}} E + eD_{\mathrm{e}} \frac{\partial N_{\mathrm{e}}}{\partial x} \tag{2.130}$$

$$J_{\mathrm{h}} = e\mu_{\mathrm{h}} N_{\mathrm{h}} E - eD_{\mathrm{h}} \frac{\partial N_{\mathrm{p}}}{\partial x} \tag{2.131}$$

where μ_{e} and μ_{h} are the mobilities of electrons and holes respectively, N_{e} and N_{h} are electron density and hole density respectively, the diffusion constants of electrons (D_{e}) and of holes (D_{h}) are according to the Einstein relations,

$$\frac{D_{\mathrm{e}}}{\mu_{\mathrm{e}}} = \frac{kJ}{e} \quad \text{and} \quad \frac{D_{\mathrm{h}}}{\mu_{\mathrm{h}}} = \frac{kJ}{e} \tag{2.132}$$

(iii) Continuity equations

$$\frac{\partial J_e}{\partial x} = e\frac{\partial N_e}{\partial t} + e\left(\frac{N_e - N_{e^0}}{\tau_e}\right) \tag{2.133}$$

$$\frac{\partial J_h}{\partial x} = -e\frac{\partial N_p}{\partial t} - e\left(\frac{N_p - N_{p^0}}{\tau_h}\right) \tag{2.134}$$

where τ_e and τ_h are the electrons and hole lifetimes; N_{e^0} and N_{p^0} are the equilibrium values of the minority carrier electron and hole densities respectively.

(iv) The bulk recombination rate per unit volume in an n-type semiconductor under low-level injection is

$$R = \frac{N_p - N_{n0}}{\tau_h} \tag{2.135}$$

and the surface recombination rate per unit area is $R_s = (N_p - N_{n0})u_r$, where u_r is the surface recombination velocity. Similar equations may be written for the p-type material when the p–n junction is subjected to a voltage V, the values of N_{e^0} and N_{p^0} change to $N_{e^0}\left[\exp\left(\frac{eV}{kT}\right) - 1\right]$ and $N_{p^0}\left[\exp\left(\frac{eV}{kT}\right) - 1\right]$, respectively. If we assume a hole mobility of μ_h, then the drift velocity of holes is

$$u_h = \mu_h E \tag{2.136}$$

Thus the conduction current density in the semiconductor is given by

$$\begin{aligned} J &= J_e + J_h = \rho_e u_e + \rho_h u_h = -\rho_e \mu_e E + \rho_h \mu_h E \\ &= (\rho_h \mu_h - \rho_e \mu_e)E = \sigma E \ \ \text{A/m}^2 \end{aligned} \tag{2.137}$$

where

$$\rho_h = eN_h \ \ \text{C/m}^3 \tag{2.138}$$

$$\rho_e = -eN_e \ \ \text{C/m}^3 \tag{2.139}$$

where N_h is the number of free holes/m^3, N_c is the number of free electron/m^3 and σ the conductivity S/m is given by

$$\begin{aligned} \sigma &= \rho_h \mu_h - \rho_e \mu_e \\ \sigma &= \begin{cases} (N_h \mu_h + N_e \mu_e)e & \text{for a semiconductor} \\ eN_e \mu_e & \text{for a good conductor such as gold} \end{cases} \end{aligned} \tag{2.140}$$

where $e = -1.6 \times 10^{-19}$ C. In dielectric materials, conductivity $\sigma = 0$ and conduction current $J = 0$. In good conductors such as copper, silver or aluminum, $\sigma \cong 10^7$, and E is very small, almost zero ($E \cong 0$) inside the conductor. In a copper conductor, for instance, $\sigma \cong 5.8 \times 10^7$ S/m, $\mu_e = 0.0032\,\mathrm{m^2/Vs}$ and $N_h = 0$. Thus the volume electric charge density $\rho_e = -\sigma/\mu_e = (-5.8 \times 10^7)/0.0032 = -1.8 \times 10^{10}\,\mathrm{C/m^{-3}}$. If the electric field inside the conductor is, say, $1\,\mathrm{mV/m}$ ($E = 1\,\mathrm{mV/m}$), then the current density $J = \sigma E = 5.8 \times 10^7 \times 10^{-3}\,\mathrm{A/m^2}$. The drift velocity of the electrons $u_e = -\mu_e E = 32 \times 10^{-7}\,\mathrm{m/s}$. The number of electrons, $N_e = -\rho_e/e = (-1.8 \times 10^{10})/(1.6 \times 10^{-19}) = 1.1 \times 10^{29}\,\mathrm{m^{-3}}$. If the conductor radius $a = 5\,\mathrm{mm}$, then the current through the conductor $I = J(\pi a^2) = 4.5\,\mathrm{A}$. We could work out the current flow using an alternative, electric circuit approach as well: the voltage (or potential) drop across the conductor $V = E\ell = \ell\,\mathrm{mV}$, where ℓ is the length of the conductor, and $E = 1\,\mathrm{mV/m}$. Resistance $R = \ell(\sigma A)$, where the cross-sectional area $A = \pi a^2$. Therefore, from Ohm's law $V = IR$, current $I = V/R = \ell(\ell/\sigma A) = \sigma A = \sigma(\pi a^2)A = 4.5\,\mathrm{A}$. Although the macroscopic electric circuit approach is an approximation of the microscopic electric field approach, it yields correct answers in many instances and hence is widely used in the analysis and design of electrical and electronic systems.

2.6.9.1 Photovoltaic effect

Semiconductor materials are used in devices other than transistors and diodes such as light emitting diodes (LED) and light dependent resistors (LDR). One such device is the photovoltaic devices as shown in Fig. 2.18.

As shown in Fig. 2.18(a), an n-type semiconductor and a p-type semiconductor are placed together, that is a silicon block is doped with n-type donors in the upper region and the lower region is doped with acceptors (holes). There will be a voltage drop across the junction of the two regions. This volt drop will drive a current I through a load connected between the n and p regions when photons (i.e. electromagnetic waves at light frequencies of around 10^{14} Hz) impinge on the n-region releasing electrons to move toward the n-region terminal. The releasing of electrons to become current carrying electric discharge is called the photoelectric effect. When the energy of the photon U_p impinging on the electrode is greater than the energy which binds the electrons to their nucleii, the electrons will be released. Note that the energy of a photon is given by $U_p = hc/\lambda = hf$, where $c = 3 \times 10^8\,\mathrm{ms^{-1}}$, the velocity of light, λ is the wavelength, f is the frequency of the light wave and h is Planck's constant.

The photovoltaic cells (PV) (Fig. 2.18(c)) are used as DC power sources. The PV cell shown in Fig. 2.18(c) is a single layer device. In order to increase the efficiency of the cell, a multilayer cell at a high cost may be made. The top layer may be a InGaP semiconductor which absorbs the high-energy carrier, such as blue light ($\lambda = 400$ nm). The next layer under the InGaP layer is an InGaAs layer which absorbs green light ($\lambda = 700$ nm), whereas the third, bottom layer is Ge, which will absorb orange light ($\lambda = 1100$ nm). Such a multilayer PV cell is more costly to produce, but its conversion efficiency is about three times higher than that of a

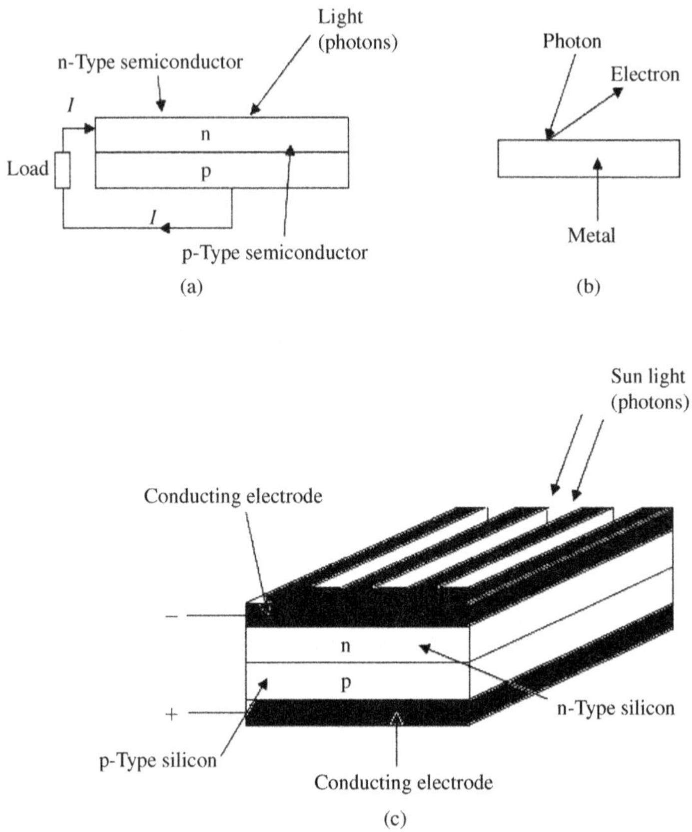

Figure 2.18: Pholtovotaic effect on (a) a p–n junction, (b) a metal and (c) a photovoltaic cell.

single layer PV cell. Several PV cells may be arranged to form, say, 4×3 PV cell module; these modules may be further cascaded to form a PV-array.

2.7 Magnetostatics: magnetic fields due to electric currents

2.7.1 Magnetic fields

Magnetic fields may be generated by DC electric fields (magnetostatics) or by time varying electric fields associated with displacement currents (electromagnetic induction). In both these fields of magnetostatics and electromagnetic induction, the frequency of the electromagntic fields is not large enough to cause electromagnetic radiation or electromagnetic wave propopagation. Hence in magnetostatics, we look at magnetic fields produced by the first term on the right-hand side of (2.3). In the electromagnetic induction, we look at magnetic fields produced by the second term in (2.3).

2.7.2 Electric currents

Electric currents are mostly associated with the movement of electric charges such as electrons under the force produced by an electric field. The displacement electric current is related to time-changing electric field and is not associated with the movement of electric charges at all. There are altogether four forms of electric current and only one type of current does not need free charges for movement.

(a) Conduction current. The electric field moves free electrons through a stationary atomic lattice.

(b) Displacement current. In displacement current no movement of electric charges is present. The time varying electric field produces a displacement current density of $\varepsilon\, \mathrm{d}E/\mathrm{d}t$.

(c) Polarization current. The polarizarion current occurs in an insulator, when electrons are transferred from one place to another in the insulator due to the straining of the atoms and alignment of molecules.

(d) Convection current. The convection current does not flow in a solid conductor, but in space. The motion of an electrically charged body carrying its electric charges with it or, as in a cathode ray tube, electrons are evaporated from a hot cathode and propagated by electric fields. Since both positive (q_+) and negative (q_-) electric charges may be present, the total current flow is found from:

$$I = \frac{\mathrm{d}(q_+ - q_-)}{\mathrm{d}t} \tag{2.141}$$

2.7.3 Moving electron and moving conductor in magnetostatic fields

2.7.3.1 Motion of an electron in magnetic field

According to Lorentz equation, the force on an electric charge, such as an electron ($q = e$), moving at a velocity v in a magnetic flux density of B Webers/m^2 is given by

$$F = evB = \frac{mv^2}{r} \tag{2.142}$$

where m is the mass of the electric charge, which yields to us the oscillating angular speed of the electron around the magnetic field

$$\omega = \frac{v}{r} = \frac{eB}{m} \text{ rads/s} \tag{2.143}$$

Consider now an electron moving inside both electric fields and magnetic fields as shown in Fig. 2.19. The force due to the magnetic field ($v \times B$) is perpendicualr to B. The force due to the electric field (qE) is along the electric field lines. As it

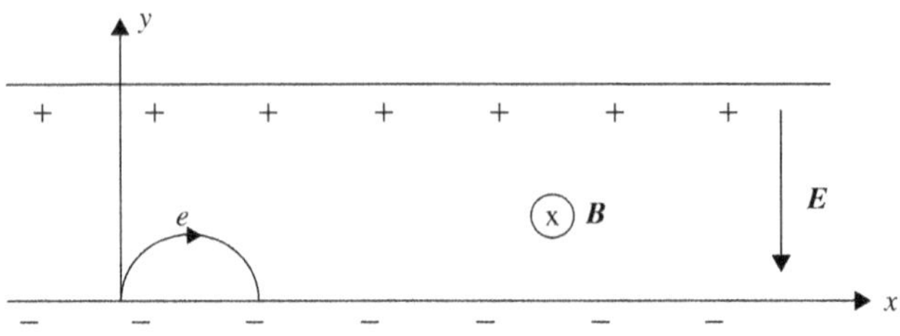

Figure 2.19: Electron motion in electric and magnetic fields.

leaves the $x = 0, y = 0$ point, its original velocity is along the y-axis only, since its inital velocity is zero and hence the force due to the magnetic field is zero at the origin. This is as shown in Fig. 2.19: the electron starts moving in the y-direction, along the electric field line. Using the Lorentz equation, the motion of the electron in E and B fields, for the geometry defined in Fig. 2.19, due to the force in the y-direction may be used in Newton's law:

$$e\mathbf{E} - e\frac{\mathrm{d}x}{\mathrm{d}t}\mathbf{B} = m\frac{\mathrm{d}^2y}{\mathrm{d}t^2} \tag{2.144}$$

Considering now the x-direction, noting that there is no electric field producing a force in the x-direction, we get

$$e\frac{\mathrm{d}y}{\mathrm{d}t}\mathbf{B} = m\frac{\mathrm{d}^2x}{\mathrm{d}t^2} \tag{2.145}$$

Integrating (2.145), we have

$$e\mathbf{B}y = m\frac{\mathrm{d}x}{\mathrm{d}t} + k \tag{2.146}$$

At the origin ($x = y = 0$), since the electron moves along the direction of the electric field only, that is along the y-axis, $\mathrm{d}x/\mathrm{d}t = 0$. Applying the conditions that at $y = 0$, $\frac{\mathrm{d}x}{\mathrm{d}t} = 0$, we have $k = 0$ in (2.146). With $k = 0$, substituting (2.146) into (2.144), we get

$$m\frac{\mathrm{d}^2y}{\mathrm{d}t^2} + \frac{e^2\mathbf{B}^2}{m}y = e\mathbf{E} \tag{2.147}$$

which yields

$$y = \frac{m\mathbf{E}}{e\mathbf{B}^2}\left(1 - \cos\frac{e\mathbf{B}t}{m}\right) \tag{2.148}$$

At $t = 0, y = 0, \frac{dy}{dt} = 0$, and solving this time for x, we have

$$x = \frac{\mathbf{E}}{\mathbf{B}}\left(t - \frac{m}{e\mathbf{B}}\sin\left(\frac{e\mathbf{B}t}{m}\right)\right) \tag{2.149}$$

Using (2.148) and (2.149), we may sketch the trajectory of the charged particle (electron), which is shown in Fig. 2.19.

2.7.3.2 A conductor moving through a *B* field

Consider now a conductor of length L moving perpendicularly to a magnetic field B at a velocity υ. The force on the conductor is given by

$$F = q\upsilon B \tag{2.150}$$

which induces a voltage along the conductor. Noting that $F = qE = q\upsilon B$, the electric field induced along the conductor by the conductor cutting the magnetic field is given by

$$E = \upsilon B \tag{2.151}$$

Therefore, the induced voltage $v = \int \mathrm{E}\,\mathrm{d}\ell = \upsilon BL$, which is typical of an electric generator where an electric charge (or current) carrying moving conductor cutting a magnetic field will experience a voltage induced across it by the static (or time-changing) magnetic field.

Consider a time-changing magnetic field. The total flux cutting the conductor as it moves a short distance x is given by

$$\psi = \iint B\,\mathrm{d}s = BLx \tag{2.152}$$

We saw above that the voltage or electromotive force (emf) induced across the conductor is

$$\mathrm{emf} = \upsilon BL \tag{2.153}$$

where the velocity of the conductor may be expressed as

$$\upsilon = -\mathrm{d}x/\mathrm{d}t \tag{2.154}$$

Substituting (2.154) into (2.153), we get

$$\begin{aligned}\mathrm{emf} &= -\frac{\mathrm{d}x}{\mathrm{d}t}BL = -\frac{\mathrm{d}}{\mathrm{d}t}(BLx) \\ &= -\frac{\mathrm{d}(BA)}{\mathrm{d}t} = -\frac{\mathrm{d}\psi}{\mathrm{d}t}\end{aligned} \tag{2.155}$$

Therefore, the voltage induced in a conductior moving across a time-changing magnetic field is equal to the rate of change of the magnetic flux ($\psi = BA$) that cuts the conductor.

The force due to a magnetic field on an electric charge, given in (2.1), may be rewritten for an element of length dL carrying a current I: $\mathbf{F} = q\mathbf{v} \times \mathbf{B} = q(\mathrm{d}\mathbf{L}/\mathrm{d}t) \times \mathbf{B} = (\mathrm{d}q/\mathrm{d}t)\mathrm{d}\mathbf{L} \times \mathbf{B} = I\,\mathrm{d}\mathbf{L} \times \mathbf{B}$ N. The Biot–Savart law, which was experimentally discovered, states that the magnetic flux density produced by a current carrying element is given by

$$\mathbf{B} = \frac{\mu_0}{4\pi} \oint_c \frac{I\,\mathrm{d}\mathbf{L} \times u_r}{r^2} \tag{2.156}$$

where the free space permeability $\mu_0 = 4\pi \times 10^{-7}$ H/m. Therefore, the force produced on a current carrying coil C_1 by another current carrying coil C_2 is given by

$$F = \frac{\mu_0}{4\pi} \int_{C1} \int_{C2} I_2\,\mathrm{d}\mathbf{L}_2 \times I_1\,\mathrm{d}\mathbf{L}_1 \tag{2.157}$$

where I_1 and I_2 are the currents flowing along the coils C_1 and C_2 respectively.

2.7.4 Magnetic dipole

As in the case of electrostatics, where randomly polarized, atomic level electric dipoles are aligned inside dielectric materials when subject to external electric fields, in magnetic materials, macroscopically aligned magnetic dipoles are formed when the materials are subject to external magnetic fields. In ferromagnetic substances, for instance, there are microscopic regions or domains within the material where the magnetic moments are aligned to each other. However from domain to domain, the microscopic dipole moment of each domain is randomly aligned with respect to others, including neighboring, domains. These macrsoscopically, randomly aligned domains, when experiencing an externally applied magnetic field, become aligned to each other. The externally applied magnetic field (e.g. through a torroidal coil) exerts a force on the net magnetic moment of each domain. Now the domains become aligned to each other, resulting in a macroscopically aligned magnetic moment. The magnetic material now is magnetized.

The electric dipole moment, for electric charges $+q$ and $-q$ separated by a distance d is defined as qd that is $m_e = qd$. the magnetic dipole moment, for a loop of area A carrying a current I, is likewise defined as $m_m = IA$. The magnetic moment m_m is directed perpendicularly to the plane on which the loop lies, in the direction of the thumb held vertically out when the other four fingers of the right hand grasp the loop in the direction of the current flow, as shown in Fig. 2.20(a).

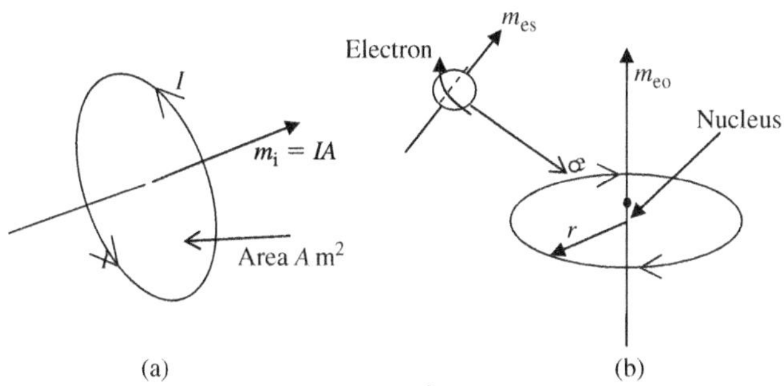

Figure 2.20: Magnetic dipole.

At the atomic level, magnetic moments are produced by the following two actions:

(i) Orbiting electron. The electron orbiting around the nucleus may be seen as a current loop of radius r. This results in an orbital magnetic moment m_{eo}. Note that to find the direction of m_{eo}, it should be remembered that the current flow direction is opposite to the direction in which the electron spins.

(ii) The spinning electron. The electron orbiting around the nucleus is also spinning in its own axis, just as the earth spins on its own axis (thus producing day and night) as it orbits around the sun. The orbiting and spinning electron is shown in Fig. 2.20(b). The magnetic moment of an atom is the summation of the orbited (m_{eo}) and spins (m_{es}) moments of the electrons. The proton dipole moments are about one-third of the electron dipole moments and may be ignored.

Consider now the orbital moment of the electron. For an electron that orbits in a circle of radius r, taking T s to complete one orbit, the current associated with its orbital motion is given by

$$I = -\frac{e}{T} = -\frac{eU_e}{2\pi r} \tag{2.158}$$

where $T = \frac{2\pi r}{U_e}$ and U_e is the orbital velocity of the electron. The area of the current loop, $A = \pi r^2$.

Therefore the orbital moment is

$$m_{eo} = -\frac{eU_e r}{2} \tag{2.159}$$

If the angular moment of the electron may be defined as $M_{ea} = m_e u_e r$, where m_e is the mass of an electron and using quantum physics, the angular momentum,

$m_{eo} = 0, \hbar, 2\hbar, \ldots$ an integer multiple of $\hbar = \frac{h}{2\pi}$, where h is the Planck's constant, we get

$$m_{eo} = -\frac{eM_{ea}}{2m_e} = -\frac{e\hbar}{2m_e} \quad (2.160)$$

Similarly, the spin magnetic moment of the electron has been found to be equal to the orbital magnetic moment of the electron and is therefore given by

$$m_{es} = m_{eo} = -\frac{e\hbar}{2\mathrm{m}_e} \quad (2.161)$$

In the absence of external magnetic fields, the electron orbital moments of atoms are randomly oriented. Hence the net magnetic moment in the absence of an external magnetic field is zero. As for spin magnetic moments, atoms with an even number of electrons have zero electrons spin moments since the paired electrons spin in opposite directions. However, atoms with odd number of electrons possess the electron spin moment of the unpaired electron.

2.7.5 Magnetization curve

Consider a closely wound toroidal coil (see Fig. 2.21(a)) with a mean radius R (or diameter $d = 2R$) and an air core. A current I A flows through the N number of turns of the coil, each coil having a diameter of a.

Outside the core (i.e. for $r < R - a$ and $r > R + a$) the magnetic field is zero, since in the closed loop integral of (2.17), the total current enclosed by the closed loop is zero when the path of integration is outside the core. Inside the core,

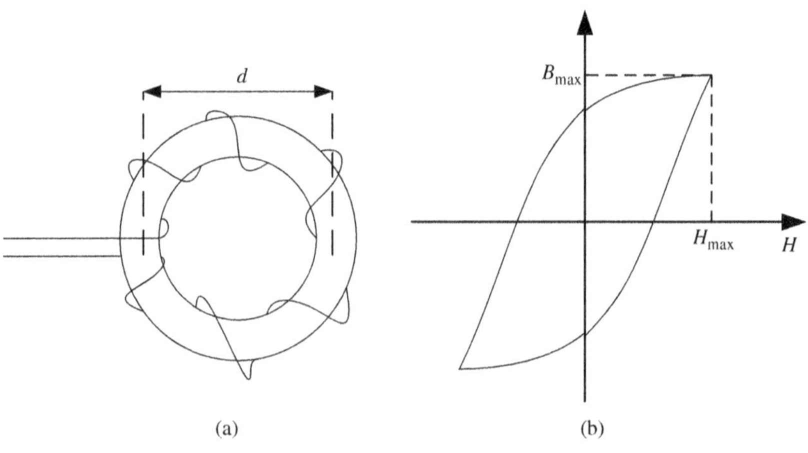

Figure 2.21: (a) Magnetic circuit and (b) magnetization curve (or hysteresis loop).

$R - a < r < R + a$, if we integrating (2.17) at $r = R$ will yield $H \cdot 2\pi R = NI$, or,

$$H = \frac{NI}{2\pi R} \text{ A/m} \tag{2.162}$$

Simlarly for an infinitely long, air-core solenoid, with n number of turns per unit length, carrying a current I A, the magnetic field inside the solenoid is $H = nI$ A/m. For both the solenoid and torroid, the magnetic flux density inside the core is $B = \mu_0 H$ W/m^2.

The relative permeabilty of certain materials in the form (name of material, its relative permeability) are as follows:

Diamagnetic materials: (silver, 0.999999976), (gold, 0.99996), (water, 0.999999901).

Paramagnetic materials: (aluminum, 1.000021).

Ferromagnetic materials: (ferrite, 1000), (nickel, 600), (steel, 2000), (iron with some impurity, 5000), (pure iron, 200,000), (supermalloy, 10^6).

If the toroidal coil is wound around a ferromagnetic material, as shown in Fig. 2.21(a), $B = \mu_0\mu_r H$ W/m^2, where μ_r is the relative permeability of the material given by $\mu_r = (1 + \chi)$ where χ is the susceptibility of the ferromagnetic material. For ferromagnetic materials such as nickel, iron and cobalt, the relative permeability is quite large, and in the range of 50–5000; for special alloys it could be as large as 10^6. Susceptibility is included to show the additional current component (J_m, the magnetization current density) that appears in the magnetic materials. In terms of the magnetization vector M(in A/m), the magentization current density defined by $J_m = \text{Curl}(M)$ A/m^2, and $M = \chi$ H. Both χ and μ_r are constants for linear, isotropic and homogeneous magnetic materials. Ferromagnetic materials are made up of magnetic domains, with the magnetic dipole moments (produced by spinning electrons) within each domain aligned with each other and producing a net magnetic moment per domain. Each domain is about 100 atoms thick, with domain walls separating each domain from the other. When an external magnetic field is applied to the material, the magnetic dipole moment of each domain tries to be aligned with the applied, external magnetic field.

What will happen when we increase the alternating current supplied to the toroidal coil? We get the hystersis characteristics as shown in Fig. 2.21(b). Initially, when the current to the toroid is zero, we are at the origin of Fig. 2.22(a), that is $H = B = 0$. As we gradually increase the current, the point moves toward ($H_{\text{max}}, B_{\text{max}}$), which is shown in Fig. 2.21(b). The curve shown in Fig. 2.21(b) is called the normal magnetization curve. At this point if we reduce the current I going into negative values, the point does not return to zero. Instead it travels back along the B–H nonlinear curve toward negative H and B. Furthermore, when increasing the

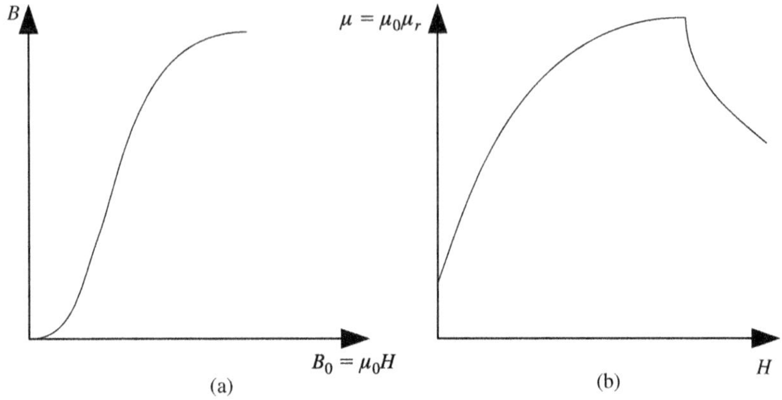

Figure 2.22: (a) Normal magnetizatuion curve and (b) change in permeability of a ferromagnetic material with increase in applied magnetc field.

current again, we return to peak $H_{\max}$, $B_{\max}$, again avoiding the origin (assuming we begin reserving the current at $-H_{\max}$ and $-B_{\max}$). This nonlinear behavior of the magentic material is due to magnetization lagging behind the field producing it. This is called hysteresis. As the field increases, the magnetization that takes place within the material becomes nonreversible. As the applied field is further increased, total magnetization occurs when the applied magnetic field and the magnetization are totally aligned to each other. The change of the material relative permeabilty is shown in Fig. 2.22(b).

There are antiferromagnetic materials such as chromium and manganese which are close to ferromagnetic materials in the periodic table, but unlike in ferromagnetic materials, despite the strong coupling forces between the atomic magnetic dipole moment, the coupling forces produce antiparallel alignments between electron spins resulting in no net magnetic moment which is found from domain to domain in ferromagnetic materials. When the temperature is increased, the antiferromagnetic materials become like a paramagnetic material which is thermally agitated as temperature is increased. Ferrimagnetic materials have a characteristics that lie between the ferromagnetic and antiferromagnetic materials.

2.7.6 Losses in iron subjected to alternating magnetic fields

2.7.6.1 Iron losses

When iron, used for instance in high-voltage power transformers, is subjected to sinusoidally varying magnetic fields, there are two kinds of power losses that it experiences. These are the hysteresis and eddy current losses.

2.7.6.2 Hysteresis loss

Consider a power generator connected to a power transformer core that is made of iron. If the emf induced in the transformer coils wrapped arond the iron core is V_e

and the current flowing from the generator to the transformer is I, the work done by the generator in time T is given by

$$W = \int_{0}^{T} - V_{\mathrm{e}} I \, \mathrm{d}t \tag{2.163}$$

According to Lenz's law, the voltage induced in the transformer coil by a flux $\psi = BA$,

$$V_{\mathrm{e}} = -N \frac{\mathrm{d}\psi}{\mathrm{d}t} = -NA \frac{\mathrm{d}B}{\mathrm{d}t} \tag{2.164}$$

Moreover, using Ampere's law,

$$\oint_{c} H \cdot \mathrm{d}l = Hl = NI \tag{2.165}$$

Using (2.164) and (2.165) in (2.163), we get

$$W = \int_{0}^{T} - \left(NA \times \frac{\mathrm{d}B}{\mathrm{d}t} \times \frac{Hl}{N} \right) \mathrm{d}t = -\int_{0}^{T} AlH \frac{\mathrm{d}B}{\mathrm{d}t} \mathrm{d}t \tag{2.166}$$

$$= -(Al) \int_{B1}^{B2} H \, \mathrm{d}B \tag{2.167}$$

where $B(0) = B_1$ and $B(T) = B_2$.

Looking at (2.167) carefully, we note that (Al) is the volume of the iron core. Moreover, the integral HB is the area of the hysteresis loop shown in Fig. 2.21(b) when we go through one cycle from B_1 back to B_1. In other words, the work done by the source supplying the transformer is the volume of the core $\times$ area of the hysteresis loop per cycle. In general, the power loss due to the hysteresis loss is,

$$P_{\mathrm{h}} = kfB^{1.6} \tag{2.168}$$

based on the shape of the curve, where B is the peak value on the hysteresis curve as excited. This loss is commenly called iron loss.

There are materials other than ferromagnetic materials ($\mu_{\mathrm{r}} >> 1$, susceptibility χ is a large positive number). These are the paramagnetic materials (where $\mu_{\mathrm{r}} > 1$ and χ is a small positive number) and diamagnetic materials (where $\mu_{\mathrm{r}} < 1$ and χ is a small negative number). Diamagnetic materials such as aluminum, titanium, magnesium and tungsten have a susceptibility of around -10^{-5}.

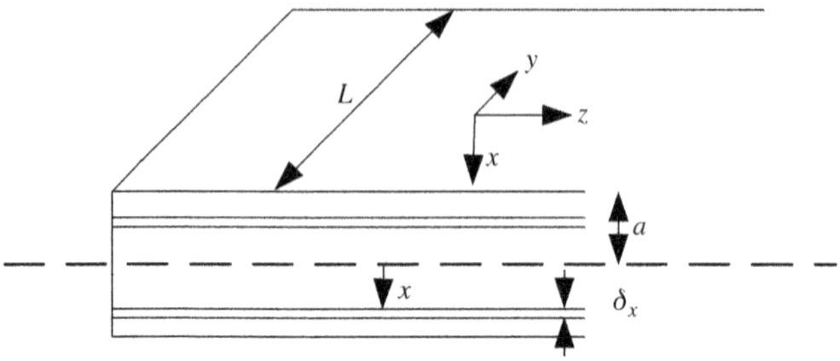

Figure 2.23: Eddy currents.

2.7.6.3 Eddy current loss

Eddy currents are currents induced inside metal substances by a time-varying magnetic field. This gives rise to two negative effects. There will be ohmic power loss and heating inside the metal substance due to eddy currents circulating inside it. To reduce eddy currents, in many kinds of high-voltage equipment, such as transformers, the iron core is laminated to reduce the path length allowed for the eddy current to flow. First this would cut down the emf induced and hence the magnitude of the eddy current and eddy current power loss. Each lamination will be insulated from its neighboring laminated plates. And second, since the direction in which the eddy currents flow, according to Lenz's law, seeks to oppose the magnetic field that induced it, the magnetic field inside the metal substance will be reduced. In some applications, such as heating up metals to melting points, eddy currents are used to produce heat deep inside the metal block to be melted.

Consider a sinusoidally time-varying magnetc field in a lamination

$$B = B_0 \sin \omega t \tag{2.169}$$

This lamination is a thin strip of rectangular conductor as shown in Fig. 2.23. Let the flux flow ψ in the z-direction inducing circularly flowing eddy currents on the xy-plane The resistivity of the material is ρ and its width is L. For a thin δ_x m strip of the lamination, the per unit length in the z-direction resistance (in ohms) to the eddy current is

$$R = \frac{\rho 2L}{1 \times \delta_x} = \frac{2\rho L}{\delta_x} \tag{2.170}$$

The emf induced in the conductor by a total flux of ψ T is

$$V_e = -\frac{d\psi}{dt} = -2xLB_0\,\omega \cos \omega t \tag{2.171}$$

Hence the mean power loss, $V_e^2/2R$, is obtained from (2.170) and (2.171), as

$$\delta p = \frac{\omega^2 B_0^2 L x^2 \delta x}{\rho} \tag{2.172}$$

Hence the total power dissipated in the conductor (in W) by the eddy currents is given by

$$P = \int_0^a \frac{\omega^2 B_0^2 L x^2}{\rho} \mathrm{d}x = \frac{\omega^2 B_0^2 L a^3}{3\rho} \tag{2.173}$$

The volume of the material is $2aL$. Hence the eddy current power loss per unit volume is

$$P_0 = \frac{\omega^2 B_0^2 a^2}{6\rho} = \frac{2\pi^2 f^2 B_0^2 a^2}{3\rho} \tag{2.174}$$

To reduce eddy current power losses, we seek to use materials that have a low conductivity at high frequencies. Ferrite substances are used, especially the magnetic spinels, which are ceramic like materials. The ferrites which are used in high frequency and microwave devices are antiferromagnetic materials which have low conductivites (e.g. 1 S/m, which is much lower than the 10^7 S/m of iron).

2.7.7 Induction heating using magnetic fields

Consider the problem of heating a conducting body by using magnetic fields. A thin metal shell lies on the z-axis; its inner radius is a and its thickness is b (i.e. its outer radius is $a + b$). A source magnetic field $\mathbf{B}_s = B_0 \cos \omega t\, \mathbf{u}_z$ T, or in phasor form $B_0 e^{j\omega t} \mathbf{u}_z$, is applied along the z-axis. According to Faraday's law, this magnetic field will induce a circulating electric field $\mathbf{E} = E\, e^{j\omega t} \mathbf{u}_\varphi$ around the outer surface of the shell, and thus a circulating current $\mathbf{J} = \sigma \mathbf{E}$. According to Ampere's law, this circulating current produces its own magnetic field $\mu_0 J e^{j\omega t} b \mathbf{u}_z$ that opposes the source field $\mathbf{B}_s$, giving rise to a total magnetic field of

$$\mathbf{B} = (B_0\, e^{j\omega t} + \mu_0 J b)\mathbf{u}_z = (B_0\, e^{j\omega t} + \mu_0 \sigma b E\, e^{j\omega t}\, \mathrm{b})\mathbf{u}_z \tag{2.175}$$

Using the integral form of Ampere's law

$$\oint_c \mathbf{E} \cdot \mathrm{d}\boldsymbol{l} = -\frac{\partial}{\partial t} \int_s \mathbf{B} \cdot \mathrm{d}\boldsymbol{s} \tag{2.176}$$

we get

$$E e^{j\omega t} \cdot 2\pi a = -\frac{\partial}{\partial t}[(B_0 e^{j\omega t} + \mu_0 \sigma b\, E\, e^{j\omega t})\pi a^2] \tag{2.177}$$

which yields, canceling the $e^{j\omega t}$ on both sides and replacing $\frac{\partial}{\partial t}$ by $j\omega t$,

$$E = \frac{-j\omega a B_0}{2 + j\omega\sigma\mu_0 ab} \tag{2.178}$$

Therefore, the heat dissipated per second in the conducting shell by the source magnetic field is

$$P_h = \frac{1}{2} Re \int_v JE \, dv = \sigma\pi ab|E|^2 = \frac{\sigma\pi ab(\omega a B_0)^2}{4 + (\omega\sigma\mu_0 ab)^2} \quad \text{W} \tag{2.179}$$

If the specific heat of the shell material is s and its mass is m, then the time taken to raise the temperature by ΔT is $((ms\Delta T)/P_h)$ s.

2.8 Further topics in electrostatics

2.8.1 An inclined plane MEMS electrostatic actuator

Consider a MEMS actuator arrangment. An electrode of length L is placed along the r-axis in the cylindrical coordinate system. The end of the conductor nearest to the origin is at a distance R from the origin; the other end of the conductor is a distance $R + L$ from the origin. A second conductor, a plate of width w, has one end of it at the origin, and it is inclined at an angle φ_a from the r-axis. A DC source V_0 is connected across the two conductors: with the positive terminal connected to the midpoint of the horizontal conductor of length L, making it the anode. The negative terminal of the DC source is connected to the inclined plate conductor at the point of origin. The Laplace's equation for the electric potential ψ simplifies to

$$\frac{d^2\psi}{d\phi^2} = 0 \tag{2.180}$$

noting that both $\frac{\partial}{\partial r}$ and $\frac{\partial}{\partial z}$ tend toward zero. Integrating the Laplace's equation twice, we have

$$\psi = C_1\phi + C_2 \tag{2.181}$$

The constants of integration C_1 and C_2 are found from two boundary consitions: (i) when $\phi = 0$ we have $\psi = V_0$. Therefore, $C_2 = V_0$. (ii) When $\phi = \phi_0, \psi = 0$. Hence $C_1 = -\frac{C_2}{\phi_0} = -\frac{V_0}{\phi_0}$. Therefore,

$$\psi = -\frac{V_0}{\phi_0}\phi + V_0 = V_0\left(1 - \frac{\phi}{\phi_0}\right) \tag{2.182}$$

Hence the electric field between the two plates is given by

$$\mathbf{E} = -\nabla\psi = -\mathbf{u}_\phi \frac{1}{r}\frac{\partial\psi}{\partial\phi} = \mathbf{u}_\phi \frac{V_0}{r\phi_0} \tag{2.183}$$

Hence the surface electric charge on the anode conductor is given by

$$\sigma_s = \mathbf{D} = \varepsilon E = \epsilon \frac{V_0}{r\phi_0} \ \mathrm{C/m^2} \tag{2.184}$$

where **D** is the electric flux density. In order to obtain the capacitance of this actuator system, we first need to get the total electric charge spread out over the inclined plate of width w:

$$Q = \iint \sigma_s \, \mathrm{d}s = \epsilon \frac{V_0}{\phi_0} \int_R^{R+L} \frac{w\,\mathrm{d}r}{r} = w\epsilon \frac{V_0}{\phi_0} \ln\left(\frac{R+L}{R}\right) \tag{2.185}$$

Therefore, the capacitance of the actuator system is given by

$$C = \frac{Q}{V_0} = \frac{w\epsilon}{\phi_0} \ln\left(\frac{R+L}{R}\right) \ \mathrm{F} \tag{2.186}$$

The inclined plane actuator has a capacitance that is a function of the angle of inclination. Consider a parallel plate arrangement where each plate has an area A m^2. The plates are parallel to the xy-plane and separated from each other by a distance d along the z-axis. The Laplace's equation for the potential between the two plates is $\nabla^2\psi = 0$, becomes $\frac{\mathrm{d}^2\psi}{\mathrm{d}z^2} = 0$. It may easily be verified that the solution for the potential is

$$\psi(z) = V_0 \frac{d-z}{z} \tag{2.187}$$

Therefore, the electric field between the two plates is

$$\mathbf{E} = -\nabla\psi = -\mathbf{u}_z \frac{\mathrm{d}\psi}{\mathrm{d}z} = \mathbf{u}_z \frac{V_0}{d} \ \mathrm{V/m}. \tag{2.188}$$

The electric displacement flux density $\mathbf{D} = \varepsilon\mathbf{E} = \mathbf{u}_z \varepsilon \frac{V_0}{d}$. Therefore, the total charge on each of the two parallel plates is $Q = \varepsilon A \frac{V_0}{d}$. Therefore, the capacitance of the two-plate arrangement is $C = Q/V_0 = \varepsilon A/d$. The energy strored in the parallel plate capacitor is $\frac{1}{2}\varepsilon E^2 \mathrm{J/m^3}$ or $\frac{1}{2}\varepsilon E^2 \times Ad = \frac{1}{2}\varepsilon\left(\frac{V_0}{d}\right)^2 Ad = \frac{1}{2}CV_0^2$ J.

The permittivity of materials used in capacitor type energy storage devices, actuators and sensors varies and determines the ability of the capacitor to store energy or respond to external signals or forces, since capacitance is directly proportional to the permittivity $\varepsilon = \varepsilon_0\varepsilon_r$, where ε_r is the relative permittivity of the material. Values of ε_r vary vastly: for air 1, for alumina 8.8, for barium titanate 1200, for epoxy 3.4, for fused quartz 3.9, for gallium arsenide 13.1, for glass 4.9, for methanol 3.2, for nylon 4.5, for paper 3, for polystyrene 2.6, for procelain 8, for rubber 6, for silicon 11.9, for sodium chloride 5.9, for styrofoam 1.03, for Teflon 2.1, for water 80 and for dry wood 2.5.

2.8.2 ESD and electric field

Electrical breakdown takes place in solids, liquids and gases when the electric field in the medium exceeds a critical breakdown value. In general, air breaks down at 3 MV/m (or 30 kV/cm). However, as defined by Paschen's curve, as the gap between two electrodes immersed in air is reduced, say to a gap distance of 50 μm and less, the field needed for the air to break down increases. For instance, the gap breakdown electric field values are at 1 cm 3 MV/m, at 1 mm 3 MV/m, at 0.2 mm 5 MV/m, at 0.1 mm 8 MV/m, at 0.05 mm 13 MV/m and at 0.01 mm 30 MV/m. At smaller gaps, the gap distance becomes comparable to the mean free collision path, resulting in reduced ionization in the gap space. This is indeed an advantage in MEMS or NEMS where electric fields in excess of 3 MV/m are required to move components made of materials such as silicon.

Electric breakdown does not occur in an instant. It takes time, and in particular, at a given onset voltage (less than the breakdown voltage) partial breakdown takes place around the conductor, and charged particles (ions and electrons) are produced. The electric charge production can gradually multiply into an avalanche leading to complete breakdown. Corona currents for any given electrode arrangement with voltage difference is given by

$$I_c = K_c(V^2 - VV_c)\ \mathrm{A} \tag{2.189}$$

where K_c is a constant depending on the electriode structure and V_c is the critical onset voltage. For a coaxial cable of an inner radius of a and outer radius b, the corona current is given by

$$I_c = \left(\frac{4\pi\epsilon_0 m_i}{b^2 \ln\left(\frac{b}{a}\right)}\right)(V^2 - VV_c)\ \mathrm{A} \tag{2.190}$$

where the ion mobility m_i for air at standard temperature and pressure is $0.00022\ \mathrm{m^2/V\,s}$. As this corona current flows out of the conductor and into the insulator, it will fill the insulator region (e.g. air or cross-linked polyethelelne insulation) outside the conductor with electric space charge q_s given by

$$q_s = \int_0^{t_0} I_c \, dt \ \text{C} \tag{2.191}$$

Therefore, the insulator gradually becomes a conductor with free electric space charges and will eventually break down to provide a very good conducting path for currents to flow into the insulator.

For a coaxial cable with air as the dielectric medium filling, the space between the inner and outer conductiors, the critical onset votage V_c is the voltage at which the electric field at the inner conductor is Peek's electric field given by

$$E_p = nE_{bd}\left(1 + \frac{0.0308}{\sqrt{a}}\right) \ \text{V/m} \tag{2.192}$$

where $E_{bd} = 3$ MV/m since the insulator is air. The surface roughness factor $n = 1$ for smooth conductiors and $n = 0.8$ for rough inner conductors. Given that the electric field at a distance r from the center of the cable inside a coaxial cable is $E(r) = \frac{V}{a \ln \frac{b}{a}}$, the onset votage is given by

$$V_c = E_p a\left[\ln\left(\frac{b}{a}\right)\right] \tag{2.193}$$

It is interesting to note that as the radius of the inner conductor is increased, the corona onset votage increases.

2.8.3 Electrostatics of floating conductors

We shall consider now the charging of a floating body. Consider a conducting sphere placed in an electric field. The radius of the sphere is R, and the electric surface charge on it is Q C. The electric field in which it is immersed is E_0. The sphere is floating, and not connected to a wired voltage source. In the spherical coordinate system, the electric field in the space outside the particle is

$$E_r(r) = E_0\left(1 + \frac{2R^3}{r^3}\right)\cos\theta + \frac{Q}{4\pi\epsilon_0 r^2} \ \text{V/m} \tag{2.194}$$

and

$$E_\theta = -E_0\left(1 - \frac{R^3}{r^3}\right)\sin\theta \tag{2.195}$$

Defining the saturatuion electric charge as the maximum charge Q_m that the charged, spherical body can hold, it may be found by equating the two terms on the right-hand side of the $E_r(r)$ electric field. Thus, for $\cos\theta = 1$ at $\theta = 0$,

$$E_0\left(1 + \frac{2R^3}{r^3}\right) = \frac{Q_m}{4\pi\epsilon_0 r^2} \ \text{V/m} \tag{2.196}$$

yielding for $r = R$,

$$Q_m = 12\pi\varepsilon_0 R^2 E_0 \ \text{C} \tag{2.197}$$

For $Q < Q_m$ the ion current flowing towards the particle is obtained by integrating the electric field over the sphere,

$$I_i = \frac{dQ}{dt} = 3\pi R^2 E_0 N q_i m_i \left(1 - \frac{Q}{Q_m}\right)^2 \ \text{A} \tag{2.198}$$

where N is the number of ions per unit volume, q_i is the electric charge of an ion and m_i is the mobility of an ion. Thus integrating (2.198), we get

$$Q(t) = Q_m\left(\frac{t/\tau}{1 + t/\tau}\right) \tag{2.199}$$

where the time constant τ is given by

$$\tau = \frac{4\varepsilon_0}{N q_i m_i} \ \text{s} \tag{2.200}$$

which is independent of both E_0 and R. In our discussion above, we have ignored the changes Q will produce in E_0.

When electric field computation (e.g. using the finite element or difference method) is carried out with floating conductors present, there are two ways to handle floating electrodes. One way is to set the electric potential at each floating electrode point (say three points with potentials that is V_4, V_5, and V_6) to be equal,

$$V_4 = V_5 = V_6 \tag{2.201}$$

at each iteration. Say at iteration number i we first calcualte V_4, then soon after calculating it, set the other two floating electrode nodes to the same value of the latest value of potential V_4 at node 4. Next in that same iteration i calculate V_5 and set the latest potential values of all three floating electrode points equal to V_5, and so on until convergence. A second technique is to set the electric permittivity of the floating point electrode as very large, for example,

$$\varepsilon \gg 1000 \tag{2.202}$$

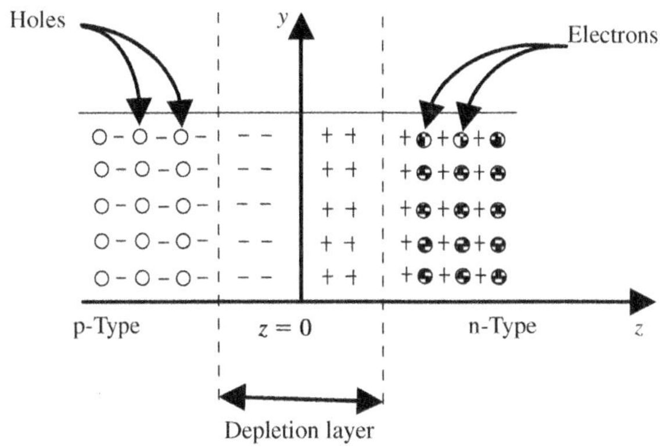

Figure 2.24: The p–n semiconductor junction.

and solve the problem as a multidielctric electrtostatic problem with a mateiral with a very large permittivity (e.g. 10,000). With finite elements, the nodes on a conductor get the same variable number but their actual coordinates.

2.8.4 Electrostatics of diode and transistor junctions

2.8.4.1 Linear electric charge distribution

Consider the p–n junction shown in Fig. 2.24. The characteristics of the junction determine the operation of both junction diodes and junction transistors. In diodes, there is only one p–n junction, whereas in transistors there are two p–n junctions. In the n-type of semiconductor, there are holes that get settled across the junction as shown through drift and thermal effects.

In the p-type of semiconductor, electrons (negative electric charge) get settled across the junction, as shown in Fig. 2.24. This region of electric charge accumulation at the junction preventing any further electric charge exchange between the two regions is called the depletion region. In Fig. 2.25 is shown the electric charge distribution at the p–n junction.

We shall first determine the potential distribution at the junction for the linear electric charge distribution shown in Fig. 2.25(a). Consider the the region between $0 < z < z_n$. The Posisson equation for this region is given by

$$\frac{\mathrm{d}^2\phi}{\mathrm{d}z^2} = -\frac{q_e N_d}{\varepsilon} \tag{2.203}$$

Integrating (2.203), we get

$$\frac{\mathrm{d}\phi}{\mathrm{d}z} = -\frac{q_e N_d}{\varepsilon} z + C_1 \tag{2.204}$$

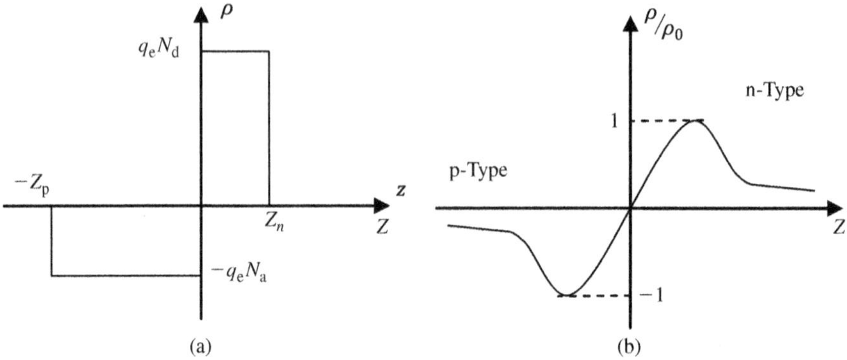

Figure 2.25: The electric charge distribution ρ C/m^3 about a p–n junction: (a) linear charge distributin and (b) nonlinear charge distributin.

Using the boundary conditions that $E_z = \frac{d\phi}{dz} = 0$ at $z = z_n$ and at $z = -z_p$, the constant of integration $C_1 = \frac{q_e N_d}{\varepsilon} z_n$.

Therefore,

$$\frac{d\phi}{dz} = \frac{q_e N_d}{\varepsilon}(z_n - z) \tag{2.205}$$

which on integration again yields

$$\phi = -\frac{q_e N_d}{2\varepsilon}(z_n - z)^2 + C_2 \tag{2.206}$$

If we now look at the region in the range of $-z_p < z < 0$, the Poisson equation is

$$\frac{d^2\phi}{dz^2} = \frac{q_e N_a}{\varepsilon} \tag{2.207}$$

which on integration is

$$\frac{d\phi}{dz} = \frac{q_e N_a}{\varepsilon} z + C_3 \tag{2.208}$$

Applying the boundary conditions as before we get $C_3 = \frac{q_e N_a Z_p}{\varepsilon}$ which on substitution yields

$$\frac{d\phi}{dz} = \frac{q_e N_a}{\varepsilon}(z_p + z) \tag{2.209}$$

integrating which we get

$$\phi = \frac{q_e N_a}{2\varepsilon}(z_p + z)^2 + C_4 \tag{2.210}$$

Since at $z = z_{\mathrm{p}}$, $\phi = 0$, $C_4 = 0$.

Now the electric potential ϕ is continuous at the boundary, that is at $z = 0$, since $E = -\mathrm{d}\phi/\mathrm{d}z$ is finite at the boundary. Thus at the boundary, the two solutions in the different regions should be equal:

$$-\frac{q_{\mathrm{e}}N_{\mathrm{d}}}{2\varepsilon}Z_{\mathrm{n}}^2 + C_2 = \frac{q_{\mathrm{e}}N_{\mathrm{a}}}{2\varepsilon}z_p^2 \tag{2.211}$$

Hence the constant of integration,

$$C_2 = \frac{q_{\mathrm{e}}}{2\varepsilon}(N_{\mathrm{d}}z_{\mathrm{n}}^2 + N_{\mathrm{a}}z_{\mathrm{p}}^2) \tag{2.212}$$

Thus for $0 < z < z_{\mathrm{n}}$ the scalar potential is

$$\phi = -\frac{q_{\mathrm{e}}N_{\mathrm{d}}}{2\varepsilon}(z_{\mathrm{n}} - z)^2 + \frac{q_{\mathrm{e}}}{2\varepsilon}(N_{\mathrm{d}}z_{\mathrm{n}}^2 + N_{\mathrm{a}}z_{\mathrm{p}}^2) \tag{2.213}$$

Likewise for $-z_{\mathrm{p}} < z < 0$, the scalar potential is

$$\phi = \frac{q_{\mathrm{e}}N_{\mathrm{a}}}{2\varepsilon}(z_{\mathrm{p}} + z)^2 \tag{2.214}$$

At the junction, that is at $z = 0$, the scalar potential is given by

$$\phi_0 = \frac{q_{\mathrm{e}}}{2\varepsilon}[N_{\mathrm{d}}z_{\mathrm{n}}^2 + N_{\mathrm{a}}z_{\mathrm{p}}^2] \tag{2.215}$$

or

$$\phi_0 = \frac{q_{\mathrm{e}}}{2\varepsilon}\frac{N_{\mathrm{a}}N_{\mathrm{d}}}{N_{\mathrm{a}} + N_{\mathrm{d}}}(z_{\mathrm{n}} + z_{\mathrm{p}})^2 \tag{2.216}$$

The total length of the depletion region $z_{\mathrm{D}} = z_{\mathrm{n}} + z_{\mathrm{p}}$.

Since

$$(z_{\mathrm{n}} + z_{\mathrm{p}})^2 = \phi_0\left(\frac{2\varepsilon}{q_{\mathrm{e}}}\right)\left(\frac{N_{\mathrm{a}} + N_{\mathrm{d}}}{N_{\mathrm{a}}N_{\mathrm{d}}}\right) \tag{2.217}$$

we have

$$z_{\mathrm{D}} = \sqrt{\left(\frac{2\varepsilon}{q_{\mathrm{e}}}\right)\phi_0\left(\frac{1}{N_{\mathrm{a}}} + \frac{1}{N_{\mathrm{d}}}\right)} \tag{2.218}$$

2.8.4.2 Nonlinear electric charge distribution in the p and n regions

Compared to the electric charge distribution shown in Fig. 2.25(a), a more realistic electric charge distribution at the p–n junction is given by

$$\rho = 2\rho_0 \operatorname{sech}\left(\frac{z}{z_\mathrm{m}}\right) \tanh\left(\frac{z}{z_\mathrm{m}}\right) \tag{2.219}$$

as sketched in Fig. 2.25(b).

At $z = 0.881z_\mathrm{m}$, we have the maximum electric charge density $\rho = \rho_0$. With an acceptor concentration of $N_\mathrm{a}\ \mathrm{m}^{-3}$ and a donor concentration of $N_\mathrm{d}\ \mathrm{m}^{-3}$, we have

$$\rho_0 = q_\mathrm{e} N_\mathrm{a} = q_\mathrm{e} N_\mathrm{d} \tag{2.220}$$

The Poisson equation in the depletion region is given by

$$\frac{\mathrm{d}^2 V}{\mathrm{d}z^2} = \frac{\rho}{\varepsilon} \tag{2.221}$$

which on substitution of the electric charge density becomes

$$\frac{\mathrm{d}^2 V}{\mathrm{d}z^2} = -\frac{2\rho_0}{\varepsilon} \operatorname{sech}\left(\frac{z}{z_\mathrm{m}}\right) \tanh\left(\frac{z}{z_\mathrm{m}}\right) \tag{2.222}$$

Integrating

$$\frac{\mathrm{d}V}{\mathrm{d}z} = \frac{2\rho_0 Z_\mathrm{m}}{\varepsilon} \operatorname{sech}\left(\frac{z}{z_\mathrm{m}}\right) + C_1 \tag{2.223}$$

The constant of integration $C_1 = 0$, since electric field, $E_z = -\frac{\mathrm{d}V}{\mathrm{d}z} = 0$, and so is the electric charge density far away from the boundary (i.e. when z is large). Hence we have

$$E_z = -\frac{2\rho_0 Z_\mathrm{m}}{\varepsilon} \operatorname{sech}\left(\frac{z}{z_\mathrm{m}}\right) \tag{2.224}$$

which on integration yields

$$V = \frac{4\rho_0 Z_\mathrm{m}^2}{\varepsilon} \tan^{-1}(\mathrm{e}^{z/z_\mathrm{m}}) + C_2 \tag{2.225}$$

Setting $V = 0$ at $z = 0$, we can get C_2, resulting in

$$V = \frac{4\rho_0 Z_\mathrm{m}^2}{\varepsilon}\left[\tan^{-1}(\mathrm{e}^{z/z_\mathrm{m}}) - \frac{\pi}{4}\right] \tag{2.226}$$

The potential difference across the p–n junction may be obtained by getting V values for a large positive value for z and a large negative value for z, and then subtracting one from the other to get a potential drop of

$$V_0 = \frac{2\pi \rho_0 Z_{\mathrm{m}}^2}{\varepsilon} \tag{2.227}$$

Using the potential drop across the junction, we may obtain the capacitance across the junction by first getting the electric charge on one side of the junction, assuming a contact area of A at the junction:

$$\begin{aligned} Q &= A \int_0^\infty 2\rho_0 \,\mathrm{sech} \frac{z}{Z_{\mathrm{m}}} \tanh \frac{z}{Z_{\mathrm{m}}} \,\mathrm{d}z \\ &= 2\rho_0 z_{\mathrm{m}} \;\; \mathrm{A} \end{aligned} \tag{2.228}$$

Substiting z_{m} as a function of V_0, we have

$$Q = A\sqrt{\frac{2\rho_0 \varepsilon V_0}{\pi}} \;\; \mathrm{C} \tag{2.229}$$

Therefore, the electric current is

$$I = \frac{\mathrm{d}Q}{\mathrm{d}t} = C\frac{\mathrm{d}V_0}{\mathrm{d}t} \tag{2.230}$$

and thus the capacitance across the junction is

$$C = \frac{\mathrm{d}Q}{\mathrm{d}V_0} = \sqrt{\frac{\rho_0 \varepsilon}{2\pi V_0}} \;\; \mathrm{A} \tag{2.231}$$

which indicates that the capacitance drops as the potential drop across the junction increases. Expressing the capacitance in another form, that is replacing the potential drop across the junction by the length of the depletion region, we have $C = \frac{\varepsilon A}{2\pi Z_{\mathrm{m}}}$, which resembles a capacitance with a plate separation of $2\pi z_{\mathrm{m}}$.

Using numerical methods, we may obtain a more accurate value for capacitance by determining first the equipotential lines and electric field lines; then the capacitance is $C = \varepsilon \frac{N_{\mathrm{Q}}}{N_{\mathrm{V}}}$, where N_{Q} = number of flux tubes joining the two conductors and N_{V} = number of potential increments between conductors. The numerical method for determining capacitance may be used for any electrostatic field device.

2.8.5 Capacitance of some simple structures

For the parallel plate conductor arrangement shown in Fig. 2.26(a), we have $C_1 = \varepsilon_2 \frac{A}{d_1}$ and $C_2 = \varepsilon_1 \frac{A}{d_2}$, where A is the cross-sectional area of both plates. Hence

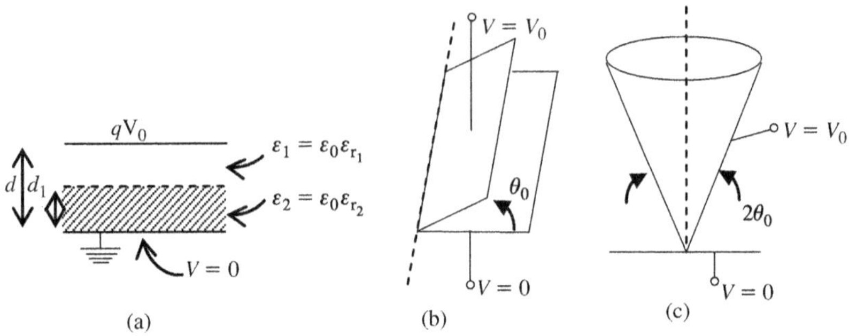

Figure 2.26: The simple arrangements of electrodes: (a) multiple dielctric capacitor (b) slanted plane capacitor and (c) cone standing above a plane. In (b) and (c) the electrodes are close without touching.

the total capacitance is $C = \frac{C_1 C_2}{(C_1+C_2)}$ and the voltage across the dielectric slab of permittivity is $V_2 = \frac{Q}{C_2}$ and $V_2 = \left(\frac{C_1}{C_1+C_2}\right) V_0$ and $V_1 + V_2 = V_0$.

For the electrode system shown in Fig. 2.26(b), $\frac{1}{r^2}\frac{\mathrm{d}^2 v}{\mathrm{d}\theta^2} = 0$. For $r \neq 0$, we have $\frac{\mathrm{d}^2 v}{\mathrm{d}\theta^2} = 0$, which has a solution of the form $V = A\theta + B$. Applying the boundary conditions $V = V_0$ when $\theta = \theta_0$, and $V = 0$, when $\theta = 0$, we get $V = \left(\frac{\theta}{\theta_0}\right) V_0$. Hence the electric field $\mathbf{E} = \left(\frac{-1}{r\theta_0}\right) V_0 \mathbf{u}_\theta$.

For the cone shown in Fig. 2.25(c), the Laplace equation in spherical coordinates is given by $\frac{1}{r^2 \sin\theta}\frac{\mathrm{d}}{\mathrm{d}\theta}\left(\sin\theta \frac{\mathrm{d}v}{\mathrm{d}\theta}\right) = 0$. For $r \neq 0, \theta \neq 0$, and $\theta \neq \pi$, this reduces to $\sin\theta \frac{\mathrm{d}V}{\mathrm{d}\theta} = A$, which on integration yields $V = A \ln\left(\tan\frac{\theta}{2}\right) + B$. Applying the boundary conditions $V = V_0$ at $\theta = \theta_0$, and $V = 0$ at $\theta = \pi/2$, we get

$$V = V_0 \frac{\ln\left(\tan\frac{\theta}{2}\right)}{\ln\left(\tan\frac{\theta_0}{2}\right)} \tag{2.232}$$

The elecric field

$$\mathbf{E} = -\frac{1}{r}\frac{\partial V}{\partial \theta}\mathbf{u}_\theta = -\frac{V_0}{r \sin\theta \ln\left(\tan\frac{\theta_0}{2}\right)}\mathbf{u}_\theta \tag{2.233}$$

The surface electric charge on the cone is, $\sigma_\mathrm{s} = -\left(\frac{\varepsilon}{\sin\theta_0 \ln\left(\tan\frac{\theta}{2}\right)}\right)\left(\frac{V_0}{r}\right)$. We have assumed that very close to the surface of the cone, the surface may be considered as a large plane carrying a surface charge of σ_s C/m^2. Applying Gauss's law at the surface, we have $\varepsilon E = \sigma_\mathrm{s}$. If the plane were open on both sides, then we get $2\varepsilon E = \sigma_\mathrm{s}$.

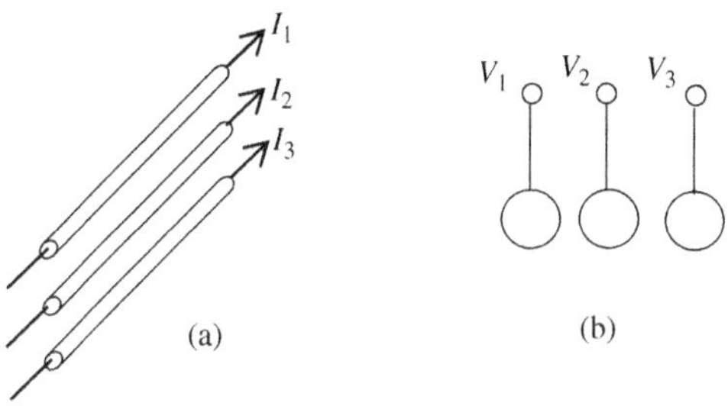

Figure 2.27: (a) Three parallel conductors carrying time-changing currents. (b) Three conductors each connected to a separate voltage source.

2.8.6 Induced electric charges and currents

When two or more conductors carry electric charges, each conductor induces electric charge on all the other conductors in the vicinity through the mutual capacitance between the conductors. Similarly when there are electric conductors carrying current, the induced votages and currents on the other conductors. Consider three electric conductors that have an inductance matrix containing self-inductance L_{nn} and mutual inductance L_{mn} (Fig. 2.27(a)):

$$[L] = \begin{bmatrix} L_{11} & L_{12} & L_{13} \\ L_{21} & L_{22} & L_{23} \\ L_{31} & L_{32} & L_{33} \end{bmatrix} \tag{2.234}$$

Then the voltage of the first conductor, for instance, due to currents flowing in it and in the two adjacent conductors is given by

$$V_1 = L_{11}\frac{\mathrm{d}I_1}{\mathrm{d}t} + L_{12}\frac{\mathrm{d}I_2}{\mathrm{d}t} + L_{13}\frac{\mathrm{d}I_3}{\mathrm{d}t} \tag{2.235}$$

And in matrix form, the voltage drop across the three conductor system is given by

$$[V(t)] = [R][I(t)] + [L]\frac{\mathrm{d}}{\mathrm{d}t}[I(t)] \tag{2.236}$$

Similarly the current in the first conductor is given by

$$I_1 = C_{11}\frac{\mathrm{d}V_1}{\mathrm{d}t} + C_{12}\frac{\mathrm{d}V_2}{\mathrm{d}t} + C_{13}\frac{\mathrm{d}V_3}{\mathrm{d}t} \tag{2.237}$$

And the current in the three conductor system is given by the matrix relationship:

$$[I(t)] = [C]\frac{\mathrm{d}}{\mathrm{d}t}[V(t)] + [G][V(t)] \tag{2.238}$$

Consider now a three conductor system where each conductor is connected to a voltage supply (Fig. 2.27(b)). Then the electric charge on each conductor will be a function of the self-capacitance C_{nn} and mutual capacitance C_{mn}:

$$Q_1 = C_{10}V_1 + C_{12}(V_1 - V_2) + C_{13}(V_1 - V_3) \tag{2.239}$$

$$Q_2 = C_{20}V_2 + C_{12}(V_2 - V_1) + C_{23}(V_2 - V_3) \tag{2.240}$$

$$Q_3 = C_{30}V_3 + C_{13}(V_3 - V_1) + C_{23}(V_3 - V_2) \tag{2.241}$$

Rearranging the three quations, we have

$$Q_1 = (C_{10} + C_{12} + C_{13})V_1 - C_{12}V_2 - C_{13}V_3 \tag{2.242}$$

$$Q_2 = -C_{12}V_1 + (C_{20} + C_{12} + C_{23})V_2 - C_{23}V_3 \tag{2.243}$$

$$Q_3 = -C_{13}V_1 - C_{23}V_2 + (C_{30} + C_{13} + C_{23})V_3 \tag{2.244}$$

which may in general be expressed in the following matrix form

$$[Q] = [C][V] \tag{2.245}$$

where

$$[C] = \begin{bmatrix} (C_{10} + C_{12} + C_{13}) & -C_{12} & -C_{13} \\ -C_{12} & (C_{20} + C_{12} + C_{23}) & -C_{23} \\ -C_{13} & -C_{23} & (C_{30} + C_{13} + C_{23}) \end{bmatrix} \tag{2.246}$$

Consider now a two conductor system, the conductors having a capacitance with reference to ground given by $C_{10} = 5\,\text{nF}$, $C_{20} = 2\,\text{nF}$ and a mutual capacitance of $C_{12} = 3\,\text{nF}$. Hence

$$[C] = \begin{bmatrix} 5+7 & -3 \\ -3 & 2+3 \end{bmatrix} = \begin{bmatrix} 12 & -3 \\ -3 & 5 \end{bmatrix} \tag{2.247}$$

Let $V_1 = 6\,\text{V}$ and V_2 is floating (i.e. with $Q_2 = 0$)

Thus,

$$\begin{bmatrix} Q_1 \\ Q_2 \end{bmatrix} = \begin{bmatrix} 12 & -3 \\ -3 & 5 \end{bmatrix} \begin{bmatrix} V_1 \\ V_2 \end{bmatrix} \tag{2.248}$$

Hence

$$Q_1 = 12 \times 6 - 3V_2 = 72 - 3V_2 \tag{2.249}$$

and

$$0 = -3 \times 6 + 5V_2 \tag{2.250}$$

Therefore, the voltage on the floating conductor is

$$V_2 = \frac{18}{5} = 3.6\,\text{V} \tag{2.251}$$

and the electric charge on the first conductor is

$$Q_1 = 72 - 3 \times 3.6 = 61.2\,\text{nC} \tag{2.252}$$

The total energy of the two conductor systems is given by:

$$\text{Energy} = \frac{1}{2}(Q_1V_1 + Q_2V_2) \tag{2.253}$$

For further reading

[1] Balanis, C.A., *Antenna Theory*, John Wiley, 1997.

[2] Bansal, R. Ed., *Fundamentals of Engineering Electromagnetics*, USA: CRC Press, 2006.

[3] Boxleitner, W., Electrostatic Discharge and Electronic Equipment: A practical Guide to Prevent ESD Problems, IEEE Press, 1989.

[4] Chen, D.K., *Fundamentals of Enginnerring Electromagnetics*, USA: Addison-Wesley, 1993.

[5] Christopoulos, C., *Principles and Techniques of Electromagnetic Compatibility*, USA: CRC Press, 1995.

[6] Cobine, *Gasoeus Condcutors*, New York, USA: Dover, 1958.

[7] Collin, R.E., *Antennas and Radio Wave Propagation*, McGraw Hill, 1985.

[8] Cross, J., *Electrtrostatics: Principles, Problems, and Applications*, Bristol, UK: IOP Publishing, 1987.

[9] Crowley, J.M., *Fundamentals of Applied Electrostatics*, USA: John Wiley, 1986.

[10] Eliot, R.S., *Electromagnetics*, IEEE Press, 1993.

[11] Greason, W., *Electrostatic Discharge in Electronics*, USA: John Wiley, 1993.

[12] Hall, M.P.M., Barclay, L.W., Hewitt, M.T., *Propagation of Radiowaves*, London, UK: IEE Press, 1996.

[13] Hall, S.H., Hall, G.W., McCall, J.A., *High Speed Digital System Design: A Handbook of Interconnect Theory and Design Practices*, USA: John Wiley, 2000.

[14] Hammond, P., *Electromagnetism for Engineers*, UK: Oxford University Press, 1997.

[15] Hansen, R.C., *Phased Array Antennas*, USA: John Wiley, 1998.

[16] Harsanyi, G., *Sensors in Biomedical Applications*, USA: Technomic Pub, 2000.

[17] Hoole, S.R.H. and Hoole, P.R.P., *A Modern Short Course in Enginneting Electromagnetics with Computer Applications*, USA: Oxford University Press, 1996.

[18] Inan, U.S. and Inan, A.S., *Electromagnetic Waves*, USA: Prentice-Hall, 2000.

[19] Johnson, R.C. Ed., *Antenna Engineering Handbook*, USA: McGraw Hill, 1993.

[20] Jordon, E.C. and Balmain, K.G., *Electromagnetic Waves and Radiating Systems*, USA: Prentice Hall, 1968.
[21] Lurtgens, G., Wilson, N., *Electrostatic Hazards*, UK: Butterworth-Heinemann, 1997.
[22] McAteer, O.J., *Electrostatics Discharge Control*, USA: McGraw Hill, 1990.
[23] Misra, D.K., *Practical Electromagnetics*, New York: John Wiley, 2007.
[24] Sampei, S., *Applications of Digital Wireless Technologies and Global Wireless Communications*, USA: Prentice Hall, 1997.
[25] Moore, A.D., *Electrostatics and Its Applications*, USA: John Wiley, 1973.
[26] Popovic, R.s. *Hall Effect Devices*, UK: IOP Press, 1991.
[27] Popovic, Z. and Popovic, B.D., *Introductory Engineering Electromagnetics*, USA: John Wiley, 1999.
[28] Smythe, W.R., *Static and Dynamic Electricity*, New York, USA: McGraw Hill, 1968.
[29] Ramo, S., Whinnery, J., Van Duzer, J., *Fields and Waves in Communication Electronics*, USA: John Wiley, 1994.
[30] Schwah, A.J., *High Voltage Measurement Techniques*, USA: MIT Press, 1972.
[31] Taylor, D.M. and Secker, P.E., *Inductrial Electrostatics*, New York, USA: John Wiley, 1994.
[32] Ulaby, F.T., *Fundamentals of Applied Electromagnetics*, USA: Prentice-Hall, 1997.
[33] Wentworth, S.M., *Fundamentals of Electromagnetics with Engineering Applications*, USA: John Wiley, 2005.

3

Dynamic fields and Maxwell's equations

3.1 Introduction

In this chapter, we shall consider the basic Maxwell's equations applied to time changing electromagnetic fields. The solution of the wave and radiation electromagnetic fields that result from manipulating the four Maxwell's equations is essential for the design and analysis of a wide variety of devices, systems and natural phenomena that are characterized by electromagnetic waves and radiation.

3.2 Time harmonic signals: why sinusoidal sources?

When a source of electromagnetic energy generates electric fields (electric current and voltage) and magnetic fields that vary sinusoidally in time, the four Maxwell's equations may be rewritten for this special case. First, let us consider the reason why sinusoidal sources, waveforms of which are illustrated in Fig. 3.1(a), are desirable for general purpose considerations.

Consider a current that flows into a radiating antenna from the source that is sinusoidal in shape, i.e., $I = I_0 \sin \omega t$, where $\omega = 2\pi f$. Then, the electric field produced by current I will also be sinusoidal. Ampere's law states that $\nabla \times (\mu^{-1}\mathbf{B}) = \sigma \mathbf{E} + \frac{\partial(\varepsilon \mathbf{E})}{\partial t}$ the first term ($\sigma \mathbf{E}$) on the right-hand side is the conduction current flowing along the antenna. The second term is the term associated with the electromagnetic fields radiated out into free space by a transmitting antenna. The second term is made up of the first time derivative of the electric field. The first derivative of the electric field produces a cosinusoidal magnetic flux density as shown in Fig. 3.1(a). Note that the waveform remains the same for both **E** and **B**, only the phase is shifted by an angle $\omega T/4$ radians. Similarly now, according to Faraday's law the time changing **B** field will produce a sinusoidal waveform **E**. Thus, the signal waveform does not change as the electromagnetic wave propagates over space. However, as shown in Fig. 3.1(b), if the source waveform I was triangular, with repeated differentiation, the waveforms of the **E** and **B** signals will be constantly changing, making detection and decoding the transmitted signal a highly complicated process at the receiver. One other reason why sinusoidal or cosoidal, time harmonic waveforms (called sinusoids) are desirable is that these will only represent a single freqency in the frequency spectrum. However, if a rectangular waveform (instead of a single sinusoid) is considered, its Fourier transform will

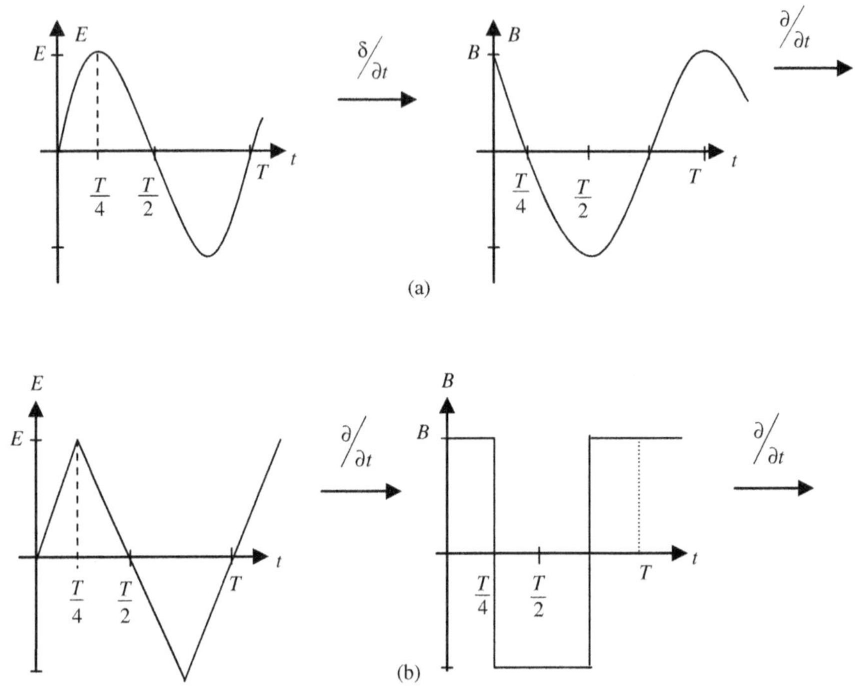

Figure 3.1: (a) Sinusoidal waveform and its first derivative and (b) triangular waveform and its first derivative.

yeild a very wide frequency bandwidth, with many cosoidal waveforms needed to make up a single rectangular waveform. Thus, sinusoidal analysis with superposition is a powerful analysis tool. Further, the sharp edges of a rectangular waveform are made up of a combination of very high frequency cosinusoidal waveforms. This means that a signal with a rectangular waveform requires a wide bandwidth of the frequency spectrum. This is a major disadvantage considering the fact that the frequency spectrum is being rapidly filled up with many telecommunication, military and other commercial and remote sensing systems which take up large portions of the frequency spectrum. Thus, the sinusoids remain the most popular form of electrical signals at all different frequencies. Digital television and radio transmission and reception, for instance, require wide bandwidths, and hence the use of wideband code division multiple access (WCDMA). For a carrier frequency of f_0, each digit of a digital system will take up $1/f_0$ seconds.

It should, however, be remembered that with a sinusoidal or cosoidal waveform starting up at a given instant in time (e.g., switching on a mobile telephone at 10.08 am to make a call), its frequency spectrum will not be a pure, single spectrum line at the operating frequency. There will be some additional spectrum lines (i.e., sinosoids), although much smaller in signal strength, in addition to the one prominent spectrum line at the main frequency. To obtain a single, pure frequency spectrum line, the waveform in the time frame, the electric and magnetic fields must exist from eternity (infinity in time) in the past to eternity in the future.

When working on the sinusoidal electromagnetic fields we may represent a signal, i.e., the electric and magnetic fields, as

$$S = S_0 \cos(\omega t + \theta) \tag{3.1}$$

which in phasor form becomes, using $\mathrm{j} = \sqrt{-1}$

$$\tilde{S} = S_0\, \mathrm{e}^{\mathrm{j}\theta} \tag{3.2}$$

where the value of the signal in real time is given by

$$\begin{aligned} S &= \mathrm{Real}[\tilde{S}\, \mathrm{e}^{\mathrm{j}\omega t}] \\ &= \mathrm{Real}[\mathrm{S}_0\{\cos(\omega t + \theta) + \mathrm{j}\sin(\omega t + \theta)\}] \\ &= S_0 \cos(\omega t + \theta) \end{aligned} \tag{3.3}$$

The first derivative of a signal S is given as

$$\frac{\partial S}{\partial t} = -\omega S_0 \sin(\omega t + \theta) \tag{3.4}$$

which in phasor form may be written as

$$\mathrm{j}\omega\tilde{S} = \mathrm{j}\omega S_0 \mathrm{e}^{\mathrm{j}\theta} \tag{3.5}$$

The first derivative in real time may be obtained from

$$\begin{aligned} \mathrm{Real}[\mathrm{j}\omega\tilde{S}\mathrm{e}^{\mathrm{j}\omega t}] &= \mathrm{Real}[\mathrm{j}\omega S_0 \mathrm{e}^{\mathrm{j}(\omega t + \theta)}] \\ &= \mathrm{Real}[S_0\{-\omega\sin(\omega \mathrm{t} + \theta) + \mathrm{j}\omega\cos(\omega \mathrm{t} + \theta)\}] \\ &= -\omega S_0 \sin(\omega \mathrm{t} + \theta) \end{aligned} \tag{3.6}$$

The integration of a signal S is given as

$$\int S \cdot \mathrm{d}t = \frac{1}{\omega} S_0 \sin(\omega \mathrm{t} + \theta) \tag{3.7}$$

which in phasor form may be written as

$$\frac{1}{j\omega}\tilde{S} = \frac{1}{j\omega}S_0\,e^{j\theta} \tag{3.8}$$

To get back the time domain form of the integration, from the phasor form, we have

$$\begin{aligned}\text{Real}\left[\frac{1}{j\omega}\tilde{S}\,e^{j\omega t}\right] &= \text{Real}\left[\frac{1}{j\omega}S_0\,e^{j(\omega t+\theta)}\right]\\ &= \frac{1}{\omega}S_0\sin(\omega t+\theta)\end{aligned} \tag{3.9}$$

In Fig. 3.2 is shown the frequency spectrum over which electromagnetic fields are found. The 0 Hz (DC) electric field is what we considered as electrostatics. In magnetostatics we looked at magnetic fields produced by 0 Hz (DC), or by very low frequency (e.g. 60 or 50 Hz) electric currents. The vast majority of telcommunications, remote sensing, and biomedical electrical systems operate at freuqencies above 10^4 Hz (or 10 kHz).

Using (3.5) and (3.8), the study of electromagnetic fields from a single frequency source or sinusoidal currents (i.e. $I = I e^{j\omega t}$, $dI/dt = j\omega I e^{j\omega t}$) requires the use of the four Maxwell equations for harmonic signals which are given below:

$$\nabla \times \mathbf{H} = \mathbf{J} + j\omega\varepsilon\mathbf{E}, \tag{3.10}$$

$$\nabla \times \mathbf{E} = -j\omega\mu\mathbf{H}, \tag{3.11}$$

$$\nabla \cdot \mathbf{B} = 0, \tag{3.12}$$

$$\nabla \cdot \mathbf{E} = \rho/\varepsilon, \tag{3.13}$$

where the term $\exp(j\omega t)$ (i.e., $e^{j\omega t}$) has been dropped from both sides of (3.10)–(3.13) since it cancels out from the right- and left-hand sides of each equation. It is important to put it back once **E** and **B** have been solved. For instance, if $\mathbf{E}_s$ is the electric field detertmined by solving (3.10)–(3.13), then the final answer will be $\mathbf{E}_s e^{j\omega t}$. Maxwell's equations relate the two basic physical parameters **E** (electric field intensity) and **B** (magnetic flux density) to the sources **J** (current density) and ρ (electric charge density). In telecommunication, radar and biomedical imaging, the presence and effects of ρ are considered negligible. Thus, **J** is the primary source of information about the text, voice and video signals being transmitted or of the object being imaged.

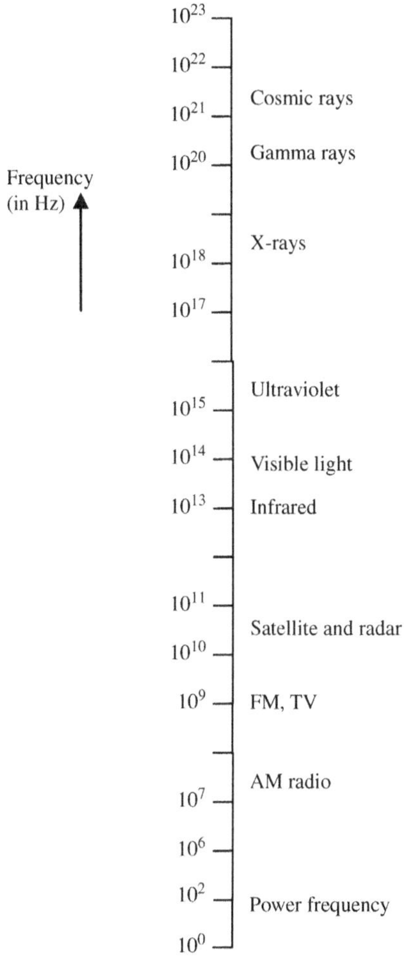

Figure 3.2: The frequency spectrum.

3.3 Electromagnetic induction: magnetic fields due to time-varying electric fields

3.3.1 General principles of induction

Faraday's law states that when the magnetic flux linking a loop changes with time, there are voltage and current induced in the loop. This means that under the following three conditions, electric current is produced in a conductor or loop that is immersed in a magnetic flux:

a) The magnetic field is time changing, whereas the conductor or loop is stationary, e.g., transformers or receiving antennas.

b) The magnetic field is static, but the conductor or loop is moving across the static magnetic field, e.g., motors and generators.

c) The magnetic field is time changing, while at the same time the area of the conductor loop that is cutting the field is also changing, e.g., induction motor or mobile communication antenna.

The general Faraday's equation for electromagnetic induction is

$$\nabla \times \mathbf{E} = -\frac{\partial \mathbf{B}}{\partial t} \tag{3.14}$$

which holds for every point in space, where the term on the right-hand side involving time-varying magnetic flux density produces the left-hand side term of time-varying electric field intensity. An electric scalar potential relation alone (e.g., $\mathbf{E} = -\nabla\phi$) is not enough to represent the electric field produced by a time-varying magnetic field; we need an additional magnetic vector potential (**A**) to represent the non-conservative electric field:

$$\mathbf{E} = -\nabla\phi + \frac{\partial \mathbf{A}}{\partial t} \tag{3.15}$$

In integral form, Faraday's law becomes for a surface bounded by a loop l

$$\oint_l \mathbf{E} \cdot \mathrm{d}l = -\frac{\partial}{\partial t}\int_s \mathbf{B} \cdot \mathbf{ds} \tag{3.16}$$

3.3.2 Transformer action

Consider the transformer coil arrangement shown in Fig. 3.3. Two coils are kept close to each other, wound on a common iron core carrying a magnetic flux ψ linking both. A source voltage V_s is connected to the primary coil of N_1 turns, and the output, or secondary, coil of N_2 turns is on open circuit or connected to a load.

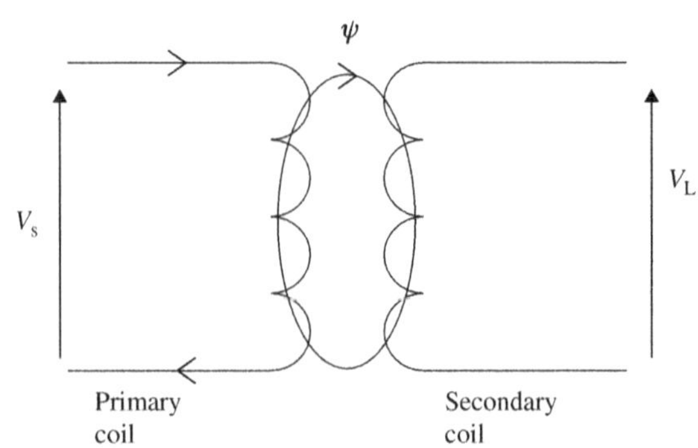

Figure 3.3: Transformer coils and electromagnetic induction.

The source voltage V_s drives a current I through the primary coil. This in turn produces a time-varying magnetic flux ψ which cuts the secondary coil as shown in Fig. 3.3. According to Faraday's law of electromagnetic induction, ψ will induce a voltage V_L across the secondary coil, where by integration over the common core cross-section **s**.

$$\psi = \int_s \mathbf{B} \cdot \mathbf{ds} \text{ Webers} \tag{3.17}$$

is the magnetic flux linking the two coils, and the load voltage is calculated from

$$V_L = \oint_l \mathbf{E} \cdot \mathbf{dl} \tag{3.18}$$

which is the voltage induced in the secondary coil by the flux.

Thus, using (3.17) and (3.18), according to Faraday's law in its integral form (3.16), we get

$$V_L = -N_2 \frac{d\psi}{dt} \tag{3.19}$$

According to Lenz's law, the induced load voltage V_L will be in such a direction as to produce a current that opposes the linking magnetic flux ψ. Since at the primary coil $V_s = -N_1 d\psi/dt$, comparing with (3.19), we have the transformer relationship

$$V_s/V_L = V_1/V_2 = N_1/N_2 \tag{3.20}$$

3.3.3 Motor action

Consider now a static (or time changing) magnetic field **B** which is cut by a conductor which moves with a velocity **v**. Then each electron in the conductor will experience a force of $ev \times \mathbf{B}$. Each unit charge will experience a force of $\mathbf{v} \times \mathbf{B}$. Therefore, the voltage produced along the conductor from $l = l_1$ to l_z is

$$V = \int_{l_1}^{l_2} (\mathbf{v} \times \mathbf{B}) \cdot \mathbf{dl} \tag{3.21}$$

If the conductor forms a closed circuit, the emf produced is

$$V = \oint_l \mathbf{v} \times \mathbf{B} \cdot \mathbf{dl} \text{ Volts} \tag{3.22}$$

3.3.4 Hall effect

Consider electric charges moving perpendicular to magnetic field lines. The Lorentz force on the electric charges gives rise to an electric field. Under steady state conditions, the net force on the electric charges will be zero under steady state conditions, where the velocity will not vary with time. This implies that there must be present an electric force due to an electric field that is equal and oppostie to the force produced by the magnetic field:

$$q\mathbf{E} + q\mathbf{v} \times \mathbf{B} = 0. \tag{3.23}$$

$$\text{i.e., } E = -vB$$

Hence, a transverse potential difference develops across the material given by

$$V = vBL \tag{3.24}$$

This is called the Hall effect. The sign of the induced voltage is determined by the sign of velocity $\mathbf{v}$. Thus in conductors and semicondcutors, the measurement of the Hall effect potential difference gives us the sign of the predominant electric charge carrier inside the material. For positive electric charges, such as the holes in a semiconductor material, the charge velocity is in the same direction as the current flow inside the material to which the magnetic field is applied. In the case of negative carriers like electrons, the charge velocity is in the opposite direction to the current flow.

3.3.5 Rotating disk in a magnetic field

In the case of the Faraday's disk generator shown in Fig. 3.4, a conducting disk rotating at ω rad/s is being cut by a magnetostatic field $\mathbf{B}$. We shall determine the voltage induced in the disk generator.

The induced voltage is given by, with $\mathbf{E} = v \times \mathbf{B}$ from (3.23) from the edge of the disk to its center

$$V = \oint (\mathbf{v} \times \mathbf{B}) \cdot \mathrm{d}\mathbf{l} \tag{3.25}$$

$$= \int_a^0 [\omega r \mathbf{u}_\phi \times B\mathbf{u_z}] \cdot (\mathrm{d}r\, \mathbf{u_r}) \tag{3.26}$$

$$= \omega B \int_a^0 r\, \mathrm{d}r \tag{3.27}$$

$$= -\frac{\omega B a^2}{2} \text{ Volts} \tag{3.28}$$

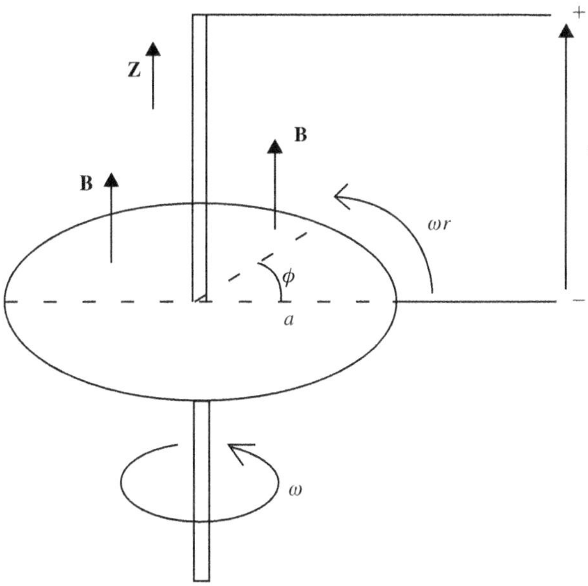

Figure 3.4: A rotating disk in a magnetic field.

The induced voltage is seen to be a strong function of the radius of the disk.

3.3.6 AC generator: a rotating coil in a magnetic field

In the case of a time-varying magnetic field cutting a moving circuit, we have in general, the induced voltage

$$V = -\frac{d\psi}{dt} = -\frac{d}{dt}\int_s \mathbf{B}\cdot d\mathbf{s} \tag{3.29}$$

Consider a rectangular conducting loop submerged in a time-varing magnetic field, $B = B_0 \sin \omega t \ \mathbf{u}_y$, as shown in Fig. 3.5.

When $\omega = 0$, that is the loop is stationary, we have

$$\phi = \int_s \mathbf{B}\cdot d\mathbf{s} = (B_0 \sin \omega t \mathbf{u}_y)\cdot(ab\ \mathbf{u}_n) \tag{3.30}$$

$$= ab\, B_0 \sin \omega t \cdot \cos\theta \tag{3.31}$$

where the plane of the coil makes an angle θ with B.

The induced voltage is given by

$$V_s = -\frac{d\phi}{dt} = -\omega ab B_0 \cos \omega t \cos\theta \tag{3.32}$$

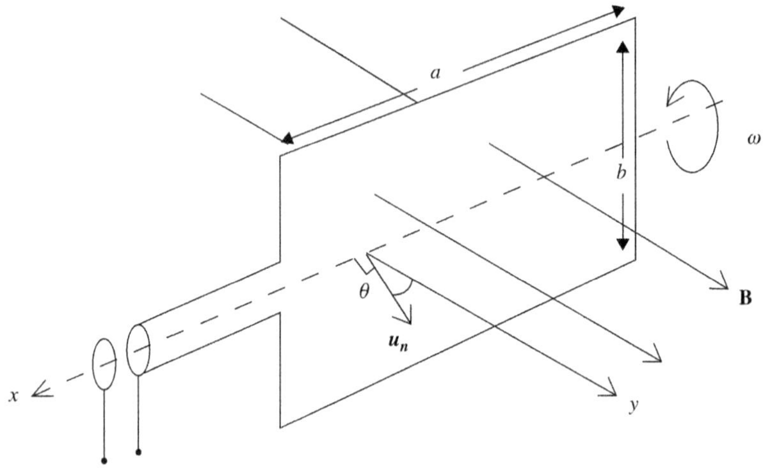

Figure 3.5: A rotating coil in a magnetic field.

Now, when the loop rotates at ω rad/s about the x-axis, we have

$$V_\mathrm{R} = \oint \mathbf{v} \times \mathbf{B} \cdot \mathrm{d}\mathbf{l} \tag{3.33}$$

$$\begin{aligned}
V_\mathrm{R} &= \int \left\{ \left[\frac{b}{2} \omega \mathbf{u}_n \right] \times B_0 \sin \omega t \; \mathbf{u}_y \right\} \cdot \mathbf{u}_x \,\mathrm{d}x + \int \left\{ \left[-\frac{b}{2} \omega \mathbf{u}_n \right] \times (\mathbf{B}_0 \sin \omega t) \right\} \cdot \mathbf{u}_x \,\mathrm{d}x \\
&= (b\omega B_0 \sin \omega t \sin \theta) a \\
&= ab\omega B_0 \sin \omega t \sin \theta \quad \text{Volts}
\end{aligned} \tag{3.34}$$

For $\theta = 0$ at $t = 0$, and $\theta = \omega t$ at instant t, thus

$$V_\mathrm{R} = \omega ab B_0 \sin \omega t \sin \omega t \tag{3.35}$$

Hence, the total induced voltage

$$\begin{aligned}
V &= V_\mathrm{S} + V_\mathrm{R} \\
&= -\omega ab B_0 \cos^2 \omega t + \omega ab B_0 \sin^2 \omega t \\
&= -\omega ab B_0 \cos 2\omega t
\end{aligned} \tag{3.36}$$

an induced voltage that oscillates at 2ω rad/s.

The flux density **B,** will yield the total flux ϕ, as

$$
\begin{aligned}
\phi(t) &= \mathbf{B}\cdot ab\,\mathbf{u}_n \\
&= abB_0 \sin\omega t \cos\theta \\
&= abB_0 \sin\omega t \cos\omega t \\
&= \frac{1}{2} abB_0 \sin 2\omega t
\end{aligned}
\tag{3.37}
$$

Therefore, the induced voltage

$$
\begin{aligned}
V &= -\frac{\mathrm{d}\phi}{\mathrm{d}t} = \frac{\mathrm{d}}{\mathrm{d}t}\left[\frac{1}{2} abB_0 \sin 2\omega t\right] \\
&= -\omega abB_0 \cos 2\omega t
\end{aligned}
\tag{3.38}
$$

3.3.7 Skin effect

We shall consider in this section the frequency dependence of electromagnetic field interaction with conductors and conducting materials. Our discussion in this section will apply for electromagnetic fields impinging on conductors (e.g., time-varying magnetic fields produced by the stator of an electric motor inducing currents in the rotor) or electric currents flowing along a conductor (e.g., as with signal generator currents along an antenna wire). It will be shown that the currents occupy only a small portion of the outer conductor for high frequency fields. Hence, the term skin effect – the time changing magnetic and electric fields occupy only the skin, or the outer layer of the conductor principally to a depth called the skin depth.

Consider a conductor placed along the x-axis as shown in Fig. 3.6(a). A current density of J A/m^2 flows in the x-direction, and associated with it is the circulating magnetic field on the yz-plane as shown in Fig. 3.6(a). In Fig. 3.6(b) is shown the magnetic field circulating around the conduction current going into the paper (in the positive x-direction). In Fig. 3.6(c) is shown the electric field being produced by the first time-domain differential of the magnetic field line coming out of the paper, as described by (3.16), i.e. the Faraday's law.

Applying Ampere's law in its intergral form to the situation shown in Fig. 3.6(b), following [9], we get,

$$
H\,\delta z - \left(H + \frac{\partial H}{\partial x}\delta x\right)\delta z = J\,\delta x\,\delta z \tag{3.39}
$$

We have from (3.39)

$$
-\frac{\partial H}{\partial x} = J \tag{3.40}
$$

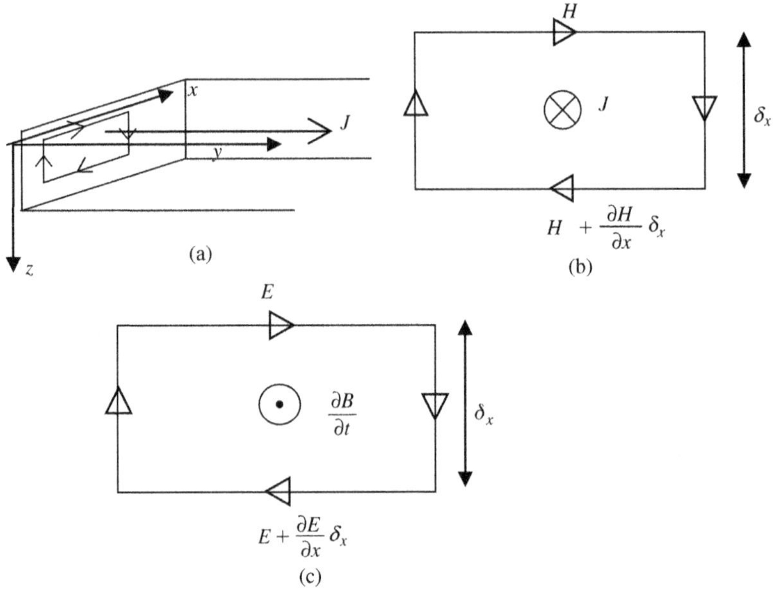

Figure 3.6: Eddy currents in laminated magnetic materials.

Applying Faraday's law in its integral form (3.16) to the loop in Fig. 3.6(c), we get

$$\left(E + \frac{\partial E}{\partial x}\delta x\right)\delta y - E\delta y = -\mu\frac{\partial H}{\partial t}(\delta x\,\delta y) \tag{3.41}$$

which reduces to

$$\frac{\partial E}{\partial x} = \mu\frac{\partial H}{\partial t} \tag{3.42}$$

Differentiating $J = \sigma E$ with respect to x, and using (3.42) we have,

$$\frac{\partial J}{\partial x} = -\mu\sigma\frac{\partial H}{\partial t} \tag{3.43}$$

Differentiating (3.43) with respect to x, we get

$$\frac{\partial^2 J}{\partial x^2} = \mu\sigma\frac{\partial J}{\partial t} \tag{3.44}$$

Expressing the sinusoidal time current in polar form,

$$J = \overline{J}\mathrm{e}^{\mathrm{j}\omega t} \tag{3.45}$$

which on substitution into (3.44) yields

$$\frac{\partial^2 \overline{J}}{\partial x^2} = \mathrm{j}\mu\sigma\omega\overline{J} \tag{3.46}$$

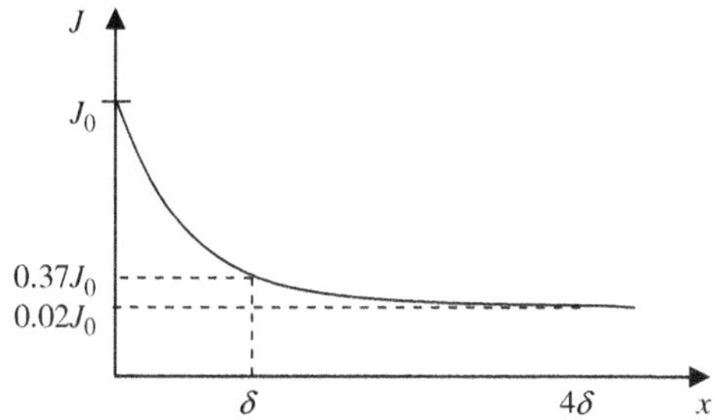

Figure 3.7: Current decay inside a conductor.

Now (3.46) has a solution of the form

$$\bar{J} = C_1 \exp\left[-(1+\mathrm{j})\sqrt{\frac{\mu\sigma\omega}{2}}x\right] + C_2 \exp\left[(1+\mathrm{j})\sqrt{\frac{\mu\sigma\omega}{2}}x\right] \tag{3.47}$$

The reader may easily verify the validity of the solution by substituting the solution in (3.47) into the left-hand side of (3.46) and seeing it yield the right-hand side of (3.46). Now $C_2 = 0$, else we have an exponential increase of current density with x, that is, an increase in current deeper into the conductor, which cannot be. Hence, setting $C_1 = J_0$, the final solution is,

$$\bar{J} = J_0 \exp\left[-(1+\mathrm{j})\sqrt{\frac{\mu\sigma\omega}{2}}x\right] \tag{3.48}$$

where J_0 is the surface (i.e., $x = 0$) current density.

Defining the skin depth (in m) as

$$\delta = \sqrt{\frac{2}{\mu\sigma\omega}} \tag{3.49}$$

the solution of (3.48) becomes

$$\bar{J} = J_0 \exp\left[-(1+\mathrm{j})\frac{x}{\delta}\right] \tag{3.50}$$

In Fig. 3.7 is given a sketch of (3.50). Equation (3.50) indicates that when $x = \delta$, the current density would have dropped by 37% to $J_0\mathrm{e}^{-1}$ A/m^2.

From Equation (3.50), the total current (in A/m) is found from

$$\bar{I} = \int_0^\infty J_0 \exp\left[-(1+\mathrm{j})\frac{x}{\delta}\right]\mathrm{d}x$$

$$= J_0 \frac{\delta}{1+\mathrm{j}} \tag{3.51}$$

The power loss in the conductor is, observing $J = J_o/\sqrt{2}$

$$P = \int_0^d \frac{\mathbf{J}^2}{2\sigma} \mathrm{d}x = \frac{\mathbf{J}_0^2 \delta}{4\sigma} \tag{3.52}$$

and using from (3.51) that $I = \frac{J_0 \delta}{\sqrt{2}}$ the equivalent resistance is given by

$$R = \frac{2P}{I^2} = \frac{1}{\sigma \delta} \tag{3.53}$$

A further application of electromagnetic heating is in the biomedical treatment of tumors. When there is local growth of tumor (or cancerous tissue), it is possible to apply heat to the tumor area to burn up the tumor. The heat is supplied by electromagnetic fields at 915 Hz carrying 60 W of power to the tumor. A transmission line in the form of a 1.5 mm-diameter coaxial line is used to carry the electromagnetic fields to the tumor. The tip of the transmission line has to be guided to the correct spot through surgery to open up the affected area, or through noninvasive electromagnetic induction of electromagnetic energy from the transmission line terminating end right on to the tumor or cancer tissues. Such treatment is possible for tumors in the liver or the lung. An alternative method, especially for skin tumors, is to make the tumor area as part of the transmission line. Applying about a 3 kV, 60 A pulse that lasts for only few hundreds of nanoseconds, the tumor may be burnt up and regrowth may be prevented.

3.4 Electromagnetic waves: general

Electromagnetic waves are produced by applying time harmonic voltages or electric fields to transmission lines, waveguides or antennas. In this section, we shall be looking at the general nature of electromagnetic waves traveling in space. Electromagnetic waves over transmission lines and waveguides, and electromagnetic waves radiated by antennas shall be separately considered in later chapters of this book. Wire antennas like dipole antennas launch out electromagnetic waves into space when time harmonic currents flow over the wires. Similarly, when time-varying electric and magnetic fields appear at a waveguide aperture (e.g., the open end of a waveguide antenna) the electromagnetic waves which have been traveling through the space inside the waveguide become waves outside the waveguide, in the open space. The time-varying currents or electric and magnetic fields at the antenna terminal will produce time-varying fields just outside the antenna. In this section, we shall show that these fields will constitute a propagating electromagnetic wave. With time-varying fields, **E** and **B** are no longer independent as in the case of static (frequency $f = 0$) or low frequency (e.g., $f = 60$ Hz) fields, but they are mutually coupled through the four Maxwell equations. We shall consider free space, that is space which is free of electric charge, where volume electric charge density $\rho = 0$, conductivity $\sigma = 0$, and conduction current densitty $J = \sigma E = 0$. Thus, the four Maxwell equations (3.10)–(3.13) reduce to

$$\nabla \cdot \mathbf{E} = 0 \tag{3.54}$$

$$\nabla \cdot \mathbf{B} = 0 \tag{3.55}$$

$$\nabla \times \mathbf{B} = \frac{1}{u^2}\frac{\partial \mathbf{E}}{\partial t} \tag{3.56}$$

$$\nabla \times \mathbf{E} = -\frac{\partial \mathbf{B}}{\partial t} \tag{3.57}$$

where we have set the velocity of electromagnetic waves as

$$u = (\mu_0 \varepsilon_0 \mu_r \varepsilon_r)^{-1/2} \tag{3.58}$$

The velocity of the wave in free space, where $\mu_r = 1$and $\varepsilon_r = 1$, becomes the velocity of light in free space, i.e., $u = c = (\mu_0\varepsilon_0)^{-1/2} = 2.9998 \times 10^8$ m/s. These four equations (3.54)–(3.57) are not only linear, but also homogeneous since there are no electric charges or electric currents present, and every term contains either **E** or **B**. One property of a linear, homogeneous equation is that the sum of two different solutions is also a solution to the equation. Thus, a general solution to these equations will consist of a sum of static as well as time-varying fields. The physical origin of such static fields would be distributions of static electric charges and steady currents. We shall assume that such a distribution of steady electric charges and currents cannot exist in free space.

Since the divergence of both **E** and **B** is zero, the flux of the time-varying **E** and **B** out of any volume is zero. Furthermore, the two curl equations (3.56) and (3.57) couple the **E** and **B** fields together. The terms on the right side of the curl equations generate the terms on the left-hand side of the equations. For a time-varying **B** field, $\partial \mathbf{B}/\partial t$ is not zero; thus, the curl of **E** (i.e., $\nabla \times \mathbf{E}$) is also not zero. A non-zero curl of **E** means that **E** is a function of position. This means that a time-varying **B** produces a spatially varying **E**. But moreover, since **B** is a function of time and the operator ∇ is independent of time, **E** must be a function of time as well. Hence, we may conclude that a time-varying **B** in space produces an **E** that changes in time and position (i.e., space). Similarly, the second curl equation (3.56) tells us that a time changing **E** produces a **B** that changes in time and space. Thus, what we have is **E** and **B** producing and supporting each other in the time and spatial domains, such that each field keeps moving (change in position or space) and each is produced by the other. Therein, we have the propagation of the electromagnetic fields **E** and **B**, due to each field sustaining, generating and nudging on the other at the velocity of light in free space.

We shall first obtain the wave equation for the electric field **E**. In order to do this, we need to use (3.56) and (3.57) to get a single equation for **E** by eliminating **B**.

The curl of (3.57) yields

$$\nabla \times (\nabla \times \mathbf{E}) = -\frac{\partial(\nabla \times \mathbf{B})}{\partial t} \tag{3.59}$$

In (3.59), the operators $\partial/\partial t$ and ∇ have been changed round since they are independent of each other. Using (3.56) in the right-hand term of (3.59) yields

$$\nabla \times (\nabla \times \mathbf{E}) = -\frac{1}{u^2}\frac{\partial^2 \mathbf{E}}{\partial t^2} \tag{3.60}$$

The curl–curl term of (3.60) may be replaced by using the vector relation

$$\nabla \times (\nabla \times \mathbf{E}) = \nabla(\nabla \cdot \mathbf{E}) - \nabla \cdot \nabla \mathbf{E} \tag{3.61}$$

Since $\nabla \cdot \mathbf{E} = 0$ for electric charge free regions, (3.61) reduces to the wave equation

$$\nabla^2 \mathbf{E} = \frac{1}{u^2}\frac{\partial^2 \mathbf{E}}{\partial t^2} \tag{3.62}$$

where we have set $\nabla \cdot \nabla$ to ∇^2.

The same two curl equations in Maxwell's equations may be used to get the wave equation for the magnetic field:

$$\nabla^2 \mathbf{B} = \frac{1}{u^2}\frac{\partial^2 \mathbf{B}}{\partial t^2} \tag{3.63}$$

Equations (3.62) and (3.63) are vector wave equations and u is the velocity of propagation of the waves. It should be noted that the displacement current term $(1/u^2)\partial E/\partial t$ on the right-hand side of (3.56) is crucial to the existence of electromagnetic waves, for without this term Maxwell's equations would not yield wave equations for **E** and **B**. Note also that the two vector wave equations are not independent of each other. This is because **E** and **B** are related by the two curl equations (3.56) and (3.57). We will employ this connection below. We should observe that the vector wave equations satisfied by **E** and **B** describe the propagation of undamped waves. This is just what we should expect as there are no charges or currents to interact with the field and hence take away energy from them.

3.5 Electromagnetic waves: plane waves

The vector wave equations (3.62) and (3.63) describe the propagation of waves in empty space in full generality, i.e., the wave fronts can be plane, spherical or cylindrical or of any shape that we care to consider. However, from now onwards we will confine our attention to plane waves as these are the simplest type of wave. Consider a wave propagating in the x-direction. A plane wave would be a wave in which at any given point along the x-direction (say $x = 55$ m), the **E** and **B** fields have constant magnitudes at each point on the yz-plane. Plane waves are also of particular, practical interest because a limited portion of a nonplanar wave front, that is far from its source approximates closely to a plane wave (where all changes are in the direction of travel with **E** and **H** perpendicular to each other and on the tranverse plane).

We will choose a rectangular Cartesian system for our discussion, because it has the appropriate symmetry. We shall consider waves that propagate in the positive z direction. Thus for a plane wave, at any given point along the z-axis, the magnitudes of the electric and magnetic fields on the xy-plane are constants. The two vector wave equations can be written as a set of six equations in terms of the Cartesian components of the two fields. For the x component of the **E** field, we have

$$\nabla^2 E_x = \frac{1}{u^2}\frac{\partial^2 E_x}{\partial t^2} \tag{3.64}$$

We may write out similar equations for the y and z components of the **E** field and for the x, y and z components of the **B** field.

Over the unbounded plane wave fronts, the magnitudes and directions of the fields will not vary, so the partial derivative of the field components with respect to x and y will be zero. Thus, for $E = E_x$ or E_y or E_z

$$\frac{\partial}{\partial x}(E) = 0 \quad \text{and} \quad \frac{\partial}{\partial y}(E) = 0 \tag{3.65}$$

The second partial derivatives of the field components with respect to x and y will also be zero, so the three-dimensional (3.62) reduces to the set of one-dimensional wave equations

$$\frac{\partial^2 E_x}{\partial z^2} = \frac{1}{u^2}\frac{\partial^2 E_x}{\partial t^2} \tag{3.66}$$

$$\frac{\partial^2 E_y}{\partial z^2} = \frac{1}{u^2}\frac{\partial^2 E_y}{\partial t^2} \tag{3.67}$$

$$\frac{\partial^2 E_z}{\partial z^2} = \frac{1}{u^2}\frac{\partial^2 E_z}{\partial t^2} \tag{3.68}$$

We get similar equations for the three components of the **B** field. Before proceeding further, we shall assign a certain vector direction for the electric field **E**, by setting,

$$\mathbf{E} = \mathbf{u}_x E_x + \mathbf{u}_z E_z \tag{3.69}$$

where $\mathbf{u}_x$ and $\mathbf{u}_z$ are unit vectors. This is legitimate since we are merely orienting the Cartesion axes so that **E** on the transverse plane is in the y direction. Now if we take the curl of this electric field defined in (3.69), remembering that $\partial/\partial x = 0$, $\partial/\partial y = 0$, we only get one nonzero component which is $\mathbf{u}_y \, dE_x/dz$, which from Maxwell's equation (3.57) is equal to $-\mu \, \partial\mathbf{H}/\partial t$. Hence, $\mathbf{u}_y \, dE_x/dz = -\mu \, \partial\mathbf{H}/\partial t$. This in turn means that the magnetic field intensity can have only a y-directed component, i.e., $\mathbf{H} = \mathbf{u}_y H_y$. And we have $-\mathbf{u}_x \, dH_y/dz$, which from Maxwell's equation (3.56), is equal to $\varepsilon \, \partial\mathbf{E}/\partial t$. Therefore, the electric field can only have the x-directed component and $E_z = 0$. Therefore, we have

$$\mathbf{E} = \mathbf{u}_x E_x \tag{3.70}$$

Consider how the two Maxwell's equations (3.56) and (3.57) have simplified for the case of plane electromagnetic waves: Faraday's law simplifies to $\mathbf{u}_y \, dE_x/dz = -\mu \, \partial\mathbf{H}/\partial t$ and the Ampere's law simplifies to $-\mathbf{u}_x \, dH_y/dz = \varepsilon \, \partial E_x/\partial t$. Differentiating this simplified Ampere's law with respect to z, and substituting from the simplified Faraday's law for dE_x/d_z, we get

$$\frac{\partial^2 H_y}{\partial z^2} = \frac{1}{u^2}\frac{\partial^2 H_y}{\partial t^2} \tag{3.71}$$

which defines a magnetic plane wave traveling in the z direction. Similarly, we get the plane wave equation for the electric field

$$\frac{\partial^2 E_x}{\partial z^2} = \frac{1}{u^2}\frac{\partial^2 E_x}{\partial t^2} \tag{3.72}$$

For an electromagnetic source oscillating at a harmonic frequency ω, setting $\partial/\partial t = j\omega$ we can show that (3.71) and (3.72) have the harmonic traveling wave solutions

$$B_y = B_0 \cos(\omega t \pm kz) \tag{3.73}$$

$$E_x = E_0 \cos(\omega t \pm kz) \tag{3.74}$$

where k is the wave number and is related to the wavelength λ by $k = 2\pi/\lambda = \omega/u$. Note that a half-wavelength long antenna designed for operation at a cyclic frequency of $f(=\omega/2\pi)$ will have a length $u/(2f)$ in a medium with relative permittivity ε_r (e.g., $\varepsilon_r = 80$ in seawater). The half wave dipole antenna will have a length of $c/2f$ for free space. Hence, the antenna designed for operation in seawater will have to be shorter by a factor of about 0.11 $(=\varepsilon_r^{-1/2})$ to that used in free space for the same frequency of operation.

Electric and magnetic waves transport electromagnetic energy. Additionally, they can carry digital or analog information in communication systems. The waves travel in the $+z$ and $-z$ directions. This is normally the case with a simple wire antenna used in mobile phones, where signals are radiated in both directions deduced from the idea of an omni-directional dipole antenna. If we want the waves to be directed in one direction only, say in the $+z$ direction, we must find a way of redirecting the energy traveling in the $-z$ direction to the $+z$ direction. This may be done by simply placing a reflecting plane just behind the antenna to turn back the $-z$ directed waves to the $+z$ direction as well; by doing this all the energy put into the antenna is radiated in the $+z$ direction instead of being divided into equal halves for transmission in both directions.

Alternative forms of expressing the solution of the wave equations (3.71) and (3.72) are

$$E_x = E^+ \exp(-kz) + E^- \exp(kz) \tag{3.75}$$

$$H_y = (E^+/\eta) \exp(-kz) - (E^-/\eta) \exp(kz) \tag{3.76}$$

where the wave or intrinsic impedance $\eta = E^+/H^+ = E^-/H^- = (\mu/\varepsilon)^{1/2}$ ohms. Note that the electric and magnetic fields of the wave are in phase with each other. For free space $\varepsilon = \varepsilon_0$ and thus $E^+/H^+ = E_0/B_0/\mu_0 = (\mu_0/\varepsilon_0)^{1/2}$. From this simplification for free space, recalling $u = c = (\mu_0\varepsilon_0)^{-1/2}$, we get

$$E_0 = cB_0 \tag{3.77}$$

The Maxwell equations are linear. Hence, the principle of superposition applies and the general solutions for spherical waves can be obtained by adding together plane-polarized electromagnetic wave solutions. An important example of an electromagnetic wave solution is the case where two plane-polarized electromagnetic waves of the same amplitude, both propagating in the positive z direction, are added together. Consider the case where one of the two waves has its E vector plane-polarized in the x direction (E_x) while the other has its E vector plane-polarized in the y direction (E_y). For a phase difference of 90° between the E_x and E_y waves, the resultant E vector and B vector have constant amplitudes and phase in the plane of the wave front as the wave advances. Moreover, the resultant electric field will rotate in space as it travels in the z-direction. Such a wave is described as circularly polarized. Such waves are found in microwave communication systems where information capacity at a particular frequency may be doubled by transmitting two orthogonal electromagnetic waves.

In the foregoing discussion, we assumed that the medium through which the electromagnetic wave travels is lossless, i.e. conductivity $\sigma = 0$. If this is not the case, our analysis for the lossless case still applies, except that we have to replace the permittivity ε by a complex permittivity ε^c. This comes about as follows. For lossy media, the conductivity is contained in the following Maxwell's equations:

$$\nabla \times \mathbf{H} = \mathbf{J} + \frac{\partial \mathbf{D}}{\partial t} \tag{3.78}$$

where $\mathbf{J} = \sigma\mathbf{E}$. Equation (3.78) may be rewritten as

$$\nabla \times \mathbf{H} = \sigma\mathbf{E} + \frac{\varepsilon\partial\mathbf{E}}{\partial t} \tag{3.79}$$

where the first term on the right-hand side is zero if the medium is lossless. Thus in the discussion so far, only the second term on the right-hand side, which may be expressed as $\mathrm{j}\omega\varepsilon E$ for a harmonic signal, existed. Now for the complete equation (3.79) the right-hand side may be rewritten as $(\sigma + \mathrm{j}\omega\varepsilon)E = \mathrm{j}\omega(\varepsilon - \mathrm{j}\sigma/\omega)E = \mathrm{j}\omega\varepsilon^{\mathrm{c}}E$, where $\varepsilon^{\mathrm{c}} = (\varepsilon - \mathrm{j}\sigma/\omega)$. Once we make this switch from permittivity ε by a complex permittivity ε^{c}, the foregoing discussion for the lossless case may be used for the lossy case. The final result for waves in the lossy medium case is given by the following two equations:

$$E_x = E_0 \exp(\pm\alpha z)\cos(\omega t \pm \beta z) \tag{3.80}$$

$$B_y = B_0 \exp(\pm\alpha z)\cos(\omega t \pm \beta z) \tag{3.81}$$

where α (in Nepers/m) is the loss attenuation constant and β (in radians/m) is the phase constant. The amplitude of the wave is attenuated as it propagates through the medium; after traveling over a distance called the skin depth $z = \delta = 1/\alpha$, the wave amplitude would have been attenuated by a factor e^{-1}. Beyond this distance δ, the signal will be too weak for a receiver system to pick up. The solution of the wave equation in lossy media may also be expressed as follows:

$$E_x = E^+ \exp(-\gamma z) + E^- \exp(\gamma z) \tag{3.82}$$

$$H_y = (E^+/\eta)\exp(-\gamma z) - (E^-/\eta)\exp(\gamma z) \tag{3.83}$$

where the propagation constant $\gamma = \alpha + \mathrm{j}\beta = [\mathrm{j}\omega\mu(\sigma + \mathrm{j}\omega\varepsilon)]^{1/2}$ (in m^{-1}) and intrinsic impedance $\eta = [\mathrm{j}\omega\mu/(\sigma + \mathrm{j}\omega\varepsilon)]^{1/2}$(ohms). E^+ and E^- are two constants associated with the amplitude of the forward (positive z directed) and backward (negative z directed) waves, respectively, at $z = 0$. When the medium is a very good conductor such that $\sigma \gg \omega\varepsilon$, we have

$$\alpha = \beta = \left(\frac{\omega\mu\sigma}{2}\right)^{1/2} \tag{3.84}$$

where the attenuation and phase constants α and β are equal to each other. If we consider the magnitude of the electric field alone at this time, the conductivity of the medium will cause the traveling wave (or signal) to decay as $E^+\mathrm{e}^{-\alpha z}$, and the power of the signal will decay as $P^+\mathrm{e}^{-2\alpha z}$. The power of the wave decays twice as fast as the electric (or magnetic) field decay since $P^+(= (E^+)^2/2\eta)\ \mathrm{W/m^2}$, giving a decay rate as $\mathrm{e}^{-2\alpha z}$.

3.6 Energy in the electromagnetic field

Electromagnetic fields may be used to both store energy, as well as to transfer energy. In this section, we shall from Maxwell's equations derive a mathematical relationship between energy stored, lost through ohmic loss and transferred through radiation. The energy radiated or transferred will be defined by the Poynting vector that shall be derived.

3.6.1 Poynting's theorem

Consider the following two of Maxwell's equations

$$\nabla \times \mathbf{H} = \mathbf{J} + \frac{\partial \mathbf{D}}{\partial t} \tag{3.85}$$

$$\nabla \times \mathbf{E} = -\frac{\partial \mathbf{B}}{\partial t} \tag{3.86}$$

Now subtract from the dot product of (3.86) with **H**, the dot product of (3.85) with **E**. We obtain

$$\mathbf{H} \cdot \nabla \times \mathbf{E} - \mathbf{E} \cdot \nabla \times \mathbf{H} = -\mathbf{H} \cdot \frac{\partial \mathbf{B}}{\partial t} - \mathbf{E} \cdot \frac{\partial \mathbf{D}}{\partial t} - \mathbf{E} \cdot \mathbf{J} \tag{3.87}$$

From vector analysis, we have

$$\nabla \cdot (\mathbf{A} \times \mathbf{B}) = \mathbf{B} \cdot \nabla \times \mathbf{A} - \mathbf{A} \cdot \nabla \times \mathbf{B} \tag{3.88}$$

which reduces (3.87) to

$$\nabla \cdot (\mathbf{E} \times \mathbf{H}) = -\mathbf{H} \cdot \frac{\partial \mathbf{B}}{\partial t} - \mathbf{E} \cdot \frac{\partial \mathbf{D}}{\partial t} - \mathbf{E} \cdot \mathbf{J} \tag{3.89}$$

For linear constitutive equations $\mathbf{B} = \mu \mathbf{H}$ and $\mathbf{D} = \varepsilon \mathbf{E}$, the volume integral of (3.89) yields

$$\int_s (\mathbf{E} \times \mathbf{H}) \cdot \mathrm{d}\mathbf{s} = -\frac{\partial}{\partial t} \int_v \frac{1}{2} (\mathbf{H} \cdot \mathbf{B} + \mathbf{E} \cdot \mathbf{D}) \mathrm{d}v - \int \mathbf{E} \cdot \mathbf{J} \mathrm{d}v \tag{3.90}$$

where the surface s bounds the volume v. To get the left-hand side term of (3.90) from the left-hand side term of (3.89), we have used Gauss's theorem for the volume integral,

$$\int_v (\nabla \cdot \mathbf{A}) \mathrm{d}v = \int_s \mathbf{A} \cdot \mathrm{d}\mathbf{s} \tag{3.91}$$

and that for non-dispersive media such as free space, μ and ε are constants and we have

$$\mathbf{H} \cdot \mathbf{B} = \mathbf{H} \cdot \mu\mathbf{H} = \mu|\mathbf{H}|^2 \tag{3.92}$$

$$\mathbf{E} \cdot \mathbf{D} = \mathbf{E} \cdot \varepsilon\mathbf{E} = \varepsilon|\mathbf{E}|^2 \tag{3.93}$$

Thus, we have the important energy conservation equation of Poynting vector, in W (J/s),

$$\int_s (\mathbf{E} \times \mathbf{H}) \cdot \mathbf{ds} = -\frac{\partial}{\partial t}\int_v \left(\frac{1}{2}\mu|\mathbf{H}|^2\right) \mathrm{d}v - \frac{\partial}{\partial \mathrm{t}}\int_v \left(\frac{1}{2}\varepsilon|\mathbf{E}|^2\right) \mathrm{d}v - \int_v \mathbf{E} \cdot \mathbf{J}\, \mathrm{d}v \tag{3.94}$$

There are four terms in the equation above.

1. The Poynting vector term

$$\mathbf{P} = \mathbf{E} \times \mathbf{H} \;\; \mathrm{W/m^2}\, (\mathrm{J/m^2 s}) \tag{3.95}$$

indicates the density of power flow. The Poynting vector **P** represents the electromagnetic energy flow per unit area per unit time across a given surface. The $\mathbf{P} = \mathbf{E} \times \mathbf{H}$ term on the left-hand side is called the Poynting vector. It denotes the flow, or radiation of electromagnetic energy out of a surface **s**. The average power in an electromagnetic wave is found from $(1/2)\mathrm{Re}(\mathbf{E} \times \mathbf{H}^*)$,where the asterisk stands for conjugate.

2. The term $\frac{1}{2}\mu|\mathbf{H}|^2$ joules/m^3 represents the energy stored in the magnetic field. The negative sign indicates that as the magnetic energy collapses with time in a given region, the energy escapes out as $\mathbf{E} \times \mathbf{H}$. The $-\frac{1}{2}\partial\mu\mathbf{H}^2/\partial t$ term indicates the rate of decay (note the negative sign) in the magnetic energy $\frac{1}{2}\mu H^2$ J/m^3 stored in a volume v. It is this decay of energy around an antenna that produces the radiated power $\mathbf{E} \times \mathbf{H}$ W/m^2 flowing out from the antenna. The term $\frac{1}{2}\mu H^2$ J/m^3 is associated with the energy stored in inductive circuits or devices.

3. The term $\frac{1}{2}\varepsilon|\mathbf{E}|^2$ J/m^3 represents the energy stored in the electric field. The negative sign implies that as the electric field energy collapses with time, energy spatially escapes the volume containing the fields. The $-\frac{1}{2}\partial\varepsilon\mathbf{E}^2/\partial t$ term indicates the rate of decay of electric energy $\frac{1}{2}\varepsilon E^2$ J/m^3 stored in a volume v. the term is what is important in the rare capacitive type of antennas. It is this term $\frac{1}{2}\varepsilon E^2$ J/m^3 that in electronic transistors provides power amplification as well as stored energy in capacitors.

4. The $\mathbf{E} \cdot \mathbf{J}$ term stands for the energy loss due to finite conductivity σ. The conduction current density $\mathbf{J} = \sigma\mathbf{E}_c$, where $\mathbf{E}_c$ is the electric field in the conductor. Moreover, $\mathbf{E} = \mathbf{E}_c + \mathbf{E}_s$, where $\mathbf{E}_s$ is the electromotive field in the source. The fourth term $E \cdot J = \sigma E^2$ (since $J = \sigma E$) is associated with the ohmic power

loss due to the conductivity of the medium through which the wave travels. Typical conductivity in S/m and relative permittivity values encountered in wireless communication channels are 4 and 80 for seawater, 0.001 and 80 for freshwater, 0.02 and 30 for swampy land, 0.004 and 13 for forest terrain, 0.002 and 14 for rocky terrain and 0.001 and 10 for sandy terrain. The values do not reflect the frequency dependence of the parameters.

3.6.2 Electromagnetic energy and material properties

For sinusoidal signals, we may replace $\partial/\partial t$ by $j\omega$ in the phasor-based frequency domain and thus the energy transported by an electromagnetic wave is $U = (1/2)\varepsilon E^2 + (1/2)\mu H^2\ \mathrm{J/m^2}$. Since $E = (\mu_0/\varepsilon_0)^{1/2} H = cB$, the energy carried by the electric field is equal to the energy carried by the magnetic field. Hence we get,

$$U_{\text{elect}} = \frac{1}{2}\varepsilon_0 E^2, \quad U_{\text{mag}} = \frac{1}{2}\varepsilon_0 c^2 B^2 \tag{3.96}$$

and the total energy $U = \varepsilon_0 E^2 = \mu_0 H^2$. The amplitude of the Poynting vector

$$P = \frac{E^2}{\eta} = c\varepsilon_0 E^2\ \mathrm{W/m^2} \tag{3.97}$$

and the average power transported by the electromagnetic wave is $(1/2)\mathrm{Re}\{\mathbf{E}\times\mathbf{H}^*) = (1/2)E^2/\eta = (1/2)c\varepsilon_0 E^2\ \mathrm{W/m^2}$.

It can be shown that for an electromagnetic wave propagating in a good conductor nearly all of the wave energy is carried by the magnetic field component. In the case of an isotropic homogeneous medium, the Poynting vector becomes

$$\mathbf{P} = \mathbf{E}\times\mathbf{H} \tag{3.98}$$

and the energy density becomes

$$U = \frac{1}{2}(\varepsilon E^2 + \mu H^2) \tag{3.99}$$

Note that these expressions reduce to those obtained previously when $\varepsilon = \varepsilon_0$, $\mu = \mu_0$ and $H = B/\mu_0$.

The ratio of electric (u_e) to magnetic (u_m) energy density is, therefore,

$$\frac{\frac{1}{2}\varepsilon E_0^2}{\frac{1}{2}\mu H_0^2} = \frac{u_\mathrm{e}}{u_\mathrm{m}} \tag{3.100}$$

Now for a good conductor $\sigma \gg \omega\varepsilon$, and the intrinsic impedance

$$\eta = [\mathrm{j}\omega\mu/\sigma + \mathrm{j}\omega\varepsilon]^{1/2} \approx [\mathrm{j}\omega\mu/\sigma]^{1/2} \tag{3.101}$$

Since $E_0/H_0 = \eta$, from (3.100) and (3.101) we get

$$\frac{\frac{1}{2}\varepsilon E_0^2}{\frac{1}{2}\mu H_0^2} = \frac{\varepsilon\omega}{\sigma} \tag{3.102}$$

Now, $\varepsilon\omega/\sigma$ is a very small quantity for a good conductor for all frequencies up to and including optical frequencies. Thus to a very good approximation, looking at (3.102), the energy is located in the magnetic component of a wave propagating in a conductor. In contrast, we earlier found that when an electromagnetic wave propagates through a vacuum its energy is shared equally between the electric and magnetic field components. Hence, for media like seawater which act like a good conductor at most undersea communication frequencies, antennas should be designed to capture the magnetic field more than the electric field. In free space, antennas such as the electric dipole are designed to capture the electric fields, whereas other antennas such as the loop antenna are designed to capture the magnetic field.

3.6.3 Microwave oven

Electromagnetic wave energy at a microwave frequency of about 2.5 GHz is used to heat up and cook food. When subject to rapidly oscillating microwave electromagnetic fields the water molecules seek to align themselves to each other as dipoles. This results in friction at molecular level and causes thermal energy dissipation. The depth to which the microwaves penetrate into the food material depends on the skin depth δ of the material at 2.5 GHz. For food materials with 20% water content, the skin depth is about 8 cm. The electromagnetic energy may penetrate to a depth of about 3δ. Magnetrons are the most commonly used microwave source. They are operated at a voltage of 4000 V (4 kV) using a transformer to step up the domestic power supply voltage. The food may be placed on plates or inside dishes made of glass, ceramic or plastic which do not absorb microwave energy.

As shown in Fig. 3.8, a metallic fan inside the microwave oven keeps the microwaves moving and bouncing inside the mesh enclosed chamber. The mesh width is kept much smaller than the wavelength (e.g., 10 cm) of the microwave. Note that as the water content of the food increases the depth of penetration decreases, and the inner part of the food could remain uncooked if a large size of such food material is placed inside. For food with 50% water content, the skin depth at 2.5 GHz is about 4 cm. Metal containers should not be placed inside the microwave oven, since the microwaves will, by dissipation of the Joule losses, exert a force on the metal plate, causing it to be rapidly thrown about inside the oven chamber.

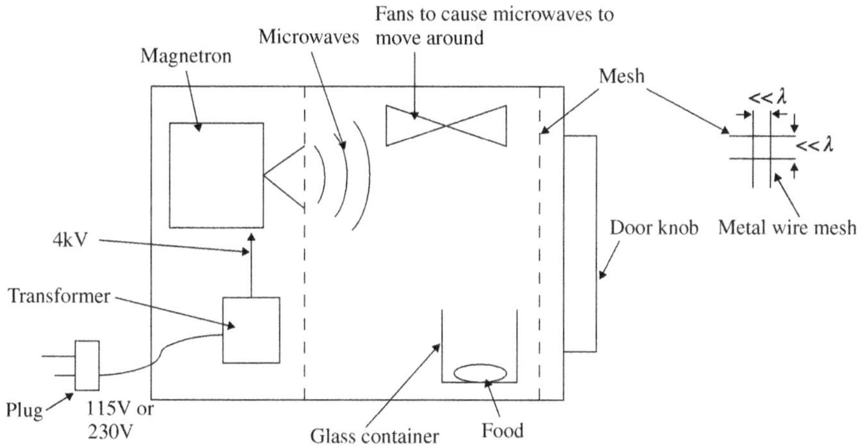

Figure 3.8: Function of a microwave oven.

3.7 Electromagnetic fields and electric circuits

Many electromagnetic field problems are solved using electric circuit theory. For instance, to determine the power transferred from an electric source, such as a signal generator, to an electronic circuit through an electric conductor, such as a copper conductor, instead of solving Maxwell's equations to obtain **E, H** and the power transferred **E** × **H**, we represent the copper conductor by the following circuit parameters: resistor R, inductor L and capacitor C, and solve circuit equations to obtain the voltage and current at the load end. Electric circuit equations are relatively easier to solve than Maxwell's equations. However, it is important to remember that R, L and C are approximate representations of devices and systems where the total energy may be conveniently divided into electrostatic energy (stored in C), magnetic energy (stored in L) and ohmic losses (in R). It is important to recognize that at higher frequencies electric circuit theory will break down, and we need to resort to electromagnetic field calculations for design purposes. Printed circuits (size of 15 cm × 15 cm boards) for instance may yield to electric circuit analysis and design up to 1 MHz or so. But at higher frequencies (e.g. 100 MHz or 1 GHz) electric circuit theory applied to printed circuit analysis will give very wrong results.

In general, we note that the circuit equations of an electromagnetic device or system are an approximation over a larger distance (e.g., over plate separation d or conductor length L) of Maxwell's equations which apply to each point in space.

Moreover, for the current at every point of a conducting line to be equal, the length of the conductor should be approximately less than a tenth of the wavelength, i.e., $\lambda/10$. Else we may have a condition where the current at different points of the conductor may be flowing in different directions or have different magnitudes. Therefore, the analysis and design of electrical devices and systems may be successfully performed using electric circuit theory if the length of the circuit element is less than about 0.1λ. In cases where there is radiation, or electromagnetic interference (EMI), designing the electronic circuit with electromagnetic compatibility (EMC) requirements is important, and electromagnetic field solutions must take precedence over electric circuit solutions.

3.7.1 Resistors

3.7.1.1 Electric fields in resistors

Consider an electric conductor of conductivity σ, cross-sectional area A, and carrying a current I. If the volt drop across a length z of the conductor is V, then the electric field may be approximated to $E = V/z$. Moreover , the current density is $J = \sigma E = \sigma(V/z)$ A/m. The power loss in the conductor is $E \cdot J$ W/m^3, or $(V/z)(\sigma(V/z)) = \sigma V^2/z^2$ W/m^3. The volume of the conductor is zA m^3. Therefore, the total power loss in length z of the conductor is $(\sigma A/z)V^2$ W. Defining the circuit resistance as $R = z/(\sigma A)$ ohms, the power loss is $P = (\sigma V^2/z^2) \times (zA) = V^2/R$ W. Here, the circuit parameter R is defined making certain assumptions such as considering the electric field along the surface of the conductor to be uniform and constant over the length z, yielding $E = V/z$.

3.7.1.2 Types of resistors

Of the three types of resistors available, the nickel wire-wound resistors perform well at DC and low frequencies. But at higher frequencies, it acts like an inductive coil and the resistance value becomes meaningless. The carbon composite resistors are the second type of resistors. These are robust and long lasting. The third type of resistor is specially designed for high frequency applications. These are films of nichrome or tantalum nitride-based chip resistors. These resistors, however, are easily damaged by electrostatic discharges (ESD). But, the advantage of these high frequency resistors is that they may be directly fabricated on an integrated electronic circuit board.

3.7.1.3 Equivalent circuit of a resistor

In the equivalent circuit of a resistor R shown in Fig. 3.9(a), the parasitic inductance L_1 is significant only in wire-wound resistors. The parasitic capacitance C_1 represents the charge separation associated with the voltage drop across a resistor and is about 2 pF for carbon composite resistors and 5 nF for chip resistors. A further parasitic inductance L_2 is 10 nH for carbon composite resistors and 0.5 nH for chip resistors with smaller leads. The equivalent impedance of the resistor circuit shown in Fig. 3.9(a) setting $L_2 = L$ and ignoring L_1, is $Z(\omega) = L\frac{\omega_2^2-\omega^2+j\omega\omega_1}{j\omega+\omega_1}$, where $\omega_1 = \frac{1}{RC}$ and $\omega_2 = \frac{1}{\sqrt{LC}}$. It can be shown that the resistor is a pure resistor, i.e., $Z(\omega) = R$, only when the signal frequency is less than ω_1. In the

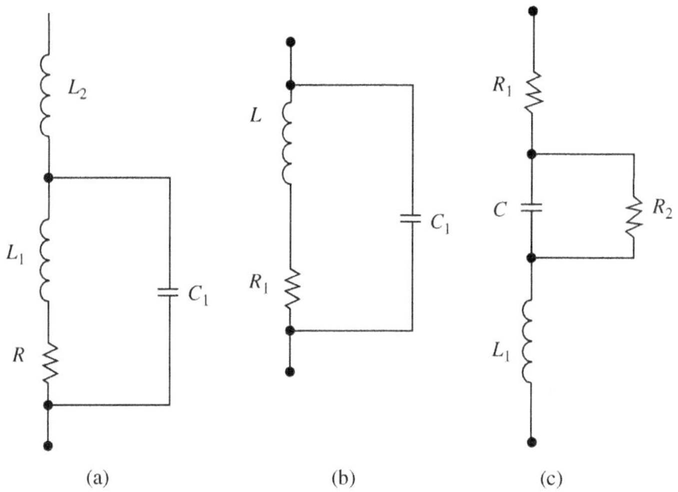

Figure 3.9: Equivalent circuits of (a) a resistor, (b) an inductor and (c) a capacitor.

frequency range $\omega_1 \leq \omega \leq \omega_2$, the impedance of the resistor becomes capacitive, i.e., $Z(\omega) = 1/(\mathrm{j}\omega C)$. For frequencies above ω_2, the impedance of the rersistor is purely inductive, i.e., $Z(\omega) = \mathrm{j}\omega L$. Thus, for instance, a resistor connected to a transistor to form a filter or amplifier, will begin to act very differently if there is high frequency noise present in the circuit, leading to the malfunction of the circuit of which the resistor is a part.

3.7.1.4 Electric plasma column

Other than the states of solid, liquid and gas, there is a fourth state of matter which is called electric plasma or ionized gases. Electric plasmas are like an intermediate state between those of (a) a perfect insulator and (b) an insulator becoming a conductor as it electrically breaks down under high electric stress when the electric field inside the insulator exceeds its breakdown electric field. Such is the case in high voltage power system insulators and cables when excessive voltages cause electric breakdown. The insulator becomes a partially conducting material where electric space charges are produced inside it, finally leading to electric breakdown when a good conducting path is created by the rapid build up of electric charges.

Consider a time harmonic field that cuts through an electric plasma. We look at a free electron of the plasma moving in the time-harmonic electric field **E**. If the electric field is in the z-direction, and the electron has an angular frequency ω, then we have from Newton's law,

$$-e\mathbf{E} = m\frac{\mathrm{d}^2 z}{\mathrm{d}t^2} \tag{3.103}$$

where m is the mass of an electron. We may replace the right-hand term of (3.103) by $-m\omega^2\mathbf{z}$. We thus have from (3.103),

$$\mathbf{z} = \frac{e}{m\omega^2}\mathbf{E} \tag{3.104}$$

This means that the displacement $\mathbf{z}$ will give rise to an electric dipole of moment $-e\mathbf{z}$, and for N number of electrons per unit volume, we get a volume density of the electric polarization vector

$$P = N(-e\mathbf{z}) = -\frac{Ne^2}{m\omega^2}\mathbf{E} \tag{3.105}$$

Therefore, the electric flux density may be written as

$$\mathbf{D} = \epsilon_0\mathbf{E} + \mathbf{P} = \varepsilon_0\left(1 - \frac{\omega_\mathrm{P}^2}{\omega^2}\right)\mathbf{E} \tag{3.106}$$

where the plasma angular frequency

$$\omega_\mathrm{P} = 2\pi f_\mathrm{P} = \sqrt{\frac{Ne^2}{m\varepsilon_0}} \tag{3.107}$$

and substituting the approriate values for e, m, and ε_0, we get a very useful formula for plasma frequency

$$f_\mathrm{P} = 9\sqrt{N} \tag{3.108}$$

where N is the number of electrons per unit volume present in the plasma or ionized gas. The permittivity of the plasma may be written as

$$\epsilon_\mathrm{P} = \varepsilon_0\left(1 - \frac{f_\mathrm{P}^2}{f^2}\right) \tag{3.109}$$

And the propagation constant is

$$\gamma = \mathrm{j}\omega\sqrt{\mu\varepsilon_0}\sqrt{1 - \left(\frac{f_\mathrm{P}^2}{f^2}\right)} \tag{3.110}$$

We observe that the propagation constant becomes zero when the signal frequency is equal to the plasma frequency. This means that at this frequency and below γ is real implying attenuation as $\mathrm{e}^{-\gamma}$; no signal can propagate through the plasma, and hence f_p is also called the cutoff frequency. Thus for instance, when a spacecraft reenters the earth's atmosphere, due to both friction heat and speed, a layer of plasma builds up around the spacecraft, with typically $N = 2 \times 10^{14}$ electrons per cubic meter. Therefore for proper communication to take place from the spacecraft, the communication frequency should be above the cutoff frequency of about 130 MHz when γ is of the form jβ.

Moreover, the earth's atmosphere contains multiple ionized gas layers at heights of about 100 km above the earth. Typically these may contain about 10^{10}–10^{12} electrons/m^3. This means that electromagnetic waves at frequencies below the cutoff frequencies bounce off back to earth from these layers. Such waves may be used to communicate all around the earth by repeated bounces between the concentric ionized layer and the earth. Moreover, communications to spacecraft and satellites placed above these layers communicate with the earth above the cutoff (plasma) frequency of about 1–10 MHz.

The intrinsic impedance of a plasma is

$$\eta_P = \frac{\eta_0}{\sqrt{1 - \frac{f_P^2}{f^2}}} \tag{3.111}$$

where the free space intrinsic impedance η_0 is 120π (or 377) ohms.

3.7.1.5 Resistance of a plasma channel

We shall consider the resistance of a lightning channel, which is an electric plasma channel, where the previously insulating air has become a conducting channel due to the high electric field stress under a thunder cloud. The resistance of the lightning return stroke column is given by $R = 1/\sigma\pi r^2$ ohms/m. The radius of the return stroke channel is assumed to remain constant throughout the presence of the return stroke, where the constant radius is maintained by equal and counteracting magnetic and kinetic forces. The conductivity of a highly ionized gas is given by $\sigma = \mathrm{j}/E = eN\mu^-$ where the electron mobility μ^- with a mass m and rms velocity c_T is given by $\mu^- = e\lambda_e/(mc_T)$. The mean free path λ_e is $(1/Nq)$ and the rms velocity $c_T = (3kT/m)^{0.5}$. The collision cross-section for strongly scattered electrons is given by $q(e^2/3kT)^2/(16\varepsilon_0^2)$. Substituting these values into $\sigma = eN\mu^-$ gives $\sigma = 1.5 \times 10^{-5} \times T^{3/2}$ (ohm-cm)$^{-1}$ where a factor of 0.1(1/1 nΛ, where Λ is the Debye length (or critical impact parameter)) is included for shielding effects. $\mathrm{Log}_e\Lambda$ varies slowly with electron density and electron temperature. Taking $r = 1$ cm we have a channel resistance of approximately 1 ohm/m for T = 20,000°K, and for T = 16,000°K the channel resistance is 5 ohm/m. Note that for T = 20,000°K, the column conductivity σ = 4242 (ohm-m)$^{-1}$ and for T = 16,000°K, σ = 3035 (ohm-m)$^{-1}$.

3.7.1.6 Earth resistance

What about the earth resistance at the point where a conductor (or plasma such as a lightning channel) is attached to the ground? The earth resistance has to be included when circuit models are used to study breakdown currents and voltages. The earth resistance represents energy dissipation in the earth. In doing this we account for the finite conductivity of the earth. In order to maintain a charge flow into the earth with a finite earth conductivity, at the contact point to earth of any earthing conductor, or of an earth flash lightning channel, the electric charge at the termination point maintains a small horizontal electric field at the point of contact. When lightning strikes the open ground, for instance, earthing is provided merely by loose attachment of the channel to the ground. After a complete flash, the channel is found to have penetrated the earth to a depth of about 0.5 m. The lightning channel radius is about 1 cm. Although, the value of the earth resistance will slightly drop for subsequent lightning strokes, its effects will not be significant. For flashes to open terrain, the earth resistance is calculated for a spherical conductor, which is given by $R_E = (1/\sigma E\pi a)$ ohms, where a is the return stroke radius. A typical value of R_E is 2000 ohms; for earth conductivity σ_E of 0.01 (ohm-m)$^{-1}$. When lightning strikes an earthed conductor, $R_E = \frac{l}{2\pi\sigma_E}\log_e(81/1.36a)$ where 1 is the length of the conductor under the earth surface. Soil conductivity in hilly areas is an order less than that of flat terrains, because of different soil constituents.

3.7.1.7 Temperature coefficient

Resistances of materials are highly dependent on temperature as well as the frequency of the current. When temperature increases, the resistance of a conductor increases. The resistance R_T at temperature T°C is related to the resistance R_0 at 0°C, according to the relationship $R_T = R_0(1 + \alpha T)$ ohms, where α is called the temperature coefficient of the material. In Table 3.1 is given a list of conductivity and temperature coefficients for different conductors.

Superconducting materials, such as lead or ceramic materials, yield zero resistivity at low temperatures: lead at 7.3°K and yttrium barium oxide at 70°K.

3.7.2 Inductors

Consider now a coaxial cable, with an inner conductor of radius a, and an outer conductor of radius b. Let the current along the inner conductor be I A. From Ampere's law, the magnetic field at a distance r from the center of the inner conductor is given by $H = I/(2\pi r)$ A/m with $a < r < b$. Therefore, the magnetic energy stored inbetween the two conductors is $(1/2)\mu H^2$ J/m^3 $= (\mu/8)I^2/(\pi^2 r^2)$ J/m^3. The volume of a cylindrical strip of width dr at distance r is $\pi[(r + dr)^2 - r^2]z$, where z is the length of the coaxial cable under consideration. The volume may be approximated to $v = 2\pi r\, dr\, z$, ignoring the dr^2 term. Hence, the total magnetic energy stored over length z of the conductor is $(\mu/8)I^2/(\pi^2 r^2)\ 2\pi r\, dr\, z = (1/4)(\mu z I^2/\pi\, r)dr$. When the stored energy is integrated from $r = a$ to $r = b$, the integration yields $(1/2)LI^2$ J, where inductance $L = (z\mu/2\pi)\ln(b/a)$ H or $(\mu/2\pi)\ln(b/a)$ H per unit length. Note that it could be shown that the per unit length capacitance, conductance and resistance of such a coaxial cable with conductivity σ and at frequency f Hz is $(z2\pi\varepsilon)/\ln(b/a)$ F, $(z2\pi\sigma)/(\ln(b/a))$ S and $(z/\pi a)(\pi f\mu/\sigma)^{1/2}$ Ω, respectively.

Table 3.1: Temperature coefficient and conductivity for some materials

	Temperature coefficient of resistance $\times 10^{-4}$	Conductivity $\times 10^{-8}$
Magnanin	0.02	45
Eureka	−0.7 to 0.4	49
Nichrome	1	108
Silicon	−75	16×10^{-12}
Gold	34	24
Platinum	37	10
Silver	38	1.62
Copper	43	1.76
Lead	43	20
Aluminum	43	2.83
Tungsten	45	5.48
Iron	55	9.4
Nickel	59	7.24

Inductors come in four basic forms: ferrite rod inductors, ferrite toroidal inductors, single loop inductors and the spiral inductor. The highest inductance is obtained by winding loops around ferromagnetic or ferrite rods. Inductors are used to tune into certain frequency signals (at resonant frequency). They may be used in series to choke off noise current through acting as filters reflecting back unwanted noise currents. Inductors are used as antennas, for instance, in Radio Frequency Identifiers (RFIDs). Such inductive antennas are formed by wrapping wires around a ferrite-rod core (or air core) to be transmitters and receivers of AM radio signals. The closed loop ferrite toroidal inductors generate and receive very little magnetic field, and are prone to saturation at high current values.

The impedance of the inductor equivalent circuit shown in Fig. 3.9(b) is

$$Z = \frac{R_1 + \mathrm{j}\omega L}{(1 - \omega^2 L C_1) + \mathrm{j}\omega R_1 C_1} \tag{3.112}$$

and it is a maximum indicating resonance when $1 - \omega^2 L C_1 = 0$, or at the self-resonance frequency of

$$f_{\mathrm{sr}} = \frac{1}{2\pi\sqrt{LC_1}} \tag{3.113}$$

At frequencies less than $\omega = R_1/L$ the inductor will act as a pure resistor, where the impedance will remain constant with changing frequency. At frequencies above

$1/\sqrt{LC_1}$ it will act like a pure capacitor, with a linearly decreasing impedance with frequency. It is only between these frequencies that the inductor will act as an inductor, with an impedance that linearly increases with frequency. A 1 μH air-core inductor has an f_{sr} value of about 100 MHz. Both inductors and capacitors should be operated at frequencies less than a half of their f_{sr} values. Inductors operating at less than 50 MHz are wire-wound inductors. For frequencies higher than 50 MHz, ferrite inductors are used.

3.7.3 Capacitors

In order to appreciate how electric circuit components are approximate parameters of the electromagnetic phenomena in insulator, magnetic and conductor materials, consider first two parallel plates, each having an area A, separated by a distance d, and at a potential difference V. Then the electric field E between the two plates is approximately V/d V/m. Then the energy stored between the two plates is $(1/2)\varepsilon E^2$ J/m^3, or $(1/2)\varepsilon(V/d)^2$ J/m^3, where ε is the permittivity of the insulatior material between the two plates. Therefore the total energy stored between the plates, noting that the volume of the capacitor is Ad m^3, is $(1/2)(\varepsilon A/d)V^2$ J. We may rewrite the stored energy as $(1/2)CV^2$ J, where $C = \varepsilon A/d$ F is the well-known expression for the capacitance of a parallel plate capacitor. But note that $(1/2)CV^2$ J which is the circuit expression of the energy stored in a capacitor is an approximation of the $(1/2)\varepsilon E^2$ J/m^3 of the electromagnetic expression of the energy. Amongst the assumptions we have made are, for instance, that there is no fringing (leakage) electric fields at the edge of the capcitor plates, an assumption allowing us to write $E = V/d$ over the entire plate. Indeed at the edge of the plates, this is not exactly correct.

Capacitors as electric charge storing devices have recently received much attention. Special capacitors have been developed as very promising energy storing super-capacitors. These capcitors may be electrically charged from the power systems at night time (when energy demand is low) and used in the day time when power demand peaks around 10 am and 4 pm. We shall first describe the foundational characteristics of a capacitor using the equivalent circuit shown in Fig. 3.9(c). The capacitor equivalent circuit is a series R, L and C circuit (assuming dielectric resistance $R_2 = \infty$) having the resonance characteristics shown in Fig. 3.10(a), in contrast to the resonance characteristics of the parallel R, L and C circuits of an inductor shown in Fig. 3.10(b).

The series parasitic resistance R_1 can be a several ohms for an electrolytic capacitor, whereas it is negligibly small for ceramic capacitors. The electrolytic capacitors are used to obtain low frequency, high capacitance values of the order of 1–1000 pF. The ceramic capacitors are used at high frequencies, within a limited value from 5 pF to 1 μF. Tuning capacitors for high frequency circuits are a third kind of capacitors, being inter-digitated capacitors. The series R, L and C equivalent

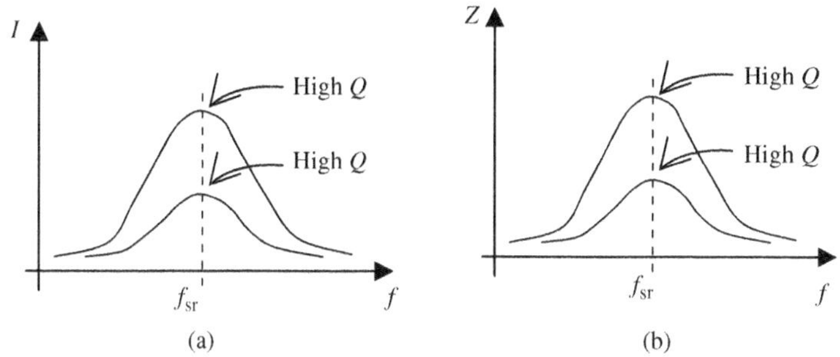

Figure 3.10: Self-resonance of (a) capacitors and (b) inductors.

of a capacitor, shown in Fig. 3.9(c), yields an impedance of

$$Z = R_1 + \frac{j(\omega^2 L_1 C - 1)}{\omega C} \tag{3.114}$$

which has a minimum value at resonance when $\omega^2 L_1 C - 1 = 0$, or at the self-resonant frequency of

$$f_{sr} = \frac{1}{2\pi\sqrt{L_1 C}} \tag{3.115}$$

where the inductance L_1 is primarily that of the leads connecting the capacitor. At resonance, the input impedance Z is a minimum and the current through the capacitor $\left(I = \frac{V}{R}\right)$ reaches a maximum. Under such a condition, the voltage across C is $V_C = IX_C = \frac{I}{\omega C}$. And the voltage across the inductor L is $V_L = IX_L = \omega LI$. The voltages V_C and V_L are both much larger than the sinusoidal supply voltage applied across the capacitor C_1. This voltage magnification exists across each L_1 and C, although the total voltage is still the supply voltage since V_L and V_C are anti-phase and therefore cancel each other out, so that the supply voltage at the resonant frequency is simply $V = V_R$, the voltage across the lead resistance ($V_R = IR$). The magnification of the voltages across C and L_1 is qualified by the Q-factor (Q for quality) defined as the ability to amplify voltage, where

$$Q = \frac{V_L}{V} = \frac{(2\pi f_{sr} L_1)}{R} \tag{3.116}$$

or

$$Q = \frac{V_C}{V} = \frac{1}{(2\pi f_{sr} CR)} \tag{3.117}$$

Both curves of Fig. 3.10 are for large Q. When Q is large, the current through the circuit is also large, as shown in Fig. 3.10(a). For an inductor with an equivalent circuit shown in Fig. 3.9(b), the impedance Z is a maximum at f_{sr} as shown in Fig. 3.10(b) and the voltage across the inductor is large.

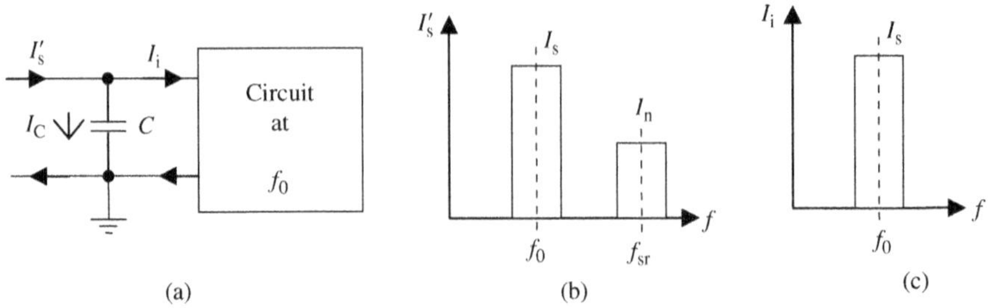

Figure 3.11: A capacitor used to eliminate noise: (a) circuit connection of a capacitor, (b) the current flowing into the circuit I_s with noise curent I_n and (c) the noise current is diverted by the capacitor, leaving only the source current I_s.

It should be noted that the impedance of the capacitor will be capacitive only at frequencies less than $1/\sqrt{LC}$, with an impedance that linearly decreases with frequency. Above this frequency, the capacitor will act like an inductor, with an impedance that linearly increases with frequency.

A capacitor connected across the input line and ground of a high-frequency circuit operating at a frequency less than half the self-resonant frequency of the capacitor shunts off to the ground any noise currents close to its resonant frequency. This is because the capacitor impedance is a minimum close to f_{sr}: thus the current through the capacitor to ground is maximum. The simple arrangement is shown in Fig. 3.11.

The current coming into the capacitor and the circuit is I_s, and it contains the necessary source current at the operating frequency f_0 of the circuit ($f_0 \ll f_{sr}$) as well as a noise current I_n at a frequency, The capacitor C is selected so that its self-resonant frequency is equal to the noise frequency. The current I_C through the capacitor will be a maximum at f_{sr} so that all the noise current will be diverted to the ground ($I_C = I_n$) and the circuit receives current I_i which is equal to the source current ($I_i = I_s$) without the noise current, as shown in Fig. 3.11(c).

Going back to the equivalent circuit of the capacitor, when the dielectric material breaks down due to high voltage across the capacitor (for instance, due to ESD), the dielectric material of the capacitor will develop a resistance associated with the movement of free electric charges through the dielectric material as pre-breakdown or corona currents. This resistance R_2 is connected in parallel to C, and is given by

$$R_2 = \frac{1}{(2\pi fC)\tan\delta} \tag{3.118}$$

where the dielectric loss tangent is given by

$$\tan\delta = \frac{\sigma_0 + \omega C_I}{\omega\varepsilon_R} = \frac{\sigma_f}{\omega\varepsilon_R} \tag{3.119}$$

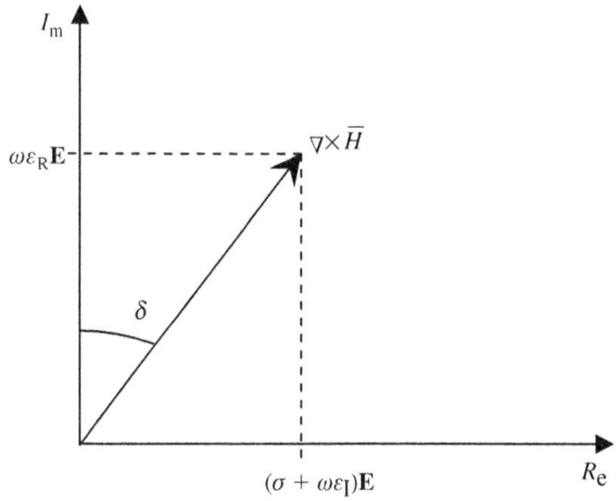

Figure 3.12: Conduction and displacement currents.

ε_R $(=\varepsilon_0\varepsilon_r)$ is the real part of the permittivity and ε_r is the relative permittivity of the capacitor dielectric material. Moreover σ_f $(= \sigma_0 + \omega\varepsilon_I)$ is the effective conductivity, ε_I being the imaginary part of the permittivity ε $(= \varepsilon_R - j\varepsilon_I)$ and σ_0 the DC conductivity of the material. In a dielectric (insulator) material, tan δ is measure of power loss in the dielectric.

Considering Ampere's law

$$\nabla \times \mathbf{H} = \mathbf{J} + \varepsilon \frac{\partial \mathbf{E}}{\partial t} \tag{3.120}$$

we may rewrite it for the capacitor insulator material as follows:

$$\begin{aligned} \nabla \times \mathbf{H} &= \sigma_0 \mathbf{E} + (\varepsilon_R - j\varepsilon_I)(j\omega\mathbf{E}) \\ &= (\sigma_0 + \omega\varepsilon_I)\mathbf{E} + j\omega\varepsilon_R\mathbf{E} \end{aligned} \tag{3.121}$$

from which we get the vector representation shown in Fig. 3.12.

Consider now the typical values of $(\sigma_0, \varepsilon_R, \varepsilon_I)$ for copper $(5.8 \times 10^7,\ \varepsilon_0,\ 0)$, seawater $(5, 72\varepsilon_0, 12\varepsilon_0)$ and glass $(10^{-12}, 10\varepsilon_0, 0.01\varepsilon_0)$ where free space permittivity $\varepsilon_0 = 8.85 \times 10^{-12}$. For a good conductor like copper, except at very high frequencies, we have $\sigma_f = \sigma_0$ and $\tan\delta >> 1$ since $\sigma_0 \gg \omega\varepsilon_R$ $(=\omega\varepsilon_0)$ and $\tan\delta = \sigma_0/\omega\varepsilon_R$. For a good dielectric like glass, $\sigma_0 \ll \omega\varepsilon_I$ and $\tan\delta = \varepsilon_I/\varepsilon_R \ll 1$. But for seawater, at low frequencies $\sigma \gg \omega\varepsilon_I, \omega\varepsilon_R$, thus making seawater a good conductor. However, at high frequencies, $\tan\delta \ll 1$, seawater is effectively a good dielectric.

Therefore, at low frequencies and for lossless dielectrics, R_2 may be considered to be like an open circuit ($R_2 \cong \infty$). But at very high frequencies, or when the dielectric material is electrically breaking down causing electric charges to appear in it as space charges, we must include R_2 in our equivalent circuit and the resonant frequency. In a lossy single core electric cable of inner diameter $2a$ and outer diameter $2b$, the resistance R_2 is represented by a conductance G, where $G = 1/R_2$ and is given by

$$G = \frac{2\pi\sigma_0}{\ln\left(\frac{b}{a}\right)} \quad \Omega^{-1} \tag{3.122}$$

And for a twin-lead cable with d being the distance between the centers of the two leads,

$$G = \frac{\pi\sigma_0}{\cosh^{-1}\left(\frac{d}{2a}\right)} \quad \Omega^{-1} \tag{3.123}$$

The per unit length inductance and capacitance for a coaxial cable with a hollow core are

$$(L, C) = \left[\left(\frac{\mu}{2\pi}\ln\left(\frac{b}{a}\right)\right) \text{ H/m}, \left(2\pi\varepsilon \ln\left(\frac{b}{a}\right)\right) \text{ F/m}\right] \tag{3.124}$$

where a is the inner conductor radius and b is the outer conductor radius of a coaxial cable.

For the twin-lead cable, the per unit length inductance and capacitance are

$$(L, C) = \left[\left(\frac{\mu}{\pi}\cosh^{-1}\left(\frac{d}{2a}\right)\right) \text{ H/m}, \left(\frac{\pi\varepsilon}{\cosh^{-1}\left(\frac{d}{2a}\right)}\right) \text{ F/m}\right] \tag{3.125}$$

where d is the distance between the centers of the two leads (or wires) of the twin-lead line and a is the radius of each lead (or wire).

Capacitors are widely used in electronic circuits like coupling capacitors in amplifiers to block off DC currents. For DC, we have $f_0 = 0, X_C = \frac{1}{\omega_0 C} = \infty$; thus the capacitor connected in series does not permit DC currents to pass through as shown in Fig. 3.13.

The DC voltage input V_0 sees the capacitor as an open circuit ($X_C \cong \infty$), through which no current can pass. But the alternating current I_ω at frequency ω passes through C to the circuit since $X_C = 1/\omega C$ which has a finite value.

In addition to being used as coupling capacitors which do not allow a DC current to pass through, capacitors are used as band-pass filters when used with an inductor.

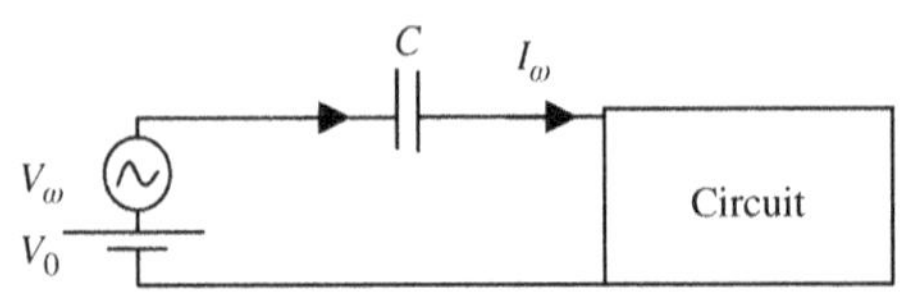

Figure 3.13: Series connection of a capacitor: blocking DC currents.

From the resonance circuit shown in Fig. 3.14(a), a capacitor used with an external inductor L will allow signals at resonant frequencies $f_r = 2\pi/\sqrt{LC}$ and signals on either side of f_r, blocking off frequencies well below or well above f_r. Moreover, such an arrangement is used to feedback a signal.

Capacitors are also used as oscillators, where with an inductor in parallel, the capacitor and the inductor will pass stored energy back and forth to each other as shown in Fig. 3.14.

The electrically charged capacitor in Fig. 3.14(a) discharges its stored energy into the inductor as magnetic energy in Fig. 3.14(b), producing positive going electric current. The next time, the discharge current will be negative going as shown in Fig. 3.14(d). The frequency of the current is

$$f_r = \frac{2\pi}{\sqrt{LC}} \tag{3.126}$$

In order to obtain an undamped oscillation of constant amplitude, a transistor circuit is used to feed the LC oscillator to make up for energy lost in the resistances of the

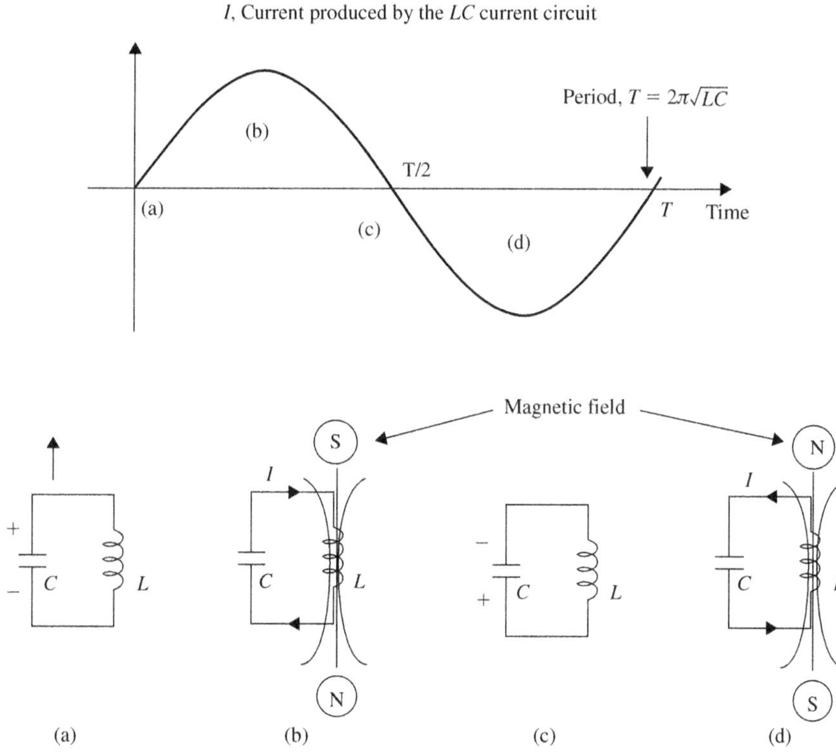

Figure 3.14: Oscillating LC circuit.

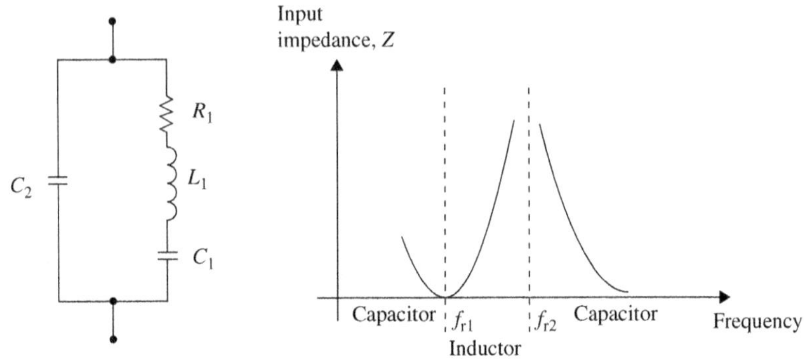

Figure 3.15: Equivalent circuit of a piezoelectric crystal.

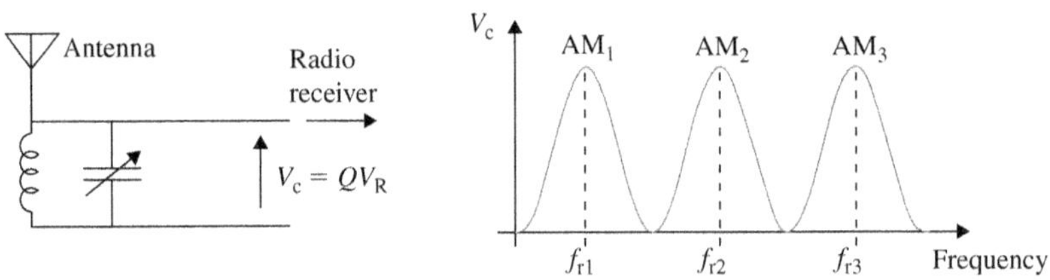

Figure 3.16: An AM radio receiver.

circuit inherent to both L and C as we see in Fig. 3.14(b) and (c). Lead lengths must be kept as short as possible to minimize resistive losses of energy.

It is of interest to note that an oscillator of very stable frequency is produced by a piezoelectric crystal oscillator which has an equivalent diagram that has the combined features of series and parallel R, L and C circuits as shown in Fig. 3.15(a). Therefore, as would be expected, it has two resonant frequencies, one f_{r1} at which the impedance $Z = 0$, or a minimum as with a series RLC circuit and another, second resonant frequency (f_{r2}) at which Z is very large as shown in Fig. 3.15(b). Typical values of R_1, L_1, C_1 and C_2 are 1000 Ω, 100 H, 0.1 pF and 10 pF, respectively.

In Fig. 3.16(a) is shown the use of an inductor L made of a coil wound on a ferrite rod and a variable capacitor C.

By changing (tuning) the values of C, we may obtain resonance frequencies f_{r1}, f_{r2} and f_{r3} for instance, of the three different AM radio frequency stations AM_1, AM_2 and AM_3, respectively. For the parallel RLC circuit of Fig. 3.16(a), the resonant frequencies are $1/(2\pi LC_1)$, $1/(2\pi LC_2)$ and $1/(2\pi LC_3)$ at which the voltages across the capacitor V_C will peak and is equal to QV_r, where V_r is the voltage at resonance and the Quality Factor,

$$Q = \frac{1}{2\pi f_r CR} \tag{3.127}$$

3.8 Electromagnetic fields and magnetic circuits

A limited amount of analysis and design on magnetic systems could be done making use of the magnetic circuit concept. Instead of circuit resistance we define magnetic reluctance $R_m = L/\mu A_m$, where L is the length of the circuit (e.g., the length of a transformer arm), A_m is the cross-sectional area of the circuit, and μ is the permeability of the material. Instead of source emf, we have the magnetomotive force (mmf), which is equal to IN for a current I flowing through a coil of N turns. Instead of electric current we have magnetic flux ψT, and for conductivity σ we have permeability μ. Thus, in the place of Kirchhoff's current law which states that at any electric circuit node

$$\sum I = 0 \tag{3.128}$$

at a magnetic circuit node, we have

$$\sum \psi = 0 \tag{3.129}$$

Similarly as in an electrtic circuit, Kirchhoff's voltage law for a closed loop states that

$$\sum \text{emf} = \sum IR = 0 \tag{3.130}$$

so in the magentic circuit, we have

$$\sum \text{mmf} = \sum \psi R_m = 0 \tag{3.131}$$

which means that the summation of magnetomotive force drives and magnetomotive force drops across reluctances in a path around a closed loop is zero.

3.9 Dielectric material and frequency-dependent permittivity

Consider first a dispersive material in which the phase velocity is a function of the frequency. In such a medium, the electromagnetic wave which carries information (e.g., as in communication or radar systems) will be unrecognizable when it is picked up by a receiver after the transmitted signal travels through a dispersive medium. The electric polarization of a dielctric material is given by

$$\mathbf{P} = \varepsilon_0 \mathbf{E} \frac{A^2}{(\omega_{oa}^2 - \omega^2 - j\omega B)} \tag{3.132}$$

where $E = E_0\, e^{-j\omega t}$ and A^2 is proportional to oscillations per unit volume of the medium through which the electrmagntic wave is traveling. The average oscillating frequency is related to the oscillating frequency ω_0 by the equation

$$\omega_{0a}^2 = \omega_0^2 - \frac{1}{3}A^2 \tag{3.133}$$

and at low frequencies the electric polarization becomes

$$\mathbf{P} = \frac{A^2}{\omega_{0a}^2}\varepsilon_0 \mathbf{E} \tag{3.134}$$

which is of a constant value, independent of frequency. We may rewrite the equation for polarization at low frequency as

$$\mathbf{P} = (\chi - 1)\varepsilon_0 \mathbf{E} \tag{3.135}$$

where the susceptibility

$$\chi = 1 + \frac{A^2}{\omega_{0a}^2} \tag{3.136}$$

Introducing the electric flux density **D**,

$$\mathbf{D} = \chi' \varepsilon_0 \mathbf{E} \tag{3.137}$$

where

$$\chi' = 1 + \frac{A^2}{\omega_{0a}^2 - \omega^2 - j\omega B} \tag{3.138}$$

we may now obtain from the Maxwell equations

$$\nabla \times \mathbf{E} + \mu_0 \frac{\partial \mathbf{H}}{\partial t} = 0 \tag{3.139}$$

and

$$\nabla \times \mathbf{H} - \varepsilon_0 \chi' \frac{\partial \mathbf{E}}{\partial t} = 0 \tag{3.140}$$

the wave equation

$$\nabla^2 \varphi - \varepsilon_0 \mu_0 \chi' \frac{\partial^2 \varphi}{\partial^2 t} = 0 \tag{3.141}$$

The wave equation has the general solution given by

$$\varphi = \varphi_0\, e^{jkz - j\omega t} \tag{3.142}$$

where the wave number

$$k = \frac{\omega}{c}\sqrt{\chi'} = \alpha + j\beta \tag{3.143}$$

and the complex susceptibility

$$\chi' = \frac{c^2}{\omega^2}(\alpha + \mathrm{j}\beta)^2 \tag{3.144}$$

The phase velocity of the wave is given by $v = \omega/\alpha = c/n$, where n is the refractive index of the material.

$$\frac{c}{\omega}(\alpha + \mathrm{j}\beta) \cong 1 + \frac{1}{2}\frac{A^2}{\omega_{oa}^2 - \omega^2 - \mathrm{j}\omega B} \tag{3.145}$$

For $\omega << \omega_0$, $\frac{\alpha c}{\omega} = n = 1 + \frac{A^2}{2\omega_0^2}$

For $|\omega_0^2 - \omega^2| >> \omega B$, we get

$$\frac{\alpha c}{\omega} = n = 1 + \frac{A^2}{2(\omega_0^2 - \omega^2)} \tag{3.146}$$

When $\omega \approx \omega_0$

$$\frac{\alpha c}{\omega} = n = 1 + \frac{1}{2}\frac{A^2(\omega_0^2 - \omega^2)}{((\omega_0^2 - \omega^2)^2 + \omega^2 B^2)} \tag{3.147}$$

and

$$\beta = \frac{1}{2c}\frac{A^2\omega^2 B}{((\omega_0^2 - \omega^2)^2 + \omega^2 B^2)} \tag{3.148}$$

The abosrption coefficient is given by $\frac{\beta c}{\omega^2}$. If we sketch the absorption coefficient against ω, we will get dumbell-shaped characteristics with the peak at $\omega = \omega_0$. Similarly, if we sketch the dispersion curve, i.e., $n - 1$ against ω, we will have $n - 1 = 0$ at $\omega = \omega_0$, and a positive peak at $\omega = \omega_1$, where $\omega_1 < \omega_0$. There will be a negative minimum, at $\omega = \omega_2$ where $\omega_2 > \omega_0$.

3.10 Conductors and frequency-dependent conductivity

Consider an electron of mass m, and electric charge e moving in an electric field $E_0 \exp(-\mathrm{j}\omega t)$ at a velocity $\mathrm{d}r/\mathrm{d}t$. Then the equation of motion is obtained applying Newton's law to the forces caused by the magnetic and electric fields:

$$m\frac{\mathrm{d}^2 r}{\mathrm{d}t^2} + mB\frac{\mathrm{d}r}{\mathrm{d}t} = eE_0\,\mathrm{e}^{-\mathrm{j}\omega t} \tag{3.149}$$

The current density associated with the movement of electrons is

$$\mathbf{J} = Ne\frac{\mathrm{d}r}{\mathrm{d}t} = \frac{N\frac{\mathrm{e}^2}{m}}{B - \mathrm{j}\omega}\mathbf{E} = \sigma_c \mathbf{E} \tag{3.150}$$

where the complex conductivity is given by

$$\sigma_c = \frac{\frac{Ne^2}{m}}{B - j\omega} \tag{3.151}$$

and the complex, frequency-dependent wave number is given by

$$k^2 = \mu\epsilon\omega^2 + j\sigma_c\mu\omega \tag{3.152}$$

We may rewrite (3.152) in the following form:

$$k^2 = (\alpha + j\beta)^2 = \frac{\omega^2}{c^2}\left(1 + j\frac{\sigma_c}{\omega\epsilon_0}\right) \tag{3.153}$$

Hence

$$\alpha^2 - \beta^2 = \frac{\omega^2}{c^2}\left(1 - \frac{Ne^2}{m\varepsilon_0}\frac{1}{\omega^2 + B^2}\right) \tag{3.154}$$

and

$$2\alpha\beta = \frac{\omega^2}{c^2}\left(\frac{Ne^2}{m\varepsilon_0\omega}\frac{B}{\omega^2 + B^2}\right) \tag{3.155}$$

At very low frequencies, $\omega \ll \omega_0, m\frac{d^2r}{dt^2} \ll \frac{dr}{dt}$, and the complex conductivity reduces to a real value given by

$$\sigma_c = \frac{Ne^2}{mB} \tag{3.156}$$

which is the case for most good conductors, such as copper or aluminum. However, in insulatiors, poor conductors and biomaterials such as body tissues, the conductivity is a frequency dependent, complex number.

For further reading

[1] Balanis, C.A., *Antenna Theory*, USA: John Wiley, New York, 2012.

[2] Bansal, R., Ed., *Fundamentals of Engineering Electromagnetics*, USA: CRC Press, Boca Raton, FL 2006.

[3] Chen, D.K., *Fundamentals of Engineering Electromagnetics*, USA: Addison-Wesley, Reading, MA 1993.

[4] Christopoulos, C., *Principles and Techniques of Electromagnetic Compatibility*, USA: CRC Press, Boca Raton, FL 1995.

[5] Collin, R.E., *Antennas and Radio Wave Propagation*, USA: McGraw Hill, New York, 1985.

[6] Elliot, R.S., *Electromagnetics*, IEEE Press, USA: New York, 1993.

[7] Hall, M.P.M., Barclay, L.W., and Hewitt, M.T., *Propagation of Radiowaves*, UK: IEE Press, London, UK, 1996.

[8] Hall, S.H., Hall, G.W., and McCall, J.A., *High Speed Digital System Design: A Handbook of Interconnect Theory and Design Practices*, USA: John Wiley, New York, 2000.
[9] Hammond, P., *Electromagnetism for Engineers*, USA: Clarendon Press, Oxford, 1997.
[10] Hansen, R.C., *Phased Array Antennas*, USA: John Wiley, New York, 1998.
[11] Harsanyi, G., Sensors in Biomedical Applications, USA: Technomic Publishers, Lancaster, PA 2000.
[12] Hoole, S.R.H., and Hoole, P.R.P., *A Modern Short Course in Engineering Electromagnetics with Computer Applications*, USA: Oxford University Press, New York, 1996.
[13] Inan, U.S., and Inan, A.S., *Electromagnetic Waves*, Addison Wesley, Menlo Park, CA, 1999.
[14] Johnson, R.C., Ed., *Antenna Engineering Handbook*, USA: McGraw Hill, New York, 1993.
[15] Jordon, E.C., and Balmain, K.G., *Electromagnetic Waves and Radiating Systems*, USA: Prentice Hall, Englewood Cliffs, NJ, 1968.
[16] Misra, D.K., *Practical Electromagnetics*, John Wiley, New York, 2007.
[17] Sampei, S., *Applications of Digital Wireless Technologies and Global Wireless Communications*, USA: Prentice Hall, Englewood Cliffo, NY, 1997.
[18] Popovic, R.S., *Hall Effect Devices*, 2nd edn, Inst of Physics, London, 2004.
[19] Popovic, Z., and Popovic, B.D., *Introductory Engineering Electromagnetics*, Prentice Hall, Englewood Cliffs, NJ, 2000.
[20] Smythe, W.R., *Static and Dynamic Electricity*, USA: McGraw Hill, NY, 1968.
[21] Ramo, S., Whinnery, J., and Van Duzer, J., *Fields and Waves in Communication Electronics*, USA: John Wiley, New York, 1994.
[22] Ulaby, F.T., *Fundamentals of Applied Electromagnetics*, USA: Prentice-Hall, 1997.
[23] Wentworth, S.M., *Fundamentals of Electromagnetics with Engineering Applications*, USA: John Wiley, New York, 2005.

4

Electromagnetic waves

4.1 The wave equation

In this chapter, we focus on various methods of energy propagation in the form of electromagnetic waves. It is appropriate to start with a study of uniform plane waves that represent the simplest form of wave propagation. The uniform plane wave is, therefore, an important fundamental phenomenon of electromagnetics. Although, we do not have ideal homogeneous media in practice, the concept of a uniform plane wave is very useful as it provides a basic understanding of electromagnetic wave propagation in a medium. Further, for media which have physical dimensions much higher than the wave length, the solution closely approximates to the uniform plane wave solutions. Therefore, there are many problems where the uniform plane wave solution is adequate and quite useful.

The wave equation that governs propagating electromagnetic waves in a linear, isotropic, homogeneous medium could be derived. We assume that the net-free charge in the region is zero ($\rho = 0$) and any currents in the region are conduction currents $\mathcal{J} = \sigma\mathcal{E}$ since these types of regions are quite general in a practical scenario. Maxwell's equations in point form for this region become

$$\nabla \times \mathcal{E} = -\mu\frac{\partial \mathcal{H}}{\partial t} \tag{4.1a}$$

$$\nabla \times \mathcal{H} = \sigma\mathcal{E} + \varepsilon\frac{\partial \mathcal{E}}{\partial t} \tag{4.1b}$$

$$\nabla \cdot \mathcal{E} = 0 \tag{4.1c}$$

$$\nabla \cdot \mathcal{H} = 0 \tag{4.1d}$$

Taking the curl of (4.1a) and substituting in (4.1b), we obtain

$$\nabla \times \nabla \times \mathcal{E} = -\mu\sigma\frac{\partial \mathcal{E}}{\partial t} - \mu\varepsilon\frac{\partial^2 \mathcal{E}}{\partial t^2} \tag{4.2}$$

We similarly obtain, by taking the curl of (4.1b) and substituting in (4.1a)

$$\nabla \times \nabla \times \mathcal{H} = -\mu\sigma\frac{\partial \mathcal{H}}{\partial t} - \mu\varepsilon\frac{\partial^2 \mathcal{H}}{\partial t^2} \tag{4.3}$$

In order to reduce (4.2) and (4.3), we use the vector identity for Cartesian coordinates

$$\nabla \times \nabla \times \mathbf{A} = \nabla(\nabla \cdot \mathbf{A}) - \nabla^2 \mathbf{A} \tag{4.4}$$

where $\nabla^2 \mathbf{A}$ is the vector laplacian. In Cartesian coordinates, the vector laplacian is evaluated as below:

$$\nabla^2 \mathbf{A} = \left(\frac{\partial^2 A_x}{\partial x^2} + \frac{\partial^2 A_x}{\partial y^2} + \frac{\partial^2 A_x}{\partial z^2}\right)\mathbf{u}_x + \left(\frac{\partial^2 A_y}{\partial x^2} + \frac{\partial^2 A_y}{\partial y^2} + \frac{\partial^2 A_y}{\partial z^2}\right)\mathbf{u}_y + \left(\frac{\partial^2 A_z}{\partial x^2} + \frac{\partial^2 A_z}{\partial y^2} + \frac{\partial^2 A_z}{\partial z^2}\right)\mathbf{u}_z \tag{4.5}$$

Substituting (4.4) into (4.2) and (4.3) as well as applying (4.1c) and (4.1d), respectively, we obtain

$$\nabla^2 \boldsymbol{\mathcal{E}} = \mu\sigma\frac{\partial \boldsymbol{\mathcal{E}}}{\partial t} + \mu\varepsilon\frac{\partial^2 \boldsymbol{\mathcal{E}}}{\partial t^2} \tag{4.6a}$$

$$\nabla^2 \boldsymbol{\mathcal{H}} = \mu\sigma\frac{\partial \boldsymbol{\mathcal{H}}}{\partial t} + \mu\varepsilon\frac{\partial^2 \boldsymbol{\mathcal{H}}}{\partial t^2} \tag{4.6b}$$

The vector differential equations in (4.6) are called the *wave equations* or *Helmholtz equations*. Each equation is composed of three scalar differential equations in terms of the components of the vectors.

The wave equations are generally derived for sinusoidal steady state variation of the field vectors. Under this condition, time varying electric and magnetic field vectors can be written in terms of their space domain vectors and the time domain function, changing over to phasors with $\frac{\partial}{\partial t}$ replaced by $\mathrm{j}\omega$:

$$\boldsymbol{\mathcal{E}}(x, y, z, t) = \mathbf{E}(x, y, z)\mathrm{e}^{\mathrm{j}\omega t} \tag{4.7a}$$

$$\boldsymbol{\mathcal{H}}(x, y, z, t) = \mathbf{H}(x, y, z)\mathrm{e}^{\mathrm{j}\omega t} \tag{4.7b}$$

Substituting (4.7a) and (4.7b), respectively, in (4.6a) and (4.6b), we obtain,

$$\nabla^2 \mathbf{E} = \mathrm{j}\omega\mu(\sigma + \mathrm{j}\omega\varepsilon)\mathbf{E} \tag{4.8a}$$

$$\nabla^2 \mathbf{H} = \mathrm{j}\omega\mu(\sigma + \mathrm{j}\omega\varepsilon)\mathbf{H} \tag{4.8b}$$

Here we use a special symbol γ^2 for the quantity $\mathrm{j}\omega\mu(\sigma + \mathrm{j}\omega\varepsilon)$:

$$\gamma^2 = \mathrm{j}\omega\mu(\sigma + \mathrm{j}\omega\varepsilon) \tag{4.9}$$

The positive square root of γ^2 is referred to as the *propagation constant* of the medium at frequency ω. Since γ^2 is a complex number, the square root of γ^2 will also be a complex number that can be written as

$$\gamma = \sqrt{\mathrm{j}\omega\mu(\sigma + \mathrm{j}\omega\varepsilon)} = \alpha + \mathrm{j}\beta \tag{4.10}$$

Expanding (4.8) in terms of components, the wave equations for the spatial components of the field vectors become

$$\frac{\partial^2 E_x}{\partial x^2} + \frac{\partial^2 E_x}{\partial y^2} + \frac{\partial^2 E_x}{\partial z^2} = \gamma^2 E_x \tag{4.11a}$$

$$\frac{\partial^2 E_y}{\partial x^2} + \frac{\partial^2 E_y}{\partial y^2} + \frac{\partial^2 E_y}{\partial z^2} = \gamma^2 E_y \tag{4.11b}$$

$$\frac{\partial^2 E_z}{\partial x^2} + \frac{\partial^2 E_z}{\partial y^2} + \frac{\partial^2 E_z}{\partial z^2} = \gamma^2 E_z \tag{4.11c}$$

$$\frac{\partial^2 H_x}{\partial x^2} + \frac{\partial^2 H_x}{\partial y^2} + \frac{\partial^2 H_x}{\partial z^2} = \gamma^2 H_x \tag{4.11d}$$

$$\frac{\partial^2 H_y}{\partial x^2} + \frac{\partial^2 H_y}{\partial y^2} + \frac{\partial^2 H_y}{\partial z^2} = \gamma^2 H_y \tag{4.11e}$$

$$\frac{\partial^2 H_z}{\partial x^2} + \frac{\partial^2 H_z}{\partial y^2} + \frac{\partial^2 H_z}{\partial z^2} = \gamma^2 H_z \tag{4.11f}$$

4.2 Uniform plane waves

Even though ideal homogeneous media do not exist in practice, the concept of a uniform plane wave that is based on homogeneous medium is very helpful for the basic understanding of wave propagation in a medium. In addition, when the physical dimensions of the medium are much higher than the wave length, the wave solution closely approximates to the uniform plane wave solutions. Therefore, the uniform plane wave solution is sufficient and quite helpful to resolve x substantial number of practical problems.

First the term *uniform plane wave* needs to be defined. The term *plane* indicates that the field vectors $\mathcal{E}$ and $\mathcal{H}$ at each point in space lie on a plane, with the planes for $\mathcal{E}$ or $\mathcal{H}$ at any two different points being parallel to each other. The term *uniform* indicates that the space field vectors are independent of position on each of these planes.

It is convention to orient our coordinate system so that the field vectors lie on the xy plane, as shown in Fig. 4.1. The Poynting vector is then in the z direction. In general agreement with this model, we may arbitrarily align the direction of the space domain field vector **E** to be in the positive x direction:

$$\mathbf{E} = E_x(z)\mathbf{u}_x \tag{4.12}$$

All field quantities are a function of only z since any field is to be uniform over the xy plane and is therefore independent of the x and y coordinates. Thus, we have

$$\frac{\partial}{\partial x} = \frac{\partial}{\partial y} = 0 \tag{4.13}$$

From Maxwell's equation for sinusoidal signals,

$$\nabla \times \mathbf{E} = -\mathrm{j}\omega\mu\mathbf{H} \tag{4.14}$$

Expanding $\nabla \times \mathbf{E}$ and substituting from (4.12) and (4.13), we obtain

$$\nabla \times \mathbf{E} = \frac{\partial E_x}{\partial z}\mathbf{u}_y \tag{4.15}$$

Therefore, by matching the components of (4.14), we see that the only nonzero component of $\mathbf{H}$ is the y component as given in (4.15). Hence $\mathbf{H}$ is given as,

$$\mathbf{H} = -\frac{1}{\mathrm{j}\omega\mu}\frac{\partial E_x}{\partial z}\mathbf{u}_y = H_y(z)\mathbf{u}_y \tag{4.16}$$

This y component of $\mathbf{H}$ is also for a uniform planar system independent of x and y. $\mathbf{E}$ and $\mathbf{H}$ for a uniform plane wave are therefore orthogonal as shown in Fig. 4.1. Note also that the Poynting vector is in the z direction: the direction of $\mathbf{E} \times \mathbf{H}$. Thus the power flow is in the z direction.

The resulting components of the wave equation given in (4.11) become simple for this case. Since $E_y = E_z = 0$ and $H_x = H_z = 0$, (4.11b), (4.11c), (4.11d) and (4.11f) become zero. The remaining two equations, (4.11a) and (4.11e),

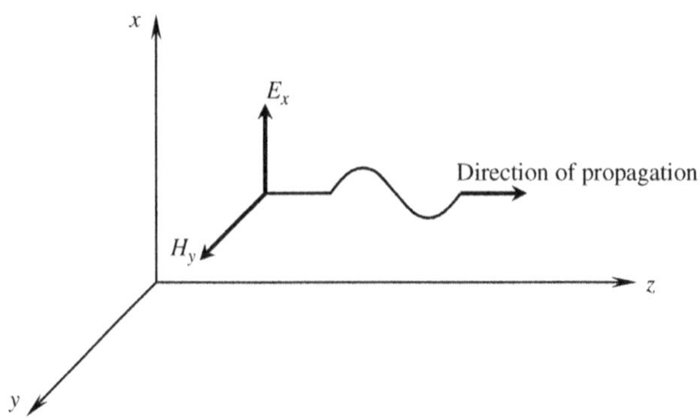

Figure 4.1: Propagation of uniform plane waves.

become

$$\frac{d^2 E_x}{dz^2} = \gamma^2 E_x \tag{4.17a}$$

$$\frac{d^2 H_y}{dz^2} = \gamma^2 H_y \tag{4.17b}$$

where the partial derivatives have been replaced by ordinary derivatives since E_x and H_y are functions of only the variable z.

The solutions to the equations in (4.17) are of the form

$$E_x = E_m^+ \,e^{-\gamma z} + E_m^- \,e^{\gamma z} = E_m^+ \,e^{-\alpha z}\, e^{-j\beta z} + E_m^- \,e^{\alpha z}\, e^{j\beta z} \tag{4.18a}$$

$$H_y = H_m^+ \,e^{-\gamma z} + H_m^- \,e^{\gamma z} = H_m^+ \,e^{-\alpha z}\, e^{-j\beta z} + H_m^- \,e^{\alpha z}\, e^{j\beta z} \tag{4.18b}$$

where E_m^+, E_m^-, H_m^+ and H_m^- are undetermined constants of integration for the second order differential equations. Substituting the solutions of E_x and H_y from (4.18) in (4.16), we obtain

$$-\gamma E_m^+ \,e^{-\gamma z} + \gamma E_m^- \,e^{\gamma z} = -j\omega\mu(H_m^+ \,e^{-\gamma z} + H_m^- \,e^{\gamma z}) \tag{4.19}$$

Matching appropriate terms in (4.19), we obtain

$$\frac{E_m^+}{H_m^+} = \frac{j\omega\mu}{\gamma} = \eta \tag{4.20a}$$

$$\frac{E_m^-}{H_m^-} = -\frac{j\omega\mu}{\gamma} = -\eta \tag{4.20b}$$

The quantity $j\omega\mu/\gamma$ has the unit of ohms (Ω) and it is called the *intrinsic impedance* of the medium and denoted by the symbol η.

$$\begin{aligned} \eta &= \frac{j\omega\mu}{\gamma} \\ &= \sqrt{\frac{j\omega\mu}{\sigma + j\omega\varepsilon}} \\ &= \bar{\eta}\, e^{j\theta_\eta} \end{aligned} \tag{4.21}$$

where the magnitude η is denoted by $\bar{\eta}$ and the phase angle by θ_η. Note that in free space or air $\gamma = j\omega\sqrt{\mu_0\varepsilon_0}$ so that $\eta = \sqrt{\frac{\mu_0}{\varepsilon_0}} = 377\ \Omega$. Substituting (4.21) and (4.20) into (4.18) and writing η as $\bar{\eta} e^{j\beta z}$ yields

$$E_x = E_m^+ \,e^{-\alpha z}\, e^{-j\beta z} + E_m^- \,e^{\alpha z}\, e^{j\beta z} \tag{4.22a}$$

$$\begin{aligned} H_y &= \frac{E_m^+}{\eta} \,e^{-\alpha z}\, e^{-j\beta z} - \frac{E_m^-}{\eta} \,e^{\alpha z}\, e^{j\beta z} \\ &= \frac{E_m^+}{\bar{\eta}} \,e^{-\alpha z}\, e^{-j\beta z}\, e^{-j\theta_\eta} - \frac{E_m^-}{\bar{\eta}} \,e^{\alpha z}\, e^{j\beta z}\, e^{-j\theta_\eta} \end{aligned} \tag{4.22b}$$

E_m^+ and E_m^- are complex constants and may be written as

$$E_m^+ = \bar{E}_m^+ \,\mathrm{e}^{\mathrm{j}\theta^+} \tag{4.23a}$$

$$E_m^- = \bar{E}_m^- \,\mathrm{e}^{\mathrm{j}\theta^-} \tag{4.23b}$$

where $\bar{E}_m^+$ and $\bar{E}_m^+$ are real numbers. Substituting (4.23) into (4.22) yields

$$E_x = \bar{E}_m^+ \,\mathrm{e}^{-\alpha z}\,\mathrm{e}^{-\mathrm{j}\beta z}\,\mathrm{e}^{\mathrm{j}\theta^+} + \bar{E}_m^- \,\mathrm{e}^{\alpha z}\,\mathrm{e}^{\mathrm{j}\beta z}\,\mathrm{e}^{\mathrm{j}\theta^-} \tag{4.24a}$$

$$H_y = \frac{\bar{E}_m^+}{\bar{\eta}}\,\mathrm{e}^{-\alpha z}\,\mathrm{e}^{-\mathrm{j}\beta z}\,\mathrm{e}^{\mathrm{j}\theta^+}\,\mathrm{e}^{-\mathrm{j}\theta_\eta} - \frac{\bar{E}_m^-}{\bar{\eta}}\,\mathrm{e}^{\alpha z}\,\mathrm{e}^{\mathrm{j}\beta z}\,\mathrm{e}^{\mathrm{j}\theta^-}\,\mathrm{e}^{-\mathrm{j}\theta_\eta} \tag{4.24b}$$

The time domain forms become

$$\begin{aligned}\mathcal{E}_x &= \mathrm{Re}(E_x\,\mathrm{e}^{\mathrm{j}\omega t})\\ &= \bar{E}_m^+\,\mathrm{e}^{-\alpha z}\cos(\omega t - \beta z + \theta^+) + \bar{E}_m^-\,\mathrm{e}^{\alpha z}\cos(\omega t + \beta z + \theta^-)\end{aligned} \tag{4.25a}$$

$$\begin{aligned}\mathcal{H}_y &= \mathrm{Re}(H_y\,\mathrm{e}^{\mathrm{j}\omega t})\\ &= \frac{\bar{E}_m^+}{\bar{\eta}}\,\mathrm{e}^{-\alpha z}\cos(\omega t - \beta z + \theta^+ - \theta_\eta) - \frac{\bar{E}_m^-}{\bar{\eta}}\,\mathrm{e}^{\alpha z}\cos(\omega t + \beta z + \theta^- - \theta_\eta)\end{aligned} \tag{4.25b}$$

We observe that functions of $\omega t - \beta z$ move in the z direction at speed ω/β and are forward-traveling waves. Similarly, functions of $\omega t + \beta z$ travel backward at the same speed.

4.2.1 Lossless medium

When the conductivity (σ) of the medium is zero, the medium is characterized as a lossless medium. For a lossless medium

$$\gamma = \sqrt{\mathrm{j}\omega\mu(\mathrm{j}\omega\varepsilon)} = \mathrm{j}\omega\sqrt{\mu\varepsilon} \tag{4.26}$$

so that

$$\alpha = 0 \tag{4.27a}$$

$$\beta = \omega\sqrt{\mu\varepsilon} \tag{4.27b}$$

Also, the intrinsic impedance becomes

$$\eta = \bar{\eta} = \sqrt{\frac{\mu}{\varepsilon}} \tag{4.28}$$

Since the intrinsic impedance is a real value

$$\theta_\eta = 0 \tag{4.29}$$

Thus, for a lossless medium, the field components in (4.25) become

$$\mathcal{E}_x = \bar{E}_m^+ \cos(\omega t - \beta z + \theta^+) + \bar{E}_m^- \cos(\omega t + \beta z + \theta^-) \tag{4.30a}$$

$$\mathcal{H}_y = \frac{\bar{E}_m^+}{\eta} \cos(\omega t - \beta z + \theta^+) - \frac{\bar{E}_m^-}{\eta} \cos(\omega t + \beta z + \theta^-) \tag{4.30b}$$

Consider the first term of $\mathcal{E}_x$ in (4.30a), $\bar{E}_m^+ \cos(\omega t - \beta z + \theta^+)$. A particular point in the waveform at time t_1 and t_2 is, respectively, located at the positions z_1 and z_2. Since it is a lossless medium, the amplitude of a specific point must be invariant. Thus, we conclude that

$$\omega t_1 - \beta z_1 + \theta^+ = \omega t_2 - \beta z_2 + \theta^+ \tag{4.31}$$

Rearranging the terms in (4.31) yields

$$\beta(z_2 - z_1) = \omega(t_2 - t_1) \tag{4.32}$$

The expression (4.32) shows that a point on the waveform must move in the positive z direction with increasing time. Hence, we state that the $\bar{E}_m^+ \cos(\omega t - \beta z + \theta^+)$ component of $\mathcal{E}_x$ represents a wave traveling in the positive z direction. In similar manner, we observe that the $\bar{E}_m^- \cos(\omega t + \beta z + \theta^-)$ component of $\mathcal{E}_x$ represents a wave traveling in the negative z direction.

Generalizing that a point in the waveform at time t located at the position z maintains constant amplitude for varying t and z, we obtain

$$\omega t - \beta z + \theta^+ = \text{a constant} \tag{4.33a}$$

$$\omega t + \beta z + \theta^- = \text{a constant} \tag{4.33b}$$

Taking the derivative of (4.33a) with respect to time yields the velocity at which the points of constant phase travel. This is referred to as the *phase velocity* of propagation of the wave, which becomes

$$u = \frac{\mathrm{d}z}{\mathrm{d}t} = \frac{\omega}{\beta} = \frac{1}{\sqrt{\mu\varepsilon}} \quad \text{m/s} \tag{4.34}$$

We remark that the *phase velocity* of propagation obtained from (4.33b) has the same amplitude but negative sign since the respective component travels in the negative z direction. Also note that similar observations hold for the magnetic field given in (4.30b). However, the magnetic field in the backward-traveling wave is in the negative y direction.

The quantity $\beta = \omega\sqrt{\mu\varepsilon}$ is referred to as the *phase constant* or *wavenumber*. The distance between corresponding adjacent points on the wave is known as the *wavelength* and is denoted by λ. Since $\beta\lambda = 2\pi$,

$$\beta = \frac{2\pi}{\lambda} \quad \text{rad/m} \tag{4.35}$$

Providing that $\omega = 2\pi f$ and using (4.34) and (4.35), we obtain

$$\lambda = \frac{u}{f} = \frac{1}{f\sqrt{\mu\varepsilon}} \tag{4.36}$$

Note that the phase velocity of propagation and wavelength depend on the properties of the medium. For typical materials $\varepsilon \geq \varepsilon_0$ and $\mu \geq \mu_0$. Thus, the phase velocity of propagation is slower than that in free space and the wavelength is shorter than that in free space.

The complete solution of (4.30) can be written as the sum of forward- and backward-traveling waves as

$$\mathcal{E}_x = \mathcal{E}_x^+ + \mathcal{E}_x^- \tag{4.37a}$$

$$\mathcal{H}_y = \mathcal{H}_y^+ + \mathcal{H}_y^- \tag{4.37b}$$

where

$$\mathcal{E}_x^+ = \bar{E}_m^+ \cos(\omega t - \beta z + \theta^+) \tag{4.38a}$$

$$\mathcal{H}_y^+ = \frac{\bar{E}_m^+}{\eta} \cos(\omega t - \beta z + \theta^+) \tag{4.38b}$$

and

$$\mathcal{E}_x^- = \bar{E}_m^- \cos(\omega t + \beta z + \theta^-) \tag{4.39a}$$

$$\mathcal{H}_y^- = -\frac{\bar{E}_m^-}{\eta} \cos(\omega t + \beta z + \theta^-) \tag{4.39b}$$

These individual waves are plotted for $t = 0$ in Fig. 4.2. The complete wave is the superposition of these forward- and backward-traveling waves. The Poynting vector for the forward–traveling components $\mathcal{E}_x^+$ and $\mathcal{H}_y^+$ is in the $+z$ direction, indicating power flow in the $+z$ direction. However, the Poynting vector for backward-traveling components $\mathcal{E}_x^-$ and $\mathcal{H}_y^-$ is in the $-z$ direction, indicating power flow in the $-z$ direction.

4.2.2 Lossy medium

When the conductivity (σ) of the medium is nonzero, the medium is characterized as a lossless medium. There are two major differences between uniform plane waves in a lossy medium and a lossless medium. The first difference is that the propagation constant of the medium has a nonzero real part.

$$\gamma = \sqrt{j\omega\mu(\sigma + j\omega\varepsilon)} = \alpha + j\beta \tag{4.40}$$

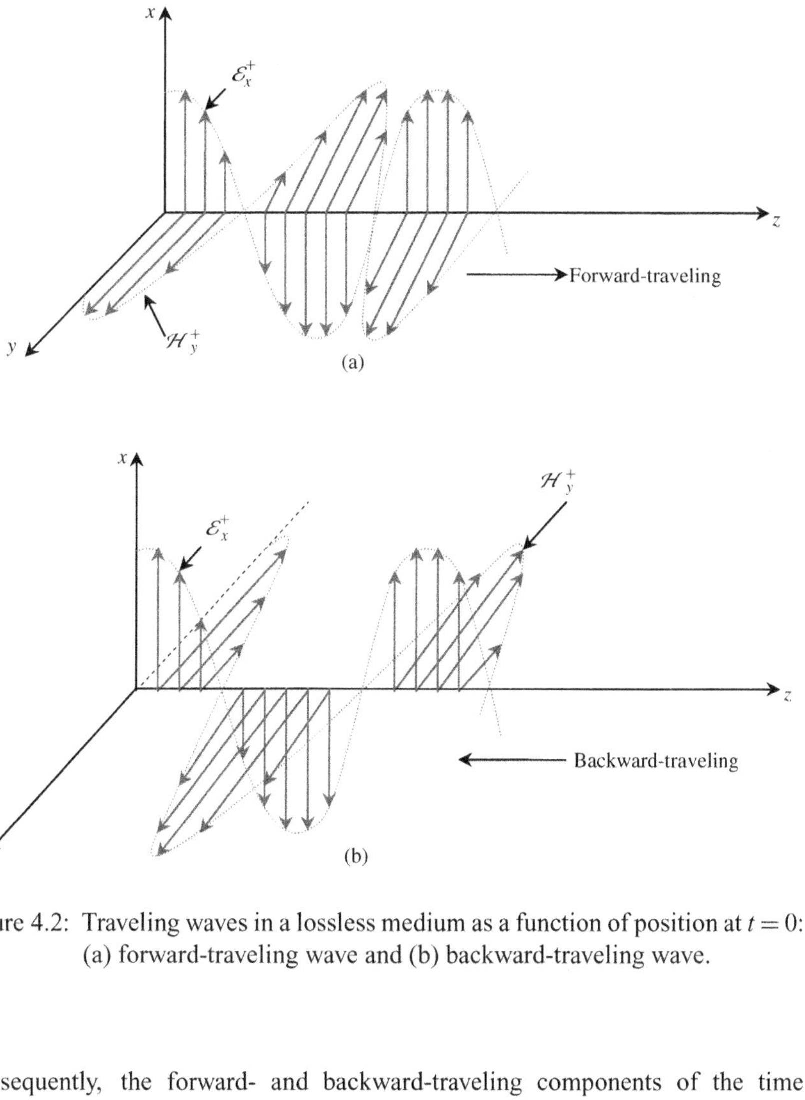

Figure 4.2: Traveling waves in a lossless medium as a function of position at $t = 0$: (a) forward-traveling wave and (b) backward-traveling wave.

Consequently, the forward- and backward-traveling components of the time domain electrical field are given, from (4.25):

$$\mathcal{E}_x^+ = \bar{E}_m^+ \, e^{-\alpha z} \cos(\omega t - \beta z + \theta^+) \tag{4.41a}$$

$$\mathcal{E}_x^- = \bar{E}_m^- \, e^{\alpha z} \cos(\omega t + \beta z + \theta^-) \tag{4.41b}$$

The amplitude of the forward-traveling wave is $\bar{E}_m^+ \, e^{-\alpha z}$, which is reduced for increasing z as shown in Fig. 4.3(a). Similarly, the amplitude of the backward-traveling wave is $\bar{E}_m^- e^{+\alpha z}$, which is also reduced since it is propagating in the $-z$ direction as shown in Fig. 4.3(b).

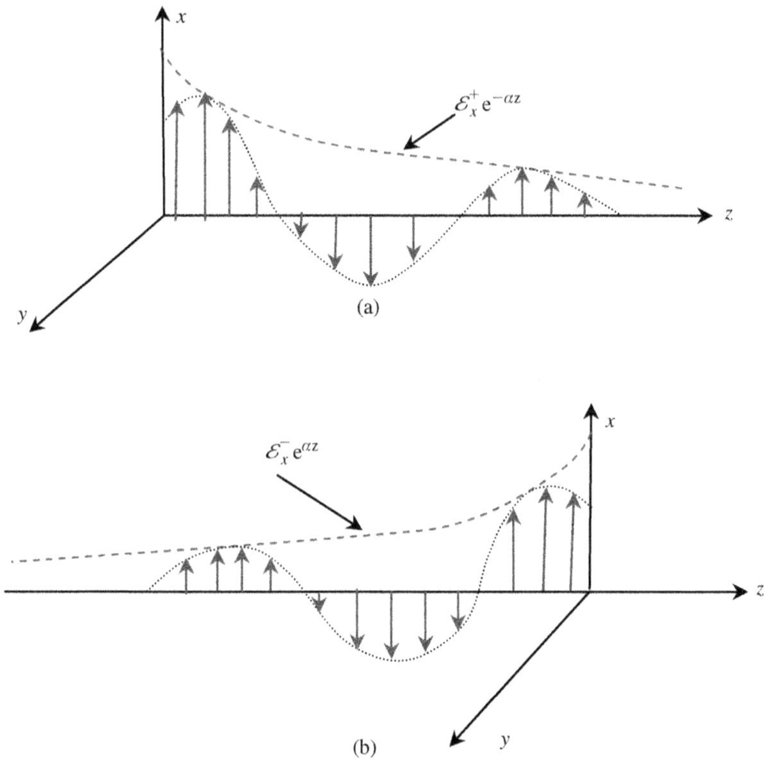

Figure 4.3: Traveling waves in a lossy medium as a function of position at $t = 0$: (a) forward-traveling wave and (b) backward-traveling wave.

The real part of γ, α, is referred to as the *attenuation constant* for the above reasons. Note that the phase constant β is the imaginary part of γ, which is not of the same value as that for a lossless medium.

$$\beta = \mathrm{Im}(\gamma) \neq \omega\sqrt{\mu\varepsilon} \quad \text{when } \sigma \neq 0 \tag{4.42}$$

The second difference is the intrinsic impedance of the medium being a complex quantity.

$$\eta = \sqrt{\frac{\mathrm{j}\omega\mu}{\sigma + \mathrm{j}\omega\varepsilon}} = \bar{\eta}\, \mathrm{e}^{\mathrm{j}\theta_\eta} \tag{4.43}$$

For a lossy medium η is complex and does not have the same value as for a lossless medium.

$$\eta \neq \sqrt{\frac{\mu}{\varepsilon}} \quad \text{when } \sigma \neq 0 \tag{4.44}$$

For a lossless medium $\theta_\eta = 0$. Therefore, we observe that the electric and magnetic fields of the forward-traveling wave are in phase, as are those in the backward-traveling wave.

However, for the lossy case the phase angle of the intrinsic impedance, θ_η, results in the electric and magnetic fields of each traveling wave being out of phase by the phase angle θ_η. On the other hand, the expressions for the phase velocity of propagation and the wavelength remain the same as for a lossless medium:

$$u = \frac{dz}{dt} = \frac{\omega}{\beta} \quad \text{m/s} \tag{4.45}$$

$$\lambda = \frac{2\pi}{\beta} = \frac{u}{f} \tag{4.46}$$

4.2.3 Power flow

We rewrite the field components of a plane wave that has been stated in (4.37)

$$\mathcal{E}_x = \mathcal{E}_x^+ + \mathcal{E}_x^- \tag{4.47a}$$

$$\mathcal{H}_y = \mathcal{H}_y^+ + \mathcal{H}_y^- \tag{4.47b}$$

where the forward- and backward-traveling wave components of (4.47) are obtained from (4.25) and given as follows:

$$\mathcal{E}_x^+ = \bar{E}_m^+ \, e^{-\alpha z} \cos(\omega t - \beta z + \theta^+) \tag{4.48a}$$

$$\mathcal{E}_x^- = \bar{E}_m^- \, e^{\alpha z} \cos(\omega t + \beta z + \theta^-) \tag{4.48b}$$

$$\mathcal{H}_y^+ = \frac{\bar{E}_m^+}{\bar{\eta}} \, e^{-\alpha z} \cos(\omega t - \beta z + \theta^+ - \theta_\eta) \tag{4.48c}$$

$$\mathcal{H}_y^- = -\frac{\bar{E}_m^-}{\bar{\eta}} \, e^{\alpha z} \cos(\omega t + \beta z + \theta^- - \theta_\eta) \tag{4.48d}$$

The electrical and magnetic fields are given in terms of the forward- and backward-traveling wave components in (4.47). The vectors of the forward and backward electric fields are in the same direction as each other, whereas those of the magnetic field are in the opposite direction to each other. This actually makes sense since $\mathcal{E}_x^+ \mathbf{u}_x \times \mathcal{H}_y^+ \mathbf{u}_y = \mathcal{E}_x^+ \mathcal{H}_y^+ \mathbf{u}_z$ and the direction of the power flow for the forward-traveling wave is the $+z$ direction as shown in Fig. 4.2(a). Similarly, the direction of power flow as shown in Fig. 4.2(b) for the backward-traveling wave is the $-z$ direction because the power flow $\mathcal{E}_x^- \mathbf{u}_x \times \mathcal{H}_y^- \mathbf{u}_y = \mathcal{E}_x^- \mathcal{H}_y^- \mathbf{u}_z$ but $\mathcal{H}_y^-$ holds a negative sign making the direction of travel $-\mathbf{u}_z$.

The time average power density for the forward and backward waves is

$$\begin{aligned}\mathbf{S}_{\text{av}} &= \frac{1}{2}\text{Re}(\mathbf{E} \times \mathbf{H}^*) \\ &= \frac{1}{2}\text{Re}(E_x\mathbf{u}_x \times H_y^*\mathbf{u}_y) \\ &= \frac{1}{2}\text{Re}(E_x H_y^*\mathbf{u}_z)\end{aligned} \tag{4.49}$$

Substituting (4.24) into (4.49)

$$E_x = \bar{E}_m^+\,\mathrm{e}^{-\alpha z}\,\mathrm{e}^{-\mathrm{j}\beta z}\,\mathrm{e}^{\mathrm{j}\theta^+} + \bar{E}_m^-\,\mathrm{e}^{\alpha z}\,\mathrm{e}^{\mathrm{j}\beta z}\,\mathrm{e}^{\mathrm{j}\theta^-}$$

$$H_y = \frac{\bar{E}_m^+}{\bar{\eta}}\,\mathrm{e}^{-\alpha z}\,\mathrm{e}^{-\mathrm{j}\beta z}\,\mathrm{e}^{\mathrm{j}\theta^+}\,\mathrm{e}^{-\mathrm{j}\theta_\eta} - \frac{\bar{E}_m^-}{\bar{\eta}}\,\mathrm{e}^{\alpha z}\,\mathrm{e}^{\mathrm{j}\beta z}\,\mathrm{e}^{\mathrm{j}\theta^-}\,\mathrm{e}^{-\mathrm{j}\theta_\eta}$$

$$\mathbf{S}_{\text{av}} = \left[\begin{aligned}&\frac{1}{2}\frac{(\bar{E}_m^+)^2}{\bar{\eta}}\,\mathrm{e}^{-2\alpha z}\cos\theta_\eta - \frac{1}{2}\frac{(\bar{E}_m^-)^2}{\bar{\eta}}\,\mathrm{e}^{2\alpha z}\cos\theta_\eta \\ &-\frac{\bar{E}_m^+\bar{E}_m^-}{\bar{\eta}}\sin\theta_\eta\,\sin(2\beta z + \theta^- - \theta^+)\end{aligned}\right]\mathbf{u}_z \tag{4.50}$$

The first term of (4.50), denoted by $\mathbf{S}_{\text{av}}^+$, is given by

$$\mathbf{S}_{\text{av}}^+ = \frac{1}{2}\frac{(\bar{E}_m^+)^2}{\bar{\eta}}\,\mathrm{e}^{-2\alpha z}\cos\theta_\eta\,\mathbf{u}_z \tag{4.51}$$

It is the time average power density in the forward-traveling wave (in the absence of the backward-traveling wave). The second term, denoted by $\mathbf{S}_{\text{av}}^-$, is given by

$$\mathbf{S}_{\text{av}}^- = -\frac{1}{2}\frac{(\bar{E}_m^-)^2}{\bar{\eta}}\,\mathrm{e}^{2\alpha z}\cos\theta_\eta\,\mathbf{u}_z \tag{4.52}$$

It is the time average power density in the backward-traveling wave (in the absence of the forward-traveling wave). The third term is the cross-coupling term that disappears for lossless media when $\theta_\eta = 0$.

For the lossless case we can write

$$\mathbf{S}_{\text{av}} = \mathbf{S}_{\text{av}}^+ + \mathbf{S}_{\text{av}}^- \quad \text{when } \sigma = 0 \tag{4.53}$$

where

$$\mathbf{S}_{\text{av}}^+ = \frac{1}{2}\frac{(\bar{E}_m^+)^2}{\bar{\eta}}\,\mathrm{e}^{-2\alpha z}\mathbf{u}_z \quad \text{when } \sigma = 0 \tag{4.54a}$$

$$\mathbf{S}_{\text{av}}^- = -\frac{1}{2}\frac{(\bar{E}_m^-)^2}{\bar{\eta}}\,\mathrm{e}^{2\alpha z}\mathbf{u}_z \quad \text{when } \sigma = 0 \tag{4.54b}$$

4.3 Conductors and dielectrics

In Maxwell's equations, the losses in the medium are introduced by Ampere's law. For $\sigma \neq 0$, Ampere's law gives

$$\nabla \times \mathbf{H} = \sigma \mathbf{E} + \mathrm{j}\omega\varepsilon \mathbf{E} = \mathrm{j}\omega\varepsilon\left(1 - \mathrm{j}\frac{\sigma}{\omega\varepsilon}\right)\mathbf{E} \tag{4.55}$$

There are two components of current involved in Ampere's law as stated in (4.55): a conduction current density

$$\mathbf{J}_c = \sigma \mathbf{E} \tag{4.56}$$

and displacement current density

$$\mathbf{J}_d = \mathrm{j}\omega\varepsilon \mathbf{E} \tag{4.57}$$

The conduction current represents an energy loss phenomenon and the displacement current represents an energy storage occurrence. The ratio of these two currents is a measure of the lossy nature of the material. Hence, the term $|\mathbf{J}_c|/|\mathbf{J}_d| = \sigma/\omega\varepsilon$ is referred to as the *loss tangent* of the material. Also the conduction and displacement phasors are 90° out of time phase.

The notation of the loss tangent as the ratio of conduction current to displacement current provides a meaningful way of distinguishing between conductors and dielectrics. We can classify materials according to whether the conduction current is much larger than the displacement current (good conductors) or the displacement current is much larger than the conduction current (good dielectrics).

$$\frac{\sigma}{\omega\varepsilon} >> 1 \quad \text{good conductor} \tag{4.58a}$$

$$\frac{\sigma}{\omega\varepsilon} << 1 \quad \text{good dielectric} \tag{4.58b}$$

It is readily recognized that whether we are dealing with good or bad conductors is a matter of frequency as well as material properties.

4.3.1 Good conductors

For good conductors $\frac{\sigma}{\omega\varepsilon} >> 1$. The expression for the propagation coefficient, γ, is

$$\begin{aligned}
\gamma &= \sqrt{\mathrm{j}\omega\mu(\sigma + \mathrm{j}\omega\varepsilon)} = \sqrt{\mathrm{j}\omega\mu\sigma\left(1 + \frac{\mathrm{j}\omega\varepsilon}{\sigma}\right)} \\
&\cong \sqrt{\mathrm{j}\omega\mu\sigma} = \sqrt{\omega\mu\sigma}\,\mathrm{e}^{\mathrm{j}\frac{\pi}{4}} \\
&= \sqrt{\frac{\omega\mu\sigma}{2}} + \mathrm{j}\sqrt{\frac{\omega\mu\sigma}{2}}
\end{aligned} \tag{4.59}$$

Therefore

$$\alpha = \beta \cong \sqrt{\frac{\omega\mu\sigma}{2}} \tag{4.60}$$

Hence the velocity of propagation is

$$\begin{aligned} u &= \frac{\omega}{\beta} \\ &\cong \sqrt{\frac{2\omega}{\mu\sigma}} \end{aligned} \tag{4.61}$$

and the intrinsic impedance becomes

$$\begin{aligned} \eta &= \sqrt{\frac{\mathrm{j}\omega\mu}{\sigma + \mathrm{j}\omega\varepsilon}} = \sqrt{\frac{\mathrm{j}\omega\mu/\sigma}{1 + \mathrm{j}\omega\varepsilon/\sigma}} \\ &\cong \sqrt{\mathrm{j}\omega\mu/\sigma} = \sqrt{\frac{\omega\mu}{2\sigma}}(1 + \mathrm{j}) \end{aligned} \tag{4.62}$$

Since $\frac{\omega\varepsilon}{\sigma} << 1$.

4.3.2 Good dielectrics

For good dielectrics $\dfrac{\sigma}{\omega\varepsilon} << 1$. The propagating constant is given as

$$\gamma = \sqrt{\mathrm{j}\omega\mu(\sigma + \mathrm{j}\omega\varepsilon)} = \mathrm{j}\omega\sqrt{\mu\varepsilon}\sqrt{\left(1 - \mathrm{j}\frac{\sigma}{\omega\varepsilon}\right)} \tag{4.63}$$

The binomial expansion of $\sqrt{\left(1 - \mathrm{j}\dfrac{\sigma}{\omega\varepsilon}\right)}$ when $\dfrac{\sigma}{\omega\varepsilon} << 1$ is given as

$$\sqrt{\left(1 - \mathrm{j}\frac{\sigma}{\omega\varepsilon}\right)} \cong 1 - \mathrm{j}\frac{\sigma}{2\omega\varepsilon} + \frac{\sigma^2}{8\omega^2\varepsilon^2} - \cdots \tag{4.64}$$

Substituting the expansion of (4.64) into (4.63)

$$\gamma \cong \mathrm{j}\omega\sqrt{\mu\varepsilon}\left(1 - \mathrm{j}\frac{\sigma}{2\omega\varepsilon} + \frac{\sigma^2}{8\omega^2\varepsilon^2}\right) \tag{4.65}$$

Hence

$$\alpha \cong \frac{\sigma}{2}\sqrt{\frac{\mu}{\varepsilon}} \tag{4.66a}$$

$$\beta \cong \omega\sqrt{\mu\varepsilon}\left(1 + \frac{\sigma^2}{8\omega^2\varepsilon^2}\right) \tag{4.66b}$$

The velocity of propagation becomes

$$u = \frac{\omega}{\beta} \cong \frac{1}{\sqrt{\mu\varepsilon}\left(1 + \dfrac{\sigma^2}{8\omega^2\varepsilon^2}\right)} \tag{4.67}$$

Thus, the velocity of propagation of the wave is reduced slightly by the losses in the medium. The intrinsic impedance using the expansion becomes

$$\eta = \sqrt{\frac{j\omega\mu}{\sigma + j\omega\varepsilon}} = \sqrt{\frac{\mu}{\varepsilon}\left(\frac{1}{1 - j\sigma/\omega\varepsilon}\right)} = \sqrt{\frac{\mu}{\varepsilon}}\left(1 - j\frac{\sigma}{\omega\varepsilon}\right)^{-\frac{1}{2}}$$
$$\cong \sqrt{\frac{\mu}{\varepsilon}}\left(1 + j\frac{\sigma}{2\omega\varepsilon} - \frac{3\sigma^2}{8\omega^2\varepsilon^2} + \cdots\right) \tag{4.68}$$

All the above expressions for the good dielectric have been approximated by considering up to the second order terms of the expansion. Depending on the required accuracy and the actual values of the loss tangent $(\sigma/\omega\varepsilon)$, the number of expansion terms can be decided.

4.4 Normal incidence of uniform plane wave

Consider uniform plane waves that are incident on a boundary between two media as shown in Fig. 4.4. The figure shows a plane boundary between two different media having material properties ε_1, μ_1 and σ_1 for medium 1 and ε_2, μ_2 and σ_2 for medium 2. A uniform plane wave traveling in medium 1 is incident on the interface, normal to the boundary. We express this incident field in space vector form as

$$\mathbf{E}_i = E_i\,e^{-\gamma_1 z}\mathbf{u}_x = E_i\,e^{-\alpha_1 z}\,e^{-j\beta_1 z}\mathbf{u}_x \tag{4.69a}$$

$$\mathbf{H}_i = \frac{E_i}{\eta_1}\,e^{-\gamma_1 z}\mathbf{u}_y = \frac{E_i}{\bar{\eta}_1}\,e^{-\alpha_1 z}\,e^{-j\beta_1 z}\,e^{-j\theta_{\eta 1}}\,\mathbf{u}_y \tag{4.69b}$$

where

$$\gamma_1 = \sqrt{j\omega\mu_1(\sigma_1 + j\omega\varepsilon_1)} = \alpha_1 + j\beta_1 \tag{4.70}$$

and

$$\eta_1 = \bar{\eta}_1\,e^{j\theta_{\eta 1}} = \sqrt{\frac{j\omega\mu_1}{\sigma_1 + j\omega\varepsilon_1}} \tag{4.71}$$

The boundary gives rise to a reflected wave that is represented as

$$\mathbf{E}_r = E_r\,e^{\gamma_1 z}\mathbf{u}_x = E_r\,e^{\alpha_1 z}\,e^{j\beta_1 z}\mathbf{u}_x \tag{4.72a}$$

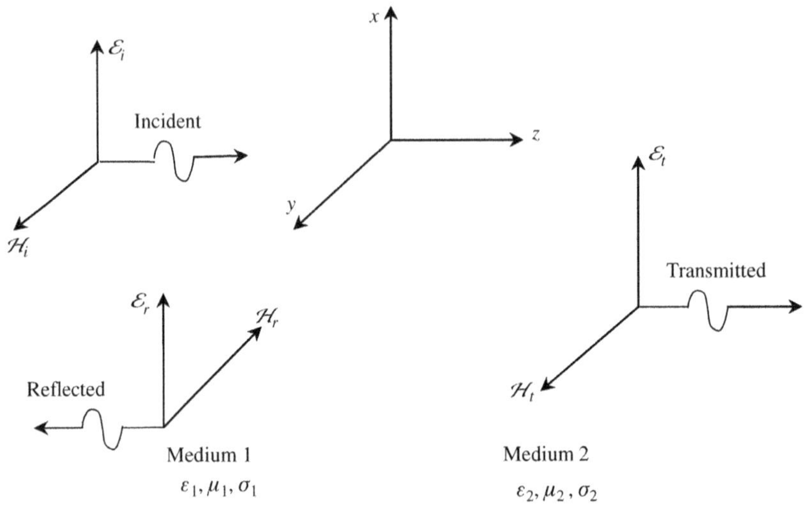

Figure 4.4: Normal incidence of a uniform plane wave on a plane boundary.

and

$$\mathbf{H}_r = -\frac{E_r}{\eta_1} e^{\gamma_1 z} \mathbf{u}_y = -\frac{E_r}{\bar{\eta}_1} e^{\alpha_1 z} e^{j\beta_1 z} e^{-j\theta_{\eta 1}} \mathbf{u}_y \tag{4.72b}$$

in accordance with the general solution of the wave equation in medium 1. A portion of the incident wave will be transmitted into the second medium, which is represented as

$$\mathbf{E}_t = E_t e^{-\gamma_2 z} \mathbf{u}_x = E_t e^{-\alpha_2 z} e^{-j\beta_2 z} \mathbf{u}_x \tag{4.73a}$$

and

$$\mathbf{H}_t = \frac{E_t}{\eta_2} e^{-\gamma_2 z} \mathbf{u}_y = \frac{E_t}{\bar{\eta}_2} e^{-\alpha_2 z} e^{-j\beta_2 z} e^{-j\theta_{\eta 2}} \mathbf{u}_y \tag{4.73b}$$

where

$$\gamma_2 = \sqrt{j\omega\mu_2(\sigma_2 + j\omega\varepsilon_2)} = \alpha_2 + j\beta_2 \tag{4.74}$$

and

$$\eta_2 = \bar{\eta}_2 e^{j\theta_{\eta 2}} = \sqrt{\frac{j\omega\mu_2}{\sigma_2 + j\omega\varepsilon_2}} \tag{4.75}$$

There will not be a backward-traveling component of the transmitted wave since the second medium is assumed to be infinite in extent. Thus, there will be no backward component as it would be created only when the forward wave impinges on something to create reflections at the surface. The unknowns in these equations,

E_r and E_t, can be obtained in terms of E_i using the boundary conditions on the continuity of the tangential field components at the boundary $z = 0$.

At the boundary $z = 0$, the boundary conditions require that the tangential components of the total electric and magnetic fields be continuous. Since $\mathbf{E}_i$, $\mathbf{E}_r$ and $\mathbf{E}_t$ are defined to be in the x direction, they are tangential to the plane $z = 0$ and

$$\mathbf{E}_i + \mathbf{E}_r = \mathbf{E}_t \quad \text{at } z = 0 \tag{4.76a}$$

and similarly for $\mathbf{H}_i$, $\mathbf{H}_r$ and $\mathbf{H}_t$, which are defined to be in the y direction,

$$\mathbf{H}_i + \mathbf{H}_r = \mathbf{H}_t \quad \text{at } z = 0 \tag{4.76b}$$

Substituting the expressions for field vectors for $z = 0$ into (4.76), we obtain

$$\frac{E_r}{E_i} = \frac{\eta_2 - \eta_1}{\eta_2 + \eta_1} = \Gamma \tag{4.77}$$

and

$$\frac{E_t}{E_i} = \frac{2\eta_2}{\eta_2 + \eta_1} = T \tag{4.78}$$

In addition, using (4.77) and (4.78), we can show an important relationship:

$$\Gamma + 1 = T \tag{4.79}$$

The quantities Γ and T are, respectively, the *reflection* and *transmission coefficients* of the boundary. By obtaining the expression for the reflection power density, it is evident that $|\Gamma| \leq 1$. However, $|T|$ may exceed unity. Usually, Γ and T are complex quantities except for the case when both media are lossless ($\sigma_1 = \sigma_2 = 0$). In general, Γ and T can be given as

$$\Gamma = \bar{\Gamma}\, e^{j\theta_\Gamma} \tag{4.80}$$

$$T = \bar{T}\, e^{j\theta_T} \tag{4.81}$$

where $\bar{\Gamma}$ and $\bar{T}$ are real quantities.

Using the expressions in (4.77) and (4.78), the space domain field vectors of reflected and transmitted components can be expressed in terms of the incident electric field component as below:

$$\mathbf{E}_i = E_i\,\mathrm{e}^{-\gamma_1 z}\mathbf{u}_x \tag{4.82a}$$

$$\mathbf{H}_i = \frac{E_i}{\eta_1}\,\mathrm{e}^{-\gamma_1 z}\mathbf{u}_y \tag{4.82b}$$

$$\mathbf{E}_r = \Gamma E_i\,\mathrm{e}^{\gamma_1 z}\mathbf{u}_x \tag{4.82c}$$

$$\mathbf{H}_r = -\Gamma\frac{E_i}{\eta_1}\,\mathrm{e}^{\gamma_1 z}\mathbf{u}_y \tag{4.82d}$$

$$\mathbf{E}_t = TE_i\,\mathrm{e}^{-\gamma_2 z}\mathbf{u}_x \tag{4.82e}$$

$$\mathbf{H}_t = T\frac{E_i}{\eta_2}\,\mathrm{e}^{-\gamma_2 z}\mathbf{u}_y \tag{4.82f}$$

Usually E_i is a complex quantity. In general, E_i can be given as

$$E_i = \bar{E}_i\,\mathrm{e}^{\mathrm{j}\theta_i} \tag{4.83}$$

where $\bar{E}_i$ is a real quantity.

We can also obtain the time domain forms of field vectors:

$$\mathcal{E}_i = \bar{E}_i\,\mathrm{e}^{-\alpha_1 z}\cos(\omega t - \beta_1 z + \theta_i)\mathbf{u}_x \tag{4.84a}$$

$$\mathcal{H}_i = \frac{\bar{E}_i}{\bar{\eta}_1}\,\mathrm{e}^{-\alpha_1 z}\cos(\omega t - \beta_1 z + \theta_i - \theta_{\eta 1})\mathbf{u}_y \tag{4.84b}$$

$$\mathcal{E}_r = \bar{\Gamma}\bar{E}_i\,\mathrm{e}^{\alpha_1 z}\cos(\omega t + \beta_1 z + \theta_i + \theta_\Gamma)\mathbf{u}_x \tag{4.84c}$$

$$\mathcal{H}_r = -\bar{\Gamma}\frac{\bar{E}_i}{\bar{\eta}_1}\,\mathrm{e}^{\alpha_1 z}\cos(\omega t + \beta_1 z + \theta_i + \theta_\Gamma - \theta_{\eta 1})\mathbf{u}_y \tag{4.84d}$$

$$\mathcal{E}_t = \bar{T}\bar{E}_i\,\mathrm{e}^{-\alpha_2 z}\cos(\omega t - \beta_2 z + \theta_i + \theta_T)\mathbf{u}_x \tag{4.84e}$$

$$\mathcal{H}_t = \bar{T}\frac{\bar{E}_i}{\bar{\eta}_2}\,\mathrm{e}^{-\alpha_2 z}\cos(\omega t - \beta_2 z + \theta_i + \theta_T - \theta_{\eta 2})\mathbf{u}_y \tag{4.84f}$$

The time-average power density vector of the transmitted wave is

$$\begin{aligned}\mathbf{S}_{\mathrm{av},t} &= \frac{1}{2}\mathrm{Re}(\mathbf{E}_t \times \mathbf{H}_t^*)\\ &= \frac{1}{2}\mathrm{Re}\left(\frac{\bar{T}^2\bar{E}_i^2}{\bar{\eta}_2}\,\mathrm{e}^{-2\alpha_2 z}\,\mathrm{e}^{\mathrm{j}\theta_{\eta 2}}\mathbf{u}_z\right)\\ &= \frac{\bar{T}^2\bar{E}_i^2}{2\bar{\eta}_2}\,\mathrm{e}^{-2\alpha_2 z}\cos(\theta_{\eta 2})\mathbf{u}_z \quad \mathrm{W/m^2}\end{aligned} \tag{4.85}$$

In medium 2, since there is only the forward-traveling component, the calculation is quite simple. However, in medium 1, both the forward- and backward-traveling wave components exist. Therefore, the Poynting vector in medium 1 contains cross-coupling term as stated in (4.50).

4.4.1 Lossless media

When both media are lossless ($\sigma_1 = \sigma_2 = 0$), the results become quite straightforward. The reflection coefficient, transmission coefficient and intrinsic impedances become real values.

$$\eta_1 = \sqrt{\frac{\mu_1}{\varepsilon_1}} = \bar{\eta}_1 \tag{4.86a}$$

$$\eta_2 = \sqrt{\frac{\mu_2}{\varepsilon_1}} = \bar{\eta}_2 \tag{4.86b}$$

$$\Gamma = \frac{\eta_2 - \eta_1}{\eta_2 + \eta_1} = \bar{\Gamma} \tag{4.87}$$

$$T = \frac{2\eta_2}{\eta_2 + \eta_1} = \bar{T} \tag{4.88}$$

Also the attenuation constants α_1 and α_2 become zero so that

$$\gamma_1 = \mathrm{j}\beta_1 = \mathrm{j}\omega\sqrt{\mu_1\varepsilon_1} \tag{4.89a}$$

$$\gamma_2 = \mathrm{j}\beta_2 = \mathrm{j}\omega\sqrt{\mu_2\varepsilon_2} \tag{4.89b}$$

The space domain forms of the field vectors in (4.82) become

$$\mathbf{E}_i = E_i\,\mathrm{e}^{-\mathrm{j}\beta_1 z}\mathbf{u}_x \tag{4.90a}$$

$$\mathbf{H}_i = \frac{E_i}{\eta_1}\,\mathrm{e}^{-\mathrm{j}\beta_1 z}\mathbf{u}_y \tag{4.90b}$$

$$\mathbf{E}_r = \Gamma E_i\,\mathrm{e}^{\mathrm{j}\beta_1 z}\mathbf{u}_x \tag{4.90c}$$

$$\mathbf{H}_r = -\Gamma\frac{E_i}{\eta_1}\,\mathrm{e}^{\mathrm{j}\beta_1 z}\mathbf{u}_y \tag{4.90d}$$

$$\mathbf{E}_t = TE_i\,\mathrm{e}^{-\mathrm{j}\beta_2 z}\mathbf{u}_x \tag{4.90e}$$

$$\mathbf{H}_t = T\frac{E_i}{\eta_2}\,\mathrm{e}^{-\mathrm{j}\beta_2 z}\mathbf{u}_y \tag{4.90f}$$

For the lossless case the time domain forms of field vectors in (4.84) simplify to

$$\mathcal{E}_i = \bar{E}_i\cos(\omega t - \beta_1 z + \theta_i)\mathbf{u}_x \tag{4.91a}$$

$$\mathcal{H}_i = \frac{\bar{E}_i}{\eta_1}\cos(\omega t - \beta_1 z + \theta_i)\mathbf{u}_y \tag{4.91b}$$

$$\mathcal{E}_r = \Gamma \bar{E}_i \cos(\omega t + \beta_1 z + \theta_i)\mathbf{u}_x \tag{4.91c}$$

$$\mathcal{H}_r = -\Gamma \frac{\bar{E}_i}{\eta_1} \cos(\omega t + \beta_1 z + \theta_i)\mathbf{u}_y \tag{4.91d}$$

$$\mathcal{E}_t = T\bar{E}_i \cos(\omega t - \beta_2 z + \theta_i)\mathbf{u}_x \tag{4.91e}$$

$$\mathcal{H}_t = T\frac{\bar{E}_i}{\eta_2} \cos(\omega t - \beta_2 z + \theta_i)\mathbf{u}_y \tag{4.91f}$$

The total space domain fields in medium 1, from (4.90), are

$$\begin{aligned}\mathbf{E}_1 &= \mathbf{E}_i + \mathbf{E}_r \\ &= E_i\, \mathrm{e}^{-\mathrm{j}\beta_1 z}(1 + \Gamma\, \mathrm{e}^{2\mathrm{j}\beta_1 z})\mathbf{u}_x \end{aligned} \tag{4.92a}$$

$$\begin{aligned}\mathbf{H}_1 &= \mathbf{H}_i + \mathbf{H}_r \\ &= \frac{E_i}{\eta_1}\, \mathrm{e}^{-\mathrm{j}\beta_1 z}(1 - \Gamma\, \mathrm{e}^{2\mathrm{j}\beta_1 z})\mathbf{u}_y \end{aligned} \tag{4.92b}$$

The magnitude of the total electric field in medium 1 varies with z. The magnitude of the electric field can be expressed as

$$|\mathbf{E}_1| = |E_i|\sqrt{(1 + \Gamma^2 + 2\Gamma\cos(2\beta_1 z))} = \bar{E}_i\sqrt{(1 + \Gamma^2 + 2\Gamma\cos(4\pi z/\lambda_1))} \tag{4.93}$$

where Γ is a real number for a lossless medium but it can be either positive or negative. The maximum and minimum amplitude of the electric field in medium 1 can be expressed by using (4.93) as

$$|\mathbf{E}_1|_{\max} = \begin{cases} \bar{E}_i(1+\Gamma) & \text{when } \Gamma > 0 \text{ and } z = -n\lambda_1/2 \\ \bar{E}_i(1-\Gamma) & \text{when } \Gamma < 0 \text{ and } z = -n\lambda_1/2 \pm \lambda_1/4 \end{cases} \tag{4.94a}$$

$$|\mathbf{E}_1|_{\min} = \begin{cases} \bar{E}_i(1-\Gamma) & \text{when } \Gamma > 0 \text{ and } z = -n\lambda_1/2 \pm \lambda_1/4 \\ \bar{E}_i(1+\Gamma) & \text{when } \Gamma < 0 \text{ and } z = -n\lambda_1/2 \end{cases} \tag{4.94b}$$

where n is a positive integer. Simplifying the conditions in (4.94), we obtain that

$$|\mathbf{E}_1|_{\max} = \bar{E}_i(1 + |\Gamma|) \tag{4.95a}$$

$$|\mathbf{E}_1|_{\min} = \bar{E}_i(1 - |\Gamma|) \tag{4.95b}$$

We notice from (4.94) that the both adjacent maxima as well as adjacent minima are separated by $\lambda_1/2$ and an adjacent maximum and minimum are separated by $\lambda_1/4$. The ratio of the maximum to minimum value of the field is referred to as the *standing wave ratio* S. From (4.95)

$$S = \frac{|\mathbf{E}_1|_{\max}}{|\mathbf{E}_1|_{\min}} = \frac{1 + |\Gamma|}{1 - |\Gamma|} \tag{4.96}$$

It can be inversely solved for $|\Gamma|$ in terms of S as

$$|\Gamma| = \frac{S-1}{S+1} \tag{4.97}$$

For lossless media ($\sigma_1 = \sigma_2 = 0$), we can also evaluate the time-average power flow density for respective incident, reflected and transmitted waveforms as

$$\mathbf{S}_{\mathrm{av},i} = \frac{1}{2}\mathrm{Re}(\mathbf{E}_i \times \mathbf{H}_i^*) = \frac{1}{2}\frac{\bar{E}_i^2}{\eta_1}\mathbf{u}_z \tag{4.98a}$$

$$\mathbf{S}_{\mathrm{av},r} = \frac{1}{2}\mathrm{Re}(\mathbf{E}_r \times \mathbf{H}_r^*) = -\frac{1}{2}\frac{\bar{E}_i^2\Gamma^2}{\eta_1}\mathbf{u}_z \tag{4.98b}$$

$$\mathbf{S}_{\mathrm{av},t} = \frac{1}{2}\mathrm{Re}(\mathbf{E}_t \times \mathbf{H}_t^*) = \frac{1}{2}\frac{\bar{E}_i^2 T^2}{\eta_2}\mathbf{u}_z \tag{4.98c}$$

In a lossless medium ($\sigma_1 = \sigma_2 = 0$), there is no power loss in the medium. In addition, there is no cross coupling power flow term between forward- and backward-traveling wave components. Thus, we can declare that the transmitted power flow must be equal to the resultant power in the positive z direction in medium 1. Hence, we write, using subscripts i, r and t to denote the three components

$$\mathbf{S}_{\mathrm{av},i} + \mathbf{S}_{\mathrm{av},r} = \mathbf{S}_{\mathrm{av},t} \quad \text{when } \sigma_1 = \sigma_2 = 0 \tag{4.99}$$

In order to satisfy (4.99), from the expression given in (4.98), the following condition needs to be satisfied.

$$\frac{1}{\eta_1} - \frac{\Gamma^2}{\eta_1} = \frac{T^2}{\eta_2} \quad \text{when } \sigma_1 = \sigma_2 = 0 \tag{4.100}$$

The expression in (4.100) can be verified by substituting (4.87) and (4.88) into (4.100) for Γ and T.

4.4.2 Incidence on perfect conductors

Suppose medium 2 is a perfect conductor ($\sigma_2 = \infty$). Since the fields in a perfect conductor are zero, there is no transmitted wave. Thus, the boundary condition at $z = 0$ becomes

$$\mathbf{E}_i + \mathbf{E}_r = 0 \tag{4.101}$$

The general solutions for the plane wave rewritten forms of (4.82) are

$$\mathbf{E}_i = E_i\,\mathrm{e}^{-\gamma_1 z}\mathbf{u}_x \tag{4.102a}$$

$$\mathbf{E}_r = \Gamma E_i\,\mathrm{e}^{\gamma_1 z}\mathbf{u}_x \tag{4.102b}$$

In order to satisfy the boundary conditions given in (4.101),

$$\Gamma = -1 \tag{4.103}$$

Thus, the transmission coefficient in (4.79) becomes

$$T = 0 \tag{4.104}$$

The total field in medium 1 is given as

$$\begin{aligned} \mathbf{E}_1 &= \mathbf{E}_i + \mathbf{E}_r \\ &= E_i(\mathrm{e}^{-\gamma_1 z} - \mathrm{e}^{\gamma_1 z})\mathbf{u}_x \end{aligned} \tag{4.105a}$$

$$\begin{aligned} \mathbf{H}_1 &= \mathbf{H}_i + \mathbf{H}_r \\ &= \frac{E_i}{\eta_1}(\mathrm{e}^{-\gamma_1 z} + \mathrm{e}^{\gamma_1 z})\mathbf{u}_y \end{aligned} \tag{4.105b}$$

If medium 1 is lossless ($\sigma_1 = 0$), the total fields in medium 1 become

$$\mathbf{E}_1 = \mathbf{E}_i + \mathbf{E}_r = E_i(\mathrm{e}^{-\mathrm{j}\beta_1 z} - \mathrm{e}^{\mathrm{j}\beta_1 z})\mathbf{u}_x = -2\mathrm{j}E_i \sin(\beta_1 z)\mathbf{u}_x \tag{4.106a}$$

$$\mathbf{H}_1 = \mathbf{H}_i + \mathbf{H}_r = \frac{E_i}{\eta_1}(\mathrm{e}^{-\mathrm{j}\beta_1 z} + \mathrm{e}^{\mathrm{j}\beta_1 z})\mathbf{u}_y = 2\frac{E_i}{\eta_1}\cos(\beta_1 z)\mathbf{u}_y \tag{4.106b}$$

The time domain expressions become

$$\mathcal{E}_1 = 2\bar{E}_i \sin(\beta_1 z)\sin(\omega t + \theta_i)\mathbf{u}_x \tag{4.107a}$$

$$\mathcal{H}_1 = 2\frac{\bar{E}_i}{\eta_1}\cos(\beta_1 z)\cos(\omega t + \theta_i)\mathbf{u}_y \tag{4.107b}$$

The total field represents *standing waves* since the power flow of the total field is zero.

4.5 Polarization of uniform plane waves

Consider the electric field vector of a uniform plane wave that is traveling in a lossless medium.

$$\mathcal{E} = E_m \cos(\omega t - \beta z)\mathbf{u}_x \tag{4.108a}$$

The magnetic field vector is given by

$$\mathcal{H} = \frac{E_m}{\eta}\cos(\omega t - \beta z)\mathbf{u}_y \tag{4.108b}$$

An instantaneous plot of these two vectors as a function of z in the medium is shown in Fig. 4.5(a). When time increases, the envelope moves to the right with velocity $u = 1/\sqrt{\mu\varepsilon}$.

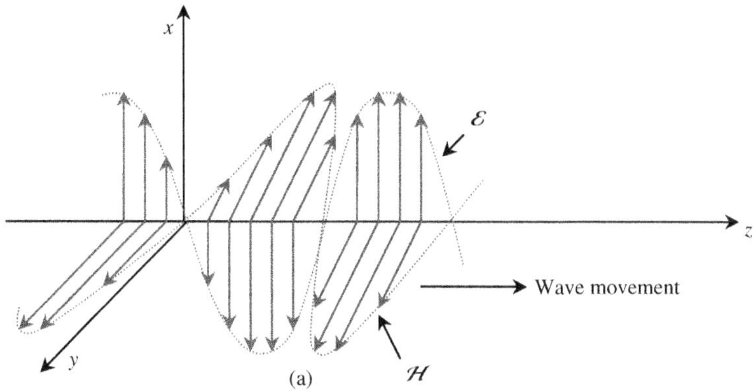

(a)

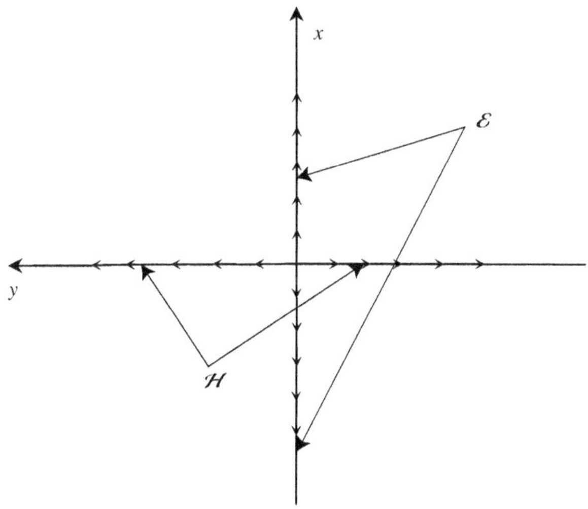

(b)

Figure 4.5: Linearly polarized uniform plane wave: (a) as a function of distance at $t = 0$ and (b) as a function of time at $z = 0$.

On the other hand, let us observe the variation of the wave at some fixed position, $z = 0$, as time varies. At $z = 0$ we obtain from (4.108),

$$\mathcal{E}|_{z=0} = E_m \cos(\omega t)\mathbf{u}_x \tag{4.109a}$$

$$\mathcal{H}|_{z=0} = \frac{E_m}{\eta} \cos(\omega t)\mathbf{u}_y \tag{4.109b}$$

In the cross-sectional view of the wave on the xy plane, we observe the magnitude of the field vectors varying according to (4.109) as shown in Fig. 4.5(b). The field vectors reverse direction with each half cycle but that variation is always along a line. This wave is said to be linearly polarized in the sense that the direction of each of the field vectors at some point in space and every point lies along a line on a plane perpendicular to the direction of propagation. Thus, the uniform plane waves that we have considered are said to be linearly polarized.

Now consider another uniform plane wave formed by the rotation of the field vectors in (4.108) by 90° in space and an addition of a phase angle θ to the argument of the cosine term.

$$\mathcal{E} = E_m \cos(\omega t - \beta z + \theta)\mathbf{u}_y \tag{4.110a}$$

$$\mathcal{H} = -\frac{E_m}{\eta} \cos(\omega t - \beta z + \theta)\mathbf{u}_x \tag{4.110b}$$

We can verify that the field vectors in (4.110) satisfy Maxwell's equations and are also linearly polarized. Further, since Maxwell's equations are linear (for linear media), the sum of field vectors in (4.108) and (4.110) is also a valid solution. Adding these solutions, we obtain

$$\mathcal{E} = E_{m1} \cos(\omega t - \beta z)\mathbf{u}_x + E_{m2} \cos(\omega t - \beta z + \theta)\mathbf{u}_y \tag{4.111a}$$

$$\mathcal{H} = \frac{E_{m1}}{\eta} \cos(\omega t - \beta z)\mathbf{u}_y - \frac{E_{m2}}{\eta} \cos(\omega t - \beta z + \theta)\mathbf{u}_x \tag{4.111b}$$

where we allow for the possibility of having different magnitudes of the components of mixed electric fields E_{m1} and E_{m2}. The combined electric field in (4.111a) at a fixed point in space, $z = 0$, as time progresses, substituting $z = 0$ into (4.111a), is

$$\mathcal{E}|_{z=0} = E_{m1} \cos(\omega t)\mathbf{u}_x + E_{m2} \cos(\omega t + \theta)\mathbf{u}_y \tag{4.112}$$

Several cases in the terms of the values of the magnitudes E_{m1} and E_{m2} and phase angle θ may be considered.

Case 1: $E_{m1} = E_{m2}$ and $\theta = 0°$. Then (4.112) reduces to

$$\mathcal{E}|_{z=0} = E_{m1} \cos(\omega t)(\mathbf{u}_x + \mathbf{u}_y) \tag{4.113}$$

which is shown in Fig. 4.6(a). The combined wave is linearly polarized with a sine variation of the wave at a fixed position and is along a line on a plane perpendicular to the *direction of propagation.*

Case 2: $E_{m1} \neq E_{m2}$ and $\theta = 0°$. Then (4.112) reduces to

$$\mathcal{E}|_{z=0} = E_{m1} \cos(\omega t)\left(\mathbf{u}_x + \frac{E_{m2}}{E_{m1}}\mathbf{u}_y\right) \tag{4.114}$$

which is also a linearly polarized wave as shown in 4.6(b).

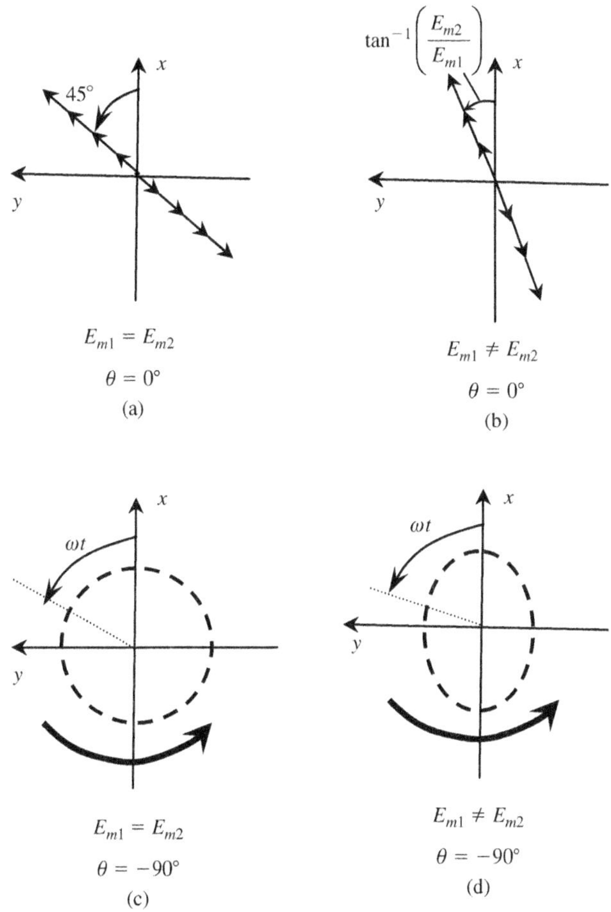

Figure 4.6: Illustration of polarization of a uniform plane wave viewed at $z = 0$ in the direction of propagation as a function of time.

Case 3: $E_{m1} = E_{m2}$ and $\theta = -90°$. Then (4.112) reduces to

$$\begin{aligned} \mathcal{E}|_{z=0} &= E_{m1}\cos(\omega t)\mathbf{u}_x + E_{m1}\cos(\omega t - 90°)\mathbf{u}_y \\ &= E_{m1}(\cos(\omega t)\mathbf{u}_x + \sin(\omega t)\mathbf{u}_y) \end{aligned} \tag{4.115}$$

The amplitude of this vector is E_{m1} at any time, and the angle between the vector and the x axis is

$$\phi = \tan^{-1}\left(\frac{\sin \omega t}{\cos \omega t}\right) = \omega t \tag{4.116}$$

Thus, the locus of the tip of the resultant vector $\mathcal{E}$ is a circle with radius E_{m1}. This is referred to as *circular polarization*. The $\mathcal{E}$ vector rotates in the counterclockwise direction and is referred to as a *left-hand circularly polarized* wave form. If $\theta = 90°$, again the locus of the tip of the resultant vector $\mathcal{E}$ is a circle but it rotates in the clockwise direction. This is referred to as a *right-hand circular polarization.*

Case 4: $E_{m1} \neq E_{m2}$ and $\theta = -90°$. For this case

$$\mathcal{E}|_{z=0} = E_{m1}\left(\cos(\omega t)\mathbf{u}_x + \frac{E_{m2}}{E_{m1}}\sin(\omega t)\mathbf{u}_y\right) \tag{4.117}$$

At any particular time, (4.117) yields the equation of an ellipse

$$\left(\frac{\mathcal{E}_x}{E_{m1}}\right)^2 + \left(\frac{\mathcal{E}_y}{E_{m2}}\right)^2 = 1 \tag{4.118}$$

Thus, the locus of the tip of the resultant vector is an ellipse and it rotates in the counterclockwise direction. The wave is said to have *left-hand elliptical polarization*. It is important to note from the above developments that a uniform plane wave has some general polarization (linear/circular/elliptical). Further, any analysis of a uniform plane wave having some general polarization can be accomplished by the superposition of the analysis of each linearly polarized orthogonal component.

4.6 Oblique incidence of uniform plane waves on plane boundaries

Previously, we considered uniform plane waves that are incident on plane boundaries, arriving normal to those boundaries. In this section, we consider uniform plane waves that are incident on plane boundaries at arbitrary angles of incidence. We again assume that the boundary between two media lies on the xy plane. This is by suitably orienting the axes in defining them. The Poynting vector of incident, reflected and transmitted waves is assumed to lie on xz plane, which is referred to as the plane of incidence. The angle of incidence θ_i is measured with respect to the normal to the boundary as shown in Fig. 4.7. The angle of reflection of the reflected wave θ_r is also measured with respect to this normal. A portion of the incident wave is transmitted into medium 2. The angle of transmission of this transmitted wave θ_t is measured with respect to the normal to the boundary as shown in Fig. 4.7.

In case of normal incidence considered previously we could assume, without loss of generality, that the electric field of the incident wave is polarized in the x direction. For oblique incidence, we have an infinite number of distinct possibilities for the polarization of the incident wave electric field. In order to consider all these cases, we decompose the incident wave into two linearly polarized waves with the electric fields of these waves orthogonal to each other. The results for a wave having some arbitrary polarization need to be achieved by the superposition of the results for these two individual waves acting severally.

Before considering these cases, we prove some general properties for this case of oblique incidence. In order to show these as well as to analyze the cases for specific

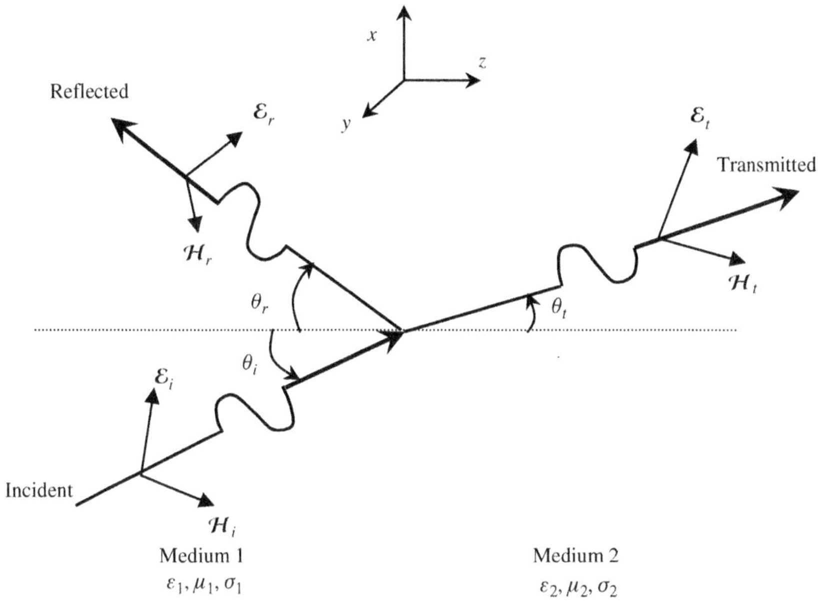

Figure 4.7: Oblique incidence of uniform plane waves on plane boundary.

polarizations, we must consider the description of uniform plane waves propagating in arbitrary spatial directions.

4.6.1 Propagation in an arbitrary spatial direction

Consider a uniform plane wave propagating in the z' direction in an orthogonal coordinate system consisting of x', y' and z' as axes as shown in Fig. 4.8. The forward-traveling space domain fields in this coordinate system are written as

$$\mathbf{E} = E_m \, \mathrm{e}^{-\gamma z'} \mathbf{u}_{x'} \tag{4.119a}$$

$$\mathbf{H} = \frac{E_m}{\eta} \, \mathrm{e}^{-\gamma z'} \mathbf{u}_{y'} \tag{4.119b}$$

We describe these in the original coordinate system consisting of the x, y and z coordinate axes. Specifically, we are interested in a rotation of the two coordinate systems. This can be obtained by using direction cosines. Knowing that

$$z' = x \cos\theta_x + y \cos\theta_y + z \cos\theta_z \tag{4.120}$$

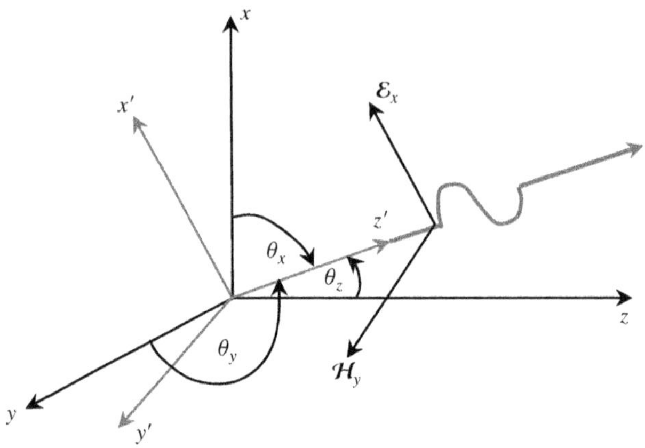

Figure 4.8: Propagation in an arbitrary spatial direction.

we can write

$$\begin{aligned} e^{-\gamma z'} &= e^{-\gamma(x\cos\theta_x + y\cos\theta_y + z\cos\theta_z)} = e^{-\gamma\cos\theta_x x}\, e^{-\gamma\cos\theta_y y}\, e^{-\gamma\cos\theta_z z} \\ &= e^{-\gamma_x x}\, e^{-\gamma_y y}\, e^{-\gamma_z z} \end{aligned} \tag{4.121}$$

where

$$\gamma_x = \gamma\cos\theta_x, \quad \gamma_y = \gamma\cos\theta_y \quad \text{and} \quad \gamma_z = \gamma\cos\theta_z \tag{4.122}$$

The remaining task is to write the electric and magnetic field vectors in terms of the original coordinate system axes. With the above result, the complete expressions in (4.119) should be in the form

$$\mathbf{E} = (E_x\mathbf{u}_x + E_y\mathbf{u}_y + E_z\mathbf{u}_z)e^{-\gamma_x x}\, e^{-\gamma_y y}\, e^{-\gamma_z z} \tag{4.123a}$$

$$\mathbf{H} = (H_x\mathbf{u}_x + H_y\mathbf{u}_y + H_z\mathbf{u}_z)e^{-\gamma_x x}\, e^{-\gamma_y y}\, e^{-\gamma_z z} \tag{4.123b}$$

where E_x, E_y and E_z and H_x, H_y and H_z are projections of the corresponding field vectors in the original coordinate system consisting of the x, y and z axes.

The unit vector in the direction of propagation in terms of the original vectors is

$$\hat{\mathbf{a}}_{z'} = \cos\theta_x\mathbf{u}_x + \cos\theta_y\mathbf{u}_y + \cos\theta_z\mathbf{u}_z \tag{4.124}$$

The magnetic field can be evaluated from the electric field as

$$\mathbf{H} = \frac{\mathbf{u}_{z'} \times \mathbf{E}}{\eta} \tag{4.125}$$

Therefore, once the electric field vector is known in the original coordinate system, it is possible to evaluate the corresponding magnetic field vectors. Hence, we obtain the relationship between E_x, E_y and E_z and H_x, H_y andH_z as

$$(H_x\mathbf{u}_x + H_y\mathbf{u}_y + H_z\mathbf{u}_z) = \frac{\mathbf{u}_{z'} \times (E_x\mathbf{u}_x + E_y\mathbf{u}_y + E_z\mathbf{u}_z)}{\eta} \tag{4.126}$$

4.6.2 Snell's law

Consider the problem of oblique incidence described in Fig. 4.7. The relation among various angles, θ_i, θ_r and θ_t, is governed by *Snell's law*. To show these relations, we write the space domain incident, reflected and transmitted field vectors as in (4.123a) and (4.123b). We assume that all three propagation vectors lie on the xz plane, which we refer to as the plane of incidence. Thus, the propagation vectors have components only in the x and z directions. The electric and magnetic field vectors may have components along all three coordinate axes. From the geometry in Fig. 4.7, we can write the space domain fields as

$$\mathbf{E}_i = (E_x^i\mathbf{u}_x + E_y^i\mathbf{u}_y + E_z^i\mathbf{u}_z)\mathrm{e}^{-\gamma_1(x\sin\theta_i + z\cos\theta_i)} \tag{4.127a}$$

$$\mathbf{H}_i = (H_x^i\mathbf{u}_x + H_y^i\mathbf{u}_y + H_z^i\mathbf{u}_z)\mathrm{e}^{-\gamma_1(x\sin\theta_i + z\cos\theta_i)} \tag{4.127b}$$

$$\mathbf{E}_r = (E_x^r\mathbf{u}_x + E_y^r\mathbf{u}_y + E_z^r\mathbf{u}_z)\mathrm{e}^{-\gamma_1(x\sin\theta_r - z\cos\theta_r)} \tag{4.127c}$$

$$\mathbf{H}_r = (H_x^r\mathbf{u}_x + H_y^r\mathbf{u}_y + H_z^r\mathbf{u}_z)\mathrm{e}^{-\gamma_1(x\sin\theta_r - z\cos\theta_r)} \tag{4.127d}$$

$$\mathbf{E}_t = (E_x^t\mathbf{u}_x + E_y^t\mathbf{u}_y + E_z^t\mathbf{u}_z)\mathrm{e}^{-\gamma_2(x\sin\theta_t + z\cos\theta_t)} \tag{4.127e}$$

$$\mathbf{H}_t = (H_x^t\mathbf{u}_x + H_y^t\mathbf{u}_y + H_z^t\mathbf{u}_z)\mathrm{e}^{-\gamma_2(x\sin\theta_t + z\cos\theta_t)} \tag{4.127f}$$

Imposing boundary conditions at the interface, the tangential electric field must be continuous at $z = 0$:

$$(E_x^i\mathbf{u}_x + E_y^i\mathbf{u}_y)\mathrm{e}^{-\gamma_1 x\sin\theta_i} + (E_x^r\mathbf{u}_x + E_y^r\mathbf{u}_y)\mathrm{e}^{-\gamma_1 x\sin\theta_r} = (E_x^t\mathbf{u}_x + E_y^t\mathbf{u}_y)\mathrm{e}^{-\gamma_2 x\sin\theta_t} \tag{4.128}$$

Similarly, the tangential magnetic field must be also continuous at $z = 0$:

$$(H_x^i\mathbf{u}_x + H_y^i\mathbf{u}_y)\mathrm{e}^{-\gamma_1 x\sin\theta_i} + (H_x^r\mathbf{u}_x + H_y^r\mathbf{u}_y)\mathrm{e}^{-\gamma_1 x\sin\theta_r} = (H_x^t\mathbf{u}_x + H_y^t\mathbf{u}_y)\mathrm{e}^{-\gamma_2 x\sin\theta_t} \tag{4.129}$$

Equations (4.128) and (4.129) satisfy the condition for all values of x. Therefore, the exponential terms must be equal:

$$\mathrm{e}^{-\gamma_1 x\sin\theta_i} = \mathrm{e}^{-\gamma_1 x\sin\theta_r} = \mathrm{e}^{-\gamma_2 x\sin\theta_t} \tag{4.130}$$

Equation (4.130) provides two conditions:

$$\theta_i = \theta_r \tag{4.131a}$$

$$\gamma_1 \sin\theta_i = \gamma_2 \sin\theta_t \tag{4.131b}$$

Equation (4.131a) is referred to as *Snell's law of reflection*. The angle of incidence must equal the angle of reflection. The second relation given by (4.170b) is referred to as *Snell's law of refraction*. For lossless media ($\sigma_1 = \sigma_2 = 0$), (4.131b) becomes

$$\mathrm{j}\beta_1 \sin\theta_i = \mathrm{j}\beta_2 \sin\theta_t \tag{4.132}$$

where $\beta = \omega\sqrt{\mu\varepsilon}$. Substituting for β in (4.132)

$$\frac{\sin\theta_i}{\sin\theta_t} = \frac{\sqrt{\mu_2\varepsilon_2}}{\sqrt{\mu_1\varepsilon_1}} \tag{4.133}$$

For typical dielectric media, which are non-magnetic $\mu_r \approx 1$. Therefore,

$$\frac{\sin\theta_i}{\sin\theta_t} = \frac{\sqrt{\varepsilon_2}}{\sqrt{\varepsilon_1}} = \frac{\sqrt{\varepsilon_{r2}}}{\sqrt{\varepsilon_{r1}}} \tag{4.134}$$

The *index of refraction* of a medium is defined as the ratio of the velocity of light in free space to that in the medium. Also if the medium is a dielectric, then $\mu_r \approx 1$. That yields

$$n = \frac{u_0}{u} = \frac{\sqrt{\mu\varepsilon}}{\sqrt{\mu_0\varepsilon_0}} = \sqrt{\varepsilon_r} \tag{4.135}$$

In terms of the refraction index, Snell's law of refraction can be stated as

$$n_i \sin\theta_i = n_t \sin\theta_t \tag{4.136}$$

The angle of incidence, when the angle of transmission is 90° (i.e., no wave is transmitted), is referred to as the *critical angle of total reflection*. From (4.136) with $\theta_t = 90°$, we obtain

$$\theta_c = \sin^{-1}\left(\frac{n_t}{n_i}\right) = \sin^{-1}\left(\frac{n_2}{n_1}\right) = \sin^{-1}\left(\sqrt{\frac{\varepsilon_2}{\varepsilon_1}}\right) \tag{4.137}$$

When the angle of incidence exceeds this critical angle ($\theta_i > \theta_c$), according to (4.136), we obtain

$$\sin\theta_t = \frac{n_1}{n_2}\sin\theta_i > 1 \tag{4.138}$$

implying that there is no real angle for θ_t. Reexamining the previous expression

$$\cos\theta_t = \sqrt{1 - \sin^2\theta_t} = \pm\mathrm{j}\sqrt{\frac{\varepsilon_1}{\varepsilon_2}\sin^2\theta_i - 1} \tag{4.139}$$

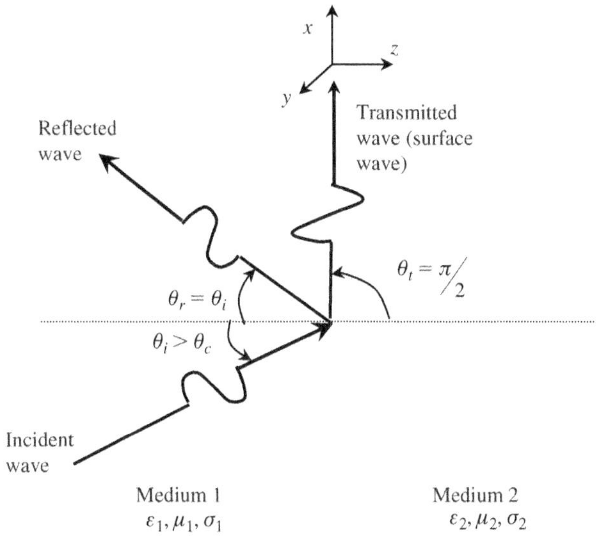

Figure 4.9: Illustration of surface waves and the critical angle of reflection.

When ($\theta_i > \theta_c$), Equation (4.138) shows that $\cos\theta_t$ is imaginary. Thus, we can write

$$\beta_2 \cos\theta_t = -\mathrm{j}\beta_2\sqrt{\frac{\varepsilon_1}{\varepsilon_2}\sin^2\theta_i - 1} = -\mathrm{j}\alpha_2 \tag{4.140}$$

Even though, $\beta_2 \cos\theta_t$ has two solutions, we have chosen the negative result in (4.140) for reasons that will soon become clear. Recalling the expression (4.127e) for the transmitted electrical field vector in a lossless medium ($\sigma_2 = 0$).

$$\mathbf{E}_t = (E_x^t \mathbf{u}_x + E_y^t \mathbf{u}_y + E_z^t \mathbf{u}_z)\mathrm{e}^{-\mathrm{j}\beta_2(x \sin\theta_t + z\cos\theta_t)} \tag{4.141}$$

Substituting (4.140) in (4.141) yields

$$\mathbf{E}_t = (E_x^t \mathbf{u}_x + E_y^t \mathbf{u}_y + E_z^t \mathbf{u}_z)\mathrm{e}^{-\alpha_2 z}\,\mathrm{e}^{-\mathrm{j}\beta_2 x \sin\theta_t} \tag{4.142}$$

This result shows that the transmitted wave is propagating in the x direction with propagation constant $\beta_2 \sin\theta_t$. However, the field is also multiplied by a term $\mathrm{e}^{-\alpha_2 z}$ that is due to the angle of incidence exceeding the critical angle. The wave is attenuated exponentially away from the boundary of medium 2. As illustrated in Fig. 4.9, waves of this type are also referred to as *surface waves* since they can be tightly bound to the surface of the interface if α_2 is large. These are important in considerations with microwave on substrates.

For further reading

[1] Balanis, C.A., *Advanced Engineering Electromagnetics*, USA: New York, 2012.
[2] Paul, C.R., Whites, K.W., and Nazar, S.A., *Introduction to Electromagnetic Fields,* USA: McGraw-Hill, New York, 1997.
[3] Griffiths, D.J., *Introduction to Electrodynamics*, USA: Prentice Hall, Englewood Cliffs, NJ, 1998.
[4] Hoole, S.R.H., Hoole, P.R.P., *A Modern Short Course in Engineering Electromagnetics*, USA: Oxford University Press, New York, 1996.
[5] Sadiku, M.N.O., *Principals of Electromagnetics*, USA: Oxford University Press, New York, 2009.
[6] Shevgaonkar, R.K., *Electromagnetic Waves*, USA: Tata McGraw-Hill Publishing Company, New York, 2006.

5

Computer-aided design

5.1 Computer-aided design – in pursuit of accuracy

In today's context, numerical methods – as distinguished from explicit closed form classical methods, hereinafter referred to as classical methods – constitute the core of computer-aided design. Classical methods rather as an almost sleight of hand have been referred to as exact methods, whereas numerical methods are called approximate methods. These claims upon closer scrutiny do not stand up.

Classical methods are limited by the mathematics by which the solutions are expressed. As shown in Fig. 5.1, for every problem, we first choose and select a model and thereupon solve the problem posed by that model chosen rather than by the actual problem. When the solution is not satisfactory, we reexamine the model with a view to improving it. For example, in the mechanics of projectiles our model may be an equation based on Newton's laws. These work most of the time. But when it comes to space travel, the solutions may not match measurements and we will incorporate Einstein's laws to improve the model with better accuracy. Likewise in electromagnetics, in classical electrodynamics modeling, say of a transmission line, we assume symmetry in the direction of the line so that the problem is two dimensional. Thereafter, we may further assume the transmission line to be circular. These assumptions permit the reduction of the problem to a simple one-dimensional differential equation of the form

$$-\frac{d\varphi}{dr} = \frac{1}{2\pi\epsilon}\frac{1}{r}. \tag{5.1}$$

This model gives us a solution which is as exact only as much as the model is exact:

$$\varphi = -\frac{1}{2\pi\epsilon}\ln\frac{r}{r_o}. \tag{5.2}$$

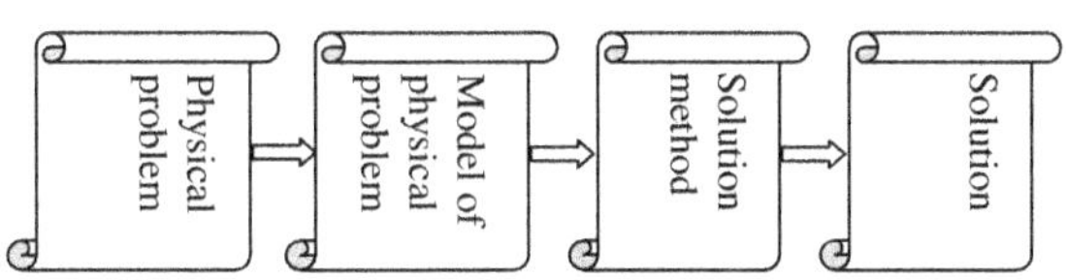

Figure 5.1: The solution process.

If we need a more accurate solution, we need to improve the model. For instance, we may model the earth above which this infinitely long line runs by considering the image of the line. This makes the problem two dimensional. With superposition, this too can be handled by our model and the corresponding solution method. If this also prove inadequate for purposes of the required accuracy, we go further and allow for nearby transmission towers and their equipment like insulator strings, and the sag of the wire from tower to tower. At this point, the model breaks down as the problem is no longer two dimensional and truly three dimensional. Although many three-dimensional problems have classical solutions, they rely upon some forms of neatness, such as spherical shapes and homogenous systems. This complex transmission line has no known classical solution. At this point a numerical solution is a must. Numerical solutions by definition are approximate insofar as derivatives and integrals assume linear variation over small discrete distances. An example of the approximation is for the derivative and integral:

$$\frac{\mathrm{d}\varphi}{\mathrm{d}x} = \lim_{x_2 \to x_1} \frac{\varphi_2 - \varphi_1}{x_2 - x_1} \approx \frac{\varphi_2 - \varphi_1}{x_2 - x_1} \tag{5.3}$$

$$\int_{x_1}^{x_2} \varphi \, \mathrm{d}x \approx \left(\frac{\varphi_1 + \varphi_2}{2}\right)(x_2 - x_1) \tag{5.4}$$

Although linear change of φ is assumed over the interval (x_1, x_2) this so-called approximation is increasingly more accurate for smaller such intervals. That is, while classical solutions are only as accurate as the models on which they are based, the numerical approximations for derivatives and integrals can be made as accurate as we want by our resorting to finer subdivisions of the relevant intervals over the domain.

That is, the accuracy of numerical methods depends on the subdivisions resorted to. This involves data preparation (called preprocessing), then solving the problems, and now faced with inordinate amounts of data corresponding to the numerous points of discretization, some form of computer-based depiction of the results to make meaning of the numbers, a stage called postprocessing.

5.2 Methods of approximate solution – differential and integral

There are two classes of numerical methods based on whether we are solving integral equations or differential equations. In electrostatics, for example, the Coulomb equation becomes the basis of the boundary element method:

$$\varphi = \iiint_{v} \frac{1}{4\pi\epsilon} \frac{\rho}{r} \, \mathrm{d}v \tag{5.5}$$

This equation is discretized and the potential φ is expressed in terms of the charge distribution ρ. However, on material boundaries there is a surface charge density σ according to the derivatives in the normal direction n at the interface separating regions 1 and 2:

$$\epsilon_1 \frac{\partial \varphi_1}{\partial n_1} - \epsilon_2 \frac{\partial \varphi_2}{\partial n_2} = \sigma \tag{5.6}$$

This simply states that the difference in normal flux density across the interface is the surface charge density.

Thus, the integral Poisson equation relates φ at each point to the known charge distribution ρ and φ's normal derivative at all material interfaces. The charge distribution too may be unknown as the charges rearrange themselves because of Coulomb forces. Thus, every discrete point has an unknown related to the unknowns at all other discrete points. The resulting matrix equation is fully populated and not necessarily symmetric. Thus although, unlike in differential methods relating every point only to others in its neighborhood thereby measuring rates of change, every point in the domain has a relationship with all other points in the domain of solution. However, here with integral methods, we have unknowns only where there are conductors and material interfaces – that is, we have fewer unknowns. Yet, the method is not too popular especially because of the non symmetric nature of the matrix associated with the solution equation, in addition to the fully populated nature of the matrix equation.

The best known differential methods are the finite difference and finite element methods. The finite difference method is best suited to homogeneous problems and therefore is very popular in solving the wave equation solving which usually involves modeling in free homogeneous space. It is not so popular in many problems where inhomogeneous materials are present. While we will focus on the finite element method due to its wide applications, here we will briefly mention the finite difference method.

We shall take as a general example for demonstrating the finite difference method, the solution of the Poisson equation which appears in many places in all specialities of engineering:

$$-\epsilon \nabla^2 \varphi = \rho \tag{5.7}$$

where in the case of electrostatic problems, ϵ is the permittivity, φ is the electric potential for which we will be solving and ρ is the space charge density. We note that the operator (in two-dimensional problems)

$$\nabla^2 = \frac{\partial^2}{\partial x^2} + \frac{\partial^2}{\partial y^2} \tag{5.8}$$

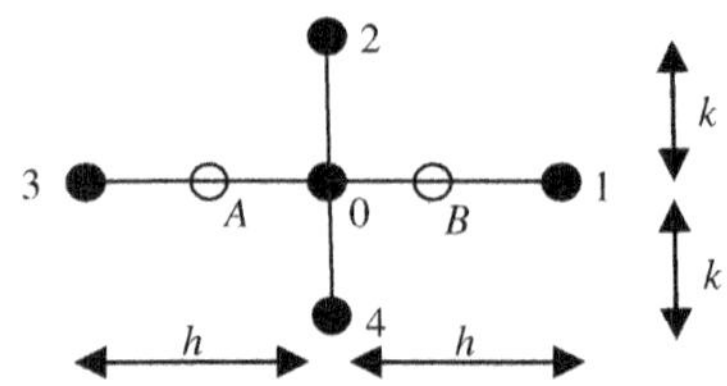

Figure 5.2: An hxk finite difference mesh with typical node 0 surrounded by nodes 1–4.

In the finite difference method, we work with a rectangular grid. Although, there are numerous papers claiming to work with triangular and quadrilateral grids, they are really re-presentations of the finite element method to make the authors' work of a life time not seem outdated. The real simplicity of the finite difference method comes from the rectangular grid as also do its weaknesses. A rectangular grid means that when a device involves fine details, the finite difference mesh needs to be fine there; but now uniformity of the mesh size demands that the same fineness be applied everywhere, including places where fineness is not required.

We need to approximate the Poisson equation at node 0. To this end, we shall estimate the first derivative at points A and B which are the midpoints between nodes 3 and 0 and nodes 0 and 1, respectively. Approximately, we may say that at the two midpoints A and B, as shown in Fig. 5.2,

$$\left.\frac{\partial\varphi}{\partial x}\right|_A = \frac{\Delta\varphi}{\Delta x} = \frac{\varphi_0 - \varphi_3}{h} \tag{5.8a}$$

$$\left.\frac{\partial\varphi}{\partial x}\right|_B = \frac{\Delta\varphi}{\Delta x} = \frac{\varphi_1 - \varphi_0}{h} \tag{5.8b}$$

Therefore, recognizing that the change in x between the points A and B is also h,

$$\frac{\partial^2\varphi}{\partial x^2} = \frac{\Delta\frac{\partial\varphi}{\partial x}}{\Delta x} = \frac{\frac{\varphi_1-\varphi_0}{h} - \frac{\varphi_0-\varphi_3}{h}}{h} = \frac{\varphi_1 + \varphi_3 - 2\varphi_0}{h^2} \tag{5.9}$$

Similarly

$$\frac{\partial^2\varphi}{\partial y^2} = \frac{\Delta\frac{\partial\varphi}{\partial y}}{\Delta y} = \frac{\varphi_2 + \varphi_4 - 2\varphi_0}{k^2} \tag{5.10}$$

so that the approximation of the Poisson equation becomes (see Fig. 5.3)

$$-\varepsilon\left(\frac{\varphi_1 + \varphi_3 - 2\varphi_0}{h^2} + \frac{\varphi_2 + \varphi_4 - 2\varphi_0}{k^2}\right) = \rho_0 \tag{5.11}$$

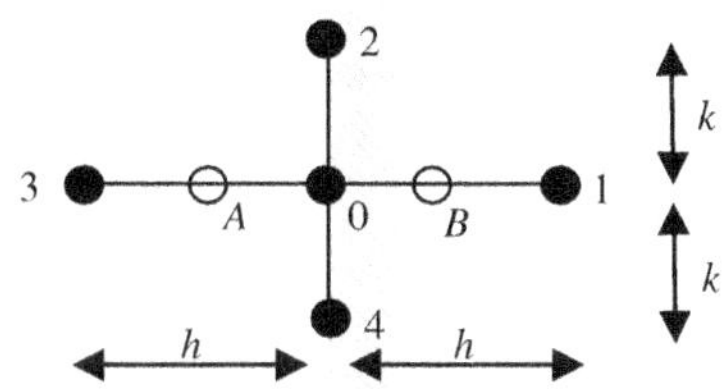

Figure 5.3: Node on left vertical boundary of solution.

Frequently, a square mesh with $h = k$ is used in which case we have

$$4\varphi_0 - \varphi_1 - \varphi_2 - \varphi_3 - \varphi_4 = \frac{h^2 \rho_0}{\varepsilon} \tag{5.12}$$

Generally an equation can be fitted to every interior node. Where the node 0 is on a boundary with the potential specified (called the Dirichlet boundary condition) there is no need to have an equation written for it. Where the node is on the boundary where the normal gradient of potential is zero (the so-called Neumann boundary condition), the node that would lie outside the boundary becomes a reflection of the interior node so that the above equation would involve the three nodes on the boundary and the interior node twice.

For example, if the node 0 is on a left vertical boundary (Fig. 5.3), the Neumann condition can also be expressed in finite differences to deal with the non existent node 3, which if it exists would lie outside the domain of solution:

$$\left.\frac{\partial \varphi}{\partial n}\right|_{\text{at } 0} = \left.\frac{\partial \varphi}{\partial x}\right|_{\text{at } 0} = \frac{\varphi_1 - \varphi_3}{h} = 0 \to \varphi_1 = \varphi_3 \tag{5.13}$$

Substituting, the relevant finite difference approximation for node 0, eliminating node 3, becomes

$$-\varepsilon\left(\frac{2\varphi_1 - 2\varphi_0}{h^2} + \frac{\varphi_2 + \varphi_4 - 2\varphi_0}{k^2}\right) = \rho_0 \tag{5.14}$$

We will stop there on the finite difference method saying that the Neumann boundary condition makes the matrix of coefficients nonsymmetric and that inhomogeneities also do that besides imposing the requirement that material interfaces must run vertically or horizontally. The latter requirement necessitates modelling slanted boundaries by short vertical and horizontal lines making the mesh very fine, thereby increasing the matrix size.

We will therefore look at the powerful finite element method that reflects the true state of the art.

5.3 A note on matrix representation of polynomials

It is convenient first to express second-order polynomials in matrix form and establish some terminology. In particular, we want to establish the rules for adding and differentiating polynomials.

The variables $x_1, x_2, \ldots, x_n$ may be expressed as the components of the column vector $\mathbf{x}$. Thus the scalar s, a first-order polynomial expression in the components of $\mathbf{x}$ becomes

$$s = c_1x_1 + c_2x_2 + \cdots + c_nx_n = \mathbf{c}^{\mathrm{t}}\mathbf{x} = \mathbf{x}^{\mathrm{t}}\mathbf{c} \tag{5.15}$$

Defining

$$\frac{\partial s}{\partial \mathbf{x}} = \frac{\partial \mathbf{x}^{\mathrm{t}}\mathbf{c}}{\partial \mathbf{x}} = \left\{ \begin{array}{c} \dfrac{\partial s}{\partial x_1} \\ \vdots \\ \dfrac{\partial s}{\partial x_n} \end{array} \right\} = \left\{ \begin{array}{c} c_1 \\ \vdots \\ c_n \end{array} \right\} = \mathbf{c} \tag{5.16}$$

A second-order polynomial p, it may be verified, is

$$\begin{aligned} p &= a_{11}x_1^2 + 2a_{12}x_1x_2 + \cdots + 2a_{1n}x_1x_n + a_{22}x_2^2 + 2a_{23}x_2x_3 + \cdots \\ &\quad + 2a_{2n}x_2x_n + \cdots + a_{nn}x_n^2 \\ &= \mathbf{x}^{\mathrm{t}}[A]\mathbf{x} \end{aligned} \tag{5.17}$$

where the symmetric matrix

$$[A] = \begin{bmatrix} a_{11} & \cdots & a_{1n} \\ \vdots & \ddots & \vdots \\ a_{n1} & \cdots & a_{nn} \end{bmatrix} \tag{5.18}$$

If we considered a second polynomial

$$\begin{aligned} q &= b_{11}x_1^2 + 2b_{12}x_1x_2 + \cdots + 2b_{1n}x_1x_n + b_{22}x_2^2 + 2b_{23}x_2x_3 + \cdots \\ &\quad + 2b_{2n}x_2x_n + \cdots + b_{nn}x_n^2 \\ &= \mathbf{x}^{\mathrm{t}}[B]\mathbf{x} \end{aligned} \tag{5.19}$$

then we have the rule for adding second-order polynomials

$$p + q = \mathbf{x}^{\mathrm{t}}[A]\mathbf{x} + \mathbf{x}^{\mathrm{t}}[B]\mathbf{x} = \mathbf{x}^{\mathrm{t}}[A + B]\mathbf{x} \tag{5.20}$$

The rule for differentiation comes from differentiating p:

$$\frac{\partial p}{\partial \mathbf{x}} = \begin{Bmatrix} \dfrac{\partial p}{\partial x_1} \\ \vdots \\ \dfrac{\partial p}{\partial x_n} \end{Bmatrix} = \begin{Bmatrix} 2a_{11}x_1 + 2a_{12}x_2 + \cdots + 2a_{1n}x_n \\ \vdots \\ 2a_{1n}x_1 + 2a_{2n}x_2 + \cdots + 2a_{nn}x_n \end{Bmatrix} = 2[A]\mathbf{x} \qquad (5.21)$$

5.4 The finite element method

The finite element method is today the most commonly used numerical method in the area of the solution of partial differential equations. Unlike integral methods and the finite difference method, it always yields symmetric sparse matrices which allow the use of the most powerful and elegant matrix solvers to be used. It allows us to use a fine mesh in parts where fine detail has to be incorporated or high accuracy is required, whereas a sparse mesh may be used in other parts. Inhomogeneities are not a problem and are handled by ensuring that an element is always entirely in a homogeneous region.

To describe the finite element method, it is good to note that it has certain ingredients and to spell them out.

5.4.1 Ingredient 1

A trial or test function on the domain of solution (the xy plane in our two-dimensional demonstration) consisting of many interpolation functions α_i with $i = 1, \ldots, n$. The test function would be a sum of all the α_is and the αs are functions that we prudently choose so as to best model the solution. This means some knowledge of the solution. In this form, the method was known as Ritz's method and was not quite successful since we do not usually know the form of the solution. So our postulate based on our choice of αs is that the solution is of the form

$$\varphi = \sum_{i=1}^{n} c_i \alpha_i \qquad (5.22)$$

Our task then would be to determine the numbers c_i so as to yield the best fit of the actual solution that can be made by the trial functions chosen.

5.4.2 Ingredient 2

A mesh of finite elements of the domain involving a tessellation of the area of solution. Commonly the domain of solution is subdivided into triangles or quadrilaterals. In this lies the brilliance of the method. Once we divide the domain into small regions, in place of the complicated functions, our αs, we may go for simple functions, such as linear or quadratic polynomials. In simple terms, to approximate a squiggly function $y(x)$ on a large domain, we may need to add several αs of different forms; but if we divide the domain into small bits, a straight line on each bit or element would suffice. The smaller the elements, the more accurate our model. Thus on a triangular linear finite element, we have

$$\varphi = a + bx + cy \tag{5.23}$$

where the functions $1, x$ and y take the place of the three αs we need for the small triangular domain. As we divide the domain into smaller and smaller triangles, the modelling of the actual shape of the unknown would become more and more accurate. This linear model is an approximation of the potential because φ may not vary linearly and may have a complex relationship with x and y. However, x and y are linear first-order functions. This is the key to determining a, b and c in a triangle.

Usually from one triangle to its neighbour we would need continuity conditions so that the trial functions match on the edge common to two adjacent elements. For this the a, b and c of one triangle need to have a relationship with the a, b and c of any adjacent triangle. This relationship being complicated, instead of a, b and c we deal with the three vertex potentials $\varphi_i, i = 1, 2, 3$. So at each vertex of a triangle (x_i, y_i) which values are known because we are solving the Poisson equation in a specific domain divided into a specific mesh of known vertices connected into a triangular mesh. Applying the linear approximation (Fig. 5.4):

$$\begin{aligned}\varphi_1 &= a + bx_1 + cy_1\\ \varphi_2 &= a + bx_2 + cy_2\\ \varphi_3 &= a + bx_3 + cy_3\end{aligned}$$

or

$$\begin{bmatrix}\varphi_1\\ \varphi_2\\ \varphi_3\end{bmatrix} = \begin{bmatrix}1 & x_1 & y_1\\ 1 & x_2 & y_2\\ 1 & x_3 & y_3\end{bmatrix}\begin{bmatrix}a\\ b\\ c\end{bmatrix} \tag{5.24}$$

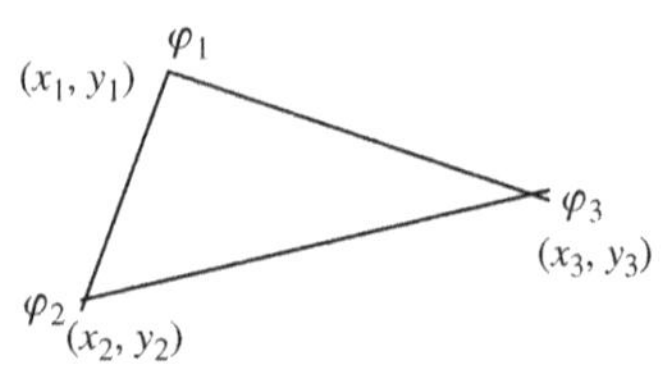

Figure 5.4: A triangle with three nodes.

Solving by Cramer's rule

$$a = \frac{\begin{vmatrix} \varphi_1 & x_1 & y_1 \\ \varphi_2 & x_2 & y_2 \\ \varphi_3 & x_3 & y_3 \end{vmatrix}}{\begin{vmatrix} 1 & x_1 & y_1 \\ 1 & x_2 & y_2 \\ 1 & x_3 & y_3 \end{vmatrix}} \tag{5.25}$$

$$\begin{aligned} b = \frac{\begin{vmatrix} 1 & \varphi_1 & y_1 \\ 1 & \varphi_2 & y_2 \\ 1 & \varphi_3 & y_3 \end{vmatrix}}{\begin{vmatrix} 1 & x_1 & y_1 \\ 1 & x_2 & y_2 \\ 1 & x_3 & y_3 \end{vmatrix}} &= \frac{1}{\Delta}\{\varphi_1(y_2 - y_3) + \varphi_2(y_3 - y_1) + \varphi_3(y_1 - y_2)\} \\ &= \sum_{i=1}^{3} \varphi_i \frac{(y_{i1} - y_{i2})}{\Delta} = \sum_{i=1}^{3} b_i \varphi_i \end{aligned} \tag{5.26}$$

$$\begin{aligned} c = \frac{\begin{vmatrix} 1 & \varphi_1 & y_1 \\ 1 & \varphi_2 & y_2 \\ 1 & \varphi_3 & y_3 \end{vmatrix}}{\begin{vmatrix} 1 & x_1 & y_1 \\ 1 & x_2 & y_2 \\ 1 & x_3 & y_3 \end{vmatrix}} &= \frac{1}{\Delta}\{\varphi_1(x_3 - x_2) + \varphi_2(x_1 - x_3) + \varphi_3(x_2 - x_1)\} \\ &= \sum_{i=1}^{3} \varphi_i \frac{(x_{i2} - x_{i1})}{\Delta} = \sum_{i=1}^{3} c_i \varphi_i \end{aligned} \tag{5.27}$$

where Δ is twice the area of the triangle provided the nodes 1, 2 and 3 go counter clockwise (negative of that if the nodes go clockwise); the numbers i, $i1$ and $i2$ are the numbers 1, 2, and 3 in cyclic order (i.e., the integers i, $i1$ and $i2$, respectively are 1, 2, 3 if i is 1, or 2, 3, 1 if i is 2 or 3, 1, 2 if i is 3; or put another way $i1 = i \bmod 3 + 1$ and $i2 = i1 \bmod 3 + 1$, where the mod 3 or modulo 3 function gives the remainder when divided by 3); and

$$b_i = \frac{y_{i1} - y_{i2}}{\Delta} \tag{5.28a}$$

and

$$c_i = \frac{x_{i2} - x_{i1}}{\Delta} \tag{5.28b}$$

5.4.3 Ingredient 3

A functional (a function of a function, in this case of φ) $\mathcal{L}(\varphi)$, minimizing which is the same as satisfying the differential equation. There is an alternative approach called the Galerkin formulation relying on the theory of function spaces going into minimizing the residual when our trial function is put into the differential equation being solved but we will not get into that here. A limitation of the approach we describe, the variational calculus approach, is that to take it we must have a functional corresponding to the solution of the differential equation whose solution is sought. Suffice it to say that the Poisson equation does have a functional corresponding to it:

$$\mathcal{L}(\varphi) = \iiint_v \left[\frac{1}{2}\epsilon(\nabla\varphi)^2 - \rho\varphi\right]\mathrm{d}v \tag{5.29}$$

We can justify this intuitively. Since the electric field strength $\mathbf{E} = -\nabla\varphi$ and the flux density $\mathbf{D} = \epsilon\mathbf{E} = -\varepsilon\nabla\varphi$, we see that the first term is the stored energy density ½$\mathbf{D}\cdot\mathbf{E}$ times elemental volume dR. Since by definition the potential φ is the work done in bringing a unit charge to where it is, the second term is the work done in bringing the charge $\rho\,\mathrm{d}R$, the charge density times elemental volume, to where it is. The essence of the functional is that it says that the difference between the two co-energies ought to be a minimum when the Poisson equation is satisfied. But let us see this more precisely in mathematical terms.

Let the potential φ take a small excursion to $\varphi + \kappa\eta(x,y,z)$, where κ is a constant and $\eta(x,y,z)$ is a small arbitrary function. Then the change in the functional is

$$\begin{aligned}\mathcal{L}(\varphi+\kappa\eta) &= \mathcal{L} + \delta\mathcal{L}\\ &= \iiint_v \left[\frac{1}{2}\epsilon(\nabla\varphi + \kappa\nabla\eta)^2 - \rho(\varphi+\kappa\eta)\right]\mathrm{d}v\\ &= \iiint_v \left[\frac{1}{2}\epsilon(\nabla\varphi\cdot\nabla\varphi + 2\kappa\nabla\varphi\cdot\nabla\eta + \kappa^2\nabla\eta\cdot\nabla\eta) - \rho(\varphi+\kappa\eta)\right]\mathrm{d}v\end{aligned} \tag{5.30}$$

Subtracting

$$\delta\mathcal{L} = \iiint_v [\epsilon(\kappa\nabla\varphi\cdot\nabla\eta) - \rho\kappa\eta]\mathrm{d}v + O^2(\kappa\eta) \tag{5.31}$$

where for small changes, the terms of order 2 (designated O^2) in $\kappa\eta$ may be dropped for being negligibly small but are retained to make a point we will encounter soon. To proceed, we need to use a vector identity expressing the divergence of a vector $\mathbf{V}$

scaled by a scalar s:

$$\nabla \cdot (s\mathbf{V}) = \nabla s \cdot \mathbf{V} + s\nabla \cdot \mathbf{V} \tag{5.32}$$

together with the divergence theorem:

$$\iiint_{v} \nabla \cdot \mathbf{V}\,\mathrm{d}R = \iint_{S} \mathbf{V} \cdot \mathbf{ds} \tag{5.33}$$

where the surface S bounds the domain R. Integrating the vector identity just encountered and applying the divergence theorem to it

$$\iiint_{v} \nabla \cdot s\mathbf{V}\,\mathrm{d}v = \iiint_{v} (\nabla s \cdot \mathbf{V} + s\nabla \cdot \mathbf{V})\mathrm{d}v$$

or

$$\iint_{S} s\mathbf{V} \cdot \mathbf{ds} = \iiint_{v} (\nabla s \cdot \mathbf{V} + s\nabla \cdot \mathbf{V})\mathrm{d}v \tag{5.34}$$

Now set $s = \eta$ and $\mathbf{V} = \nabla\varphi$

$$\iint_{S} \eta\nabla\varphi \cdot \mathbf{ds} = \iiint_{v} (\nabla\eta \cdot \nabla\varphi + \eta\nabla^2\varphi)\mathrm{d}v \tag{5.35}$$

We note further that the elemental surface vector points in the normal direction so that $\mathbf{ds} = \mathrm{d}s\mathbf{u}_n$ where $\mathbf{u}_n$ is a unit vector in the normal direction. Further, since $\nabla\varphi = \mathbf{u}_x\frac{\partial\varphi}{\partial x} + \mathbf{u}_y\frac{\partial\varphi}{\partial y} + \mathbf{u}_z\frac{\partial\varphi}{\partial z}$ and therefore $\mathbf{u}_x \cdot \nabla\varphi = \frac{\partial\varphi}{\partial x}$, and since the direction x is arbitrary according to how we orient the axes, we may write $\mathbf{u}_n \cdot \nabla\varphi = \frac{\partial\varphi}{\partial n}$. So the above equation upon rearranging becomes

$$\iiint_{v} (\nabla\eta \cdot \nabla\varphi)\mathrm{d}v = \iint_{S} \eta\frac{\partial\varphi}{\partial n}\mathrm{d}s - \iiint_{v} (\eta\nabla^2\varphi)\mathrm{d}v \tag{5.36}$$

Using this in the expression for the change in the Lagrangian functional

$$\delta\mathcal{L} = \left\{\iint_{S} \epsilon\kappa\eta\frac{\partial\varphi}{\partial n}\mathrm{d}s - \iiint_{v} EK(\eta\nabla^2\varphi)\mathrm{d}v\right\}\iiint_{v} [-\rho\kappa\eta]\mathrm{d}v + O^2(\kappa\eta) \tag{5.37}$$

Now the boundary S has either Dirichlet or Neumann conditions on each point of it. Where the Dirichlet condition applies $\eta = 0$ since the potential is specified and cannot vary. On those parts of S where the Neumann condition is specified,

$\frac{\partial\varphi}{\partial n} = 0$. Therefore, the surface integral vanishes because the integrand is zero at every point on S, thereby leading to

$$\delta\mathcal{L} = \left\{ \iiint_v \kappa\eta(-\varepsilon\nabla^2\varphi - \rho)\mathrm{d}v \right\} + O^2(\kappa\eta) \tag{5.38}$$

Now we are in a position to say

a) If the functional is at an extremum, whether a minimum or a maximum, the left-hand side is zero since changes are flat at the point of extremum. Further the second-order terms $O^2(\kappa\eta)$ may be neglected so that

$$\iiint_v \kappa\eta(-\epsilon\nabla^2\varphi - \rho)\mathrm{d}v = 0 \tag{5.39}$$

But since the change in potential $\kappa\eta$ is arbitrary, this is possible only if the other factor of the integrand

$$-\epsilon\nabla^2\varphi - \rho = 0$$

or

$$-\epsilon\nabla^2\varphi = \rho \tag{5.40}$$

meaning that the Poisson equation is satisfied at the extremum.

b) We need one more consideration to show that the extremum is a minimum. At that extremum since the residual of the Poisson equation $-\epsilon\nabla^2\varphi - \rho = 0$, we must have

$$\delta\mathcal{L} = O^2(\kappa\eta) \tag{5.41}$$

which is a strictly positive quantity so that any changes in φ about the extremum of $\mathcal{L}$ will make $\mathcal{L}$ rise so that that extremum has to be a minimum.

5.5 Uniqueness

For completeness, we will show that the Poisson equation has a unique solution when the potential φ or its normal gradient is defined at every point on the boundary s and φ at least at one point. Practically this is a requirement for the well-posedness of a problem. If a problem is not well-posed, the solution will not be unique and the matrix equation we form will not have a solution. To this end of proving the conditions for uniqueness, consider two solutions to the Poisson equation φ_1 and φ_2 satisfying the boundary conditions:

$$-\epsilon\nabla^2\varphi_1 = \rho \text{ in } v$$
$$-\epsilon\nabla^2\varphi_2 = \rho \text{ in } v$$

with

$$\varphi_1 = f \text{ on } s_1 \quad \text{and} \quad \frac{\partial \varphi_1}{\partial n} = g \text{ on } s_2$$

$$\varphi_2 = f \text{ on } s_1 \quad \text{and} \quad \frac{\partial \varphi_2}{\partial n} = g \text{ on } s_2$$

where s_1 is the Dirichlet part of s, s_2 is the Neumann part and f and g are the specified values. Let us now examine $\psi = \varphi_1 - \varphi_2$ to see under what circumstances it will be zero, the condition for uniqueness – that is, there being no difference between two solutions. Subtracting the equations in v and s

$$-\epsilon\nabla^2\psi = 0 \text{ in } v$$

or simply

$$\nabla^2\psi = 0 \text{ in } v \tag{5.42}$$

with

$$\psi = 0 \text{ on } s_1 \quad \text{and} \quad \frac{\partial \psi}{\partial n} = 0 \text{ on } s_2$$

In the identity

$$\nabla \cdot (s\mathbf{V}) = \nabla s \cdot \mathbf{V} + s\nabla \cdot \mathbf{V}$$

Simple if we let $s = \psi$ and $\mathbf{V} = \nabla\psi$

$$\nabla \cdot (\psi\nabla\psi) = \nabla\psi \cdot \nabla\psi + \psi\nabla^2\psi$$

Noting that $\nabla^2\psi$ is zero, integrating over v

$$\iiint_v \nabla\psi \cdot \nabla\psi \, \mathrm{d}v = \iiint_v \nabla \cdot (\psi\nabla\psi)\mathrm{d}v \tag{5.43}$$

And applying the divergence theorem noting that the dot product of the vector $\nabla\psi$ with itself is the square of its magnitude:

$$\iiint_v (\nabla\psi)^2 \, \mathrm{d}v = \iint_S \psi\nabla\psi \cdot \mathbf{ds} = \iint_S \psi\frac{\partial \psi}{\partial n} \, \mathrm{d}s = 0 \tag{5.44}$$

since the surface integral is zero because ψ or $\frac{\partial \psi}{\partial n}$ is zero on every point on S. And now since $(\nabla\psi)^2$ is either positive or zero, for its integral to be zero we must have

$$\nabla\psi = 0 \text{ everywhere in } v \tag{5.45}$$

which means that ψ is a constant. So we make the additional stipulation that φ ought to be specified at least at one point on S, giving us a reference potential. At that point $\psi = 0$ so that constant is zero giving us the requisite condition for uniqueness:

> The potential or its normal gradient should be specified at every point on the boundary S with the potential specified at least at one point on S.

5.6 Natural boundary conditions

It may be shown that the variational formulation of the finite element method has the zero Neumann boundary condition implicit to it – that is, on boundaries where no condition is imposed, it will naturally turn out to be imposed in the solution.

Consider the problem shown in Fig. 5.5. Say that the problem is prepared for a finite element solution with free potentials $\phi_1, \ldots, \phi_n$. The functional $\mathcal{L} = \mathcal{L}(\phi_1, \ldots, \phi_n)$. The left-hand side end at $x = x_1$ has a Dirichlet condition whereas the right-hand side end at $x = x_2$ is free. As of now, we do not know if this problem has a unique solution if tackled by the finite element method.

Let us now double the problem by symmetrically reflecting it about the right end x_2. All nodes in the left half too are reflected and those in the right half have the same variable values as the corresponding nodes in the left half. That is, the unknowns are still $\phi_1, \ldots, \phi_n$ as before. The variables ε and ρ too are symmetrically reflected in the extended half of the problem. The new right end of the doubled problem will be at coordinate location $x_2 + (x_2 - x_1) = 2x_2 - x_1$ and has the same value of ϕ, ϕ_1, imposed as at the right end. Now this problem satisfies the condition that the ends ought to have either the Neumann or Dirichlet condition imposed so it has a unique solution. But the new functional by the way the doubled problem has been formulated is twice $\mathcal{L}(\phi_1, \ldots, \phi_n)$. Therefore the original problem also has not only a unique solution but also, in the left half, the same solution as the doubled problem. By symmetry, at the mid-point, $\frac{\mathrm{d}\varphi}{\mathrm{d}x} = 0$. Therefore, we may assert that when we ignore boundary conditions, the variational formulation automatically imposes a Neumann condition and yields unique solutions.

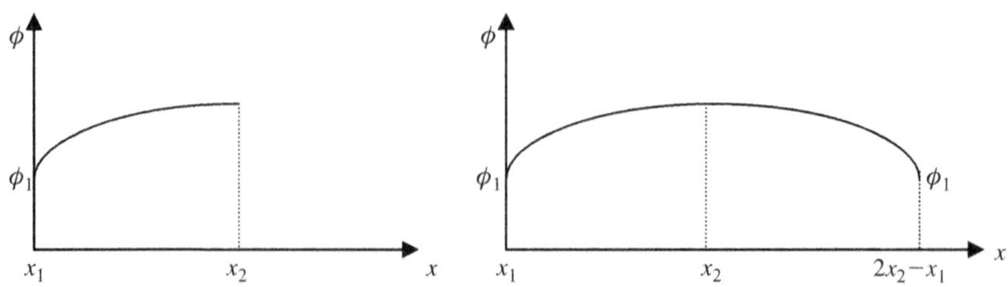

Figure 5.5: A finite element problem with the right-end free and a symmetric doubled problem.

This argument may be repeated in two dimensions with some limitations because multiple portions of the Neumann boundary or curved portions cannot be reflected as done here in one dimension. Getting into that, however, is beyond the scope of this work.

Often we have the freedom to impose the Neumann condition rather than ignoring it. This is called a strongly imposed condition as opposed to what we call a weakly imposed condition when it is allowed to occur naturally by ignoring the condition. In a one-dimensional situation, say imposing it strongly on the right boundary would mean making the potential values in the last two nodes equal while being free. This would ensure that the φ curve would be flat. But then it would be flat throughout the last element on the right which means that the equal potential values in the last two nodes cannot bend so as to accommodate the differential equation being solved. This then is the classic poser: Is it more important to satisfy the equation set (consisting of the differential equation and boundary conditions) at the end point by strong imposition or overall through weak imposition? Clearly the latter is the better and indeed more implementationally convenient approach.

5.7 One-dimensional linear finite elements

Let us approach the implementation of the finite element method through a simple one-dimensional problem in demonstration of the principles. In the domain [0, 10] let the electric potential be governed by the one-dimensional Poisson equation

$$-\varepsilon \frac{\mathrm{d}^2\varphi}{\mathrm{d}x^2} = \rho$$

with $(\varepsilon, \rho) = (2,\ 0)$ in the interval (0, 5) and (1,1) in the interval (5, 10). In addition φ is 30 V at the two end points $x = 0$ and $x = 10$ (i.e., Dirichlet conditions at both boundaries). Solve for φ.

The element in one dimension x is a straight line (Fig. 5.6). A linear function in x, $\varphi = a + bx$, has two constants a and b which need to be replaced by two nodes.

Applying the interpolation to the two nodes and solving

$$a + bx_1 = \varphi_1$$

$$a + bx_2 = \varphi_2$$

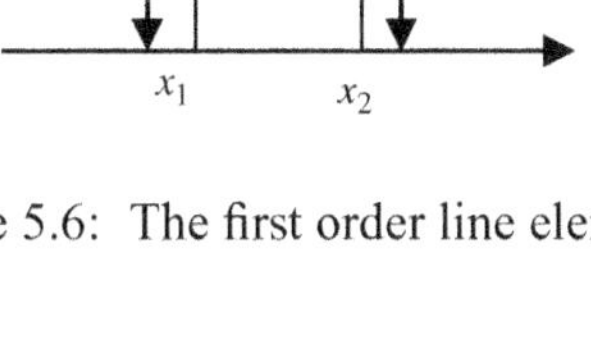

Figure 5.6: The first order line element.

or

$$\begin{bmatrix} 1 & x_1 \\ 1 & x_2 \end{bmatrix} \begin{Bmatrix} a \\ b \end{Bmatrix} = \begin{Bmatrix} \varphi_1 \\ \varphi_2 \end{Bmatrix} \tag{5.46}$$

yielding

$$\begin{Bmatrix} a \\ b \end{Bmatrix} = \frac{1}{x_2 - x_1} \begin{bmatrix} x_2 & -x_1 \\ -1 & 1 \end{bmatrix} \begin{Bmatrix} \varphi_1 \\ \varphi_2 \end{Bmatrix} \tag{5.47}$$

Therefore, noting that $x_2 - x_1 = L$, the length of the element. Thus

$$\begin{aligned} \varphi = a + bx &= \frac{1}{L}[x_2\varphi_1 - x_1\varphi_2 + x(\varphi_2 - \varphi_1)] = \varphi_1 \frac{x - x_1}{L} + \varphi_2 \frac{x_2 - x}{L} \\ &= \alpha_1\varphi_1 + \alpha_2\varphi_2 \end{aligned} \tag{5.48}$$

where the interpolation functions

$$\alpha_1 = \frac{x - x_1}{L} \quad \text{and} \quad \alpha_2 = \frac{x_2 - x}{L}$$

We can also see that

$$\frac{\mathrm{d}\varphi}{\mathrm{d}x} = \frac{\varphi_2 - \varphi_1}{L} \tag{5.49}$$

The functional in one dimension is now

$$\begin{aligned} \mathcal{L} = \int_R \left\{ \frac{1}{2}\varepsilon \left[\frac{\mathrm{d}\varphi}{\mathrm{d}x} \right]^2 - \rho\varphi \right\} \mathrm{d}x &= \sum_{\text{All elements}} \int_{\text{Element}} \left\{ \frac{1}{2}\varepsilon \left[\frac{\mathrm{d}\varphi}{\mathrm{d}x} \right]^2 - \rho\varphi \right\} \mathrm{d}x \\ &= \sum_{\text{All elements}} \int_{x_1}^{x_2} \left\{ \frac{1}{2}\varepsilon \left[\frac{\varphi_2 - \varphi_1}{L} \right]^2 - \rho\varphi \right\} \mathrm{d}x \\ &= \sum_{\text{All elements}} \left\{ \frac{1}{2}\varepsilon \left[\frac{\varphi_2 - \varphi_1}{L} \right]^2 L - \rho_0 \frac{\varphi_1 + \varphi_2}{2} L \right\} \end{aligned} \tag{5.50}$$

where we have used the fact that the first integral term is constant so that taking it out, the integral is simply x which becomes L when limits are applied and for the second integral we have assumed that ρ is a constant so that the integral of φ,

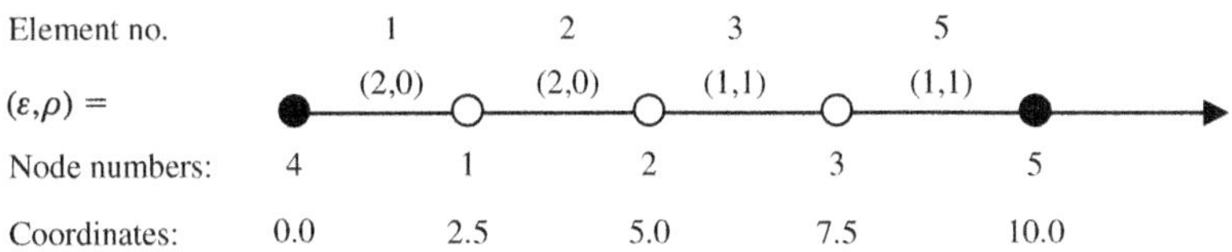

Figure 5.7: Data for sample problem and its mesh.

varying as a straight line (Fig. 5.6), is the area under the curve, which is the area of a trapezoid. Expressing this polynomial expression for energy in matrix terms

$$\mathcal{L} = \sum_{\text{All elements}} \left\{ \frac{1}{2}\underline{\varphi}^{t} \begin{bmatrix} \frac{\varepsilon}{L} & -\frac{\varepsilon}{L} \\ -\frac{\varepsilon}{L} & \frac{\varepsilon}{L} \end{bmatrix} \underline{\varphi} - \underline{\varphi}^{t} \left\{ \begin{matrix} \frac{\rho_0 L}{2} \\ \frac{\rho_0 L}{2} \end{matrix} \right\} \right\} = \sum_{\text{All elements}} \left\{ \frac{1}{2}\underline{\varphi}^{t}[P]\underline{\varphi} - \underline{\varphi}^{t}\underline{Q} \right\} \tag{5.51}$$

where what we call the Dirichlet matrix

$$[P] = \begin{bmatrix} \frac{\varepsilon}{L} & -\frac{\varepsilon}{L} \\ -\frac{\varepsilon}{L} & \frac{\varepsilon}{L} \end{bmatrix}$$

and

$$\underline{Q} = \left\{ \begin{matrix} \frac{\rho_0 L}{2} \\ \frac{\rho_0 L}{2} \end{matrix} \right\}$$

Let us divide the domain into four elements of equal length as shown in Fig. 5.7. The two known nodes are shown in black and the three unknowns in white. Note well that the two known nodes (i.e., whose potential values are known) have been numbered last. The reason for this will become apparent as we go through the example.

There being five potentials $\varphi_1, \varphi_2, \ldots, \varphi_5$, the functional will be a polynomial in these, formed by a 5×5 matrix. Let us say that the final form of the functional is

$$\mathcal{L} = \frac{1}{2}\varphi_{g}^{t}[P_{g}]\varphi_{g} - \varphi_{g}^{t}\{Q_{g}\} \tag{5.52}$$

These matrices $[P_g]$ and $\{Q_g\}$ are referred to as global matrices because they contain the functional with the contributions of the local matrices from the elements.

The so-called local matrices $[P]$ and $\{Q\}$ for element no. 1 of length = 2.5 are

$$[P] = \begin{bmatrix} \frac{2}{2.5} & -\frac{2}{2.5} \\ \frac{2}{2.5} & -\frac{2}{2.5} \end{bmatrix} = \begin{bmatrix} 0.8 & -0.8 \\ -0.8 & 0.8 \end{bmatrix} \quad \text{and} \quad \{Q\} = \begin{Bmatrix} 0 \\ 0 \end{Bmatrix}$$

because ρ is zero. The same goes for the second element. But for the third and fourth elements,

$$[P] = \begin{bmatrix} \frac{1}{2.5} & -\frac{1}{2.5} \\ -\frac{1}{2.5} & \frac{1}{2.5} \end{bmatrix} = \begin{bmatrix} 0.4 & -0.4 \\ -0.4 & 0.4 \end{bmatrix} \quad \text{and} \quad \{Q\} = \begin{Bmatrix} \frac{1*2.5}{2} \\ \frac{1*2.5}{2} \end{Bmatrix} = \begin{Bmatrix} 1.25 \\ 1.25 \end{Bmatrix}$$

Fig. 5.8 shows the global matrices being built up from the four local matrices and four local vectors.

Therefore,

$$\mathcal{L} = \frac{1}{2}\varphi_g^t \begin{bmatrix} 1.6 & -0.8 & 0 & -0.8 & 0 \\ -0.8 & 1.2 & -0.4 & 0 & 0 \\ 0 & -0.4 & 0.8 & 0 & -0.4 \\ -0.8 & 0 & 0 & 0.8 & 0 \\ 0 & 0 & -0.4 & 0 & 0.4 \end{bmatrix} \varphi_g - \varphi_g^t \begin{Bmatrix} 0 \\ 1.25 \\ 2.5 \\ 0 \\ 1.25 \end{Bmatrix} \tag{5.53}$$

The three potentials φ_1, φ_2 and φ_3 are free to vary so as to make this as small as possible, but the two potentials φ_4 and φ_5 are fixed and cannot vary. Therefore of the derivative vector $\frac{\partial \mathcal{L}}{\partial \emptyset_g}$ only the first three rows may be set to zero:

$$\begin{bmatrix} 1.6 & -0.8 & 0.0 & -0.8 & 0.0 \\ -0.8 & 1.2 & -0.4 & 0.0 & 0.0 \\ 0.0 & -0.4 & 0.8 & 0.0 & -0.4 \end{bmatrix} \begin{Bmatrix} \varphi_1 \\ \varphi_2 \\ \varphi_3 \\ \varphi_4 \\ \varphi_5 \end{Bmatrix} - \begin{Bmatrix} 0 \\ 1.25 \\ 2.5 \end{Bmatrix} = \begin{Bmatrix} 0 \\ 0 \\ 0 \end{Bmatrix} \tag{5.54}$$

Now we see why the unknown nodes were numbered first. This ensured that we could simply drop the last few rows of *PG* corresponding to known nodes and shift the last columns of *PG* corresponding to known nodes to the right. Otherwise, we would have had rows in the middle of the matrix needing to be eliminated. Of course in modern programs where we progressively refine the mesh as a solution is found to be too crude to be acceptable, nodes are numbered arbitrarily but then mapped on to variables that are structured unknowns first and knowns last.

Elem. 1	PLocal		QLocal
Vertices	4	1	
4	0.8	−0.8	0
1	−0.8	0.8	0

PGlobal	1	2	3	4	5	QGlobal
1	0.8			−0.8		0
2						
3						
4	−0.8			0.8		0
5						

Elem. 2	PLocal		QLocal
Vertices	1	2	
1	0.8	−0.8	0
2	−0.8	0.8	0

PGlobal	1	2	3	4	5	QGlobal
1	0.8 + 0.8 = 1.6	**−0.8**		**−0.8**		0 + 0 = 0
2	−0.8	0.8				
3						
4	−0.8			0.8		0
5						

Elem. 3	PLocal		QLocal
Vertices	2	3	
2	0.4	−0.4	1.25
3	−0.4	0.4	1.25

PGlobal	1	2	3	4	5	QGlobal
1	1.6	−0.8		**−0.8**		0
2	**−0.8**	0.8 + 0.4	**−0.4**			0 + 1.25
3		−0.4	0.4			+1.25
4	**−0.8**			0.8		0
5						

Elem. 4	PLocal		QLocal
Vertices	3	5	
3	0.4	−0.4	1.25
5	−0.4	0.4	1.25

PGlobal	1	2	3	4	5	QGlobal
1	1.6	**−0.8**		**−0.8**		0
2	**−0.8**	1.2	−0.4			1.25
3		**−0.4**	0.4 + 0.4		**−0.4**	1.25 + 1.25
4	**−0.8**			0.8		0
5			−0.4		0.4	+1.25

Figure 5.8: Contributions of the local matrices **P** and **Q** as they add to the Global Matrices **PG**, **QG**.

Returning to our problem, recognizing that the last two values, φ_4 and φ_5, are each 30 V, these may be shifted to the right-hand side of the three equations:

$$\begin{bmatrix} 1.6 & -0.8 & 0.0 \\ -0.8 & 1.2 & -0.4 \\ 0.0 & -0.4 & 0.8 \end{bmatrix} \begin{Bmatrix} \varphi_1 \\ \varphi_2 \\ \varphi_3 \end{Bmatrix} = \begin{Bmatrix} 0 + 0.8 \times 30 \\ 1.25 \\ 2.5 + 0.4 \times 30 \end{Bmatrix} = \begin{Bmatrix} 24 \\ 1.25 \\ 14.5 \end{Bmatrix} \tag{5.55}$$

Solving, we have $\varphi_1 = 32.0433, \varphi_2 = 34.1667, \varphi_3 = 35.2083$.

We have seen how the polynomials are added to the global matrices using the corresponding node numbers. But hand calculations are prone to error and tedious. Now, therefore, let us look at a problem to see how this process is implemented on the computer.

Defining the demonstrative example, the domain is [0, 10]. The left and right ends are at potential 10. ε and ρ are 1 and 2. This is a test problem for learning purposes and has an analytical solution so it is a problem for which we would normally not need a computer:

$$-\frac{d^2\varphi}{dx^2} = 2$$

Integrating twice

$$\frac{d\varphi}{dx} = -2x + c$$

$$\varphi = -x^2 + cx + d$$

Applying the boundary conditions, $d = 10$ and $c = 10$. Therefore, the exact solution is

$$\varphi = -x^2 + 10x + 10 \tag{5.56}$$

To see if we get the same solution by the finite element method, the domain length is divided into 100 elements so we have 101 nodes with all elements of equal length = 10/100 = 0.1 m. Since the known nodes are numbered last, the numbering from left to right is $101, 1, 2, \ldots, 100$. So element 1 has nodes 101-1, element 2 is 1–2; element 3 is 2–3, etc. to element 100 which connects nodes 99–100. All elements have the same local matrices:

$$P = \frac{\varepsilon}{L}\begin{bmatrix} 1 & -1 \\ -1 & 1 \end{bmatrix} = \begin{bmatrix} 10 & -10 \\ -10 & 10 \end{bmatrix} \quad Q = \begin{Bmatrix} \dfrac{\rho L}{2} \\ \dfrac{\rho L}{2} \end{Bmatrix} = \begin{Bmatrix} 0.1 \\ 0.1 \end{Bmatrix} \tag{5.57}$$

The MATLAB code is shown in box as Code 5.1. Note that in placing the local matrices P and Q in the global matrices PG and QG, if the node number is of one of the knowns, the row of the global matrix is not formed. If the row is unknown, the corresponding Q goes into QG; and if the column of P is also of an unknown, then the corresponding P value goes into PG. But if the row of P corresponds to an unknown while the column of P to a known, then that P value times the known potential value is moved to the right-hand side of the equation (i.e., the multiple is taken out of QG). Upon the formation of PG and QG, the solution is obtained through inversion here (although in more sophisticated programs a more efficient matrix solution scheme will be used.

The analytical and numerical solutions are shown in Fig. 5.9 and they are seen to match exactly as best as the eye can see. It is easy to see how the method is very powerful when different values of ε and ρ come into different segments of the problem where an analytical solution would be very difficult.

```
clc; clear all; %Clear memory and screen
n=100;
%n= No. of elements. No. of nodes=n-2
m=n-1; %no. of unknowns
PG=zeros(m,m); %Initializing
QG=zeros(m,1);
L=10/n; %Domain Length/No. of Elem
y=zeros(n,1);
%Solution in y. Setting last 2 known values
y(n+1)=10;
y(n)=10;
P=[1,-1;-1,1]/L;
%Same Local P,Q for all elements
Q=[1;1]*L;

for e=1:n %e is element number
    if e==1
       v(1)=n+1;
       v(2)=1;
    else
       v(1)=e-1;
       v(2)=e;
    end;%if
    for r=1:2
        if v(r)<=m %Otherwise do nothing
            QG(v(r))=QG(v(r))+Q(r);
            for c=1:2
                if v(c)<=m %P(r,c) goes into PG
                   PG(v(r),v(c))= PG(v(r),v(c))+P(r,c);
                else %P(r,c)needs to go into QG
                   QG(v(r))=QG(v(r))-P(r,c)*y(v(c));
                end;
            end;
        end;
    end;
end;
xx=inv(PG)*QG;
for i=1:n+1
    xaxis(i,1)=(i-1)*L;
    yexact(i,1)=10+10*xaxis(i,1)-(xaxis(i,1))^2;
    if i==1
       yaxis(i,1)=10;
    else
        if i==n+1
           yaxis(i,1)=10;
        else
           yaxis(i,1)=xx(i-1,1);
        end;
    end;
end;
%plot (xaxis, yexact);
plot (xaxis, yaxis,'r', xaxis, yexact, 'b');
```

Code 5.1: One-D FEM with Dirichlet conditions.

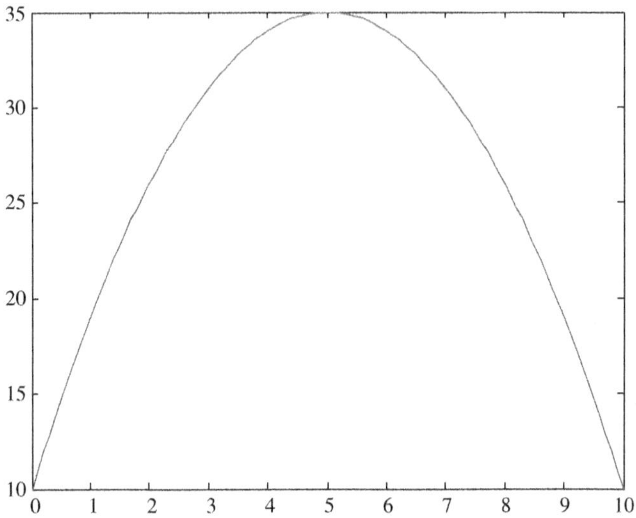

Figure 5.9: Explicit and numerical solutions to one-dimensional problem.

To test and see for ourselves how naturally the Neumann boundary condition works when ignored, let us revisit the one-dimensional finite element problem with the right-end boundary condition turned into a Neumann condition

$$-\frac{d^2\varphi}{dx^2} = 2$$

Integrating twice

$$\frac{d\varphi}{dx} = -2x + c$$

$$\varphi = -x^2 + cx + d$$

Applying the boundary conditions on the right, $c = 20$, and now on the left, $d = 10$, so that the exact solution is

$$\varphi = -x^2 + 20x + 10 \tag{5.58}$$

Now, however, we have 100 unknowns, the right-end node also corresponding to an unknown potential. The MATLAB code is given in Code 5.2. Fig. 5.10 gives the explicit and numerical solutions overlapping as best as the eye can see with the right end of the curve flat, that is the zero Neumann boundary condition turning out naturally.

```
clc; clear all;
n=100;
m=n;%no. of unknowns
PG=zeros(m,m);
QG=zeros(m,1);
L=10/n;
y=zeros(n,1);
y(n+1)=10;
P=[1,-1;-1,1]/L;
Q=[1;1]*L;
for e=1:n
    if e==1
       v(1)=n+1;
       v(2)=1;
    else
       v(1)=e-1;
       v(2)=e;
    end;%if
    for r=1:2 %
        if v(r)<=m
           QG(v(r))=QG(v(r))+Q(r);
           for c=1:2
               if v(c)<=m
                  PG(v(r),v(c))= PG(v(r),v(c))+P(r,c);
               else
                  QG(v(r))=QG(v(r))-P(r,c)*y(v(c));
               end;
            end;
        end;
    end;
end;
xx=inv(PG)*QG;
for i=1:n+1
    xaxis(i,1)=(i-1)*L;
    yexact(i,1)=10+20*xaxis(i,1)-(xaxis(i,1))^2;
    if i==1
        yaxis(i,1)=10;
    else
        yaxis(i,1)=xx(i-1,1);
    end;
end;
%plot (xaxis, yexact);
plot (xaxis, yaxis,'r', xaxis, yexact, 'b');
```

Case 5.2: One-D FEM with Dirichlet and Neumann conditions.

5.8 Two-dimensional linear triangular finite elements

Just as we did this in one dimension, we will broach this too through an example. Figure 5.11 shows a two-dimensional system governed by the Poisson equation $-\sigma\nabla^2 T = q$ with zero Neumann conditions along the upper and lower boundaries and Dirichlet conditions on the left and right boundaries; and $\sigma = 4$ and $q = 1$

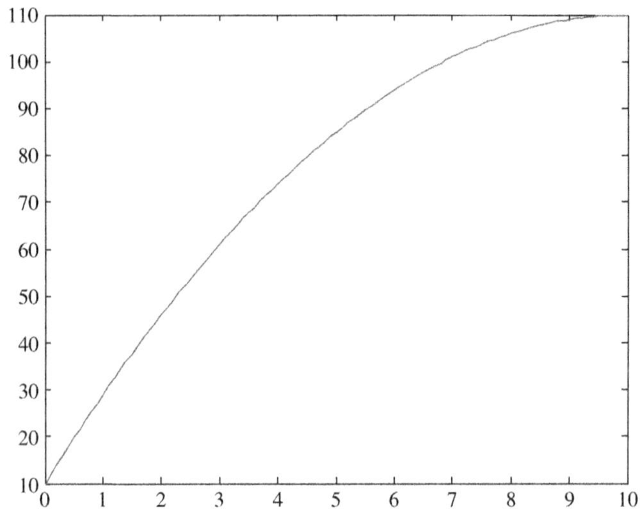

Figure 5.10: Explicit and numerical solutions to one-dimensional problem with natural boundary condition.

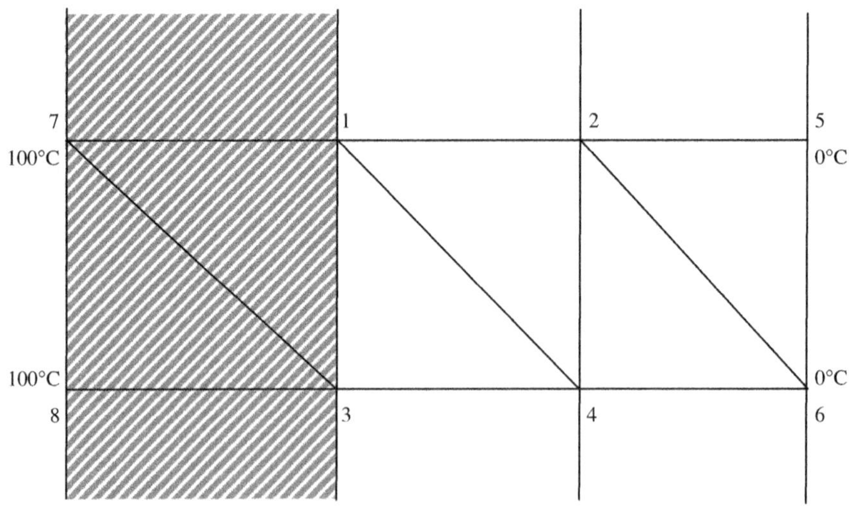

Figure 5.11: Heat flow problem.

in the shaded region and $\sigma = 1$ and $q = 0$ elsewhere. Although this is a problem governing the flow of heat, the equations are similar to those developed for an electrostatic system with $\sigma = \varepsilon$ and $q = \rho$. Solve for T everywhere using hand calculations.

We have by our linear approximations or trial functions

$$T = a + bx + cy$$

We have already seen that the first-order approximation $\varphi = a + bx + cy$ on a triangle leads to

$$b = \sum_{i=1}^{3} b_i \varphi_i; \quad c = \sum_{i=1}^{3} c_i \varphi_i$$

Now

$$\nabla T = \mathbf{u}_x \frac{\partial T}{\partial x} + \mathbf{u}_y \frac{\partial T}{\partial y} = \mathbf{u}_x b + \mathbf{u}_y c \tag{5.59}$$

The functional

$$\begin{aligned}\mathcal{L}(T) &= \iint_v \left(\left[\frac{1}{2}\sigma(\nabla T)^2 - qT \right] \right) \mathrm{d}v \\ &= \sum_{\text{Elements}} \iint_{\text{Triangle}} \left[\frac{1}{2}\sigma(b^2 + c^2) - qT \right] \mathrm{d}v \\ &= \sum_{\text{Elements}} \left\{ \frac{1}{2}\sigma(b^2 + c^2)A - q_0 \frac{T_1 + T_2 + T_3}{3} A \right\}\end{aligned} \tag{5.60}$$

where, as before in one dimension, we have used the fact that $\sigma(b^2 + c^2)$ is a constant to pull it out of the integral sign so that the integral of dR is simply the triangle area A. Likewise, by assuming that q is a constant at q_0 in a triangle, that too comes out of the integral sign so that the integral of the linearly varying T is the average of T at the three vertices times a measure of the integration area which is A; this is a variant of Simpson's rule in one dimension for two dimensions. Thus we have, substituting for b and c, while noting that

$$b = \sum_{i=1}^{3} b_i T_i = \mathbf{T}^{\mathrm{t}}\mathbf{b}; \quad c = \sum_{i=1}^{3} c_i T_i = \mathbf{T}^{\mathrm{t}}\mathbf{c} \tag{5.61}$$

With notation for T, b and c defined by correspondence for terms, and noting that a scalar b is its own transpose, we have

$$b^2 = b \cdot b = b \cdot b^{\mathrm{t}} = (\mathbf{T}^{\mathrm{t}}\mathbf{b})(\mathbf{b}^{\mathrm{t}}\mathbf{T}) = \mathbf{T}^{\mathrm{t}}\mathbf{b}\mathbf{b}^{\mathrm{t}}\mathbf{T} = \mathbf{T}^{\mathrm{t}} \begin{bmatrix} b_1^2 & b_1 b_2 & b_1 b_3 \\ b_2 b_1 & b_2^2 & b_2 b_3 \\ b_3 b_1 & b_3 b_2 & b_3^2 \end{bmatrix} \mathbf{T} \tag{5.62}$$

where the brackets have been dispensed with in the absence of there being multiplicative compatibility. Similarly

$$c^2 = c \cdot c = c \cdot cb^{\mathrm{t}} = (\mathbf{T}^{\mathrm{t}}\mathbf{c})(\mathbf{cb}^{\mathrm{t}}\mathbf{T}) = \mathbf{T}^{\mathrm{t}}\mathbf{cc}^{\mathrm{t}}\mathbf{T} = \mathbf{T}^{\mathrm{t}}\begin{bmatrix} c_1^2 & c_1c_2 & c_1c_3 \\ c_2c_1 & c_2^2 & c_2c_3 \\ c_3c_1 & c_3c_2 & c_3^2 \end{bmatrix}\mathbf{T} \quad (5.63)$$

Therefore

$$\begin{aligned} \mathcal{L}(T) &= \sum_{\text{Elements}} \left\{ \frac{1}{2}\sigma(b^2+c^2)A - q_0\frac{T_1+T_2+T_3}{3}A \right\} \\ &= \sum_{\text{Elements}} \left\{ \frac{1}{2}\mathbf{T}^{\mathrm{t}}\sigma A \begin{bmatrix} b_1^2+c_1^2 & b_1b_2+c_1c_2 & b_1b_3+c_1c_3 \\ b_2b_1+c_2c_1 & b_2^2+c_2^2 & b_2b_3+c_2c_3 \\ b_3b_1+c_3c_1 & b_3b_2+c_3c_2 & b_3^2+c_3^2 \end{bmatrix}\mathbf{T} - \mathbf{T}^{\mathrm{t}}\begin{Bmatrix} \frac{q_0A}{3} \\ \frac{q_0A}{3} \\ \frac{q_0A}{3} \end{Bmatrix} \right\} \\ &= \sum_{\text{All elements}} \left\{ \frac{1}{2}\mathbf{T}^{\mathrm{t}}[P]\mathbf{T} - \mathbf{T}^{\mathrm{t}}\mathbf{Q} \right\} \end{aligned} \quad (5.64)$$

where, in corresponding notation, the local matrices

$$[P] = \sigma A \begin{bmatrix} b_1^2+c_1^2 & b_1b_2+c_1c_2 & b_1b_3+c_1c_3 \\ b_2b_1+c_2c_1 & b_2^2+c_2^2 & b_2b_3+c_2c_3 \\ b_3b_1+c_3c_1 & b_3b_2+c_3c_2 & b_3^2+c_3^2 \end{bmatrix} \quad (5.65)$$

and

$$\mathbf{Q} = \begin{Bmatrix} \frac{q_0A}{3} \\ \frac{q_0A}{3} \\ \frac{q_0A}{3} \end{Bmatrix} \quad (5.66)$$

From here onwards, the procedure is exactly as in the one-dimensional example we worked on before. Consider the first triangle of the mesh, 8–3–7. Where we start numbering the nodes, whether 8–3–7 or 3–7–8 or 7–8–3 does not matter but that the order ought should be counter clockwise is important (Fig. 5.12).

For this triangle

$$\Delta = \begin{vmatrix} 1 & x_1 & y_1 \\ 1 & x_2 & y_2 \\ 1 & x_3 & y_3 \end{vmatrix} = \begin{vmatrix} 1 & 0 & 0 \\ 1 & 1 & 0 \\ 1 & 0 & 1 \end{vmatrix} = 1 \quad (5.67)$$

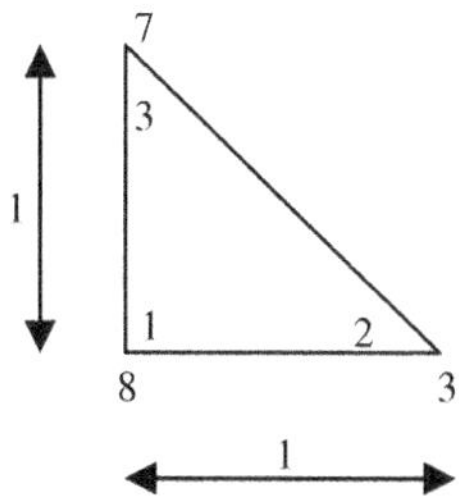

Figure 5.12: Triangle with internal generic numbers 1, 2, 3 and corresponding actual numbers 8, 3, 7.

Therefore area $A = \frac{1}{2}$. We could have said area $= \frac{1}{2} \times$ base $\times$ height $= \frac{1}{2}$ but this is not general enough an approach for triangles whose sides are not parallel to the Cartesian axes. We have

$$\mathbf{b} = \begin{Bmatrix} \dfrac{y_2 - y_3}{\Delta} \\ \dfrac{y_3 - y_1}{\Delta} \\ \dfrac{y_1 - y_2}{\Delta} \end{Bmatrix} = \begin{Bmatrix} \dfrac{0-1}{1} \\ \dfrac{1-0}{1} \\ \dfrac{0-0}{1} \end{Bmatrix} = \begin{Bmatrix} -1 \\ 1 \\ 0 \end{Bmatrix} \tag{5.68}$$

$$\mathbf{c} = \begin{Bmatrix} \dfrac{x_3 - x_2}{\Delta} \\ \dfrac{x_1 - x_3}{\Delta} \\ \dfrac{x_2 - x_1}{\Delta} \end{Bmatrix} = \begin{Bmatrix} \dfrac{0-1}{1} \\ \dfrac{0-0}{1} \\ \dfrac{1-0}{1} \end{Bmatrix} = \begin{Bmatrix} -1 \\ 0 \\ 1 \end{Bmatrix} \tag{5.69}$$

Observe that only relative coordinates are important since we are dealing with differences in coordinate values which makes the actual values of the coordinates irrelevant. Now

$$\begin{aligned} [P] &= \sigma A \begin{bmatrix} b_1^2 + c_1^2 & b_1 b_2 + c_1 c_2 & b_1 b_3 + c_1 c_3 \\ b_2 b_1 + c_2 c_1 & b_2^2 + c_2^2 & b_2 b_3 + c_2 c_3 \\ b_3 b_1 + c_3 c_1 & b_3 b_2 + c_3 c_2 & b_3^2 + c_3^2 \end{bmatrix} \\ &= 4 \times 0.5 \begin{bmatrix} 1+1 & -1+0 & 0-1 \\ -1+0 & 1+0 & 0+0 \\ 0-1 & 0+0 & 0+1 \end{bmatrix} \\ &= \begin{bmatrix} 4 & -2 & -2 \\ -2 & 2 & 0 \\ -2 & 0 & 2 \end{bmatrix} \end{aligned} \tag{5.70}$$

Ele. 1	PMatrix				QMatrix	Ele. 2	PMatrix				QMatrix
Verts	8	3	7			Verts	1	7	3		
8	4	−2	−2		1/6	1	4	−2	−2		1/6
3	−2	2	0		1/6	7	−2	2	0		1/6
7	−2	0	2		1/6	3	−2	0	2		1/6

Ele. 3	PMatrix				QMatrix	Ele. 4	PMatrix				QMatrix
Verts	3	4	1			Verts	2	1	4		
3	1	−0.5	−0.5		0	2	1	−0.5	−0.5		0
4	−0.5	0.5	0		0	1	−0.5	0.5	0		0
1	−0.5	0	0.5		0	4	−0.5	0	0.5		0

Ele. 5	PMatrix				QMatrix	Ele. 6	PMatrix				QMatrix
Verts	4	6	2			Verts	5	2	6		
4	1	−0.5	−0.5		0	5	1	−0.5	−0.5		0
6	−0.5	0.5	0		0	2	−0.5	0.5	0		0
2	−0.5	0	0.5		0	6	−0.5	0	0.5		0

Figure 5.13: The local matrices of the six elements of Fig. 5.11.

$$\mathbf{Q} = \begin{Bmatrix} \dfrac{1 \times 0.5}{3} \\[2ex] \dfrac{1 \times 0.5}{3} \\[2ex] \dfrac{1 \times 0.5}{3} \end{Bmatrix} = \begin{Bmatrix} \dfrac{1}{6} \\[2ex] \dfrac{1}{6} \\[2ex] \dfrac{1}{6} \end{Bmatrix} \tag{5.71}$$

We observe (and leave it as an exercise to the reader to check this out) that if we number the next triangle as 1–7–3 to have shape correspondence to the previous triangle, the local matrices will be exactly the same. This is a common way to reduce hand computations although most computations are done today on the computer so that how we number triangles does not matter. We may do likewise for the next two triangles except for differences in σ and q, which can be easily accounted for through scaling. So the local matrices are as shown in Fig. 5.13.

The four rows and columns of the matrix $[P]$ are filled up as shown in Fig. 5.14. For clarity, each position of the global matrix PG is divided into six, each sub-location corresponding to the contribution from one of the six elements. Similarly, the column vector $\mathbf{Q}$ shows its contributions from the six elements as it is built up.

The numbers 200 in the right-hand side column vector $\mathbf{Q}$ comes from the two additional columns of $[P]$ corresponding to the known T values at 100. The numbers

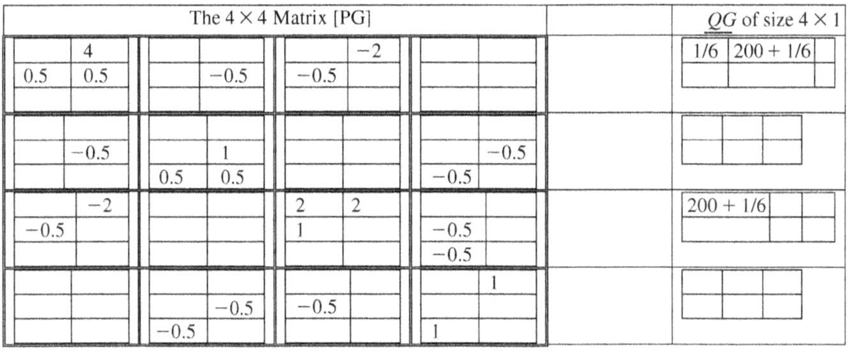

Figure 5.14: The local matrices of Fig. 5.13 added into global matrices (each matrix location has six boxes for each contributing element).

zero when they need to be added are not reflected. Thus we have, from Fig. 5.14,

$$\begin{bmatrix} 5 & -0.5 & -2.5 & 0 \\ -0.5 & 2 & 0 & -1 \\ -2.5 & 0 & 5 & -0.5 \\ 0 & -1 & -0.5 & 2 \end{bmatrix} \begin{Bmatrix} T_1 \\ T_2 \\ T_3 \\ T_4 \end{Bmatrix} = \begin{Bmatrix} \frac{601}{3} \\ 0 \\ \frac{1201}{3} \\ 0 \end{Bmatrix} \tag{5.72}$$

Solving $\mathbf{T}^{t} = [200 \quad 0 \quad 200 \quad 0]$.

5.9 A two-dimensional finite element program through an example

We will consider without too much explanation a two-dimensional problem and a MATLAB program to solve it – not too much because the principles are an extension of the work done in one dimension.

Figure 5.15 shows an eighth of a square cable with a square core surrounded by a sheath at 100 V. It has been prepared for finite element analysis with 16 elements and 15 nodes. The sheath is a Dirichlet boundary and the two lines of symmetry are Neumann boundaries. So the known nodes, nodes 11–15, are numbered last. There are 16 elements and 15 nodes of which 10 need their potential values computed. Left of the line 4–5–6, $(\epsilon, \rho) = (2, 1)$. To the right of the line, the corresponding value of the number pair is $(1, 0)$.

Codes 5.3–5.7 give the MATLAB solution to this problem. In the main program (Code 5.3), the coordinates of the 15 nodes are defined in XG and YG. ϕ contains dummy values for the potentials of the 10 nodes for which it is to computed and the last 5 values of ϕ are the known values. DatE contains the element data – for each element the three vertices and its ε and ρ values. The program then, after

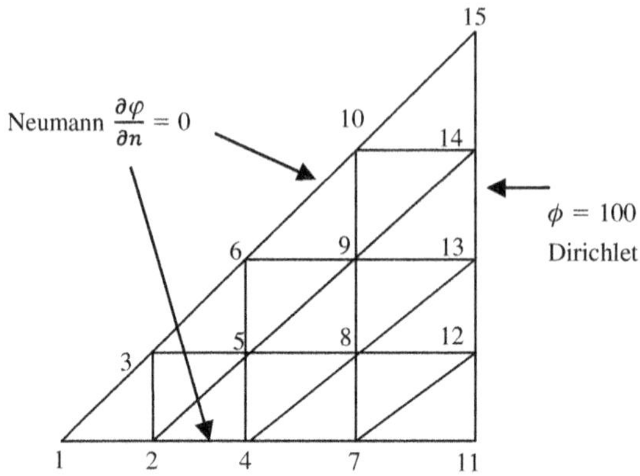

Figure 5.15: An eighth of a square cable core prepared for finite element analysis.

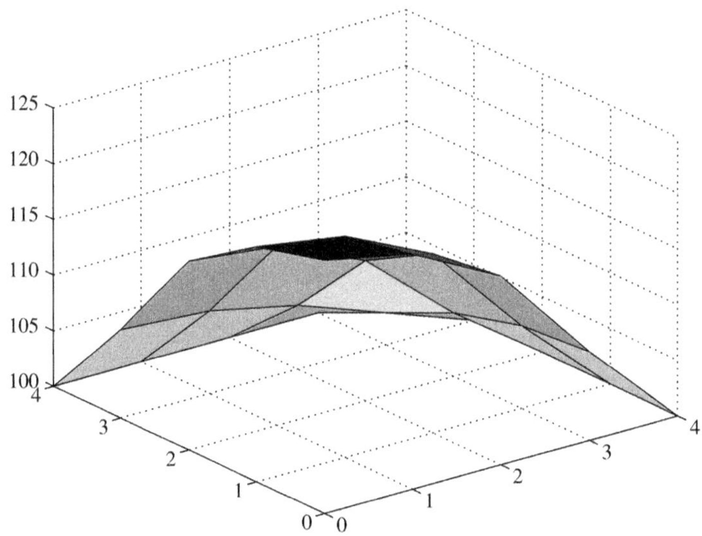

Figure 5.16: The potential in square cable.

initializing PG and QG, the global matrices, takes up each element, forms its local matrices P and q using the function Triangle (Code 5.6) and then places them in the global matrices using the function GlobalPlace (Code 5.5). Once all elements' contributions are done, the function ElYoo (Code 5.4) solves the equation using Cholesky's factorization (also called LU-decomposition), forward elimination and back-substitution. Now the solution is ready. To process it, the MATLAB function surf is used. It plots a vector $\mathbf{z}$ defined at coordinate locations in the vectors $\mathbf{x}$ and $\mathbf{y}$ on a rectangular grid. The function Nearest (Code 5.7), of no general value but written for this problem, identifies the nodes of the finite element grid nearest to the nodes in x, y and gets the solution plotted. The output of code 53 is shown in Fig. 5.16.

```
clc; clear all;
n= 10; %matrix size
NNodes = 15; %No. of Nodes
NElems = 16;
PG = zeros (10,10);
QG = zeros(10,1);
XG = [0;1;1;2;2;2;3;3;3;3;4;4;4;4;4]; %These are the x-coordinates
                                          of nodes 1 to 10
YG = [0;0; 1;0;1;2;0;1;2;3;0;1;2;3;4];%These are the y-coordinates
                                          of nodes 1 to 10
DatE= [1,2,3, 2, 10; %v1,v2,v3,eps,rho
    2, 4, 5, 2, 10;
    2, 5, 3, 2, 10;
    3, 5, 6, 2, 10;
    4, 7, 8, 1, 0;
    4, 8, 5, 1, 0;
    5, 8, 9, 1, 0;
    5, 9, 6, 1, 0;
    6, 9, 10, 1, 0;
    7, 11, 12, 1, 0;
    7, 12, 8, 1, 0;
    8, 12, 13, 1, 0;
    8, 13, 9, 1, 0;
    9, 13, 14, 1, 0;
    9, 14, 10, 1, 0;
    10, 14, 15, 1, 0];
Phi = [0.0; 0.0; 0.0; 0.0; 0.0; 0.0; 0.0; 0.0; 0.0; 0.0;100.0;100.0;
      100.0;100.0;100.0];
%Last 5 positions known
for Element = 1: NElems
    for i=1:3
        V(i)=DatE(Element,i);
    end;
    eps=DatE(Element,4);
    rho=DatE(Element,5);
    for i = 1:3
        x(i) = XG(V(i));% x-coordinates of 3 vertices
        y(i) = YG(V(i));% y-coordinates of 3 vertices
    end;
    [P,q] = Triangle(x,y,eps,rho);
    [PG,QG] = GlobalPlace(PG,QG,P,q,n,V,Phi);
end; %for Element
Solution=ElYooFunction(PG,QG,n);
for i=1:n
    Phi(i)=Solution(i);
end;
for i=1:5
    for j=1:i
        xx(i,1)=i-1;
        yy(i,1)=j-1;
        k=Nearest(xx(i,1),yy(i,1), XG,YG,NNodes);
        zz(i,j)=Phi(k);
        if j<i
            zz(j,i)=Phi(k);
        end;
    end;
end;
surf(xx,yy,zz);
```

Code 5.3: For problem of Fig. 5.15 by 2-D FEM.

```
function x= ElYooFunction(A,B,n)
L(1,1)= sqrt(A(1,1));
for i=2:n
    %First compute all Ls up to the diagonal
    for j=1:i-1
        sum=A(i,j);
        for k=1:j-1
            sum=sum-L(j,k)*L(i,k);
        end;
        L(i,j)=sum/L(j,j);
    end;
    %Then compute L at the diagonal
    sum=A(i,i);
    for k=1:i-1
        sum=sum-L(i,k)*L(i,k);
    end;
    L(i,i)=sqrt(sum);
end;
L
U=L';
P=L*U
Z(1)=B(1)/L(1,1);
for i=2:n
    sum=B(i);
    for k=1:i-1
        sum=sum-L(i,k)*Z(k);
    end;
    Z(i)=sum/L(i,i);
end;
Q=L*Z'
x(n,1)=Z(n)/L(n,n);
for i=1:n-1
    ir=n-i;
    sum=Z(ir);
    for ic=ir+1:n
        sum=sum-L(ic,ir)*x(ic);
    end;
    x(ir,1)=sum/L(ir,ir);
end
```

Code 5.4: The function ElYoo for LU-decomposition.

5.10 Other equations

In addition to the Poisson equation, two other equations commonly appear in electrical engineering. These are the wave equation and the diffusion equation for eddy current problems. Functionals need to be found for these.

The wave equation for waveguides is for the component of the non-transverse electric or magnetic field φ,

$$\nabla^2\varphi = -k^2\varphi \tag{5.73}$$

```
GlobalPlace.m
function [PG,QG] = GlobalPlace(PGI,QGI,P,q,MatSize,V,Phi)
   PG = PGI;
   QG = QGI;
   for i = 1:3
     if V(i)<= MatSize %Then Goes into PG
        NR = V(i);
        QG(NR)= QG(NR)+q(i);
        for j = 1:3
            NC = V(j);
            if NC <= MatSize
               PG(NR,NC)=PG(NR,NC)+P(i,j);
            else
               QG(NR)= QG(NR)-P(i,j)*Phi(NC);
            end; %if statement end;
        end;% Loop for j ends
  end;
end;
```

Code 5.5: Function GlobalPlace placing local matrices in global matrices.

```
function [P,q] = Triangle(x,y,eps,rho)
Del=(x(2)*y(3)-x(3)*y(2)) + (x(3)*y(1)-x(1)*y(3))+ ...
(x(1)*y(2)-x(2)*y(1));
 Area=Del/2;
 for i=1:3
     i1=mod(i,3)+1;
     i2=mod(i1,3)+1;
     b(i)=(y(i1)-y(i2))/Del;
     c(i)=(x(i2)-x(i1))/Del;
 end;
 const1=rho*Area/3;
 const2=Area*eps;
 for i=1:3
   q(i,1)=const1;
   for j=1:3
     P(i,j)=const2*(b(i)*b(j)+c(i)*c(j));
   end;
end;
```

Code 5.6: Function triangle computes local matrices for a triangle.

```
function k=Nearest(x,y, XG,YG,n)
Error=10000;
k=0;
for i=1:n
    SquareDistance=(XG(i)-x)^2+(YG(i) -y)^2;
    if Error > SquareDistance
       Error=SquareDistance;
       k=i;
    end;
end;
```

Code 5.7: Function nearest identified nodes nearest to coordinate (x,y).

where ∇^2 operates on the transverse plane and k is defined in terms of the propagation constant γ, frequency ω and velocity of light c:

$$k^2 = \gamma^2 + \frac{\omega^2}{c^2} \tag{5.74}$$

From our considerations, we know that

$$\mathcal{L} = \frac{1}{2}\iint_v (\nabla\varphi)^2 \, dv \tag{5.75}$$

when minimized will satisfy $-\nabla^2\varphi = 0$. We also can see that $\frac{k^2}{2}\varphi^2$ upon differentiation with respect to φ will yield $k^2\varphi$. Therefore, the functional we seek is

$$\mathcal{L} = \frac{1}{2}\iint_v \left\{(\nabla\varphi)^2 - \frac{k^2}{2}\varphi^2\right\} dv \tag{5.76}$$

Putting in the approximations it can be shown that the relevant equation is

$$\sum_{\text{All elements}} \left\{\frac{1}{2}\varphi^t[P]\varphi - \frac{k^2}{2}\varphi^t[Q]\varphi\right\} = 0 \tag{5.77}$$

where the local matrices are

$$[P] = A\begin{bmatrix} b_1^2 + c_1^2 & b_1b_2 + c_1c_2 & b_1b_3 + c_1c_3 \\ b_2b_1 + c_2c_1 & b_2^2 + c_2^2 & b_2b_3 + c_2c_3 \\ b_3b_1 + c_3c_1 & b_3b_2 + c_3c_2 & b_3^2 + c_3^2 \end{bmatrix} \tag{5.78}$$

and

$$[Q] = \frac{A}{12}\begin{bmatrix} 2 & 1 & 1 \\ 1 & 2 & 1 \\ 1 & 1 & 2 \end{bmatrix} \tag{5.79}$$

where A is the area of the triangular element. This leads to the linearized eigen value problem of the form

$$[P]\varphi = \lambda[Q]\varphi \tag{5.80}$$

In the case of eddy current problems in two dimensions, the governing equation is

$$-\frac{1}{\mu}\nabla^2\mathbf{A} = \mathbf{J}_0 - j\omega\sigma\mathbf{A} \tag{5.81}$$

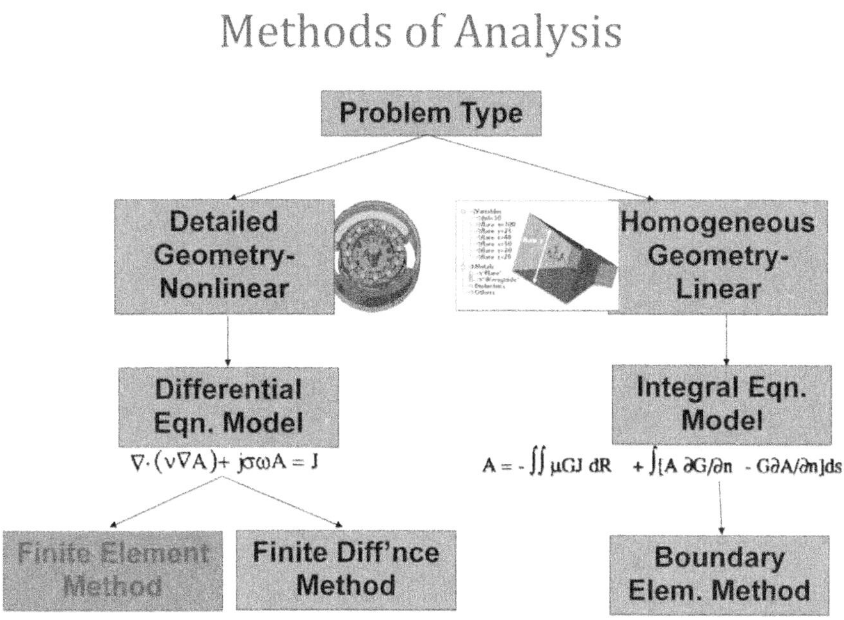

Figure 5.17: The Finite Element, Finite Difference and Finite Element Methods.

where **A** is the complex, single-component vector potential. By similar considerations as to the wave equation, we may arrive at the functional

$$\mathcal{L} = \iint_v \left\{ \frac{1}{2\mu}(\nabla \mathbf{A})^2 + \frac{1}{2}\mathrm{j}\omega\sigma\mathbf{A}^2 - \mathbf{A}\mathbf{J}_0 \right\} \mathrm{d}v \tag{5.82}$$

where the first and third terms together are from the Poissonian part of the diffusion equation and the middle term arises from like considerations to the wave equation. Figure 5.17 shows the three different electromagnetic solution methods used for the linear and non-linear problems. The differential equations that model the non linear problems may be solved using the finite difference method or the finite element method. The linear problem which is modelled by integral equation may be solved using the boundary element method.

For further reading

[1] Garg, R., *Analytical and Computational Methods in Electromagnetics*, Artech House, Norwood, MA, 2008.

[2] Hoole, S.R.H., *Computer Aided Analysis and Design of Electromagnetic Devices*, Elsevier, New York, 1986.

[3] Hoole, S.R.H., and Hoole, P.R.P., *A Modern Short Course in Engineering Electromagnetics*, Oxford University Press, New York, 1996.

[4] Poljak, D., and Brebbia, C.A., *Boundary Element Methods for Electrical Engineers*, WIT Press, Southampton, UK, 2005.

[5] Poljak, D., Doric, V., and Antonijevic, S., *Computer Aided Design of Wire Structures*, WIT Press, Southampton, UK, 2007.

[6] Sadiku, M., *Numerical Techniques in Electromagnetics*, 3rd Edition, CRC Press, Bota Racon, FL, 2009.

6

Transmission lines and waveguides

6.1 TEM waves on transmission line

Wave propagation is mainly characterized into two as unguided wave propagation and guided wave of propagation. Unguided wave propagation indicates that once waves are launched in an infinite block of material, they continue to propagate in the same direction.

On the other hand, transmission lines and waveguides serve to guide the propagation of energy from one point to another. Common examples of transmission lines are parallel-wire lines, coaxial lines and microstrip lines as shown in Fig. 6.1. The transmission lines mentioned here are said to be *uniform* lines in the sense that cross-sectional views of the line at any two points along the line are the same.

A transverse electromagnetic (TEM) field structure is one in which the electric and magnetic field vectors at each point in space have no components in the direction of propagation. Stated another way if we assume the direction of propagation for a TEM wave to be the z direction, we require that $\mathcal{E}_z = \mathcal{H}_z = 0$ for a TEM field structure.

To illustrate the generality of this statement, we consider general two conductor uniform transmission lines as shown in Fig. 6.2. The axis of the line is the z axis of a rectangular coordinate system and the line cross section at $z = 0$ lies on the xy plane. Since the two conductors of the line are infinite in length, we may neglect fringing of the fields at the conductor endpoints which technically do not exist. The electric and magnetic field lines lie on the xy plane, transverse to the line axis, as shown in Fig. 6.3.

When fields are time varying, it is generally not possible to define a unique voltage between two points since a time varying electrical field is not a conservative field. However, if we assume a TEM field structure, it is possible to define a unique voltage difference between the two conductors of the transmission line in Fig. 6.3 even when time-varying. Similarly, we can define the current on the conductors

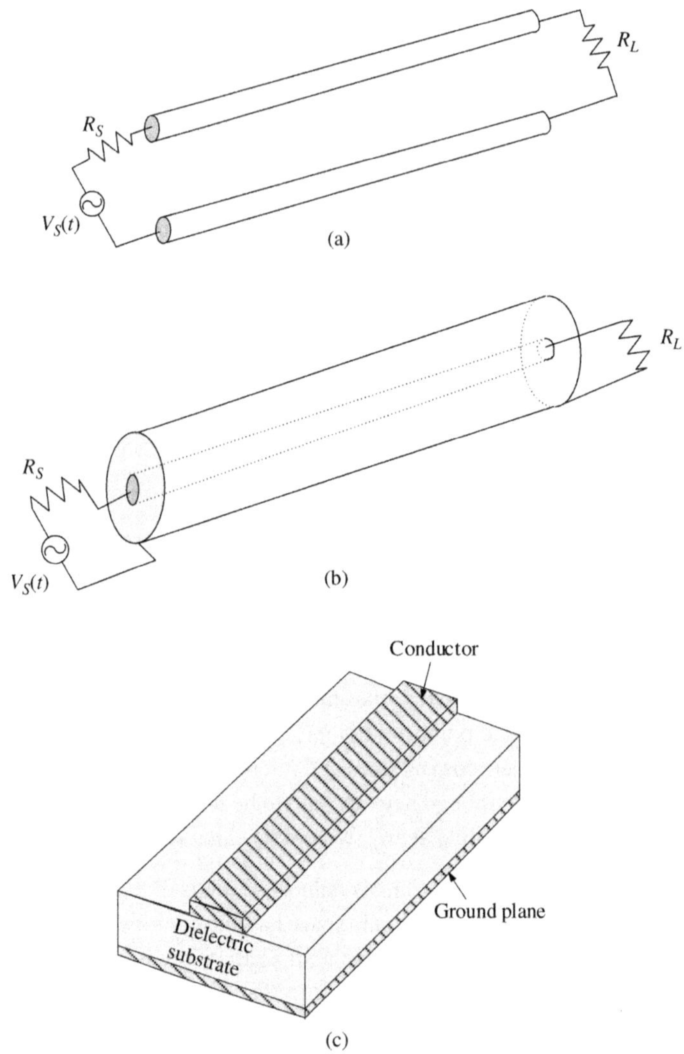

Figure 6.1: Transmission line configurations: (a) parallel-wire line, (b) coaxial line and (c) microstrip line.

similar to that for the case of static excitation. To illustrate this, we recall Faraday's law and Ampere's law in integral form in a homogeneous medium surrounding the conductors.

$$\oint_{C_{xy}} \mathcal{E} \cdot \mathbf{dl} = -\mu \frac{\mathrm{d}}{\mathrm{d}t} \int_{S_{xy}} \mathcal{H} \cdot \mathbf{ds} \tag{6.1a}$$

$$\oint_{C_{xy}} \mathcal{H} \cdot \mathbf{dl} = \int_{S_{xy}} \mathcal{J} \cdot \mathbf{ds} + \varepsilon \frac{\mathrm{d}}{\mathrm{d}t} \int_{S_{xy}} \mathcal{E} \cdot \mathbf{ds} \tag{6.1b}$$

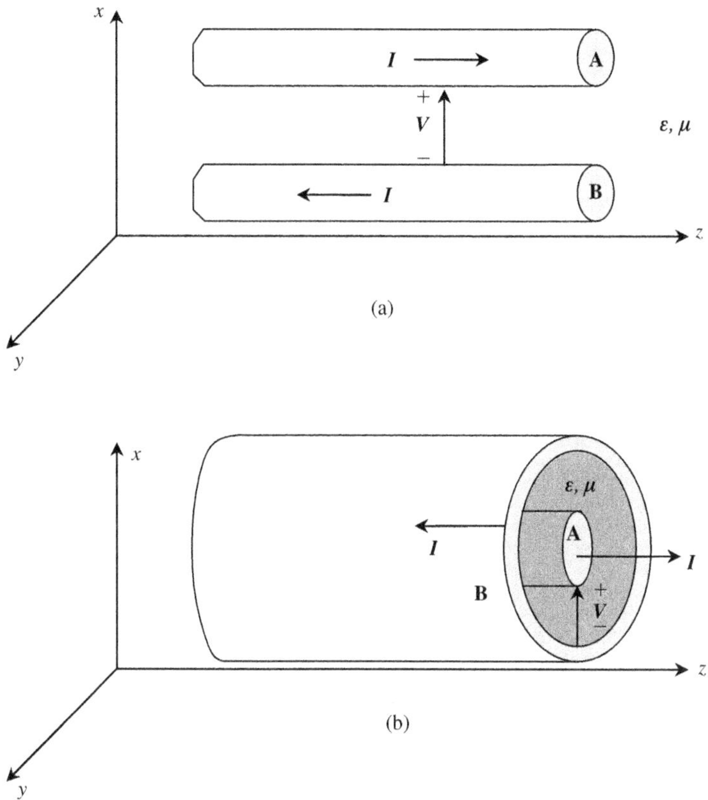

Figure 6.2: Two-conductor transmission lines: (a) open-conductor line and (b) coaxial line.

Expanding the expression by substituting for the field components in (6.1) yields

$$\oint_{C_{xy}} (\mathcal{E}_x \,\mathrm{d}x + \mathcal{E}_y \,\mathrm{d}y) = -\mu \frac{\mathrm{d}}{\mathrm{d}t} \int_{S_{xy}} \mathcal{H}_z \,\mathrm{d}x \,\mathrm{d}y \tag{6.2a}$$

$$\oint_{C_{xy}} (\mathcal{H}_x \,\mathrm{d}x + \mathcal{H}_y \,\mathrm{d}y) = \int_{S_{xy}} \mathcal{J}_z \,\mathrm{d}x \,\mathrm{d}y + \varepsilon \frac{\mathrm{d}}{\mathrm{d}t} \int_{S_{xy}} \mathcal{E}_z \,\mathrm{d}x \,\mathrm{d}y \tag{6.2b}$$

For the TEM mode of propagation $\mathcal{E}_z = \mathcal{H}_z = 0$. Thus, we obtain

$$\oint_{C_{xy}} (\mathcal{E}_x \,\mathrm{d}x + \mathcal{E}_y \,\mathrm{d}y) = 0 \tag{6.3a}$$

$$\oint_{C_{xy}} (\mathcal{H}_x \,\mathrm{d}x + \mathcal{H}_y \,\mathrm{d}y) = \int_{S_{xy}} \mathcal{J}_z \,\mathrm{d}x \,\mathrm{d}y \tag{6.3b}$$

Applying the results obtained in (6.3a) to the two-conductor lines in Fig. 6.2, we conclude that it is possible to uniquely define a voltage between the conductors at each point along the line integral of $\boldsymbol{\mathcal{E}}$ along any path on the transverse xy plane as

$$V(z,t) = -\int_{c_V} \boldsymbol{\mathcal{E}}_T \cdot \mathbf{dl} \tag{6.4}$$

where contour c_V is shown in Fig. 6.3 and $\boldsymbol{\mathcal{E}}_T$ is the transverse electric field.

$$\boldsymbol{\mathcal{E}}_T = \mathcal{E}_x \mathbf{u}_x + \mathcal{E}_y \mathbf{u}_y \tag{6.5}$$

Similarly, from (6.3b) we conclude that we can uniquely define the current on each conductor surface as the line integral of $\boldsymbol{\mathcal{H}}_T$ around any closed contour on the

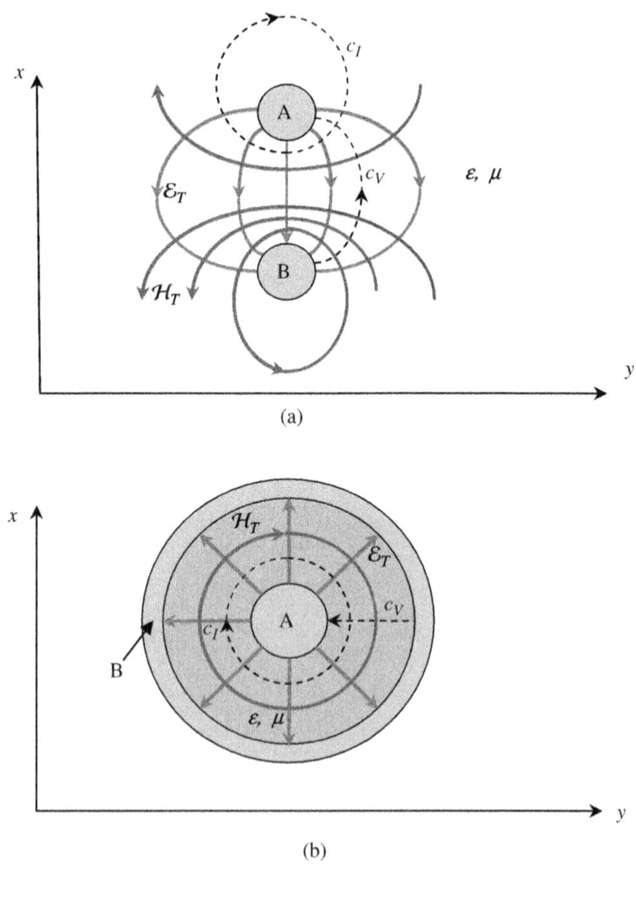

Figure 6.3: Transmission-line cross-sectional field distribution: (a) open-conductor line and (b) coaxial line.

teraservers xy plane that encircles the conductor as

$$I(z,t) = \int_{c_I} \mathcal{H}_T \cdot \mathbf{dl} \tag{6.6}$$

where contour c_I is shown in Fig. 6.3 and $\mathcal{H}_T$ is the transverse magnetic field.

$$\mathcal{H}_T = \mathcal{H}_x \mathbf{u}_x + \mathcal{H}_y \mathbf{u}_y \tag{6.7}$$

Note that the voltage and current in (6.4) and (6.6) are shown as being functions of position along the line, z, as well as time, t.

6.2 Transmission line equations

6.2.1 Model and equations

As mentioned in the above Section 6.1, an important property of TEM waves is that the fields $\mathcal{E}_T$ and $\mathcal{H}_T$ are uniquely related to voltage $V(z,t)$ and current $I(z,t)$, respectively. In view of this, we use circuit quantities $V(z,t)$ and $I(z,t)$ in solving the transmission line problems instead of solving for the field quantities $\mathcal{E}_T$ and $\mathcal{H}_T$ since the voltage and current circuit model is simpler and more convenient.

We propose the equivalent circuit of a portion of the line by examining an incremental portion of length Δz of a two-conductor transmission line as shown in Fig. 6.4. Further, we intend to find an equivalent circuit for this line and derive the line equations. The model is in terms of the line parameters, such as resistance per unit length (R), inductance per unit length (L), conductance per unit length (G) and capacitance per unit length (C), to represent the two-conductor line. In the model shown in Fig. 6.4, we assume that the wave propagates along the positive z direction, from the generator to the load.

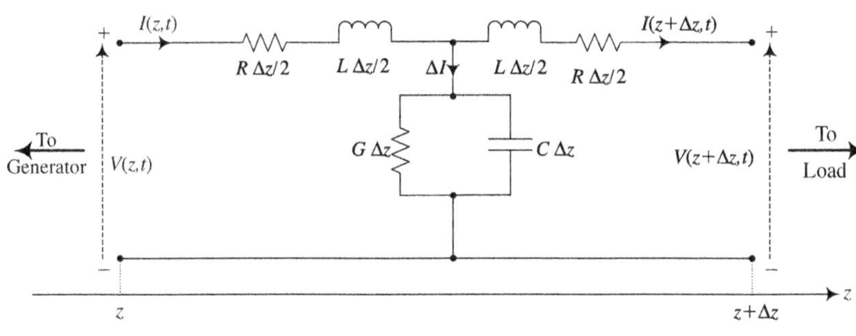

Figure 6.4: A T-type equivalent circuit model of a two-conductor transmission line of differential length Δz.

By applying Kirchhoff's current law to the main node of the circuit in Fig. 6.4, we obtain

$$
\begin{aligned}
I(z,t) &= I(z+\Delta z,t) + \Delta I \\
&= I(z+\Delta z,t) + G\Delta z(V(z,t)+\Delta V) + C\Delta z\frac{\partial(V(z,t)+\Delta V)}{\partial t}
\end{aligned}
\tag{6.8}
$$

Equation (6.8) can be further written as

$$
\frac{I(z+\Delta z,t)-I(z,t)}{\Delta z} = -G(V(z,t)+\Delta V) - C\frac{\partial(V(z,t)+\Delta V)}{\partial t}
\tag{6.9}
$$

When $\Delta z \to 0$, Equation (6.9) becomes

$$
\frac{\partial I(z,t)}{\partial z} = -GV(z,t) - C\frac{\partial V(z,t)}{\partial t}
\tag{6.10}
$$

By applying Kirchhoff's voltage law to the main loop of the circuit in Fig. 6.4, we obtain

$$
\begin{aligned}
V(z,t) = {} & R\frac{\Delta z}{2}I(z,t) + L\frac{\Delta z}{2}\frac{\partial I(z,t)}{\partial t} \\
& + R\frac{\Delta z}{2}(I(z,t)-\Delta I) + L\frac{\Delta z}{2}\frac{\partial(I(z,t)-\Delta I)}{\partial t} + V(z+\Delta z,t)
\end{aligned}
\tag{6.11}
$$

Equation (6.11) can be further written as

$$
\begin{aligned}
\frac{V(z+\Delta z,t)-V(z,t)}{\Delta z} = {} & -\frac{R}{2}I(z,t) - \frac{L}{2}\frac{\partial I(z,t)}{\partial t} \\
& -\frac{R}{2}(I(z,t)-\Delta I) - \frac{L}{2}\frac{\partial(I(z,t)-\Delta I)}{\partial t}
\end{aligned}
\tag{6.12}
$$

When $\Delta z \to 0$, Equation (6.12) becomes

$$
\frac{\partial V(z,t)}{\partial z} = -R\,I(z,t) - L\frac{\partial I(z,t)}{\partial t}
\tag{6.13}
$$

If we assume harmonic time dependence so that

$$
V(z,t) = \mathrm{Re}[V_s(z)\mathrm{e}^{\mathrm{j}\omega t}]
\tag{6.14a}
$$

$$
I(z,t) = \mathrm{Re}[I_s(z)\mathrm{e}^{\mathrm{j}\omega t}]
\tag{6.14b}
$$

where $V_s(z)$ and $I_s(z)$ are the phasor forms of $V(z,t)$ and $I(z,t)$, respectively. Equations (6.10) and (6.13) become

$$\frac{dI_s}{dz} = -(G + j\omega C)V_s \tag{6.15}$$

$$\frac{dV_s}{dz} = -(R + j\omega L)I_s \tag{6.16}$$

The differential equations (6.15) and (6.16) are coupled. To separate them, we take the derivative of (6.16) with respect to z and employ (6.15) to be rid of I_s. Hence, we obtain

$$\begin{aligned}\frac{d^2 V_s}{dz^2} &= (R + j\omega L)(G + j\omega C)V_s \\ &= \gamma^2 V_s \end{aligned}\tag{6.17}$$

where $\gamma^2 = (R + j\omega L)(G + j\omega C)$. Similarly, taking the derivative of (6.15) and employing (6.16), we obtain

$$\begin{aligned}\frac{d^2 I_s}{dz^2} &= (R + j\omega L)(G + j\omega C)I_s \\ &= \gamma^2 I_s \end{aligned}\tag{6.18}$$

Equations (6.17) and (6.18) are, respectively, the wave equations for voltage and current. In our usual notation, γ is named the ***propagation constant*** (in per meter). Since γ is a complex term, it can be written as

$$\gamma = \sqrt{(R + j\omega L)(G + j\omega C)} = \alpha + j\beta \tag{6.19}$$

where α is the ***attenuation constant*** (in nepers per meter) and β is the ***phase constant*** (in radians per meter). The wavelength λ and wave velocity u are, respectively, given by

$$\lambda = \frac{2\pi}{\beta} \tag{6.20}$$

$$u = f\lambda = \frac{\omega}{\beta} \tag{6.21}$$

The solutions of the linear homogeneous differential equations (6.17) and (6.18) are given as

$$V_s(z) = V_0^+ e^{-\gamma z} + V_0^- e^{\gamma z} \tag{6.22}$$

$$I_s(z) = I_0^+ e^{-\gamma z} + I_0^- e^{\gamma z} \tag{6.23}$$

where V_0^+, V_0^-, I_0^+ and I_0^- are voltage and current amplitudes. The + and − signs, respectively, denote waves traveling along $+z$ and $-z$ directions. We obtain the instantaneous expression for voltage as

$$\begin{aligned} V(z,t) &= \mathrm{Re}\lfloor V_s(z)\,\mathrm{e}^{\mathrm{j}\omega t}\rfloor \\ &= V_0^+\,\mathrm{e}^{-\alpha z}\cos(\omega t-\beta z)+V_0^-\,\mathrm{e}^{\alpha z}\cos(\omega t+\beta z) \end{aligned} \tag{6.24}$$

The ***characteristic impedance*** Z_0 of the line is the ratio of the positively traveling voltage wave to current wave at any point of the line. By substituting (6.22) and (6.23) in (6.16), we obtain

$$\frac{\mathrm{d}(V_0^+\,\mathrm{e}^{-\mathrm{j}\gamma z}+V_0^-\,\mathrm{e}^{\mathrm{j}\gamma z})}{\mathrm{d}z} = -(R+\mathrm{j}\omega L)(I_0^+\,\mathrm{e}^{-\mathrm{j}\gamma z}+I_0^-\,\mathrm{e}^{\mathrm{j}\gamma z}) \tag{6.25}$$

Equating the coefficients of the terms $\mathrm{e}^{-\mathrm{j}\gamma z}$ and $\mathrm{e}^{\mathrm{j}\gamma z}$ in (6.25), we obtain

$$Z_0 = \frac{V_0^+}{I_0^+} = -\frac{V_0^-}{I_0^-} = \frac{R+\mathrm{j}\omega L}{\gamma} = \frac{\gamma}{G+\mathrm{j}\omega C} = \sqrt{\frac{R+\mathrm{j}\omega L}{G+\mathrm{j}\omega C}} \tag{6.26}$$

The propagation constant γ and the characteristic impedance Z_0 are important properties of the line because both depend on the line parameters R, L, G and C and the operating frequency.

The general transmission line considered so far in this section is the lossy transmission line since conductors comprising the line are not perfect conductors ($\sigma_c \neq \infty$) and the dielectric by which the conductors are surrounded is not perfect ($\sigma_d \neq 0$). We are now going to consider two special cases, namely the lossless line and the distortion-less line.

6.2.2 Lossless line

A transmission line is said to be lossless if the conductors of the line are perfect ($\sigma_c \approx \infty$) and the dielectric medium separating them is lossless ($\sigma_d \approx 0$). Hence, we can state that $R = G = 0$. This is a necessary condition for a line to be lossless. Thus for such a line, (6.19) yields

$$\gamma = \mathrm{j}\omega\sqrt{LC} = \alpha + \mathrm{j}\beta \tag{6.27}$$

Equating the real and imaginary term in (6.27),

$$\begin{aligned} \alpha &= 0 \\ \beta &= \omega\sqrt{LC} \end{aligned} \tag{6.28}$$

Further, (6.20), (6.21) and (6.26), respectively, become

$$\lambda = \frac{2\pi}{\beta} = \frac{1}{f\sqrt{LC}} \tag{6.29}$$

$$u = f\lambda = \frac{1}{\sqrt{LC}} \tag{6.30}$$

$$Z_0 = \sqrt{\frac{L}{C}} \tag{6.31}$$

6.2.3 Distortion-less line

Wave amplitudes of different frequency components are attenuated differently in a lossy line because the attenuation constant is frequency dependent. Moreover, in general, the phase velocity of each frequency component is also frequency dependent. Frequency dependent attenuation constants and phase velocities result in distortion in a signal since a signal normally consists of a band of frequencies. A distortion-less line is one in which the attenuation constant and phase velocity are frequency independent while the phase constant is linearly dependent on frequency.

Rewriting the general expression for the propagation constant

$$\gamma = \sqrt{(R + \mathrm{j}\omega L)(G + \mathrm{j}\omega C)} = \sqrt{RG\left(1 + \mathrm{j}\omega\frac{L}{R}\right)\left(1 + \mathrm{j}\omega\frac{C}{G}\right)} = \alpha + \mathrm{j}\beta \tag{6.32}$$

It is evident from (6.32), that when $L/R = C/G$, α is frequency independent whereas β is linearly dependent on frequency.

$$\alpha = \sqrt{RG} \tag{6.33a}$$

$$\beta = \omega\sqrt{LC} \tag{6.33b}$$

The phase velocity is obtained as

$$u = \frac{\omega}{\beta} = \frac{1}{\sqrt{LC}} \tag{6.34}$$

Since α and u are frequency independent, the band of frequencies in a signal have these same attenuation and phase velocity, and the signal will not be distorted. Hence the condition for the distortion-less line is given as

$$\frac{R}{L} = \frac{G}{C} \tag{6.35}$$

The characteristic impedance of the distortion-less line is

$$Z_0 = \sqrt{\frac{R(1+j\omega L/R)}{G(1+j\omega C/G)}} = \sqrt{\frac{R}{G}} = \sqrt{\frac{L}{C}} \tag{6.36}$$

The characteristic impedance of a distortion-less line is a real quantity similar to lossless lines despite the fact that there are power losses in distortion-less lines.

6.3 Reflection in transmission lines

6.3.1 Input impedance, standing wave ratio and power

Consider a transmission line of length l, characterized by α and Z_0, connected to a load Z_L as shown in Fig. 6.5. Looking into the line, the generator sees the line with the load as an input impedance Z_{in}. Let the transmission line extend from $z = 0$ at the generator to $z = l$ at the load. Thus, the voltage and current waves can be given from (6.22) and (6.23) incorporated with (6.26), as

$$V_s(z) = V_0^+ \, e^{-\gamma z} + V_0^- \, e^{\gamma z} \tag{6.37}$$

$$I_s(z) = \frac{V_0^+}{Z_0} e^{-\gamma z} - \frac{V_0^-}{Z_0} e^{\gamma z} \tag{6.38}$$

To find V_0^+ and V_0^-, the terminal conditions must be given. Let us say the voltage and current at $z = l$ are

$$\begin{aligned} V_s|_{z=l} &= V_L \\ I_s|_{z=l} &= I_L \end{aligned} \tag{6.39}$$

Substituting these into (6.37) and (6.38) results in

$$V_0^+ = \frac{1}{2}(V_L + Z_0 I_L) e^{\gamma l} \tag{6.40a}$$

$$V_0^- = \frac{1}{2}(V_L - Z_0 I_L) e^{-\gamma l} \tag{6.40b}$$

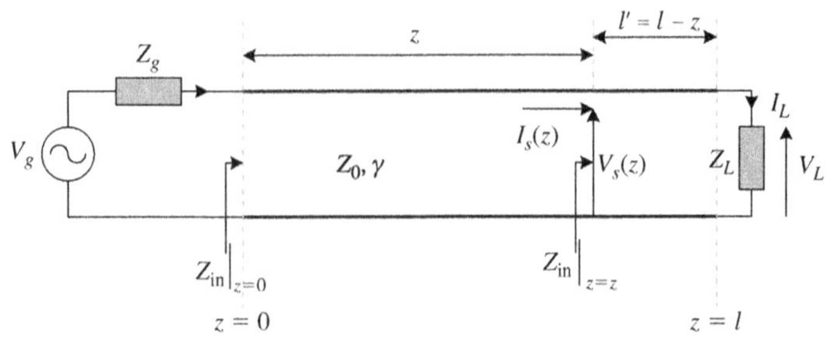

Figure 6.5: Transmission line terminated by a load and fed by a source.

The input impedance at any point in the line can be obtained from (6.37) and (6.38) and is given by

$$Z_{\text{in}}(z) = \frac{V_s(z)}{I_s(z)} = \frac{Z_0(V_0^+ \,\mathrm{e}^{-\gamma z} + V_0^- \,\mathrm{e}^{\gamma z})}{(V_0^+ \,\mathrm{e}^{-\gamma z} - V_0^- \,\mathrm{e}^{\gamma z})} \tag{6.41}$$

Substituting (6.40) in (6.41) and utilizing the fact that

$$\tanh \gamma(l - z) = \frac{(\mathrm{e}^{\gamma(l-z)} - \mathrm{e}^{-\gamma(l-z)})}{(\mathrm{e}^{\gamma(l-z)} + \mathrm{e}^{\gamma(l-z)})} \tag{6.42}$$

we get

$$Z_{\text{in}}(z) = Z_0 \left[\frac{Z_L + Z_0 \tanh \gamma(l - z)}{Z_0 + Z_L \tanh \gamma(l - z)}\right] = Z_0 \left[\frac{Z_L + Z_0 \tanh \gamma l'}{Z_0 + Z_L \tanh \gamma l'}\right] \tag{6.43}$$

where l' is the distance of the point from the load as shown in Fig. 6.5. For a lossless line, $\gamma = \mathrm{j}\beta$, $\tanh \mathrm{j}\beta l' = \mathrm{j} \tan \beta l'$. Hence equation (6.43) becomes

$$Z_{\text{in}}(z) = Z_0 \left[\frac{Z_L + \mathrm{j}Z_0 \tan \beta l'}{Z_0 + \mathrm{j}Z_L \tan \beta l'}\right] \tag{6.44}$$

showing that the input impedance varies periodically with distance l' from the load. The quantity $\beta l'$ in (6.44) is usually referred to as the *electrical length* of the line and can be expressed in radians.

We define Γ_L as the voltage reflection coefficient at the load. The reflection coefficient at load Γ_L is the ratio of the reflected wave voltage to the incident wave voltage at the load. That is,

$$\Gamma_L = \frac{V_0^- \,\mathrm{e}^{\gamma l}}{V_0^+ \,\mathrm{e}^{-\gamma l}} \tag{6.45}$$

Substituting V_0^+ and V_0^- from (6.40) into (6.45) and incorporating $V_L = Z_L I_L$ gives

$$\Gamma_L = \frac{Z_L - Z_0}{Z_L + Z_0} \tag{6.46}$$

The *voltage reflection coefficient* at any point on the line is the ratio of the voltage reflection wave to incident wave. It is given as

$$\Gamma(z) = \frac{V_0^- \,\mathrm{e}^{\gamma z}}{V_0^+ \,\mathrm{e}^{-\gamma z}} = \frac{V_0^-}{V_0^+} \mathrm{e}^{2\gamma z} \tag{6.47}$$

Combining (6.47) with (6.45), we get

$$\Gamma(z) = \Gamma_L \,\mathrm{e}^{-2\gamma(l-z)} = \Gamma_L \,\mathrm{e}^{-2\gamma l'} \tag{6.48}$$

The *current reflection coefficient* at any point on the line is the negative of the voltage reflection coefficient at that point.

The *standing wave ratio* (SWR) is defined as

$$s = \frac{V_{\rm max}}{V_{\rm min}} = \frac{I_{\rm max}}{I_{\rm min}} = \frac{1+|\Gamma_L|}{1-|\Gamma_L|} \tag{6.49}$$

The input impedance $Z_{\rm in}(z)$ in (6.41) has maxima and minima that occur at the maximum voltage and minimum current, and minimum voltage and maximum current, respectively. It can be shown that

$$|Z_{\rm in}(z)|_{\rm max} = \frac{V_{\rm max}}{I_{\rm min}} = sZ_0 \tag{6.50}$$

and

$$|Z_{\rm in}(z)|_{\rm min} = \frac{V_{\rm min}}{I_{\rm max}} = \frac{Z_0}{s} \tag{6.51}$$

To verify the above results, consider a lossless line with characteristic impedance Z_0 and phase constant β. The voltage and current in the line at any point can be given from (6.37) and (6.38) incorporated with (6.47) and (6.48), and are

$$V_s(z) = V_0^+\, {\rm e}^{-{\rm j}\beta z}(1+\Gamma_L\, {\rm e}^{-2{\rm j}\beta l'}) \tag{6.52}$$

$$I_s(z) = \frac{V_0^+}{Z_0}\, {\rm e}^{-{\rm j}\beta z}(1-\Gamma_L\, {\rm e}^{-2{\rm j}\beta l'}) \tag{6.53}$$

The maxima and minima of $V_s(z)$ can be obtained from (6.52) as

$$V_s(z)|_{\rm max} = |V_0^+|(1+|\Gamma_L|) \tag{6.54a}$$

$$V_s(z)|_{\rm min} = |V_0^+|(1-|\Gamma_L|) \tag{6.54b}$$

Similarly, the maxima and minima of $I_s(z)$ can be obtained from (6.53) as

$$I_s(z)|_{\rm max} = \frac{|V_0^+|}{Z_0}(1+|\Gamma_L|) \tag{6.55a}$$

$$I_s(z)|_{\rm min} = \frac{|V_0^+|}{Z_0}(1-|\Gamma_L|) \tag{6.55b}$$

Using the expressions given in (6.54) and (6.55), we can verify the expressions given in (6.49), (6.50) and (6.51). Also the voltage and current standing wave patterns on a lossless line, respectively expressed in (6.53) and (6.54), are given in Fig. 6.6.

The transmission line is used in transferring power from the source to load. The average power flow at a distance z from the generator is given by

$$P_{\rm ave} = \frac{1}{2}{\rm Re}(V_s(z)I_s^*(z)) \tag{6.56}$$

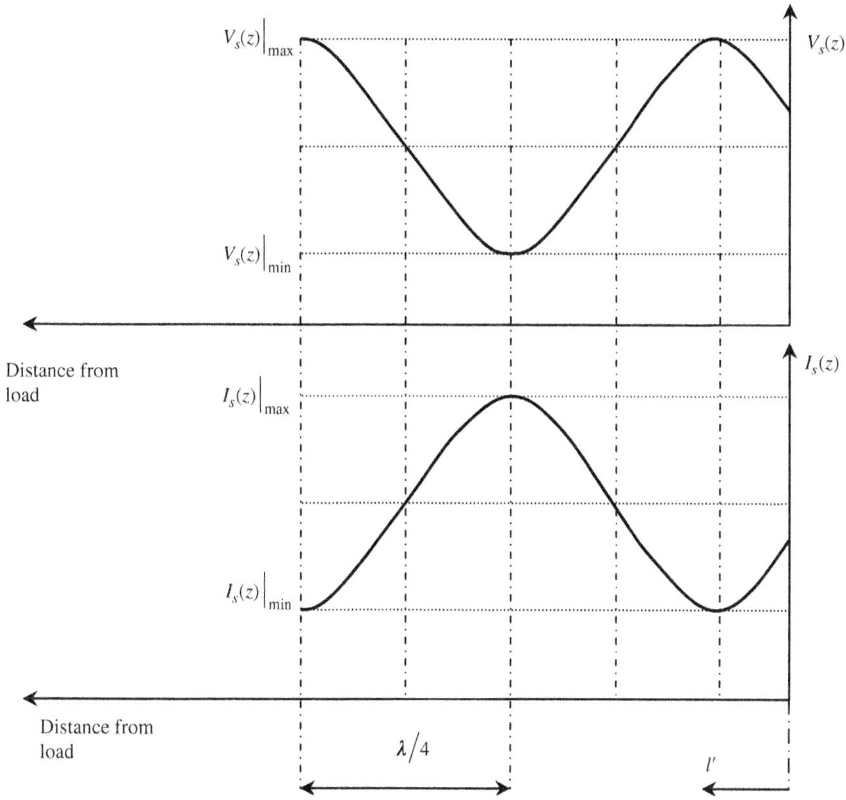

Figure 6.6: Voltage and current standing wave patterns on a lossless line terminated by a load.

Assuming a lossless line, we obtain the expressions for voltage and current from (6.37) and (6.38) incorporated with (6.47) as

$$V_s(z) = V_0^+ \,\mathrm{e}^{-\mathrm{j}\beta z}(1 + \Gamma(z)) \tag{6.57a}$$

$$I_s(z) = \frac{V_0^+}{Z_0}\mathrm{e}^{-\mathrm{j}\beta z}(1 - \Gamma(z)) \tag{6.57b}$$

Substituting (6.57) into (6.56) yields

$$\begin{aligned} P_{\text{ave}} &= \frac{1}{2}\mathrm{Re}\left[V_0^+ \,\mathrm{e}^{-\mathrm{j}\beta z}(1 + \Gamma(z))\frac{V_0^{+^*}}{Z_0}\,\mathrm{e}^{\mathrm{j}\beta z}(1 - \Gamma^*(z))\right] \\ &= \frac{1}{2}\mathrm{Re}\left[\frac{|V_0^+|^2}{Z_0}(1 - |\Gamma(z)|^2 + \Gamma(z) - \Gamma^*(z))\right] \end{aligned} \tag{6.58}$$

Since the term $[\Gamma(z) - \Gamma^*(z)]$ is purely imaginary,

$$P_{\text{ave}} = \frac{1}{2}\left[\frac{|V_0^+|^2}{Z_0}(1 - |\Gamma(z)|^2)\right] \tag{6.59}$$

For lossless line $\Gamma(z) = \Gamma_L\,\mathrm{e}^{-2\beta l'}$. Therefore

$$|\Gamma(z)| = |\Gamma_L| \tag{6.60}$$

Substituting (6.60) into (6.59) yields

$$P_{\text{ave}} = \frac{1}{2}\left[\frac{|V_0^+|^2}{Z_0}(1 - |\Gamma_L|^2)\right] \tag{6.61}$$

It is noted from (6.61) that the power flow is constant and does not depend on z, since it is a lossless line. Moreover, in lossless lines, the power flow is equal to the power delivered to the load. Thus, it is also noted that the maximum power is delivered to the load when $\Gamma_L = 0$, as expected.

We now consider three special cases when the line is connected to load $Z_L = 0$, $Z_L = \infty$ and $Z_L = Z_0$.

6.3.2 Short-circuit line ($Z_L = 0$)

For this case, (6.44) becomes

$$Z_{\text{sc}} = Z_{\text{in}}(z)|_{Z_L=0} = Z_0\left[\frac{Z_L + \mathrm{j}Z_0 \tan\beta l'}{Z_0 + \mathrm{j}Z_L \tan\beta l'}\right]\Bigg|_{Z_L=0} = \mathrm{j}Z_0 \tan\beta l' \tag{6.62a}$$

The reflection coefficient and SWR from (6.46) and (6.49) are obtained as

$$\Gamma_L = \left.\frac{Z_L - Z_0}{Z_L + Z_0}\right|_{Z_L=0} = -1 \tag{6.62b}$$

$$s = \frac{1 + |\Gamma_L|}{1 - |\Gamma_L|} = \infty \tag{6.62c}$$

It is noted from (6.62a) that $Z_{\text{in}}(z)$ is a pure reactance, which could be capacitive or inductive depending on the distance.

6.3.3 Open-circuit line ($Z_L = \infty$)

For this case, Equation (6.44) becomes

$$Z_{\text{oc}} = Z_{\text{in}}(z)|_{Z_L=\infty} = Z_0\left[\frac{Z_L + \mathrm{j}Z_0 \tan\beta l'}{Z_0 + \mathrm{j}Z_L \tan\beta l'}\right]\Bigg|_{Z_L=\infty} = -\mathrm{j}Z_0 \cot\beta l' \tag{6.63a}$$

The reflection coefficient and SWR from (6.46) and (6.49) are obtained as

$$\Gamma_L = \left.\frac{Z_L - Z_0}{Z_L + Z_0}\right|_{Z_L=\infty} = 1 \tag{6.63b}$$

$$s = \frac{1 + |\Gamma_L|}{1 - |\Gamma_L|} = \infty \tag{6.63c}$$

It is also noted from (6.63a) that $Z_{\text{in}}(z)$ could be capacitive or inductive depending on the distance. Further, from (6.62a) and (6.63a), we obtain

$$Z_{\text{sc}}Z_{\text{oc}} = Z_0^2 \tag{6.64}$$

6.3.4 Matched line ($Z_{\text{L}} = Z_0$)

For this case, Equation (6.44) becomes

$$Z_{\text{matched}} = Z_{\text{in}}(z)|_{Z_L=Z_0} = Z_0\left.\left[\frac{Z_L + \mathrm{j}Z_0\tan\beta l'}{Z_0 + \mathrm{j}Z_L\tan\beta l'}\right]\right|_{Z_L=Z_0} = Z_0 \tag{6.65a}$$

The reflection coefficient and SWR from (6.46) and (6.49) are obtained as

$$\Gamma_L = \left.\frac{Z_L - Z_0}{Z_L + Z_0}\right|_{Z_L=Z_0} = 0 \tag{6.65b}$$

$$s = \frac{1 + |\Gamma_L|}{1 - |\Gamma_L|} = 1 \tag{6.65c}$$

It is noted that the whole wave is transmitted, and there is no reflection. The incident power is fully absorbed by the load. Thus, maximum power transfer is possible when a transmission line is matched to the load.

6.4 The Smith Chart

Before the invention of computers and scientific calculators, to avoid tedious calculations involved in obtaining the characteristics of transmission lines, graphical means were developed. The Smith Chart is the most commonly used graphical technique to evaluate transmission line parameters. It is basically a graphical representation of the impedance of a transmission line and of the corresponding reflection coefficient as one moves along the line. We will first examine how the Smith Chart is constructed and that will help us to utilize it in the calculation of transmission line characteristics, such as Γ_L, s and Z_{in}. We generally assume that the transmission line to which the Smith Chart will be applied is lossless, although this is not fundamentally required. Today the Smith Chart is computerized and widely used easily if its principles are understood.

The construction of the Smith Chart is based on the relation in (6.46), that is,

$$\Gamma_L = \frac{Z_L - Z_0}{Z_L + Z_0} = \Gamma_L^r + \mathrm{j}\Gamma_L^i \tag{6.66}$$

where Γ_L^r and Γ_L^i are the real and imaginary parts of the reflection coefficient Γ_L. Instead of having separate Smith Charts for transmission lines with different characteristic impedances, we prefer to have just one that can be used for any line. We achieve this by using a normalized chart in which all impedances are normalized with respect to the characteristic impedance Z_0. Thus, for the load impedance Z_L, the normalized impedance $\bar{Z}_L$ is given by

$$\bar{Z}_L = \frac{Z_L}{Z_0} = r + \mathrm{j}x \tag{6.67}$$

Substituting (6.67) into (6.66) yields, upon separation of Γ_L intio its real and imaginary parts,

$$\Gamma_L = \frac{\bar{Z}_L - 1}{\bar{Z}_L + 1} = \Gamma_L^r + \mathrm{j}\Gamma_L^i = \frac{(r + \mathrm{j}x) - 1}{(r + \mathrm{j}x) + 1} \tag{6.68a}$$

or

$$\bar{Z}_L = r + \mathrm{j}x = \frac{(1 + \Gamma_L^r) + \mathrm{j}\Gamma_L^i}{(1 - \Gamma_L^r) - \mathrm{j}\Gamma_L^i} \tag{6.68b}$$

Equating real and imaginary components in (6.68b), we obtain

$$r = \frac{1 - (\Gamma_L^r)^2 - (\Gamma_L^i)^2}{(1 - \Gamma_L^r)^2 + (\Gamma_L^i)^2} \tag{6.69a}$$

$$x = \frac{2\Gamma_L^i}{(1 - \Gamma_L^r)^2 + (\Gamma_L^i)^2} \tag{6.69b}$$

Rearranging terms in (6.69a) and (6.69b), respectively, leads to

$$\left(\Gamma_L^r - \frac{r}{1+r}\right)^2 + (\Gamma_L^i)^2 = \left(\frac{1}{1+r}\right)^2 \tag{6.70}$$

and

$$(\Gamma_L^r - 1)^2 + \left(\Gamma_L^i - \frac{1}{x}\right)^2 = \left(\frac{1}{x}\right)^2 \tag{6.71}$$

Each of (6.70) and (6.71) is similar to

$$(\Gamma_L^r - a)^2 + (\Gamma_L^i - b)^2 = (r_0)^2 \tag{6.72}$$

which is the general equation of a circle of radius r_0, centered at (a, b) on the (Γ_L^r, Γ_L^i) plane axes. Therefore, Equation (6.70) can be called the resistance circle (*r-circle*) which is centered at $\left(\frac{r}{1+r}, 0\right)$ with radius $\left(\frac{1}{1+r}\right)$.

Table 6.1: Radii and Centers of x-Circles for Typical Value of x.

Normalized Resistance (r)	Radius $\left(\dfrac{1}{1+r}\right)$	Center $\left(\dfrac{r}{1+r}, 0\right)$
0	1	(0, 0)
0.2	5/6	(1/6, 0)
0.5	2/3	(1/3, 0)
1	1/2	(1/2, 0)
2	1/3	(2/3, 0)
5	1/6	(5/6, 0)
10	1/11	(10/11, 0)
∞	0	(1, 0)

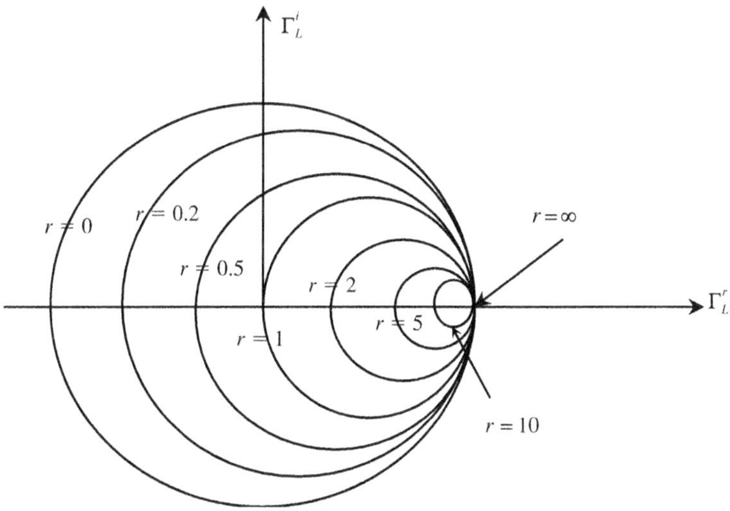

Figure 6.7: Typical r-circles for different values of r.

For typical values of the normalized resistance r, the corresponding centers and radii of the r-circle are presented in Table 6.1. Typical examples of the r-circles based on the data in Table 6.1 are shown in Fig. 6.7.

Similarly, Equation (6.71) is an x-circle (*reactance circle*) centered at $\left(1, \dfrac{1}{x}\right)$ with the radius $\left(\dfrac{1}{x}\right)$. Table 6.2 presents centers and radii of the x-circle for the typical values of x whereas Fig. 6.8 shows the corresponding illustrations. Note that x can be positive (for inductive impedance) or negative (for capacitive impedance).

Table 6.2: Radii and Centers of x-Circles for Typical Value of x.

Normalized Reactance (x)	Radius $\left(\frac{1}{x}\right)$	Centre $\left(1, \frac{1}{x}\right)$
0	∞	$(1, \pm\infty)$
± 0.2	5	$(1, \pm 5)$
± 0.5	2	$(1, \pm 2)$
± 1	1	$(1, \pm 1)$
± 2	1/2	$(1, \pm 1/2)$
± 5	1/5	$(1, \pm 1/5)$
± 10	1/10	$(1, \pm 1/10)$
$\pm\infty$	0	$(1, 0)$

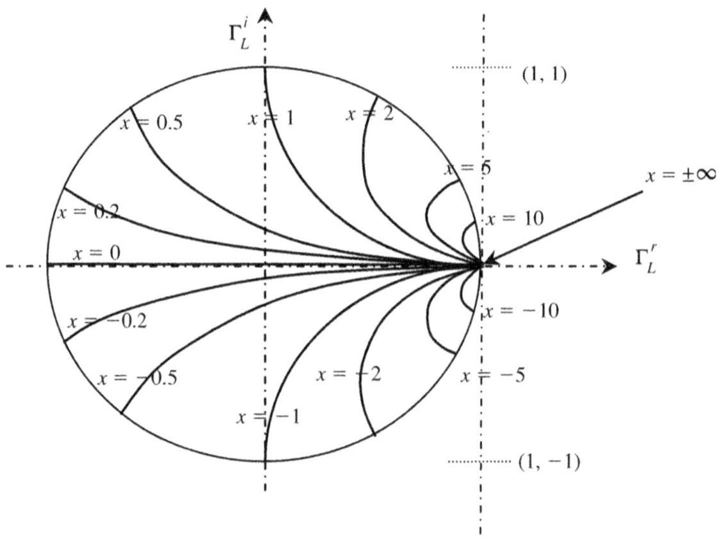

Figure 6.8: Typical x-circles for different values of x.

If we superimpose the r-circles and x-circles, we have the Smith Chart shown in Fig. 6.9.

Apart from the r-circles and x-circles, we can draw the s-circles (*constant SWR circles*) which are generally not shown in the Smith Chart. s-Circles are centered at the origin with s varying from 1 to ∞. The value of the SWR s is determined by locating where an s-circle crosses the Γ_L^r axis. Typical examples of s-circles are shown in Fig. 6.10.

Some important points to note are:

- At point P_{SC} (indicated in Fig. 6.9) on the Smith Chart $r=0$ and $x=0$, which is showing that P_{SC} represents a short circuit on the transmission line.

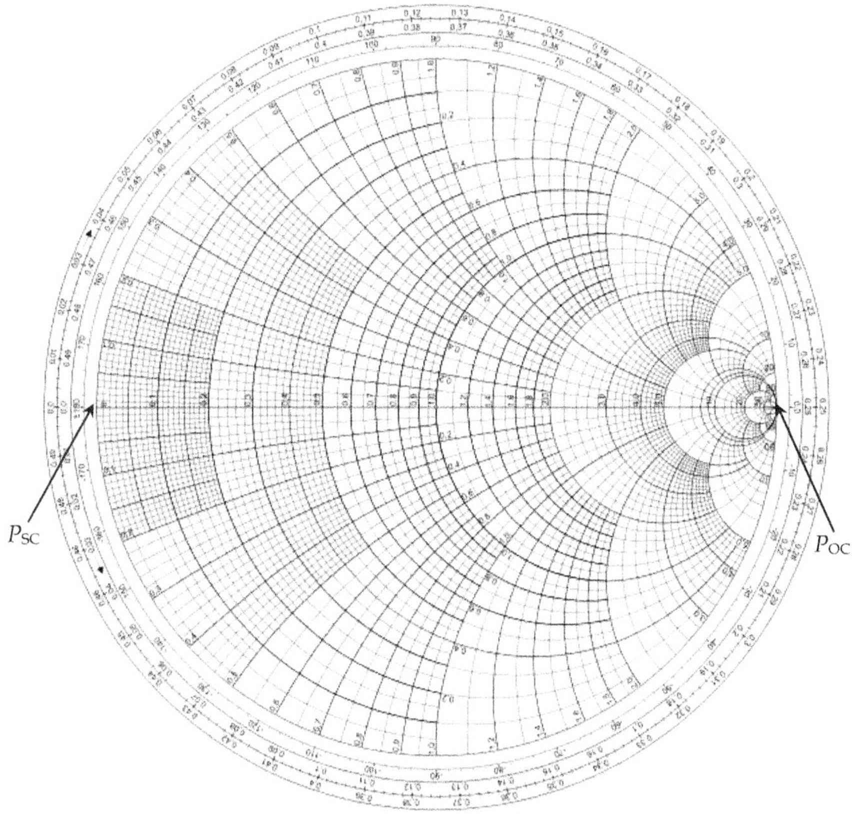

Figure 6.9: Illustration of r-circles and x-circles on the Smith Chart.

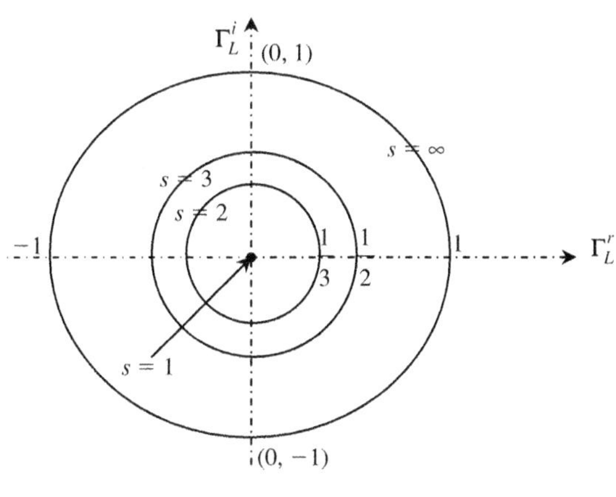

Figure 6.10: Typical s-circles for typical values of s.

At point P_{OC}, $r = \infty$ and $x = \infty$, which implies that P_{OC} corresponds to an open circuit on the line.

- For a lossless line, Equation (6.48) can be given as

$$\Gamma(z) = \Gamma_L \, \mathrm{e}^{-2\mathrm{j}\beta l'} = \Gamma_L \, \mathrm{e}^{-\mathrm{j}4\pi \frac{l'}{\lambda}} \tag{6.73}$$

where the distance l' is measured from the load towards the generator. The negative sign of the phase angle in (6.73) indicates that the phase angle changes in the clockwise direction. In addition, a complete revolution in the Smith Chart occurs when the phase angle changes by 2π. Thus, it is clear that

$$4\pi \frac{l'}{\lambda} = 2\pi$$

$$l' = \frac{\lambda}{2} \tag{6.74}$$

That is, a complete revolution (360°) around the Smith Chart represents a distance $\lambda/2$ on the line. Clockwise movement on the chart is regarded as moving towards the generator (away from load) as shown in Fig. 6.11. Similarly, counter-clockwise movement on the chart corresponds to moving towards the load (away from the generator).

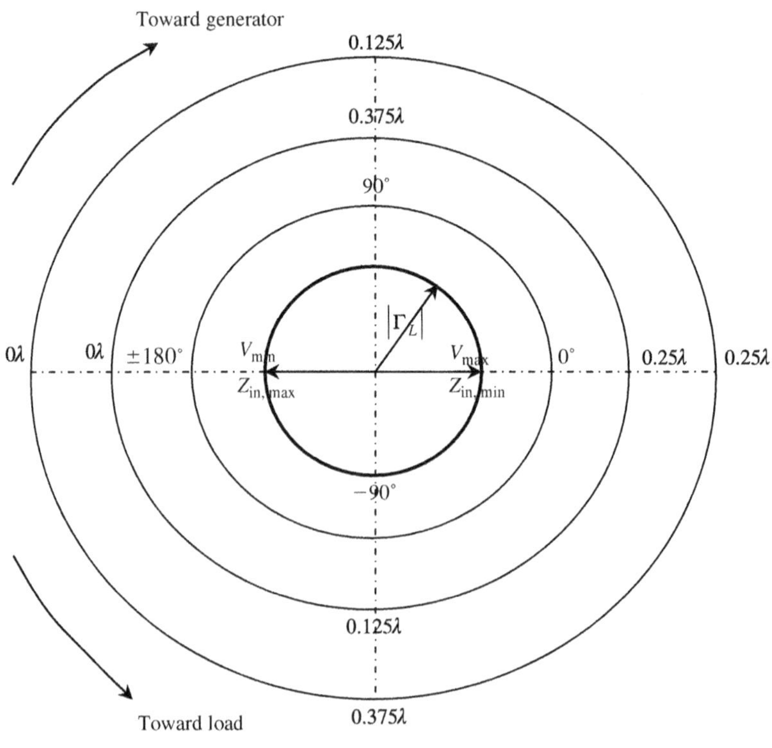

Figure 6.11: Smith Chart illustrating scales around the periphery and movements around the chart.

- There are three scales around the periphery of the Smith Chart as shown in Fig. 6.11. The three scales are included for the sake of convenience but they are actually meant to serve the same purpose. The scales are used in determining the distance from the load or generator in degrees or wavelengths. Since $\lambda/2$ distance on the line corresponds to a movement of 360° on the chart, λ distance on the line corresponds to 720° movement on the chart.
- Based on (6.50), (6.51), (6.54) and (6.55), the voltage V_{max} occurs where $Z_{\text{in,max}}$ is located on the chart, which is on the positive Γ_L^r-axis as shown in Fig. 6.9 and 6.11. The voltage V_{min} is located at the same point where we have $Z_{\text{in,min}}$ on the chart which is on the negative Γ_L^r-axis. Also note that V_{max} and V_{min} or ($Z_{\text{in,max}}$ and $Z_{\text{in,min}}$) are $\lambda/4$ apart.
- The normalized admittance at load obtained from (6.68b) is

$$\bar{Y}_L = \frac{1}{\bar{Z}_L} = \frac{1}{r + \mathrm{j}x} = \frac{(1 - \Gamma_L^r) - \mathrm{j}\Gamma_L^i}{(1 + \Gamma_L^r) + \mathrm{j}\Gamma_L^i} = g + jb \tag{6.75}$$

Thus, comparing expressions for normalized impedance and normalized admittance in terms of the reflection coefficient they are

$$\begin{aligned} r + \mathrm{j}x &= \frac{(1 + \Gamma_L^r) + \mathrm{j}\Gamma_L^i}{(1 - \Gamma_L^r) - \mathrm{j}\Gamma_L^i} = \frac{1 + \Gamma_L}{1 - \Gamma_L} \\ g + jb &= \frac{(1 - \Gamma_L^r) - \mathrm{j}\Gamma_L^i}{(1 + \Gamma_L^r) + \mathrm{j}\Gamma_L^i} = \frac{1 - \Gamma_L}{1 + \Gamma_L} \end{aligned} \tag{6.76}$$

Looking at (6.76), we can say that if a given normalized impedance is represented by a reflection coefficient Γ in the Smith Chart, then the corresponding normalized admittance must be represented by $-\Gamma$ in the Smith Chart.

Based on the important properties given above, the Smith Chart may be used to determine reflection coefficient, SWR, input impedance, input admittance and the locations V_{max} and V_{min} provided that we are given the characteristic impedance, load impedance, wavelength and the length of the line.

6.5 Applications of transmission lines

Transmission lines are used to serve different purposes. Here, we consider how transmission lines are used for load matching and impedance measurements.

6.5.1 Quarter-wave transformer (matching)

When $Z_0 = Z_L$, it is said that the load is matched and no reflected wave exists on the line as $\Gamma_L = 0$. There is no reflection component thus maximum power is transmitted. Hence, for maximum power transmission in the line, the load impedance must

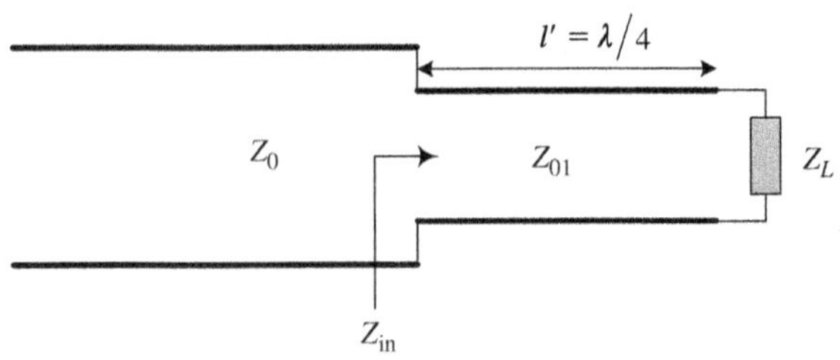

Figure 6.12: Load matching using a λ/4 transformer.

be matched to the characteristic impedance of the line. It is also possible to achieve matching of a mismatched load by placing a short section of transmission line between the load and the end of the line as shown in the Fig. 6.12. Thus by adding a $\lambda/4$ line with characteristic impedance (Z_{01}) between load and the line, a mismatched load (Z_L) can be properly matched to a line with characteristic impedance (Z_0). Substituting $l' = \lambda/4$ or $\beta l' = \pi/2$ in (6.44), we obtain

$$Z_{\text{in}} = Z_{01}\left[\frac{Z_L + \text{j}Z_{01}\tan\pi/2}{Z_{01} + \text{j}Z_L\tan\pi/2}\right] = \frac{Z_{01}^2}{Z_L} \tag{6.77}$$

To satisfy the matching condition,

$$\begin{aligned} Z_{\text{in}} &= \frac{Z_{01}^2}{Z_L} = Z_0 \\ Z_{01} &= \sqrt{Z_0 Z_L} \end{aligned} \tag{6.78}$$

The $\lambda/4$ section of the transmission line is called a *quarter-wave transformer* because it is used for impedance matching like an ordinary transformer.

The voltage standing wave pattern with the $\lambda/4$ transformer is shown in Fig. 6.13. It is observed that although a standing wave still exists in the quarter-wave transformer line, there is no standing wave to the left of the transformer due to the matching. However, there will be reflection at a slightly different wavelength (or frequency). It must be noted that this is not a practicable implementation since for a given load Z_L at the end of the line Z_0, a line of characteristic impedance Z_{01} may not be readily available.

6.5.2 Single-stub tuner (matching)

The major drawback of using a quarter-wave transformer is finding an appropriate quarter wave line with the required characteristic impedance $Z_{01} = \sqrt{Z_0 Z_L}$. This problem is evaded by using a *single-stub* tuner where the stub line has the same characteristic impedance of the main line. The tuner consists of an open or shorted section of transmission line of length l connected in parallel (stub) with the main line at some distance d from the load as in Fig. 6.14.

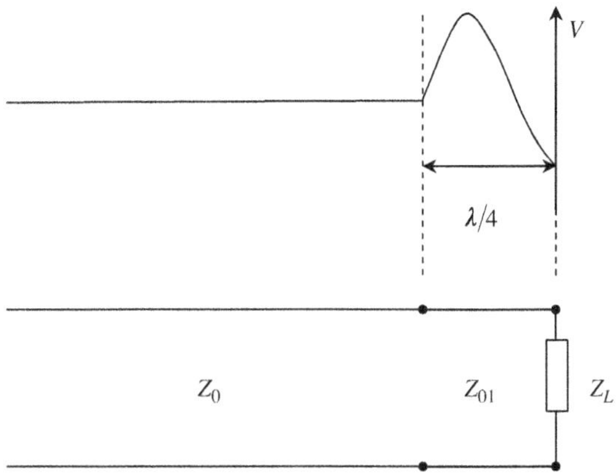

Figure 6.13: Voltage standing wave pattern of mismatched load with a $\lambda/4$ transformer.

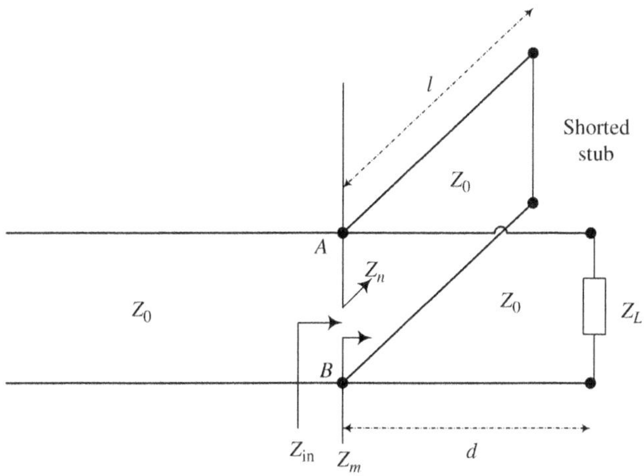

Figure 6.14: Matching with single-stub tuner.

Assuming that the line is lossless, so that Z_0 is real. We obtain the normalized impedances of Z_n and Z_m for the short circuited stub and the line of length d as below from (6.62a).

$$\bar{Z}_n = \frac{Z_n}{Z_0} = \mathrm{j}\tan\beta l \tag{6.79a}$$

$$\bar{Z}_m = \frac{Z_m}{Z_0} = \left[\frac{\bar{Z}_L + \mathrm{j}\tan\beta d}{1 + \mathrm{j}\bar{Z}_L\tan\beta d}\right] \tag{6.79b}$$

In order to satisfy the matching condition in the line, the normalized impedance at terminal AB needs to be 1. Thus, we write

$$\bar{Z}_{\text{in}} = \frac{Z_{\text{in}}}{Z_0} = 1 = \left[\frac{1}{\bar{Z}_m} + \frac{1}{\bar{Z}_n}\right]^{-1} \tag{6.80}$$

Equation (6.80) can be rewritten in terms of the respective normalized admittance as

$$\bar{Y}_{\text{in}} = 1 = \bar{Y}_m + \bar{Y}_n \tag{6.81}$$

Substituting from (6.79) into (6.81) yields

$$1 = -\text{j}\cot\beta l + \frac{1 + \text{j}\bar{Z}_L \tan\beta d}{\bar{Z}_L + \text{j}\tan\beta d} \tag{6.82}$$

Equating real and imaginary parts

$$\text{Re}\left[\frac{1 + \text{j}\bar{Z}_L \tan\beta d}{\bar{Z}_L + \text{j}\tan\beta d}\right] = 1$$

$$\text{Im}\left[\frac{1 + \text{j}\bar{Z}_L \tan\beta d}{\bar{Z}_L + \text{j}\tan\beta d}\right] = \cot\beta l \tag{6.83}$$

The lengths l and d can be evaluated either numerically or using the Smith Chart to satisfy (6.83) since Z_L is known.

6.6 Waveguides

6.6.1 Introduction

A waveguide can be used to guide electromagnetic (EM) energy from one point to another point, the same as in a transmission line. However, a waveguide differs from transmission line since a transmission line supports only a TEM wave whereas a waveguide supports many possible field configurations that are called modes. Higher order modes in waveguides do not begin to propagate until the frequency of excitation is more than a certain frequency called the cut off frequency but TEM waves do not suffer this restriction and are able to propagate at all frequencies. Further, at microwave frequencies (3–300 GHz), transmission lines become inefficient as a result of the skin effect and dielectric losses. However, waveguides are used at the microwave range of frequencies to achieve a larger bandwidth and lower signal attenuation.

Generally, a waveguide may assume any uniform cross section and common waveguides have either rectangular or circular cross sections. Analysis of circular waveguides is involved and requires familiarity with Bessel functions. For simplicity, we consider only rectangular waveguides here. Cavity resonators may be thought of as rectangular waveguides of finite length having conducting planes at each end. This will allow us to extend the results obtained for rectangular waveguides to cavity resonators in a simple manner.

6.6.2 Separable characteristic of the wave equation

The wave equation obtained in Chapter 4 is a general form equation for a propagating EM wave. Therefore, waves in wave guiding structures are also governed by the wave equations. The following general development applies to any structure whose cross section is uniform along the z axis.

In order to characterize wave propagation on a general structure, we examine the wave equation for a structure composed of perfect conductors ($\sigma_c = \infty$) surrounded by lossless dielectrics ($\sigma_d = 0$) characterized by ε and μ.

In the surrounding lossless dielectric medium, the wave equation (4.8) of Chapter 4 reduces to

$$\nabla^2\mathbf{E} + \omega^2\mu\varepsilon\mathbf{E} = 0 \tag{6.84a}$$

$$\nabla^2\mathbf{H} + \omega^2\mu\varepsilon\mathbf{H} = 0 \tag{6.84b}$$

Since the wave propagates in the z direction, the field components could be written as

$$\begin{aligned} E_x(x,y,z) &= \hat{E}_x(x,y)\,\mathrm{e}^{-\gamma z} \\ E_y(x,y,z) &= \hat{E}_y(x,y)\,\mathrm{e}^{-\gamma z} \\ &\vdots \\ H_z(x,y,z) &= \hat{H}_z(x,y)\,\mathrm{e}^{-\gamma z} \end{aligned} \tag{6.85}$$

The constant γ is referred to as the propagation constant. To represent propagation, γ must be either complex or imaginary. The field vectors given in (6.85) are in the form of a forward traveling wave (traveling in the $+z$ direction). If backward-traveling waves are also of interest, the general form of the field vectors is to be the summation of both forms as

$$\begin{aligned} E_x(x,y,z) &= \hat{E}_x^+(x,y)\,\mathrm{e}^{-\gamma z} + \hat{E}_x^-(x,y)\,\mathrm{e}^{+\gamma z} \\ E_y(x,y,z) &= \hat{E}_y^+(x,y)\,\mathrm{e}^{-\gamma z} + \hat{E}_y^-(x,y)\,\mathrm{e}^{+\gamma z} \\ &\vdots \\ H_z(x,y,z) &= \hat{H}_z^+(x,y)\,\mathrm{e}^{-\gamma z} + \hat{H}_z^-(x,y)\,\mathrm{e}^{+\gamma z} \end{aligned} \tag{6.86}$$

Substituting the assumed forms of the field vectors given in (6.86) into the wave equation given in (6.84) and matching components yields

$$\begin{aligned}
&\frac{\partial^2 E_x}{\partial x^2} + \frac{\partial^2 E_x}{\partial y^2} + (\gamma^2 + \omega^2 \mu\varepsilon)E_x = 0 \\
&\frac{\partial^2 E_y}{\partial x^2} + \frac{\partial^2 E_y}{\partial y^2} + (\gamma^2 + \omega^2 \mu\varepsilon)E_y = 0 \\
&\qquad\qquad\vdots \\
&\frac{\partial^2 H_z}{\partial x^2} + \frac{\partial^2 H_z}{\partial y^2} + (\gamma^2 + \omega^2 \mu\varepsilon)H_z = 0
\end{aligned} \tag{6.87}$$

Combining (6.87) into vector form gives the reduced wave equation

$$\begin{aligned}
&\nabla_T^2 \mathbf{E} + (\gamma^2 + \omega^2 \mu\varepsilon)\mathbf{E} = 0 \\
&\nabla_T^2 \mathbf{H} + (\gamma^2 + \omega^2 \mu\varepsilon)\mathbf{H} = 0
\end{aligned} \tag{6.88}$$

where ∇_T^2 is the Laplacian operator transverse to the z axis

$$\nabla_T^2 = \frac{\partial^2}{\partial x^2} + \frac{\partial^2}{\partial y^2} \tag{6.89}$$

Now consider Faraday's and Ampere's laws in the dielectric medium ($\sigma_\mathrm{d} = 0$) of the guide.

$$\begin{aligned}
&\nabla \times \mathbf{E} = -\mathrm{j}\omega\mu\mathbf{H} \\
&\nabla \times \mathbf{H} = \mathrm{j}\omega\varepsilon\mathbf{E}
\end{aligned} \tag{6.90}$$

Substituting the field vectors given for forward traveling waves in (6.85) into (6.90) yields

$$\frac{\partial E_z}{\partial y} + \gamma E_y = -\mathrm{j}\omega\mu H_x \tag{6.91a}$$

$$-\gamma E_x - \frac{\partial E_z}{\partial x} = -\mathrm{j}\omega\mu H_y \tag{6.91b}$$

$$\frac{\partial E_y}{\partial x} - \frac{\partial E_x}{\partial y} = -\mathrm{j}\omega\mu H_z \tag{6.91c}$$

and

$$\frac{\partial H_z}{\partial y} + \gamma H_y = \mathrm{j}\omega\varepsilon E_x \tag{6.92a}$$

$$-\gamma H_x - \frac{\partial H_z}{\partial x} = \mathrm{j}\omega\varepsilon E_y \tag{6.92b}$$

$$\frac{\partial H_y}{\partial x} - \frac{\partial H_x}{\partial y} = -\mathrm{j}\omega\varepsilon E_z \tag{6.92c}$$

because $(\partial/\partial z)e^{-\gamma z} = -\gamma\, e^{-\gamma z}$. For backward traveling wave we replace γ in (6.91) and (6.92) with $-\gamma$ since $(\partial/\partial z)e^{-\gamma z} = -\gamma\, e^{-\gamma z}$. Combining (6.91) and (6.92) into vector form

$$
\begin{aligned}
(\nabla_T - \gamma \mathbf{u}_z) \times \mathbf{E} &= -j\omega\mu\mathbf{H} \\
(\nabla_T - \gamma \mathbf{u}_z) \times \mathbf{H} &= j\omega\varepsilon\mathbf{E}
\end{aligned}
\tag{6.93}
$$

where ∇_T is the operator transverse to the z axis

$$\nabla_T = \frac{\partial}{\partial x}\mathbf{u}_x + \frac{\partial}{\partial y}\mathbf{u}_y \tag{6.94}$$

Rearranging (6.91a) and (6.92b), we can write the E_y and H_x components in terms of E_z and H_z. Similarly rearranging (6.92a) and (6.91b), we can write the E_x and H_y components in terms of E_z and H_z. Therefore, we recognize that we only need to solve for E_z and H_z components of **E** and **H** from which all the other field components can easily be evaluated. The E_x, E_y, H_x and H_y components are derived in terms of E_z and H_z as follows:

$$E_x = -\frac{1}{\gamma^2 + \omega^2\mu\varepsilon}\left(j\omega\mu\frac{\partial H_z}{\partial y} + \gamma\frac{\partial E_z}{\partial x}\right) \tag{6.95a}$$

$$E_y = \frac{1}{\gamma^2 + \omega^2\mu\varepsilon}\left(j\omega\mu\frac{\partial H_z}{\partial x} - \gamma\frac{\partial E_z}{\partial y}\right) \tag{6.95b}$$

$$H_x = \frac{1}{\gamma^2 + \omega^2\mu\varepsilon}\left(j\omega\varepsilon\frac{\partial E_z}{\partial y} - \gamma\frac{\partial H_z}{\partial x}\right) \tag{6.95c}$$

$$H_y = -\frac{1}{\gamma^2 + \omega^2\mu\varepsilon}\left(j\omega\varepsilon\frac{\partial E_z}{\partial x} + \gamma\frac{\partial H_z}{\partial y}\right) \tag{6.95d}$$

Note that Equation (6.95) is obtained for the forward traveling components. Replacing γ with $-\gamma$ in (6.95), we can obtain the relationship for backward traveling waves. Finally, we note that we only need to solve for the z components of the wave equations within the guide.

$$\frac{\partial^2 E_z}{\partial x^2} + \frac{\partial^2 E_z}{\partial y^2} + (\gamma^2 + \omega^2\mu\varepsilon)E_z = 0 \tag{6.96a}$$

$$\frac{\partial^2 H_z}{\partial x^2} + \frac{\partial^2 H_z}{\partial y^2} + (\gamma^2 + \omega^2\mu\varepsilon)H_z = 0 \tag{6.96b}$$

In order to solve for E_z given in (6.96a), we further assume that the field vector's spatial dependence can be separated as

$$E_z(x, y, z) = F(x)G(y)e^{-\gamma z} \tag{6.97}$$

We assume that the spatial electric field component E_z and magnetic field component H_z can be expressed as the product of three functions, each being a function of only one of the coordinate variables. This solution technique for partial differential equations is referred to as the method of separation of variables. In differential equation theory, this is legitimate because we are merely investigating if there is such a solution. If we find such a solution with two constants of integration, then at that point we assert that it is the solution. This method is only applicable to metallic waveguides where a conducting surface is used to guide the wave in a particular direction. Substituting (6.97) into (6.96a) and rearranging terms, we obtain

$$\frac{1}{F(x)}\frac{\mathrm{d}^2F(x)}{\mathrm{d}x^2}+\frac{1}{G(y)}\frac{\mathrm{d}^2G(y)}{\mathrm{d}y^2}=-(\gamma^2+\omega^2\mu\varepsilon) \tag{6.98}$$

Since $F(x)$ is a function only of x, and $G(y)$ is a function only of y, each term on the left hand side must be constant. Otherwise, the left-hand side of the equation cannot be equal to a constant as on the right for all values of x and y. Thus, we state

$$\frac{1}{F(x)}\frac{\mathrm{d}^2F(x)}{\mathrm{d}x^2}=-\beta_x^2 \tag{6.99a}$$

$$\frac{1}{G(y)}\frac{\mathrm{d}^2G(y)}{\mathrm{d}y^2}=-\beta_y^2 \tag{6.99b}$$

The sign of these *separation constants* by their squaring is arbitrarily chosen to simplify the solution. We can show that for a wave to exist, these constants cannot be zero or positive. Substituting (6.99) into (6.98) yields the relationship between the separation constants and the propagation constant.

$$\beta_x^2+\beta_y^2=(\gamma^2+\omega^2\mu\varepsilon) \tag{6.100}$$

The solutions to (6.99) are of the form

$$F(x)=F_1\cos\beta_x x+F_2\sin\beta_x x \tag{6.101a}$$

$$G(y)=G_1\cos\beta_y y+G_2\sin\beta_y y \tag{6.101b}$$

where F_1, F_2, G_1 and G_2 are undetermined constants. Hence, E_z becomes

$$E_z(x,y,z)=(F_1\cos\beta_x x+F_2\sin\beta_x x)(G_1\cos\beta_y y+G_2\sin\beta_y y)\mathrm{e}^{-\gamma z} \tag{6.102}$$

Similarly, H_z can be written as

$$H_z(x,y,z)=(F_3\cos\beta_x x+F_4\sin\beta_x x)(G_3\cos\beta_x x+G_4\sin\beta_x x)\mathrm{e}^{-\gamma z} \tag{6.103}$$

where F_3, F_4, G_3 and G_4 are undetermined constants.

Once the separation constants β_x and β_y are determined for a particular waveguide, the propagation constant is determined from (6.100) as

$$\gamma = \sqrt{\beta_x^2 + \beta_y^2 - \omega^2 \mu \varepsilon} \tag{6.104}$$

In order to have propagation in the z direction, γ should be purely imaginary. Hence, we can write that

$$\gamma = \mathrm{j}\beta_z \tag{6.105}$$

Thus, the waves have z dependence as $\mathrm{e}^{-\mathrm{j}\beta_z z}$. Also, to satisfy the condition given in (6.105), it requires from (6.104) that $\omega^2 \mu \varepsilon > \beta_x^2 + \beta_y^2$. Further, substituting (6.105) into (6.100), we obtain the *dispersion relation*:

$$\beta_x^2 + \beta_y^2 + \beta_z^2 = \omega^2 \mu \varepsilon \tag{6.106}$$

Therefore, the phase constant in the z direction is derived as

$$\beta_z = \sqrt{\omega^2 \mu \varepsilon - (\beta_x^2 + \beta_y^2)} \quad \text{provided that } \omega^2 \mu \varepsilon > \beta_x^2 + \beta_y^2 \tag{6.107}$$

When $\omega^2 \mu \varepsilon < \beta_x^2 + \beta_y^2$, then the propagation constant is purely real as

$$\gamma = \alpha_z \tag{6.108}$$

and the wave will have z dependence as $\mathrm{e}^{-\alpha_z z}$ representing attenuation of the wave amplitude, where

$$\alpha_z = \sqrt{(\beta_x^2 + \beta_y^2) - \omega^2 \mu \varepsilon} \quad \text{provided that } \omega^2 \mu \varepsilon < \beta_x^2 + \beta_y^2 \tag{6.109}$$

Thus, we can say that the wave is about to propagate when $\omega^2 \mu \varepsilon = \beta_x^2 + \beta_y^2$. Hence, the cutoff frequency is defined as

$$\begin{aligned} \omega_c^2 \mu \varepsilon &= \beta_x^2 + \beta_y^2 \\ f_c &= \frac{1}{2\pi} \sqrt{\frac{\beta_x^2 + \beta_y^2}{\mu \varepsilon}} \end{aligned} \tag{6.110}$$

There are several other quantities of interest that can be derived from the above results and written in terms of the cutoff frequency given in (6.110). First the phase constant in the z direction obtained from (6.107):

$$\beta_z = \beta \sqrt{1 - (f_c/f)^2} \quad \text{provided that } f > f_c \tag{6.111}$$

where $\beta = \omega \sqrt{\mu \varepsilon}$.

The second is the phase velocity in the z direction:

$$u_z = \frac{\omega}{\beta_z} = \frac{u}{\sqrt{1-(f_c/f)^2}} \quad \text{provided that } f > f_c \tag{6.112}$$

where $u = 1/\sqrt{\mu\varepsilon}$ and the third is the wavelength in the z direction:

$$\lambda_z = \frac{2\pi}{\beta_z} = \frac{u_z}{f} = \frac{\lambda}{\sqrt{1-[f_c/f]^2}} \quad \text{provided that } f > f_c \tag{6.113}$$

where $\lambda = u/f$.

The separation constants β_x and β_y can be determined by applying appropriate boundary conditions to the general form of the solutions for E_z and H_z given in (6.102) and (6.103), respectively. Once β_x and β_y are determined, the range of frequencies for which wave propagation is possible for the particular waveguide can be determined from (6.110). Once the form of the solution for E_z and H_z is known by applying appropriate boundary conditions, the solutions for the other field components are obtained from (6.95), which can be rewritten by substituting (6.100) in (6.105) as

$$E_x = -\frac{1}{\beta_x^2+\beta_y^2}\left(\mathrm{j}\omega\mu\frac{\partial H_z}{\partial y} + \mathrm{j}\beta_z\frac{\partial E_z}{\partial x}\right) \tag{6.114a}$$

$$E_y = \frac{1}{\beta_x^2+\beta_y^2}\left(\mathrm{j}\omega\mu\frac{\partial H_z}{\partial x} - \mathrm{j}\beta_z\frac{\partial E_z}{\partial y}\right) \tag{6.114b}$$

$$H_x = \frac{1}{\beta_x^2+\beta_y^2}\left(\mathrm{j}\omega\varepsilon\frac{\partial E_z}{\partial y} - \mathrm{j}\beta_z\frac{\partial H_z}{\partial x}\right) \tag{6.114c}$$

$$H_y = -\frac{1}{\beta_x^2+\beta_y^2}\left(\mathrm{j}\omega\varepsilon\frac{\partial E_z}{\partial x} + \mathrm{j}\beta_z\frac{\partial H_z}{\partial y}\right) \tag{6.114d}$$

There are two separate components (E_z and H_z) contributing to the E_x, E_y, H_x and H_y components as given in (6.114). Therefore, we can categorize the system as consisting of two major field configurations as follows

1. $E_z = 0$, $H_z \neq 0$: In this case, the electrical fields are transverse to the guide axis, and these structures are called *transverse electric* (TE) *modes*, or H-modes since H_z determines all other components.

2. $E_z \neq 0$, $H_z = 0$: In this case, magnetic fields are transverse to the guide axis, and these structures are called *transverse magnetic* (TM) *modes*, or E-modes since E_z determines all other components.

Note that setting $E_z = H_z = 0$ (the TEM mode), the numerator and the denominator of all expressions in (6.114) become zero. Thus, we cannot conclude from these equations that the TEM mode is not a possible mode of operation. From Ampere's law, given in (6.2b), we observe that the integral of this transverse magnetic field around a closed path must yield the axial conduction current or displacement current. Since $E_z = 0$ for the TEM mode, no axial displacement current can exist. Also, since there is no interior conductor to carry a conduction current, we therefore conclude that the axial conduction current is zero. Therefore, it is the conclusion that the TEM mode cannot exist in a hollow waveguide.

6.7 Rectangular waveguides

6.7.1 Application of guided wave equations

General results for the guiding structures in the Cartesian coordinate system have been derived in the previous section. In this section, we consider the application of those results to the rectangular waveguide shown in Fig. 6.15. For the analysis of this structure, we need to enforce the boundary conditions at the perfectly conducting walls of the guide in order to determine the constants in the general equations given in (6.102) and (6.103) as well as the separation constants β_x and β_y for the specified (TM/TE) mode solutions. The general solution is the superposition of these TE and TM solutions as given in (6.114).

6.7.2 TM modes

For TM modes $H_z = 0$ and the general solution for E_z is given in (6.102). The boundary conditions are that the tangential electric field must be zero at the walls

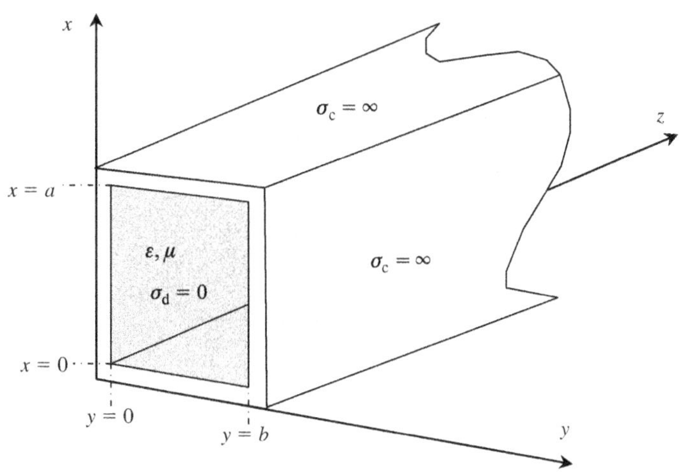

Figure 6.15: Schematic diagram of a rectangular waveguide.

of the guide. The boundary conditions are given as

$$E_z = 0;\ E_y = 0 \quad \text{at } x = 0 \tag{6.115a}$$

$$E_z = 0;\ E_y = 0 \quad \text{at } x = a \tag{6.115b}$$

$$E_z = 0;\ E_x = 0 \quad \text{at } y = 0 \tag{6.115c}$$

$$E_z = 0;\ E_x = 0 \quad \text{at } y = b \tag{6.115d}$$

Since the expression for the E_z component is available, we can directly apply the conditions for E_z given in (6.115a) and (6.115c) into (6.102). This yields $F_1 = G_1 = 0$. Hence (6.102) can be simplified as

$$E_z(x, y, z) = E_0 \sin \beta_x x \sin \beta_y y\, \mathrm{e}^{-\gamma z} \tag{6.116}$$

where $E_0 = F_2 G_2$. Applying (6.115b) and (6.115d) to (6.116) yields

$$\sin \beta_x a = 0 \tag{6.117a}$$

$$\sin \beta_y b = 0 \tag{6.117b}$$

This results in the separation constants as

$$\beta_x = \frac{m\pi}{a} \quad m = 1, 2, 3, \ldots \tag{6.118a}$$

$$\beta_y = \frac{n\pi}{a} \quad n = 1, 2, 3, \ldots \tag{6.118b}$$

Every possible combination of the integers m and n ends up with numerous TM_{mn} modes. Substituting (6.118) into (6.110) to (6.113) yields the cutoff frequency

$$f_{c,mn} = \frac{1}{2\pi\sqrt{\mu\varepsilon}} \sqrt{\left(\frac{m\pi}{a}\right)^2 + \left(\frac{n\pi}{b}\right)^2} \tag{6.119}$$

the phase constant in the z direction

$$\beta_{z,mn} = \beta\sqrt{1 - (f_{c,mn}/f)^2} \tag{6.120}$$

the phase velocity in the z direction:

$$u_{z,mn} = \frac{u}{\sqrt{1 - (f_{c,mn}/f)^2}} \tag{6.121}$$

and the wavelength in the z direction:

$$\lambda_{z,mn} = \frac{\lambda}{\sqrt{1 - [f_{c,mn}/f]^2}} \tag{6.122}$$

Substituting (6.118) into (6.116) and replacing the propagation constant (γ) in terms of the phase constant ($\gamma = \mathrm{j}\beta_{z,mn}$ when $f > f_c$), we obtain the expression for E_z as

$$E_z = E_0 \sin\left(\frac{m\pi}{a}x\right)\sin\left(\frac{n\pi}{b}y\right) \mathrm{e}^{-\mathrm{j}\beta_{z,mn}z} \tag{6.123}$$

The solution for other field components is obtained by substituting (6.118) and (6.123) into (6.114), resulting in

$$E_x = -\mathrm{j}\beta_{z,mn}E_0 \frac{(m\pi/a)}{[(m\pi/a)^2 + (n\pi/b)^2]} \cos\left(\frac{m\pi}{a}x\right)\sin\left(\frac{n\pi}{b}y\right) \mathrm{e}^{-\mathrm{j}\beta_{z,mn}z} \tag{6.124a}$$

$$E_y = -\mathrm{j}\beta_{z,mn}E_0 \frac{(n\pi/b)}{[(m\pi/a)^2 + (n\pi/b)^2]} \sin\left(\frac{m\pi}{a}x\right)\cos\left(\frac{n\pi}{b}y\right) \mathrm{e}^{-\mathrm{j}\beta_{z,mn}z} \tag{6.124b}$$

$$H_x = \mathrm{j}\omega\varepsilon E_0 \frac{(n\pi/b)}{[(m\pi/a)^2 + (n\pi/b)^2]} \sin\left(\frac{m\pi}{a}x\right)\cos\left(\frac{n\pi}{b}y\right) \mathrm{e}^{-\mathrm{j}\beta_{z,mn}z} \tag{6.124c}$$

$$H_y = -\mathrm{j}\omega\varepsilon E_0 \frac{(m\pi/a)}{[(m\pi/a)^2 + (n\pi/b)^2]} \cos\left(\frac{m\pi}{a}x\right)\sin\left(\frac{n\pi}{b}y\right) \mathrm{e}^{-\mathrm{j}\beta_{z,mn}z} \tag{6.124d}$$

In the expressions given in (6.123) and (6.124), setting either m or n to zero leads to all field components being zero. Hence, the lowest-order TM mode is TM_{11}.

For a propagating EM wave, the transverse electrical field vector ($\mathbf{E}_T = (E_x\mathbf{u}_x + E_y\mathbf{u}_y)$), transverse magnetic field vector ($\mathbf{H}_T = (H_x\mathbf{u}_x + H_y\mathbf{u}_y)$) and propagation (power flow) direction vector ($\mathbf{u}_z$) are related by the wave impedance ($\eta_{\mathrm{TM},mn}$) as below:

$$\eta_{\mathrm{TM},mn} = \frac{\mathbf{u}_z \times \mathbf{E}_T}{\mathbf{H}_T} \tag{6.125}$$

Expanding the field components given in (6.125) and substituting from (6.124), we obtain

$$\eta_{\mathrm{TM},mn} = \frac{E_x}{H_y} = -\frac{E_y}{H_x} = \frac{\beta_{z,mn}}{\omega\varepsilon} = \eta\sqrt{1 - \left(\frac{f_{c,mn}}{f}\right)^2} \quad f > f_c \tag{6.126}$$

where $\eta = \mu/\varepsilon$ is the intrinsic impedance of the medium.

The time domain forms of the fields are obtained by multiplying (6.123) and (6.124) by $e^{j\omega t}$ and taking the real part of the result:

$$\mathcal{E}_z = E_0 \sin\left(\frac{m\pi}{a}x\right)\sin\left(\frac{n\pi}{b}y\right)\cos(\omega t - \beta_{z,mn}z) \tag{6.127a}$$

$$\mathcal{E}_x = \beta_{z,mn}E_0\frac{(m\pi/a)}{[(m\pi/a)^2 + (n\pi/b)^2]}\cos\left(\frac{m\pi}{a}x\right)\sin\left(\frac{n\pi}{b}y\right)\sin(\omega t - \beta_{z,mn}z) \tag{6.127b}$$

$$\mathcal{E}_y = \beta_{z,mn}E_0\frac{(n\pi/b)}{[(m\pi/a)^2 + (n\pi/b)^2]}\sin\left(\frac{m\pi}{a}x\right)\cos\left(\frac{n\pi}{b}y\right)\sin(\omega t - \beta_{z,mn}z) \tag{6.127c}$$

$$\mathcal{H}_x = -\omega\varepsilon E_0\frac{(n\pi/b)}{[(m\pi/a)^2 + (n\pi/b)^2]}\sin\left(\frac{m\pi}{a}x\right)\cos\left(\frac{n\pi}{b}y\right)\sin(\omega t - \beta_{z,mn}z) \tag{6.127d}$$

$$\mathcal{H}_y = \omega\varepsilon E_0\frac{(m\pi/a)}{[(m\pi/a)^2 + (n\pi/b)^2]}\cos\left(\frac{m\pi}{a}x\right)\sin\left(\frac{n\pi}{b}y\right)\sin(\omega t - \beta_{z,mn}z) \tag{6.127e}$$

where constant E_0 is assumed to be real.

6.7.3 TE modes

The TE modes ($E_z = 0$) field components of the rectangular waveguide can be obtained by applying the boundary conditions in (6.114). Since $E_z = 0$, we need to apply the boundary conditions on the tangential electric fields at the walls of the waveguide. Thus, we state the conditions as

$$E_y = 0 \quad \text{at } x = 0 \text{ and } x = a \tag{6.128a}$$

$$E_x = 0 \quad \text{at } y = 0 \text{ and } y = b \tag{6.128b}$$

Applying the boundary conditions given in (6.128) into (6.114a) and (6.114b), we can obtain the boundary conditions in terms of H_z as

$$\frac{\partial H_z}{\partial x} = 0 \quad \text{at } x = 0 \tag{6.129a}$$

$$\frac{\partial H_z}{\partial x} = 0 \quad \text{at } x = a \tag{6.129b}$$

$$\frac{\partial H_z}{\partial y} = 0 \quad \text{at } y = 0 \tag{6.129c}$$

$$\frac{\partial H_z}{\partial y} = 0 \quad \text{at } y = b \tag{6.129d}$$

Differentiating the general solution given in (6.103) with respect to x and with respect to y and applying conditions given in (6.129a) and (6.129c) results in

$$H_z = H_0 \cos(\beta_x x)\cos(\beta_y y)e^{-\gamma z} \tag{6.130}$$

where $H_0 = F_3 G_3$.

The same as with the TM mode, applying (6.129b) and (6.129d) to the solution obtained in (6.130) yields

$$\beta_x = \frac{m\pi}{a} \quad m = 1, 2, 3, \ldots \tag{6.131a}$$

$$\beta_y = \frac{n\pi}{a} \quad n = 1, 2, 3, \ldots \tag{6.131b}$$

Thus $f_{c,mn}$, $\beta_{z,mn}$, $u_{z,mn}$ and $\lambda_{z,mn}$ for the TE_{mn} mode remain the same as for the TM_{mn} mode. Replacing the propagation constant (γ) in terms of the phase constant ($\gamma = j\beta_{z,mn}$ when $f > f_c$), we obtain the expression for H_z from (6.130)

$$H_z = H_0 \cos\left(\frac{m\pi}{a}x\right)\cos\left(\frac{n\pi}{b}y\right)e^{-j\beta_{z,mn}z} \tag{6.132a}$$

Substituting (6.131a) into (6.114) yields

$$E_x = j\omega\mu H_0 \frac{(n\pi/b)}{[(m\pi/a)^2 + (n\pi/b)^2]} \cos\left(\frac{m\pi}{a}x\right)\sin\left(\frac{n\pi}{b}y\right)e^{-j\beta_{z,mn}z} \tag{6.132b}$$

$$E_y = -j\omega\mu H_0 \frac{(m\pi/a)}{[(m\pi/a)^2 + (n\pi/b)^2]} \sin\left(\frac{m\pi}{a}x\right)\cos\left(\frac{n\pi}{b}y\right)e^{-j\beta_{z,mn}z} \tag{6.132c}$$

$$H_x = j\beta_{z,mn} H_0 \frac{(m\pi/a)}{[(m\pi/a)^2 + (n\pi/b)^2]} \sin\left(\frac{m\pi}{a}x\right)\cos\left(\frac{n\pi}{b}y\right)e^{-j\beta_{z,mn}z} \tag{6.132d}$$

$$H_y = j\beta_{z,mn} H_0 \frac{(n\pi/b)}{[(m\pi/a)^2 + (n\pi/b)^2]} \cos\left(\frac{m\pi}{a}x\right)\sin\left(\frac{n\pi}{b}y\right)e^{-j\beta_{z,mn}z} \tag{6.132e}$$

Using the basic definition for the wave impedance given in (6.125), we can evaluate the wave impedance as

$$\eta_{TE,mn} = \frac{E_x}{H_y} = -\frac{E_y}{H_x} = \frac{\omega\mu}{\beta_{z,mn}} = \frac{\eta}{\sqrt{1 - \left(\frac{f_{c,mn}}{f}\right)^2}} \quad f > f_c \tag{6.133}$$

where η is the intrinsic impedance of the medium. Note that when the frequency increases beyond the cutoff frequency, the wave impedance of the TE modes decreases and asymptotically approaches the intrinsic impedance η. Further, carefully studying (6.132), we can identify that the lowest order TE mode is TE_{10} provided that $a > b$.

6.7.4 Power flow

The time average power density vector for any TM or TE mode in the z direction is given by

$$\begin{aligned}\mathbf{S}_{\text{av},z} &= \frac{1}{2}\text{Re}[\mathbf{E}_T \times \mathbf{H}_T^*] \\ &= \frac{1}{2}\text{Re}(E_x H_y^* - E_y H_x^*)\mathbf{u}_z\end{aligned} \tag{6.134}$$

Expression given in (6.134) can further be simplified depending on whether we can deal with TE modes or TM modes. The power flow density vector for TM modes,

$$\begin{aligned}\mathbf{S}_{\text{av},z} &= \frac{1}{2}\text{Re}[\mathbf{E}_T \times \mathbf{H}_T^*] \\ &= \frac{1}{2}\frac{|E_x|^2 + |E_y|^2}{\eta_{\text{TM},mn}}\mathbf{u}_z\end{aligned} \tag{6.135a}$$

and for TE modes

$$\begin{aligned}\mathbf{S}_{\text{av},z} &= \frac{1}{2}\text{Re}[\mathbf{E}_T \times \mathbf{H}_T^*] \\ &= \frac{1}{2}\frac{|E_x|^2 + |E_y|^2}{\eta_{\text{TE},mn}}\mathbf{u}_z\end{aligned} \tag{6.135b}$$

Thus, the total time-average power transmitted in the $+z$ direction across a cross section of the guide

$$P_{\text{av},z} = \int_{y=0}^{b}\int_{x=0}^{a} \mathbf{S}_{\text{av},z} \cdot \mathbf{u}_z \,\mathrm{d}x\,\mathrm{d}y \tag{6.136}$$

6.8 Cavity resonators

It is necessary to construct resonant devices at microwave frequencies. Lumped circuit elements, such as inductors and capacitors, are virtually useless due to their nonlinear behavior and their huge losses at these frequencies.

Consider the rectangular waveguide. Previously, we assumed a z dependence of the form $e^{-\gamma z}$, indicating only forward-traveling waves. The possibility of the forward- and backward-traveling waves suggests the possibility of standing waves

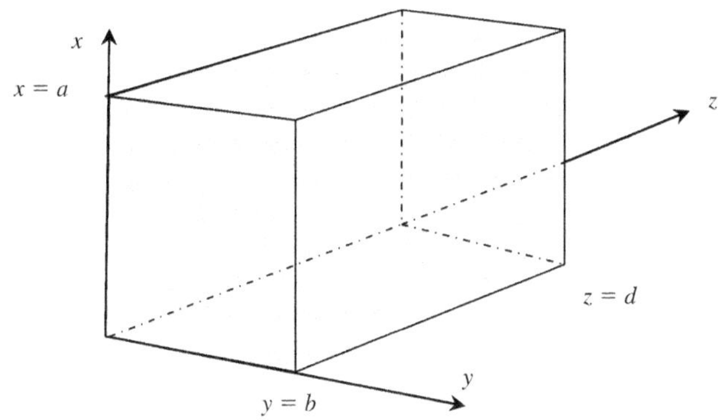

Figure 6.16: Schematic diagram of a rectangular waveguide.

and resonant behavior. Suppose we insert conducting plates at $z = 0$ and $z = d$, as shown in Fig. 6.16. These conducting plates give rise to reflections of the waveguide fields, resulting in standing waves. Here, we will investigate the resulting fields and the resonant frequencies of this cavity.

6.8.1 TM modes

The total field is the sum of forward- and backward-traveling waves obtained in Section 6.7.1:

$$\begin{aligned} E_z &= E_z^+ + E_z^- \\ &= \sin(\beta_x x)\sin(\beta_y y)(E_0^+\,\mathrm{e}^{-\gamma z} + E_0^-\,\mathrm{e}^{+\gamma z}) \end{aligned} \tag{6.137a}$$

Substituting this field individually into (6.114) and remembering that $\gamma(=\mathrm{j}\beta_z)$ is to be replaced with $-\gamma(=-\mathrm{j}\beta_z)$ in these equations for the backward-traveling wave, yields

$$E_x = \frac{\beta_x}{[\beta_x^2+\beta_y^2]}\cos(\beta_x x)\sin(\beta_y y)(-\gamma E_0^+\,\mathrm{e}^{-\gamma z} + \gamma E_0^-\,\mathrm{e}^{\gamma z}) \tag{6.137b}$$

$$E_y = \frac{\beta_y}{[\beta_x^2+\beta_y^2]}\sin(\beta_x x)\cos(\beta_y y)(-\gamma E_0^+\,\mathrm{e}^{-\gamma z} + \gamma E_0^-\,\mathrm{e}^{\gamma z}) \tag{6.137c}$$

$$H_x = \mathrm{j}\omega\varepsilon\frac{\beta_y}{[\beta_x^2+\beta_y^2]}\sin(\beta_x x)\cos(\beta_y y)(E_0^+\,\mathrm{e}^{-\gamma z} + E_0^-\,\mathrm{e}^{\gamma z}) \tag{6.137d}$$

$$H_y = -\mathrm{j}\omega\varepsilon\frac{\beta_x}{[\beta_x^2+\beta_y^2]}\cos(\beta_x x)\sin(\beta_y y)(E_0^+\,\mathrm{e}^{-\gamma z} + E_0^-\,\mathrm{e}^{\gamma z}) \tag{6.137e}$$

where

$$\beta_x = \frac{m\pi}{a} \qquad m = 1, 2, 3, \ldots \tag{6.138a}$$

$$\beta_y = \frac{n\pi}{a} \qquad n = 1, 2, 3, \ldots \tag{6.138b}$$

as earlier. The boundary conditions imposed by the plates at $z = 0$ and $z = d$ require

$$E_x = 0 \quad \text{at } z = 0 \text{ and } z = d \tag{6.139a}$$

$$E_y = 0 \quad \text{at } z = 0 \text{ and } z = d \tag{6.139b}$$

From (6.137b) and (6.137c), it requires $(-\gamma E_0^+ \, e^{-\gamma z} + \gamma E_0^- \, e^{\gamma z}) = 0$ at $z = 0$ and $z = d$. Therefore,

$$E_0^+ = E_0^- = E_0 \tag{6.140}$$

$$e^{\gamma d} - e^{-\gamma d} = 0 \tag{6.141}$$

To satisfy Equation (6.141), γ must be imaginary. Thus, we can write

$$\gamma = \mathrm{j}\beta_z \tag{6.142}$$

Hence, the solution of β_z is given as

$$\beta_z = \frac{p\pi}{d} \qquad p = 1, 2, 3 \ldots \tag{6.143}$$

Substituting (6.142) and (6.140) into (6.137) we obtain

$$E_z = 2E_0 \sin(\beta_x x) \sin(\beta_y y) \cos \beta_z z \tag{6.144a}$$

$$E_x = -\frac{2\beta_x \beta_z E_0}{[\beta_x^2 + \beta_y^2]} \cos(\beta_x x) \sin(\beta_y y) \sin(\beta_z z) \tag{6.144b}$$

$$E_y = -\frac{2\beta_y \beta_z E_0}{[\beta_x^2 + \beta_y^2]} \sin(\beta_x x) \cos(\beta_y y) \sin(\beta_z z) \tag{6.144c}$$

$$H_x = \frac{2\mathrm{j}\omega\varepsilon\beta_y E_0}{[\beta_x^2 + \beta_y^2]} \sin(\beta_x x) \cos(\beta_y y) \cos(\beta_z z) \tag{6.144d}$$

$$H_y = \frac{-2\mathrm{j}\omega\varepsilon\beta_x E_0}{[\beta_x^2 + \beta_y^2]} \cos(\beta_x x) \sin(\beta_y y) \cos(\beta_z z) \tag{6.144e}$$

The time domain forms of the fields can be easily obtained by multiplying each spatial expression by $e^{j\omega t}$ and taking the real part of the result. Since the phasors do not contain isolated complex exponentials, the time domain fields will vary with time as a sinusoidal function. Thus, the fields are oscillatory.

Each set of integers m, n and p is said to be associated with a TM_{mnp} mode. The resonant frequency of each mode is calculated by substituting (6.143) into (6.106):

$$f_{c,mnp} = \frac{1}{2\pi\sqrt{\mu\varepsilon}}\sqrt{\left(\frac{m\pi}{a}\right)^2 + \left(\frac{n\pi}{b}\right)^2 + \left(\frac{p\pi}{d}\right)^2} \tag{6.145}$$

Equations (6.129a)–(6.129e) show that neither m nor n can be zero; otherwise, all field components vanish. However, p can be zero. Thus, the lowest order TM mode is the TM_{110} mode.

6.8.2 TE modes

A similar development can be obtained for the TE modes. Taking the general solution from (6.130) and applying backward-traveling wave components yields

$$H_z = \cos(\beta_x x)\cos(\beta_y y)(H_0^+ \,\mathrm{e}^{-\mathrm{j}\gamma z} + H_0^- \mathrm{e}^{\mathrm{j}\gamma z}) \tag{6.146a}$$

Substituting this field individually into (6.114) and remembering that $\gamma(=\mathrm{j}\beta_z)$ is to be replaced with $-\gamma(=-\mathrm{j}\beta_z)$ in these equations for the backward-traveling wave yields

$$E_x = \frac{\mathrm{j}\omega\mu\beta_y}{[\beta_x^2+\beta_y^2]}\cos\beta_x x \sin\beta_y y(H_0^+ \,\mathrm{e}^{-\mathrm{j}\gamma z} + H_0^- \,\mathrm{e}^{\mathrm{j}\gamma z}) \tag{6.146b}$$

$$E_y = \frac{\mathrm{j}\omega\mu\beta_x}{[\beta_x^2+\beta_y^2]}\sin\beta_x x \cos\beta_y y(H_0^+ \,\mathrm{e}^{-\mathrm{j}\gamma z} + H_0^- \,\mathrm{e}^{\mathrm{j}\gamma z}) \tag{6.146c}$$

$$H_x = \frac{\beta_x}{[\beta_x^2+\beta_y^2]}\sin\beta_x x \cos\beta_y y(\gamma H_0^+ \,\mathrm{e}^{-\mathrm{j}\gamma z} - \gamma H_0^- \,\mathrm{e}^{\mathrm{j}\gamma z}) \tag{6.146d}$$

$$H_y = \frac{\beta_y}{[\beta_x^2+\beta_y^2]}\cos\beta_x x \sin\beta_y y(\gamma H_0^+ \,\mathrm{e}^{-\mathrm{j}\gamma z} - \gamma H_0^- \,\mathrm{e}^{\mathrm{j}\gamma z}) \tag{6.146e}$$

The boundary conditions imposed by conductors at $z = 0$ and $z = d$ are again given by (6.139), once again resulting in (6.142), (6.143) and

$$H_0^+ = -H_0^- = H_0 \tag{6.147}$$

Therefore, substituting (6.142), (6.143) and (6.147) in (6.146) yields

$$H_z = -2jH_0 \cos(\beta_x x)\cos(\beta_y y)\sin(\beta_z z) \tag{6.148a}$$

$$E_x = \frac{2\omega\mu\beta_y H_0}{[\beta_x^2 + \beta_y^2]}\cos(\beta_x x)\sin(\beta_y y)\sin(\beta_z z) \tag{6.148b}$$

$$E_y = -\frac{2\omega\mu\beta_x H_0}{[\beta_x^2 + \beta_y^2]}\sin(\beta_x x)\cos(\beta_y y)\sin(\beta_z z) \tag{6.148c}$$

$$H_x = \frac{2j\beta_z\beta_x H_0}{[\beta_x^2 + \beta_y^2]}\sin(\beta_x x)\cos(\beta_y y)\cos(\beta_z z) \tag{6.148d}$$

$$H_y = \frac{2j\beta_z\beta_y H_0}{[\beta_x^2 + \beta_y^2]}\cos(\beta_x x)\sin(\beta_y y)\cos(\beta_z z) \tag{6.148e}$$

The resonant frequency is again given in (6.145). Equations (6.148a–6.148e) show that either m or n (but not both) can be zero and that p cannot be zero. Hence, the lowest order mode is the TE101 mode provided that $a > b$ and $a \geq d$.

For further reading

[1] Balanis, C., *Advanced Engineering Electromagnetics*, USA: Wiley, 2012.
[2] Chrisitopoulos, C., *The Transmission Line Modeling Method*, USA: IEEE Press, 1995.
[3] Griffiths, D.J., Introduction to Electrodynamics, USA: Prentice Hall, 1998.
[4] Hoole, S.R.H., and Hoole, P.R.P., *A Modern Short Course in Engineering Electromagnetics with Computer Applications*, Oxford University Press, USA, 1996.
[5] Jordon, E.C., Balmain, K.G., *Electromagnetic Waves and Radiating Systems*, USA: Prentice Hall, 1968.
[6] Magnussen, P.C., Alexander, G.C., Tripathi, V.K., Weishaar, A., *Transmission Lines and Wave Propagation*, 4th Edn, USA: CRC Press, 2001.
[7] Paul, C.R., Whites, K.W., Nazar, S.A., *Introduction to Electromagnetic Fields*, USA: McGraw-Hill, 1997.
[8] Rachidi, F., Tkachenko, S., *Electromagnetic Field Interaction with Transmission Lines*, UK: WIT Press, 2008
[9] Sadiku, M.N.O., *Principals of Electromagnetics*, Oxford University Press, 2009.
[10] Shevgaonkar, R.K., *Electromagnetic Waves*, Tata McGraw-Hill Publishing Company, 2006.
[11] Ulaby, F.T., *Fundamentals of Applied Electromagnetics*, USA: Prentice Hall, 1997.

7

Antennas I: Electromagnetic radiation and interference

7.1 Antennas for telecommunications and remote sensing

Antennas are conductors made in various geometrical shapes and sizes to effectively launch elelctromagnetic signals into space (transmitter antennas), as well as to receive electromagnetic signals (receiver antennas) transmitted by other antennas or electromagnetic reflectors. When designing a wired communication line to be constructed, it is designed so as to minimize radiation from the wires. But in a wireless communication systems, an antenna is designed so as to maximize radiation power and that in a given preferred direction. The shape of an antenna depends on the direction in which an electromagnetic signal (or wave) is to be launched, as well as on the frequency of operation.

In a receiving antenna, the antenna geometric shape depends on the direction of travel and the frequency of the electromagnetic signal which is to be picked up relative to the receiver antenna. Often the same antenna is used as the transmitter and receiver antenna (sometimes called a transceiver antenna). Moreover, the antenna bandwidth must encompass the overall bandwidth of the signals or channels that must be received by the antenna. The transmitting and receiving center frequency and bandwidth of an antenna must be such that there is no overlap or interference between the transmitted and received signals. In telecommunications, this is achieved by having the transmitted wave and received wave operating at two different frequencies, but close to each other. For instance, elelctromagnetic signals at 1600 MHz (with a 1 MHz bandwidth) may be used to transmit data from a low-earth-orbit (LEO) satellite–antenna to a handheld telephone antenna unit (downlink). For transmission from the handheld antenna unit to the same LEO satellite-antenna communication (uplink), the carrier frequency would be 1601 MHz (also with a 1 MHz bandwidth).

In a radar system, on the other hand, the same signal at a frequency of 10 GHz is transmitted and received in an L-band monostatic, weather radar; the 10 GHz antenna transmits out a short duration pulse of electromagnetic waves, and before the next pulse is transmitted out it waits to receive the reflected 10 GHz signal from a target such as a thundercloud or tornado. Only after the reflected signal at 10 GHz is received, will it transmit out the next pulse at 10 GHz.

One of the simplest and most widely used antennas is the half-wave ($\lambda/2$) dipole antenna, which looks like a two-wire transmission line with wires bent out at the end perpendicular to the incoming transmission line. Each of the two bent out wires has a length of a quarter wavelength ($\lambda/4$) making up a total of half-wavelength ($\lambda/2$) for the long antenna. In this case, the two parallel transmission line wires guide electric and magnetic communication signals from a signal generator electronic circuit to the bent two wire, $\lambda/2$ long antenna. At the dipole antenna, the high-frequency signals will no longer cling to the antenna conductor as they did over the trabsmission line wires. Instead the signals will travel out into free spece as a wireless wave. The wave will be traveling out at the velocity of light (for free space). But the free space area the traveling waves will occupy will be constrained by the radiation pattern (or beam pattern) of the given antenna. Moreover, the maximum signal power will be radiated in one given direction (or more), determined by the direction in which the antenna radiation pattern peaks. The radiation pattern indicates the magnitude of the electric field strength observed at a constant distance r from the antenna in different directions. Most antennas, as indicated, are designed to transmit and receive maximum signal in one particular direction at a specified center frequency and bandwidth.

The size of the antenna depends mostly on the wavelength (or frequency) of operation. The size of a half-wave dipole antenna (where the length of the antenna is half the operating wavelength), for instance, will reduce with any increase in operating frequency. Moreover, the wavelength of a signal reduces with the dielectric properties of the medium. If the free space wavelength of a signal is λ, then the wavelength of the signal is $\lambda/10$ if the medium in which the antenna is operated has a relative permittivity $\varepsilon_r = 100$. This reduction of the velocity of the electromagnetic signal inside a dielctric medium is $c/(\varepsilon_r)^{1/2}$. Hence the size of an antenna is also a function of the relative permittivity of the medium; the length of a half-wave dipole antenna, for instance operating in a medium characterized by $\varepsilon_r = 100$, should be constructed to be 10 times smaller than the free space ($\varepsilon_r = 1$) length required. The fundamental theory of antennas discussed herein is based on Ref. [12].

7.2 Broadcast frequency bands

Maxwell's four equations for electromagnetic fields form the functional basis, and define the basic parameters of the electric field intensity and magnetic flux density, of every single device or system ranging from a direct current (DC) power system, or alternating current (e.g. 50 or 60 Hz AC) power system to communication and radar systems that may operate from 1 MHz to 100 GHz. All electromagnetic waves from say about 1 kHz to gamma rays carry energy (or information) at a velocity of 3×10^8 m/s (which is equal to c, the velocity of light), irrespective of frequency.

In phase modulation (PM) communication systems, for instance, the digital bit 1 may be represented by a sinusoidal signal frequency f at phase 0°, whereas bit 0 may be represented by a sinusoidal signal of the same frequency but at phase 45°. Thus

Table 7.1: Typical frequencies of operation and applications.

Application	Frequency (Hz)	Wavelength (m)
Power	0 (DC)	∞
Power	50 (AC)	3×10^{6}
Radio/TV		
AM	10^{6}	300
FM	10^{8}	3
Microwave	10^{11}	3×10^{-3}
Infrared/heat	10^{12}	3×10^{-4}
Visible light	10^{14}	3×10^{-6}
Ultraviolet light	10^{16}	3×10^{-8}
X rays	10^{18}	3×10^{-10}
Gamma rays	10^{20}	3×10^{-12}

in either case, the signal to be handled by the antenna is a sinusoidal electromagnetic signal. Now to pack more data (i.e. more 1s and 0s) into a short wave packet, higher frequencies have to be used, since one bit typically takes $1/f$ s. This is the reason why communication systems carrying large amounts of information (e.g. two TV channels and 3000 telephone channels) must use the very high microwave frequency bands (e.g. 4 GHz, with a period of 0.25 ns). Typical mobile communication frequency bands of four different service providers, operating in the same service area (thus the need to avoid interference with each other), may be as follows: 2140–2155 MHz (15 MHz bandwidth), 2125–2140 MHz (15 MHz bandwidth), GSM 900/1800 MHz, GSM 1800 MHz. However, low-data-rate AM and FM radio stations use the lower frequency spectra. Table 7.1 gives the general layout of the frequencies at which electromagnetic systems such as power, communication, remote sensing and medical systems operate.

7.3 Basic characteristics and definitions of terms

Wavelength. Wavelength is the distance traveled within one time cycle of a radiated electric signal. The frequency of the signal is in cycles per second. It follows that frequency f is inversely proportional to the wavelength. Both wavelength and frequency are related to the speed of light c by the formula

$$c = f \times \lambda \tag{7.1}$$

where $c = 3 \times 10^{8}$ m/s (speed of light), $f =$ frequency (Hz) and $\lambda =$ wavelength (m).

Sinusoidal waveform. Why do we mostly use sinusoidal signals? There are two simple reasons for this. First, consider the differentiation of a sine term:

$$\frac{\mathrm{d}(\sin \omega t)}{\mathrm{d}t} = \omega \cos \omega t \tag{7.2}$$

Although we get a cosinusoidal term from a sinusoidal waveform, the shape of the resultant waveform is the same as that of the sinusoidal waveform. Communication electronic circuits are full of capacitors C and inductors L, which integrate (C) or differentiate (L) current signals. This means that the waveform shape will not change if we use sinusoidal signals. Furthermore, we shall see that when high-frequency signals are launched into free space, the signals propagate by means of the electric field intensity (E, V/m) and magnetic flux density (B, Webers/m^2) maintaining each other through a process of spatial and time differentiation as defined by Maxwell's equations. Here again, using sinusoidal signals ensures that the basic waveforms of the signals carried by the electromagnetic wave do not change. If we had used a triangular waveform, for example, the resultant waveform after differentiation would be rectangular in shape!

There is a second reason why a pure sinusoidally shaped signal is attractive for communication signals: minimization of frequency bandwidth. A single pure sinusoidal wave will only occupy a single-frequency (spectrum) line in the frequency spectrum. Consider now a rectangular wave, say switching between 6 V (bit "1" in digital terms) and 0 V (bit "0"). If its Fourier transform is taken the resulting frequency spectrum will contain a collection of sinusoidal signals at different signal strengths. This means that to correctly receive or transmit a square wave pulse, a much larger frequency bandwidth is required.

The electromagnetic frequency spectrum is rapidly becoming filled up with many communications and remote sensing channels and systems. This is also one reason why a digital communication system is more attractive than the analog frequency modulation (FM) system. In the digital system, bits 1 and 0 may be modulated onto a single-frequency carrier with a phase shift to differentiate bit 1 from bit 0. However, in an FM system, to carry a TV channel, for instance, a 6 MHz bandwidth (with a carrier frequency, say, of 1800 MHz) is required to transmit the image frames and another 100 kHz or so for the voice.

Most of the antenna analysis presented in this book will assume a single-frequency electromagnetic signal. Of course, all the results we obtain apply to a multiple-frequency signal as well (e.g. an FM signal), if we analyze the signal frequency by frequency (in its frequency domain) and then sum them all up to get back the resultant signal.

Radiation. Radiation is the emission of coherently modulated electromagnetic waves, into free space, from a single or a group of radiating antenna elements.

In order to get directive radiation beams (patterns), a group of radiating antenna elements is used to maximize signal radiation in a given direction.

Array antenna. An array antenna consists of a group of radiating or receiving antenna elements.

Beamforming (or beam steering). In beamforming, we seek to determine the geometry of an antenna or the phase and current amplitudes (weights) for each element of an array antenna in order to obtain a prescribed antenna beam (radiation pattern). This is also known as antenna *synthesis*. In analog beamforming (i.e. switched-beam antennas), hardware signal phase shifters are used for each element of an array antenna to get the required, overall beam for the array antenna. In digital beam forming (e.g. adaptive antennas), a microprocessor is used to impose the magnitude and phase changes on currents supplied to each element of an array antenna to obtain the required radiation patterns.

Adaptive antenna. An adaptive antenna is an array of antennas that is able to change its radiation pattern to minimize the effects of noise, interference and multipaths and to maximize signal reception (and transmission) in a given direction. It may also be used to transmit signals to a mobile receiver such that the incident signal strength at the mobile unit is always maximum. The adaptive nature of the antenna is due to the fact that it uses powerful signal processing techniques to do online, real-time adjustments of the radiation pattern to maximize its performance.

Switched-beam antenna. In a switched-beam antenna, a series of nonadaptive antenna beams are available at the transmitter/receiver site. By selecting a combination of beams, reception and transmission in a given direction may be maximized. This is not as dynamic or flexible as the adaptive antennas.

Smart antenna. In general, a smart antenna combines the adaptive antenna and switched-beam antenna technologies. In communication systems, a smart antenna is essentially an adaptive antenna since it intelligently adjusts the radiation pattern to maximize the performance of the antenna. In imaging systems, it may also include processing the signals to identify targets and to perform diagnostics on the object under observation.

Decibel units. The power carried by a signal of electric field strength E_1 (V/m) with reference to a signal with field strength E_2 is given by

$$P(\text{dB}) = 20\log\left(\frac{E_1}{E_2}\right) \tag{7.3}$$

The formula for dB calculation when directly dealing with power levels is given by

$$P(\text{dB}) = 10\log\left(\frac{P_1}{P_2}\right) \tag{7.4}$$

where both P_1 and P_2 are in Watts (W). If the reference power P_2 is 1 mW, then power is represented by P (dBm). In other words, if power P is measured in W, then $P(\text{dB}) = 10\log_{10}(P)$ dB. If power is measured in mW, then $P(\text{dB}) = 10\log_{10}(P)$ dBm. Since power is proportional to (voltage)^2 or (electric field)^2 we have 20 instead of 10 when the electric field strength (or the voltage) is used to determine P(dB). If power P is in W, then gain $G = 10\log_{10}(P)$ dBW, and if it is in mW, then $G = 10\log_{10}(P)$ dBmW.

Radiation resistance. Radiation resistance is defined in terms of the power radiated out (P_r) of the antenna, $P_r = I_A^2 R_r$, where R_r is radiation resistance of the antenna and I_A is the rms value of the driving current at the antenna terminals. From the perspective of the transmitter circuit connected to the antenna, the power radiated by the antenna is a power loss, which is identified as the power dissipated in the radiation resistance. The additional power loss is the heat dissipation $I_A^2 R_0$, the ohmic loss in the ohmic resistance R_0. Thus the efficiency of an antenna (η_e), which is defined as the power radiated divided by power delivered to the antenna terminals, is given by $R_r/(R_0 + R_r)$.

Polarization. Polarization is the angle of the radiated field vector deviation from the direction of maximum radiation. Polarization is classified into

(a) Linear polarization of electric and magnetic fields. There are two kinds of linearly polarized signals. These are vertical and horizontal polarizations.

If the vector of the electric field is parallel to the ground, it is known as a horizontally polarized signal.

If the vector of radiated electric field from a vertically held wire antenna is perpendicular to ground, then the wave is called a vertically polarized signal. When the receiving antenna is located on the same plane as the transmitting antenna, the received signal strength will be maximum.

(b) Circular polarization. If the radiated or tranmitted signal is rotating at the operating frequency in space, the rotation being acheived by electrical means in feeding the signal to the transmitting antenna, the radiated signal is said to be circularly polarized. At the receiver antenna, the circularly polarized signal produces equal received signal strengths for both a horizontally and a vertically polarized receiving antenna. Moreover, if the rotation of the electric field in space is in the same direction as the right-hand stretched out from the body and swung left into the body, the electronagnetic signal is called a left-hand circularly polarized signal. Similarly if the rotation of the electric field is in the same direction as that of the left-hand swung right into the body, it is called a left-hand circularly polarized signal.

(c) Eliptical polarization. A circularly polarized signal is made up of two orthogonal, linearly polarized signals: these are vertically and horizontally polarized signals. A circulary polarized signal becomes an elliptically polarized signal when the two orthogonal (perpendicualr to each other) electric fields are not equal in magnitide. This occurs, for instance, when a satellite antenna radiates circularly polarized signals through rain. The rain drops absorb most power from the transmission signals which are polarized along the elongated length of a rain drop. This makes the two (vertically and horizontally polarized) electric fields unequal to each other.

Note that for effective land to satellite communication in a wireless, mobile communication system that uses satellite base stations (BS) instead of a land based BS, the handheld telephone unit should use antennas such as the helical antennas (instead of using some variation of linear wire antennas) to effectively capture circularly polarized signals from satellites. Using circularly polarized signals, the system information capacity may be doubled, since the two electric fields that are perpendicular (orthogonal) to each other will not interfere with each other though they are at the same carrier frequency. That means, for instance, two TV channels may be transmitted and received at the same carrier frequency, if one channel is transmitted as a vertically polarized signal and the other as a horizontally polarized signal. The resultant signal will be a circularly polarized signal, each carrying the text, voice and image information of two TV channels.

Antenna power gain, directivity and beamwidth. The antenna power gain G is the signal level produced (or radiated) by an antenna relative to the total power input to the antenna. The directivity D of an antenna is the maximum power radiated by an antenna relative to the maximum power radiated by a short, Hertzian dipole antenna or any other standard reference dipole antenna. Power gain is frequently used as a figure of merit. Gain is closely related to directivity, which in turn is dependent upon the radiation pattern of a given antenna. Higher values of gain are usually obtained at the cost of reduction in beamwidth. Gain can be calculated only for simple antenna configurations. Consequently, it is determined by measuring the antenna performance relative to a reference dipole antenna. The reference dipole is either an omnidirectional, isotropic antenna or a half-wave dipole antenna.

The directivity, gain and efficiency of an antenna are scalar quantities which are used to express the antenna's radiation characteristics. All these are defined for a particular antenna with respect to a reference isotropic antenna. If the total power radiated is P_{r} W, then the density of power radiated by an equivalent isotropic antenna at a distance r from the antenna is $P_{\mathrm{r}}/4\pi r^2$ W/m^2. The directivity (D) of an antenna may alternatively be expressed as

$$D = \frac{P_{\mathrm{m}}}{P_{\mathrm{r}}/4\pi r^2} = \frac{r^2 P_{\mathrm{m}}}{P_{\mathrm{r}}/4\pi} = \frac{S_{\mathrm{m}}}{P_{\mathrm{r}}/4\pi} \tag{7.5}$$

where P_{m} is the maximum power density radiated by the antenna in W/m^2 and $S_{\mathrm{m}} = r^2 P_{\mathrm{m}}$, which is the maximum value of radiation intensity in a unit solid angle in W/sr (where the unit of solid angle is steradians or sr). The quantity $P_{\mathrm{r}}/4\pi$ is the radiation intensity of an isotropic antenna. For an isotropic antenna which has a spherical radiation pattern, the radiation power P_r will be equally distributed on the surface of the sphere, thus yielding $P_m = P_r/4\pi r^2$. Therefore the directivity of an isotropic antenna is $D_i = 1$. The directivity of the antenna will be entirely determined by the radiation pattern of the antenna. A narrower beam with fewer side lobes will mean a highly directive antenna. The power gain (G) of the antenna is given by

$$G = \frac{S_m}{P_T/4\pi} < D \quad \text{for } P_r < P_T \tag{7.6}$$

where P_T (W) is the power delivered to the transmitting antenna (e.g. 0.5–10 W for cell phones and as high as 500 kW in the case of radar antennas).

The power gain of an antenna is usually less than the directivity of the antenna due to losses in the antenna. Sometimes the power gain is specified in decibels: $G_{dB} = 10\log_{10}(G)$ dBi, the i in units dBi indicating that the gain is measured with reference to an isotropic antenna. The efficiency of an antenna is given by

$$\eta_e = \frac{P_r}{P_T} = \frac{G}{D} \tag{7.7}$$

When there is no loss in the antenna, that is $R_r/(R_0 + R_r) = 1$, $R_0 << R_r$ and $P_r = P_T$, the efficiency of the antenna is 100%. The antenna power gain G, sometimes called the isotropic gain does not mean that the antenna amplifies the overall power input. If ohmic losses in the antenna are neglibibly small, then the power delivered to the antenna is equal to the power radiated out. The power delivered to the antenna will also be equal to the power appearing at a transmission line (or waveguide)–antenna termination, when there is no reflected energy at the termination due to good impedance matching at that transmission line–antenna termination.

What then is the power gain G of an antenna? The power gain of an antenna denotes the increase in maximum power density (W/m^2) radiated by a given antenna A_l compared to the power density radiated by an isotropic antenna A_i. An isotropic antenna A_i radiates equal power density over the entire surface of a sphere surrounding it, with the antenna at the center of the sphere. Thus at a distance r from an isotropic antenna, the power density is $P_r/4\pi r^2$ W/m^2, where P_r is the total power radiated (in W) by the isotropic antenna. When the same amount of power P_r is radiated by another antenna A_l, the maximum power density it radiates will be $GP_r/4\pi r^2$ W/m^2, where G_l is the gain of antenna A_l. Some manufacturers define the gain of an antenna with reference to a half-wave dipole antenna, instead of an isotropic antenna. In this case, the beam area of $4\pi r^2$ over which an isotropic antenna radiates has to replaced by the smaller beam area over which a dipole antenna radiates, which means that the dipole gain G_d of a given antenna A_l will be smaller than its isotropic gain G_i. Throughout the rest of this chapter the gain of an antenna will be assumed to be the isotropic gain of the antenna.

The total efficiency of an antenna η_e is defined by taking into account the power losses at the antenna terminals and within the antenna structure itself. Power losses may be due to both reflections at the antenna terminals because of poor impedance matching between the antenna and the transmission line, as well as the I^2R_0 losses due to the ohmic resistance R_0 of the antenna conductor and dielectric materials. Hence the efficiency of an antenna is

$$\eta_e = \eta_r \eta_c \eta_d \tag{7.8}$$

where η_r is the reflection efficiency, η_c is the conduction efficiency and η_d is the dielectric efficiency. When an antenna is well matched with the transmission line, waveguide or coaxial cable, and the dielectric loss is neglibily small in antennas using dielctric materials, then $\eta_e = \eta_c = R_r/(R_0 + R_r)$.

Half-power beam width (HPBW $= \theta_B$) of an antenna on the plane of the antenna is the angular width of the radiation pattern, where the power level of the received signal is down by 50% (i.e. the 3 dB point) from the maximum power density. Let Ω_B be the HPBW solid angle (sr) measured between the 3 dB points on the radiation pattern. We assume that all radiated power P_r is confined to the half-power beam; that is we ignore the radiation outside the HPBW angle (θ_B radians) as being small and negligible. For most antennas, the HPBW is approximately equal to FNBW/2, where FNBW is the first null beamwidth, the angle between two straight lines drawn through the first two null points that appear at each end of the main beam. The FNBW is thus the angular width between the two null points on either side of the main beam of the radiation pattern. The FNBW is always greater than the HPBW.

The half-power beam area is given by

$$A_b = r^2\Omega_B = r^2\frac{\pi}{4}\theta_B^2 \text{ m}^2 \tag{7.9}$$

where the solid angle is

$$\Omega_B = \frac{\pi}{4}\theta_B^2 \tag{7.10}$$

If the total radiated power is P_r, then the maximum power density P_m, assuming that the power density is uniform and constant over the beam area A_b, is given by,

$$P_m = \frac{P_r}{A_b} = \frac{P_r}{r^2\Omega_B} \text{ W/m}^2 \tag{7.11}$$

Therefore, the directivity of the antenna is given by

$$D = \frac{P_m}{P_r/4\pi r^2} \tag{7.12}$$

which could be approximated to

$$D = \frac{4\pi}{\Omega_B} \tag{7.13}$$

If the antenna is lossless, then $P_T = P_r$, and thus

$$G = D = \frac{4\pi}{\Omega_B} \tag{7.14}$$

where Ω_B is the HPBW solid angle (in sr) measured between the 3 dB points. Equation (7.14) is an approximate formula and applies to all four types of antennas. For an exact calculation of D and G, we must not assume that the power density over the beam area is constant and uniform. Since maximum power density $P_m = (E_m^2/2\eta)$, the maximum electric field radiated to a distance r from the antenna is given by $E_m = (1/r)(DP_r\eta/2\pi)^{\frac{1}{2}}$ V/m.

In order to obtain the power gain $G(\theta, \phi)$, or directivity $D(\theta, \phi)$, in a given direction (θ, ϕ) other than in the direction of maximum radiation, P_m in (7.5) and (7.6) must be replaced by $P_i(\theta, \phi)$, which is the power density radiated in the direction (θ, ϕ). $P_i(r, \theta, \phi) = E(r, \theta, \phi)^2/(2\eta)$, where $P_i(r, \theta, \phi)$ is the radiated power density at the observation point (r, θ, ϕ), and $E(r, \theta, \phi)$ is the electric field at the observation point (r, θ, ϕ). Similarly S_m W/sr must be replaced by $S_i(r, \theta, \phi)$ W/sr.

7.4 Antenna input impedance

The antenna input impedance is the terminating resistance into which a receiving antenna will deliver maximum power. As with the gain of an antenna, the input impedance of an antenna can be calculated only for very simple formats and is determined by actual measurement (Fig. 7.1).

The antenna input impedance

$$Z_A = R_A + jX_A, \tag{7.15}$$

where X_A is the antenna reactance that is associated with the reactive energy (both capacitive and inductive) stored in the near-field zone surrounding the antenna. This reactance is largely capacitve and may be associated with the capacitance that will form between the upper and lower condctors of a dipole antenna. Moreover, in installing antennas, it is important to ensure that the capacitance that will be formed between the antenna conductor and the surrounding conductor (e.g. the metallic body of a motor vehicle) must be minimized; else a part of the transmitter antenna signal will be stored in this capacitance and will not be released into free space.

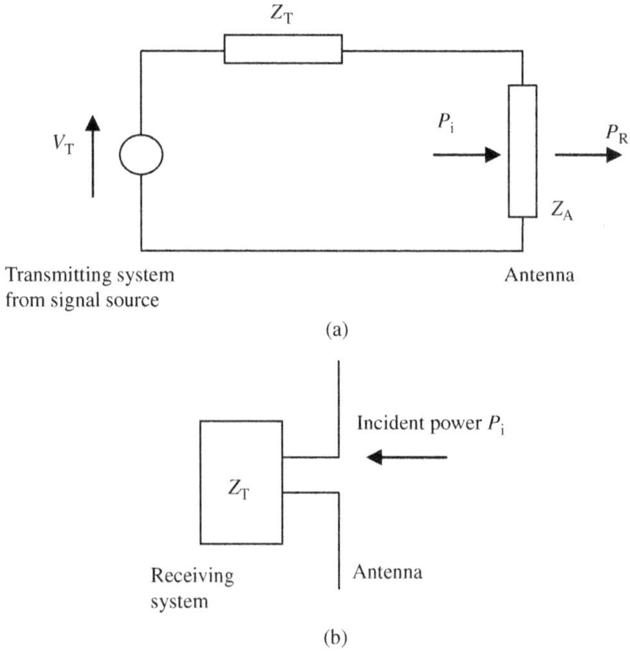

Figure 7.1: The Thevenin equivalent circuits of (a) transmitting and (b) receiving antennas.

The antenna resistance R_A combines the so-called radiation resistance (R_r) of the antenna and the ohmic resistance of the finite-conductivity antenna (R_0). R_0 is related to the ohmic power loss in the antenna. The radiation resistance is a measure of an antenna's ability to radiate power. The power dissipated in R_r is the radiated power P_r; hence for an antenna we seek to maximize R_r, whereas for other circuits like printed circuit boards (PCBs) we seek to minimize R_r since in PCBs we want the information and energy to flow along the copper transmission line instead of getting radiated out and interfering with other lines and devices. Such electromagnetic interference (EMI), that induces signals in other connecting conductors, circuit components and integrated chips, thus degrades the performance of electronic devices and circuits.

If the antenna is transmitting and the source current that flows through the antenna is I_{rms}, then the radiated power $P_r = I_{rms}^2 R_r$ W. If the source voltage is $V_{s,rms}$ and the source impedance Z_s is properly matched to the antenna impedance Z_A (i.e. $Z_s = Z_A^*$), $I_{rms} = V_{s,rms}/2R_A$. If the antenna is receiving and its impedance is properly matched to the load (transmission line/receiver electronics) impedance Z_L (i.e. $Z_L = Z_A^*$), then the curent in the receiver antenna is $I_{rms} = V_{s,rms}/2R_A$, where $V_{s,rms}$ is the voltage induced in the antenna by the incident electric field E. For a short dipole of length L, an approximate expression for the induced voltage is

$V_{s,rms} = EL$. The received power or power captured by the antenna $P_r = I_{rms}^2 R_r$W. For an efficient antenna (i.e. $R_r \gg R_O$), the antenna resistance is equal to the radiation resistance, that is $R_A = R_r$. Thus, in general, power dissipated in R_r is equal to the power radiated by a transmitting antenna. In a receiving antenna, the power dissipated in R_r is the power captured by the receiving antenna which is the power (P_R) that is delivered to the receiver electronics. For a small wire antenna of length L, we have

$$R_r = 800 \left(\frac{L}{\lambda}\right)^2 \quad \Omega \tag{7.16}$$

For a small circular loop antenna (magnetic dipole) of area A ($=\pi$(radius of loop)^2), the radiation resistance is

$$R_r = 31{,}200 \left(\frac{A}{\lambda}\right)^2 \quad \Omega \tag{7.17}$$

As highlighted above, the antenna reactance X_A is related to the electromagnetic energy stored in the vicinity of the antenna. Note that the input impedance of the antenna is a function of frequency, since X_A increases linearly with frequency if the antenna is inductive and decreases inversely with frequency if it is capacitive. Similarly the resistance is also frequency dependent. Hence when we match an antenna to the impedance Z_T of a two-wire transmission line or coaxial cable, it is matched at a particular center frequency and bandwidth. If the frequency of operation is changed, unless matching is done again, then power transfer to and from the antenna will be poor.

7.5 Effective aperture

The concept of an effective aperture captures the idea of an antenna being open to capture power radiated by another (transmitting) antenna, which falls within its transmission or radiation beam. If P_i is the incident power density (in W/m^2) at the receiver antenna and P_R (W) is the power received or captured by the antenna, the effective area (in m^2) is defined by

$$A_{em} = \frac{P_R}{P_i} = \frac{\frac{1}{2} I^2 R_T}{P_i} \tag{7.18}$$

For maximum power transfer, ensuring that all the power captured by the reciever antenna is passed on to the receiver low-noise amplier (LNA),

$$R_A = R_r + R_O = R_T \tag{7.19}$$

$$X_A = -X_T \tag{7.20}$$

where R_A is the antenna resistance, R_r is the antenna radiation resistance, R_0 is the antenna ohmic resistance, R_T is the termination resistance, X_A is the antenna reactance and X_T is the terminatiing reactance. Hence the current in the receiver circuit of the antenna is given by

$$I_T = \frac{V_T}{(Z_A + Z_T)} = \frac{V_T}{2R_T} = \frac{V_T}{[2(R_r + R_0)]} \tag{7.21}$$

Hence the maximum effective aperture area of the antenna is given by

$$A_{em} = \frac{V_T^2}{8P_i(R_r + R_0)} = \frac{V_T^2}{8P_iR_r} \quad \text{for } R_0 \ll R_r \tag{7.22}$$

For a short dipole of length L and an incident electric field E, using the following parameters: the radiation resistance $R_r = 80(\pi L/\lambda)^2$; the directivity $D = 1.5$; the incident power density at the receiver antenna $P_i = E^2/2\eta_0 = E^2/2(377)\,\text{W/m}^2$; and the voltage induced along the antenna wire $V_T = EL$, we get $A_{em} = 0.12\lambda^2\,\text{m}^2$ for a short or infinitesimal dipole. In general, for any antenna, the maximum effective aperture area is given by

$$A_{em} = \frac{\lambda^2}{4\pi}D \tag{7.23}$$

It should be noted that if the receiver is not directly pointed toward the transmitting antenna and is tilted at an angle α measured from the line connecting the transmitter and receiver antennas, then the effective aperture area is $A_{em}\sin\alpha$. Maximum reception occurs when $\alpha = \pi/2$.

A useful equation for the gain of an antenna in terms of its maximum effective aperture area A_{em} is $G = D = (4\pi A_{em})/\lambda^2$. Hence $G \propto A_{em}$ and $G \propto f^2$. Note that in general the gain of an antenna increases as the operating frequency f is increased, where $f = c/\lambda$, c is the velocity of light (approximately equal to 3×10^8 m/s) and λ is the wavelength. For an elemental dipole, $G = D = 1.5$, and for a half-wave dipole $G = D = 1.67$.

Now, because the aperture and directivity are proportionate quantities, for any two antennas, we might say that the ratio of D to A_e is constant. To get this constant, we use results from the short dipole antenna

$$\frac{A_e}{D} = \left.\frac{A_e}{D}\right|_{\text{Dipole antenna}} = \frac{0.125\lambda^2}{1.5} = \frac{\lambda^2}{4\pi} \tag{7.24}$$

For a parabolic reflector-type antenna with a dish diameter d_0, the physical aperture A_p is $\pi d_0^2/7$. The aperture efficiency η_a is the ratio of the effective aperture to the physical aperture

$$\eta_a = \frac{A_e}{A_p} \tag{7.25}$$

7.6 Antenna temperature

An antenna receives radiation from a variety of objects other than the transmitter antenna, and its temperature T_b is usually above absolute zero $T_b > 0°C\,(-273\,K)$.

This temperature has adverse effects on communication and radar systems. The electrons in conductor and semiconductor materials are agitated by the temperature and radiate broad band electromagnetic signals as they oscillate. The broad band signals associated with the antenna temperature (as well as the temperature of other nearby objects) appear as noise at the receiver electronics. However in remote sensing systems such as radar systems, the radiation due to the finite temperature of objects other than the antenna can be captured and used to determine the material properties of the object. This is called passive remote sensing. Passive remote sensing is used in astronomy to study the properties of stars and far-off hot bodies, as well as in studying the geography of land through a passive sensing antenna attached to the fuselage of an aircraft or to a satellite.

However, in telecommunication systems, the physical temperature of both the antenna and other objects produces electromagnetic noise which interferes with the signal received. The noise power is given by $p = kT_b\,W/Hz$. Hence for a bandwidth of Δf Hz, the total noise is $\Delta f k T_b$ W. This noise is specified in terms of power per unit frequency over the bandwidth. It is given by

$$p = kT_b \tag{7.26}$$

where by definition, the unit is W/Hz and k is Boltzmann's constant $1.38 \times 10^{-23}\,J\,K^{-1}$. Now if the entire pattern of the antenna over its bandwidth Δf sees an object at temperature T_b, it receives a total "noise" of power

$$P = p\,\Delta f = kT_b\,\Delta f \text{ W} \tag{7.27}$$

T_b is often termed the *brightness temperature* of the object. Each one of the following objects is at a specific temperature as shown and will radiate an electromagnetic signal over a broad spectrum:

Sky: 3 K; Mars: 164 K; Earth: 290 K; Man: 310 K

Thus the *noise temperature* at an antenna associated with temperature produced noise results from a summation of temperatures at the antenna due to many hot objects, both near and far away, and is given by the total noise temperature:

$$T_{\mathrm{A}} = T_{\mathrm{B1}} + T_{\mathrm{B2}} + \cdots \tag{7.28}$$

Transmission line theory applied to the conduction of noise along the line or waveguide, which connects the antenna to the receiver electronics, will give

$$T_{\mathrm{a}} = T_{\mathrm{A}}\,\mathrm{e}^{-2\alpha d} + T_0(1 - \mathrm{e}^{-2\alpha d}) \tag{7.29}$$

where $T_0 = 273\,\mathrm{K}$ is the physical temperature of the system (0°C), and α is the attenuation coefficient of the electrical circuit.

Hence the total noise power at the receiver electronics is the sum of noise coming from the antenna (T_{a}) and the noise produced by the hot electronic circuitry (at a temperature T_{R}) itself. The total noise power at the receiver electronics operating at a bandwidth Δf Hz is given by,

$$P_{\mathrm{N}} = k(T_{\mathrm{a}} + T_{\mathrm{R}})\Delta f \tag{7.30}$$

For very high-frequency antennas, with frequency $f > 10\,\mathrm{GHz}$, the temperature is given by $T = hf/k$, where Planck's constant $h = 1.055 \times 10^{-34}$ J s and Boltzmann's constant $k = 1.38 \times 10^{-23}$ J/K. The density of power received at an earth station antenna from a 12 GHz satellite downlink is of the order of 10^{-13} W/m^2. It should be appreciated that with such weak signals arriving at the earth station antenna, it is important to minimize temperature produced additive white Gaussian noise (AWGN). Moreover, powerful computer software based filters with LNA must be used to separate the signal from the noise.

7.7 Antenna operating zones

Three zones are defined for an antenna. These zones depend on the size of the antenna L (e.g. L = length of antenna for a wire type antenna) and the wavelength λ.

The near-field zone is the distance r from the antenna that satisfies

$$r < r_1 = 0.62L\sqrt{L/\lambda} \tag{7.31}$$

The near-field region of an antenna is the region immediately surrounding the radiating near-field (Fresnel) zone or the intermediate zone, defined by distance r which lies in the range

$$r_1 < r < r_2 = 2L^2/\lambda \tag{7.32}$$

In the radiating zone, that is beyond the near-field or intermediate field regions, the radiation field dominates and the angular field distribution is dependent on the distance from the surface of the antenna. For a short or infinitesimal antenna where the size of the antenna is very small, compared to the wavelength, the intermediate zone does not exist. In the far-field (Fraunhofer) zone in which most antennas operate, the receiving antenna is far away from the transmitting antenna:

$$r > r_2 \tag{7.33}$$

In the far-field region of an antenna, the angular field distribution may be assumed to be dependent on the distance from the antenna. Although most communication antennas operate in the far-field zone, in biomedical imaging, the antennas (or coils) operate at frequencies of the order of 160 MHz and are only about 0.5 m away from the radiating source. In mobile communications, it is possible for the receiver antenna to pick up near fields when it is close to the BS, as in the case of indoor picocells of the order of 10 m × 10 m.

7.8 Antenna operating principles

We shall consider now antenna operation using Faraday's law. How does an antenna work? The principle of antenna operation may be best explained by considering Faraday's law, which defines the induced voltage V in a coil as given by

$$V = \frac{\mathrm{d}(N\phi)}{\mathrm{d}t} = \frac{\mathrm{d}(NBA)}{\mathrm{d}t} = N\left[B\frac{\mathrm{d}A}{\mathrm{d}t} + A\frac{\mathrm{d}B}{\mathrm{d}t}\right] \tag{7.34}$$

where N is the number of turns, B is the flux density (W/m^2) and A (m^2) is the area of the coil. We have two terms in the expression for the induced voltage. For the first term to be effective, while the magnetic flux density may be time varying or constant, the area cut by the flux density must vary with time. In the second term, while the area of the coil or antenna may be constant, the magnetic flux density should be a function of time. A receiving antenna may make use of one or both of these terms. When a time-varying magnetic field cuts a conductor, it will induce a voltage, which will be proportional to $\mathrm{d}B/\mathrm{d}t$. This is how a wire antenna of a mobile telephone picks up signals. Alternatively, if a coil antenna is rotated in an electromagnetic field, there will be an induced voltage due to $\mathrm{d}A/\mathrm{d}t$, where the area of the coil cutting the signal is changing. Such antennas are used in direction finding. In a coil antenna rotated in an electromagnetic field, when the induced voltage is maximum, the antenna is directly facing (when $\mathrm{d}A/\mathrm{d}t$ is maximum) the source that is radiating the electromagnetic signal. In military applications, such rotating antennas may be used to detect hidden transmitters and radar. In a transmitting antenna, a wire to which a time-varying voltage (e.g. an FM radio signal) is applied will produce time-varying magnetic flux density B in the space around it. At high frequencies, this B will induce an electric field intensity E. Both E and B together will now propagate through free space at the velocity of light.

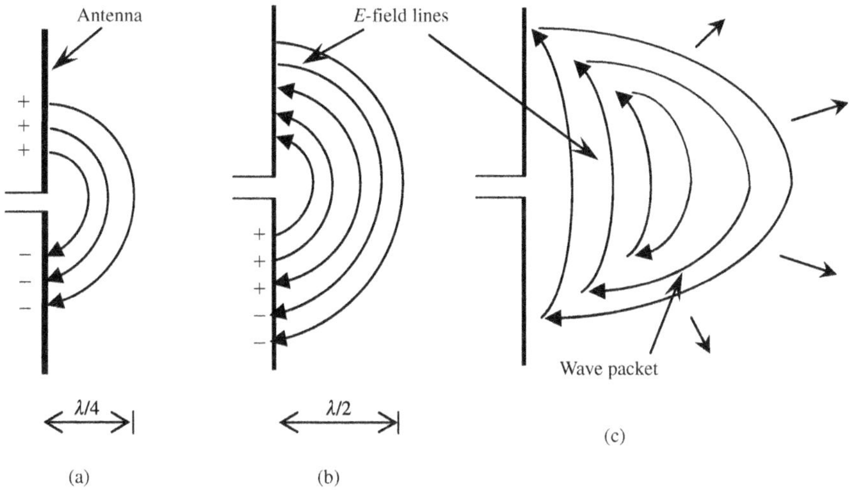

Figure 7.2: Formation of free space electromagnetic information packets: (a) $t = \frac{T}{4}$, (b) $\frac{T}{2}$ and (c) $t = T$ (T = Period).

7.9 Antenna radiation (or beam) patterns

7.9.1 Waves along conductors and in free space

The launching or radiating of an electromagnetic wave from a dipole antenna is shown in Fig. 7.2. A two-wire transmission line carrying high-frequency signals (waves) is connected to the antenna wires which are opened out vertically. The two vertical lines (wires) shown in Fig. 7.2 form the dipole antenna. When each wire is of length $\lambda/4$, where λ is the wavelength of the electromagnetic signal to be transmitted, the wire antenna of total length $\lambda/2$ is called a half-wave dipole. The voltages are shown on the transmission line (going from a negatively charged conductor to the positively charged line) and the electric field intensity around the antenna is also shown going from positive to negative electric charges). The voltage waves arrive from the source at the transmission line and antenna connection in packets of pulses, reversing direction every half wavelength. The velocity of the voltage (and current) waves along the transmission line is close to the velocity of light. We may look at the radiaiton process in three stages.

Stage I. When the top positive and bottom negative voltage pulse arrive at the antenna, there is a buildup of positive electric charges on the top wire of the antenna and negative electric charges on the bottom wire.

Stage II. As the next voltage pulse arrives, this time with negative electric charges on top half of the two wires, the negative electric charges gradually push out the previously present positive electric charges and form electric field lines in the reverse direction on the antenna. Hence the bottom antenna wire is progressively positively charged and the top is progressively negatively charged. During this time, the previous collection of electric field lines going from the wire to the bottom wire all get pushed out and crowded on the edge of the wires.

Stage III. When the next voltage pulse arrives (top positive and bottom negative), the electric field lines of phase I and phase II (which are pointing in the opposite directions) join together at both ends of the antenna conductor and form one whole loop of displacement current in the free space. Now these closed electric field packets of half wavelength width will be pushed out into space and travel out as radiatioin fields. This can be roughly pictured as a bubble (electric field line packets) flowing out of a tube (the antenna). With each time-varying electric field packet, there will also be a magnetic flux line circulating around the electric field lines, as stipulated by Maxwell's equations, where magnetic and electric fields generate each other at high frequencies. A whole sequence of such electromagnetic field packets form each packet containing information coded from the source connected to the transmission line. Now these self-contained, self-generating electromagnetic wave packets travel out at the velocity of light.

7.9.2 Antenna radiation pattern

The radiation pattern of an antenna gives us an idea of the propagation of the radiated signal around the antenna. For an isotropic antenna (i.e. point radiating element), the radiation pattern will be spherical. The radiation pattern changes as the configuration of the radiating element changes. Figure 7.3 shows the radiating patterns of a directive antenna.

The radiation pattern of an antenna may be obtained by using the mathematical expression for the electric (or magnetic) field radiated by a given antenna. If we draw a circle around the antenna, with the antenna at the center of the circle, and calculate the electric field magnitude along the circumference of the circle drawn, we will get different values for the electric (or magnetic) field at each point on the circle. We may also obtain these electric (or magnetic) field values by measuring the electric field strength at each point on the circle, with the antenna radiating from the origin. Measurements are necessary in the case of antennas for which getting an accurate analytical or numerical solution for the radiation fields is not possible. If we now place points around the antenna where the distance of each point from the antenna at the origin is proportional to the electric field calcualted (or measured) for that point on the circle, we will get the antenna radiation pattern as shown in Fig. 7.3.

Note that the maximum electric and magnetic fields E_{m} and H_{m} (and maximum power density P_{m} W/m^2) radiated is along the direction of the radiation pattern where the distance between the origin and the radition pattern point is maximum.

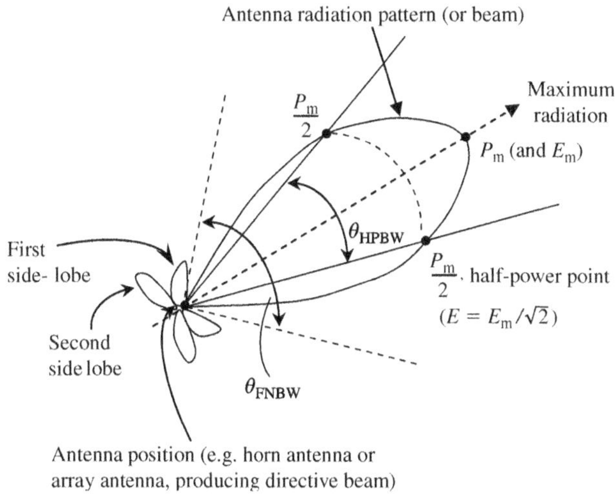

Figure 7.3: An antenna radiation pattern.

Note that since we have drawn the radiation pattern for the electric field, if we draw the radiation pattern for the power radiated, we would need to square the distance to each point, since power P is proportional to the square of electric field (E^2).

Furthermore, note that the radiated power is half the maximum power when the electric field magnitude is the electric field when the power is maximum divided by square root two, that is the eletric field is $0.707E_m$, Now E_m is the maximum electric field measured in a circle of radius r around the antenna: $E_m = \text{maximum } \{E(r, \theta, \phi)\}$. The time domain electric field at position (r, θ, ϕ) is given by $E(r, \theta, \phi)\cos(\omega t - kr)$. Therefore, the power at (r, θ, ϕ) is given by $E(r, \theta, \phi) = P_m f^2(\theta, \phi)$, where $P_m = E_m^2/2\eta\ W/m^2$.

As we move down from the peak point on either side of the radiation beam pattern, we come to a point where the distance between that point and the origin is 0.707 times that of the distance between the origin and the maximum point. These two points, on either side of the maximum point, are called the half-power points. Along the direction of these two points, the power density radiated will be $(P_m/2)$ W/m^2. The angle between these two points, as shown in Fig. 7.4, is the half-power point beamwidth (θ_{HPBW}). As we move further down the radiation pattern, we come to the first set of null points along which the antenna will neither radiate as a transmitting antenna nor receive signals as a receiving antenna. The angle between these first two null points is called the first null beamwidth (θ_{FNBW}). It is obvious that if we have an isotropic antenna which radiates signals of equal magnitudes of power density, the beam pattern will be a perfect circle around the antenna. The antenna that comes closest to such an isotropic antenna is the short (or Hertzian) dipole antenna, where the size or length of the antenna is much smaller than the signal wavelength.

After the first null, we encounter the side lobes of the antenna beam. In the case of a transmitting antenna, these side lobes are regions in which the antenna radiation is wasteful. Similarly, in directional antennas, we want signals received to be within the main beam's half-power points. Hence signals that get into the antenna electronics through the side lobes are usually unwanted, interference signals. The major lobe (main beam) is the radiation lobe containing the maximum radiation power. Minor lobes are any lobes other than the main lobe. The minor (or side) lobes represent radiation power lost.

Any signal transmitted out through the side lobes is unwanted since it radiates information (or electromagnetic energy) in unwanted and unintended directions. A back lobe is a minor lobe that occupies a direction opposite to that of the major lobe. Minor lobes should be minimized at antenna design stage in order to get the antenna to radiate and receive only in the prescribed or intended direction.

Narrow or pencil beams are preferred for antennas that must be highly directive. In some applications like radio broadcasts where the radio program should be transmited in all directions, the antenna radiation pattern should have a wide main beam that spans the entire 360° in the ground (i.e. H-) plane, that is a beam that resembles the isotropic radiation pattern of a circle (in two dimensions) or a sphere (in three dimensions). For a fictitious isotropic antenna (point radiating element), the radiation pattern will be perfectly spherical. In reality what comes closest to an isotropic antenna is a short infinitesimal dipole antenna. The radiation pattern depends on the antenna used and the combination of a number of antenna elements (as in array antennas). Let us say that the radiation pattern of a transmitting antenna and a distance r from the antenna is defined by $E(r, \theta, \phi) = E_{\mathrm{m}} f(\theta, \phi)$, where E_{m} is the maximum electric field at distance r from the antenna, and the function $f(\theta, \phi)$ defines the radiating (or elemental) pattern. The transmitting antenna is assumed to be placed at the origin. Although we discuss the radiation pattern of a transmitting antenna, it is the same pattern when the antenna is receiving signals at the same frequency as when transmitting.

7.9.3 Polarization and polarization diversity

There are three types of polarizations used in electromagnetic fields used in communication and remote sensing systems. These are shown in Fig. 7.5.

(a) Linear polarization of electric and magnetic fields. There are two kinds of linearly polarized signals. These are vertical and horizontal polarizations. If the plane of the field is parallel to the ground, it is known as a horizontally polarized signal. If the plane of radiated electric field from a vertically held wire antenna is perpendicular to ground and the electric field is perpendicular to the ground, the wave is called a vertically polarized signal. When the receiving antenna is

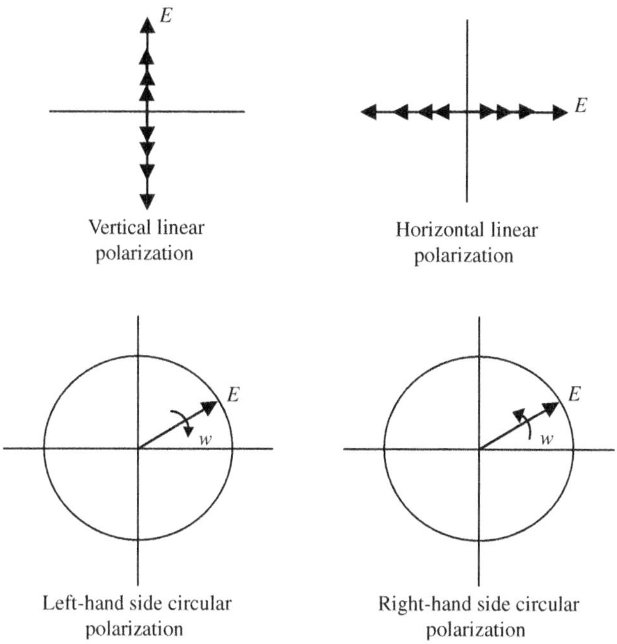

Figure 7.4: Linear, Circular and elliptical polarization of the electric field vector.

located in the same plane as the transmitting antenna, and the two are directly facing each other, the received signal strength will be maximum.

(b) Circular polarization. If the radiated or tranmitted signal is rotating at its operating frequency in space, the rotation being acheived by electrical means in feeding the signal to the transmitting antenna, the radiated signal is said to be circularly polarized. At the receiver antenna, the circularly polarized signal produces equal received signal strengths, whether for both a horizontally or a vertically polarized receiving antenna. Moreover, if the rotation of the electric field in space is in the same direction as the right-hand stretched out from the body and swung into the body, the electromagnetic signal is called a left-hand circularly polarized signal. Similarly if the rotation of the electric field is in the same direction as that of the left-hand swung into the body, it is called a left-hand circularly polarized signal.

(d) Elliptical polarization. A circularly polarized signal is made up of two orthogonal, linearly polarized signals; these are vertically and horizontally polarized signals. A circularly polarized signal becomes an elliptically polarized signal when the two orthogonal (perpendicular to each other) electric fields are not equal in magnitude. This occurs, for instance, when a satellite antenna radiates circularly polarized signals through rain. The rain drops absorb most power from the transmission signals which are polarized along the elongated length of a rain drop. This makes the two (vertically and horizontally polarized) electric fields unequal to each other.

Note that for effective land to satellite communication in a wireless, mobile communication system that uses satellite BS instead of a land based BS, the hand-held telephone unit should use antennas such as the helical antenna (instead of using some variation of linear wire antennas) to effectively capture circularly polarized signals from satellites. Using a circularly polarized signal the system information capacity may be doubled, since the two electric fields that are perpendicular (orthogonal) to each other will not interfere with each other though they are at the same carrier frequency. That means, for instance, two TV channels may be transmitted and received at the same carrier frequency, if one channel is transmitted as a vertically polarized signal, and the other as a horizontally polarized signal. The resultant signal will be a circularly polarized signal, carrying the text, voice and image information of two TV channels.

Since most mobile antennas are mounted on handheld units, most mobile transmitters radiate both vertically and horizontally polarized signals. These horizontally and vertically polarized signals arrive at the BS antenna together. The path and signal attenuation of the horizontally and vertically polarized signals are very different. This diversity which is inherent in most mobile transmistters may be used to improve link margin and capacity by getting the BS receiver antenna to operate on the vertically and harizontally polarized signals separately before combining them to get the final received signal. This technique makes good use of polarization diversity, which is present in most mobile communication and radar signals.

7.10 Antennas as a temporal filter: frequency bandwidth

An antenna is a temporal filter with its response peaking at a certain resonant frequency and not allowing signals at other frequencies to be transmitted or received.The frequency bandwidth of an antenna, as in other electrical devices, is a measure of the frequency range over which the antenna will perform satisfactorily. The bandwidth may be expressed as the ratio of the highest frequency at which the antenna performs acceptably to the lowest frequency at which it does so. For example, in a 10:1 bandwidth broadband antenna, the upper frequency of acceptable performance is 10 times greater than the lower frequency of acceptable performance. The bandwidth is also expressed as a percentage of the frequency difference. An example of such a classification is the 5% bandwidth narrowband antenna, where the frequency range of acceptable operation is 5% about the mid-frequency of the bandwidth. Thus:

i. Broadband antennas perform over a center frequency with a bandwidth ratio of 10:1, which is the ratio of the maximum-to-minimum frequencies of acceptable operation. Antennas operating at 5:1 bandwidth are generally suitable for most applications, including mobile communications, which require a bandwidth of about 70 MHz.

ii. If the bandwidth is about 1:0.05, the antenna is called a narrowband antenna. In a narrowband antenna, the antenna conductor diameter to length ratio must be kept small. If the diameter of the antenna is increased, then it becomes a wideband antenna. Narrowband antennas tend to detune under ice wind loading.

The antenna can be characterized as a temporal filter. Since an antenna can be considered as an RLC (resistor–inductor–capacitor) circuit, it has a certain frequency bandwidth. The current I induced in the antenna by an incoming, incident signal is maximum at a given, resonant frequency. The bandwidth of an antenna may be conceptually understood by considering it as an RLC circuit. The capacitance from between the upper and lower wire of a half-wave dipole antenna will result in the bell shaped frequency response of a bandpass filter. The wires will have both resistance and inductance.

We define the Q-factor of the antenna which is represented by a circuit to be the ratio of the voltage V_c across the capacitor of the circuit to the total voltage applied V:

$$Q = \frac{V_c}{V} = \frac{1}{(2\pi f_{sr} CR)} \tag{7.35}$$

When the Q-factor is large, the peak current increases. In other words, for maximum reception, the antenna Q-factor should be designed to be large at the center frequency at which it is designed to operate. Now any series RLC circuit will have a resonant frequency at which the current peaks. The resonant frequency in terms of the inductance and capacitance of the antenna is given by

$$f_0 = \frac{1}{2\pi \sqrt{L_A C_A}} \tag{7.36}$$

This is the best frequency to operate the particular antenna. It is known that the response of an RLC circuit is degraded as the operational frequency moves up or down from the resonant frequency. At given points on either side of the peak, the resonant-frequncy point, the current induced along a receiver wire antenna will become $0.707 I_m$, where I_m is the maximum current that will be induced by the incident signal at resonant frequency. At these two (3dB) points, the received power will be half the maximum power that could have been received. If we operate the antenna beyond the 3 dB point, then very little of the signal will get into the antenna to be radiated out. Hence we have the concept of bandwidth, which specifies the permissible frequency fluctuations for a given antenna. Given the resonant frequency and the bandwidth of an antenna, we will know the frequencies at which the antenna will satisfactorily perform. For a given communication or remote sensing system, the bandwidth of the antenna must be matched with the bandwidth of the signals.

7.11 Antenna as a spatial filter: angular beamwidth

Figure 7.3 shows the antenna characteristics when an antenna is considered as a spatial, or angular, filter. The antenna transmits effectively within the HPBW (θ_{HPBW}), angle, and, beyond the first-null beamwidth (FNBW, θ_{FNBW}), we get the undesirable radiation through side lobes. The fact that the antenna radiation pattern has nulls in certain (θ, ϕ) directions and peaks in other directions implies that the antenna acts like a filter in the spatial (r, θ, ϕ) domain, a concept that will be used in antenna heamforming in adaptive array antennas. Such antennas which improve the spatial resolution of the beam (i.e. narrower beams with small beamwidths) is expected to make a significant impact on future systems such as wireless mobile communications, magnetic resonance imaging in medicine, and safety and security systems. The part of the radiation pattern plot that is the longest radiation (beam) pattern is called the *major lobe*. This antenna is called a *directional antenna* because it radiates more in one direction than in other directions. The azimuth spatial resolution of the beam, at a distance R from the antenna, is approximately $R\theta_{\text{HPBW}}$ m.

The power density may be obtained from the electric field intensity in W/m^2:

$$P_{\text{av}} = \frac{1}{2}\text{Re}\{E \times H^*\} \tag{7.37}$$

Assuming plane-wave propagation, the intrinsic impedance of the medium through which the wave travels, for example $\eta = 120\pi$ for free space, is defined by

$$\eta = \frac{E}{H} \tag{7.38}$$

Thus from (7.37) and (7.38), we have

$$P = \frac{1}{2\eta}E^2 \tag{7.39}$$

When $E/E_{\text{max}} = 1$, we have $P/P_{\text{max}} = 1$. When $E/E_{\text{max}} = 0.707$, we have $P/P_{\text{max}} = 0.5$. Thus we see that the polar plots of E and P are very similar, except that the plots of P would be narrower at the middle of the lobes, the HPBW was defined above as the angular difference between the half of the peak values (see Fig. 7.4), or $10\log_{10}(1/2) = -3$ dB on the decibel scale. The side lobes depict the undesired signals radiated in any direction other than the intended or desired direction. The ratio of the largest side (minor) lobe solid angle to the major-lobe solid angle represents the ratio of the power density in the largest minor lobe to that of the major lobe. Acceptable values of side lobe levels are −20 dB or less. The larger side lobes could be wasteful because they contribute to significant power loss. In a radar system, the large side lobes could create problems by reflecting undesired signals back into the detection electronics. The beam efficiency η_b is defined as the ratio of the main beam area in solid angle (Ω_M) to the total beam solid angle, Ω_T,

$$\eta_b = \frac{\Omega_M}{\Omega_T} \tag{7.40}$$

The minor-lobe area Ω_m is given by $\Omega_m = \Omega_T - \Omega_M$.

7.12 Reciprocity

7.12.1 The Friis equation

The reciprocity theorem tells us that if a transmitter is connected to an antenna A_x and the signal is received by another receiver antenna A_y, then the power received at the output terminal of the receiver antenna will be the same if the transmitter is now connected to A_y and the receiver to A_x. This also implies that the transmit and receive beam (radiation) patterns of an antenna are identical.

Thus, for example, if antenna A_x transmits maximum signal strength along the z-axis ($\theta = 90^\circ, \phi = 90^\circ$) and zero radiation along the direction $\theta = 30^\circ, \phi = 90^\circ$, then if A_x is used to receive signals, it will best receive signals coming toward it along the z-axis and will not pick up any signal coming from the $\theta = 30^\circ, \phi = 90^\circ$ direction.

Let the gain of the transmitting antenna in the direction of the receiver antenna be G_x and that of the receiver antenna in the direction of the transmitter antenna be G_y; and P_T be the total power input to the transmitting antenna. Referring to Fig. 7.5, if we ignore the losses in the antenna ($G = D$), then the total power radiated is P_T W. The power density at the receiver antenna $P_i = G_x P_T/(4\pi r^2)\,\text{W/m}^2$. The effective aperture of the receiving antenna in the direction of the transmitting antenna is $A_{em} = D_y \cdot \lambda^2/(4\pi) = G_y \cdot \lambda^2/(4\pi)$. Hence the total power received by the second antenna

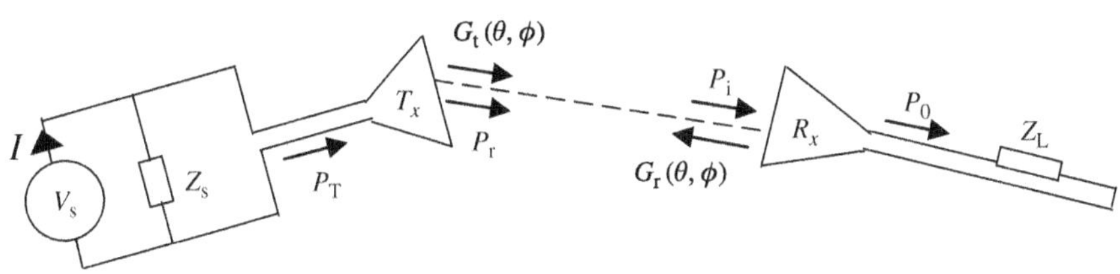

Figure 7.5: The main and side lobes of an antenna

at a distance r from the transmitting antenna is given by

$$P_R = \left(\frac{G_x P_T}{4\pi r^2}\right) \cdot A_{em} = \left(\frac{G_x P_T}{4\pi r^2}\right) \cdot \left(\frac{D_y \cdot \lambda^2}{4\pi}\right) \text{ W} \tag{7.41}$$

or,

$$\frac{P_R}{P_T} = \frac{G_x G_y \lambda^2}{(4\pi r)^2} \tag{7.42}$$

This is called the Friis transmission formula, and it relates the power P_R that appears at the communication receiver amplifier to the total power P_T transmitted to the communication transmitter antenna. We have assumed that both antennas are lossless, that the impedances are all properly matched and that both antennas are correctly polarized.

The gains G_x and G_y are, in general, given by $G_x = G_{x0} f_x^2(\theta, \phi)$ and $G_y = G_{y0} f_x^2(\theta, \phi)$, where G_{x0} and G_{y0} are the maximum gains of the two antennas (e.g. $G_{x0} = G_{y0} = 1.5$ if both antennas are elemental, short dipole or $G_{x0} = G_{y0} = 1.64$ if both antennas are half-wave dipoles). The functions $f_x(\theta, \phi)$ and $f_y(\theta, \phi)$ are the normalized electric (and magnetic) field radiation patterns of the two antennas in spherical coordinates in the direction (θ, ϕ). The maximum value of the elemental radiation pattern function is 1.0.

In mobile or wireless communication systems, the antennas are considered short dipoles that may be approximated to isotropic (omnidirectional) antennas. For omnidirectional antennas, $G = D = 1$. This means that the received power is

$$P_R = P_T \left(\frac{\lambda}{4\pi r}\right)^2 \text{ W} \tag{7.43}$$

or

$$P_{RdB}(r_0) = 10 \log_{10} P_T \left(\frac{\lambda}{4\pi r}\right)^2 \text{ dB} \tag{7.44}$$

The $1/r^2$ term causes a decrease in signal strength with distance and may be considered as a type of power loss or noise. It is termed a multiplicative noise since the $1/r^2$ term is multiplied with the transmitted power term P_T. In (7.43) for power, we have ignored the signal loss over the path traveled by the signal and shadowing effects. Path loss is loss in the finite-conductivity ground over which the signal travels. Shadowing is due to objects such as trees and buildings that tend to obstruct a part of the signal. Shadowing too is a multiplicative noise, since the signal attenuation constant of a wall or tree is multiplied to the transmitted power to cause a decrease in the radiated power. Shadowing causes slow signal variation. There is also multipath fading due to reflection of the transmitted signals resulting in mutual interference between the directly received signals and incoming reflected signals from all directions. It is also called fast noise, since there is fast or deep fluctuations of the signal strength. Thus the received power will be less than the ideal received P_R. Expressions for path loss have been obtained from measurements and include the following:

(a) The path loss for an open zone system without buildings on trees:

$$L(\text{dB}) = 32.44 + 20\log_{10}(f_c) - 20\log_{10}(d), \text{ units} \tag{7.45a}$$

where d is the distance between transmitter and receiver and f_c is the carrier frequerncy in MHz.

(b) The path loss for urban zones:

$$L_u(\text{dB}) = 69.55 + 26.16\log_{10}(f_c) + (47.9 + 6.55\log_{10}(h_b)) - 13.82\log_{10}(h_b) - C(h_m), \text{ units} \tag{7.45b}$$

where h_b, h_m are the heights of the BS and mobile station antennas, respectively.

(c) For suburban zones:

$$L_s(\text{dB}) = L_u - 2\log_{10}(f_c/28)^2 - 5.4 \text{ units} \tag{7.45c}$$

(d) For rural areas:

$$L_r(\text{dB}) = L_p - 7.78\log_{10}(f_c)^2 + 18.33\log_{10}(f_c) - 40.94 \text{ units} \tag{7.45d}$$

where the correction factor for mobile station height h_m, is given by

$$C(h_m) = \{1.1\log_{10}(f_c) - 0.7\}h_m - \{1.5\log_{10}(f_c) - 0.8\} \tag{7.45e}$$

The path loss component typically varies from 3 to 4 units in an urban macrocellular environment, and from 2 to 8 units for a microcellular environment. The cell size is 1 km for macrocellular systems, 100 m for outdoor microcellular systems and 1 m for indoor picocelluar systems. The loss component due to shadowing is a zero mean square Gaussian random variable. The transmitted power P_T for a typical mobile phone is less then 2 W or even down to 1 W. There is now an attempt to take this power level down to 0.1 or 0.01 mW by reducing the cell size. Moreover, future semiconductor devices may operate at a reduced voltage of 1.2 V, instead of 5 V. By using high-frequency electromagnetic signals for wireless mobile communications, the path loss is increased and thus it is possible to reuse frequency (frequency reuse) in another adjacent cell. Thus to reduce cell size, the terrestrial mobile communication operating frequency of 800/900 MHz was increased, toward 1885–2000 MHz and beyond.

7.12.2 The Radar equation

Consider now a radar system. A transmitting antenna of gain G_x radiates a pulse of sinusoidal waves (continuous wave, or CW, radar) or linear-frequency modulated waves (Chirrp radar) that impinge on a target of radar cross-sectional area (RCS) of σ m^2. The transmitting antenna itself or another antenna of gain G_y captures some portion of the signal reflected by the target. When the same antenna is used to capture the reflected signal, the radar is called a monostatic radar. When a second antenna is used at a slightly different position to capture the reflected signal, it is called a bistatic radar.

If the target is at a distance r_1 from the transmitting antenna, then the power density at the target is $G_x P_T / 4\pi r_1^2$ W/m^2. Hence the power reflected by the target is $G_x P_T / 4\pi r_1^2$ W.

If the second antenna is at a distance r_2 from the target, then the power density appearing at the receiving antenna is $G_x P_T / (4\pi r_1 r_2)^2$ W/m^2. If the maximum effective aperture of the receiving antenna is expressed as

$$A_{\text{em}} = G_y \cdot \frac{\lambda^2}{4\pi} \tag{7.46}$$

the received power is given by

$$P_R = \left[G_y \cdot \frac{\lambda^2}{4\pi} \right] \left[\frac{\sigma G_x P_T}{(4\pi r_1 r_2)^2} \right] \tag{7.47}$$

As before,

$$\frac{P_R}{P_T} = \left[\sigma G_x G_y \cdot \frac{\lambda^2}{(4\pi)^3 (r_1 r_2)^2} \right] \tag{7.48}$$

When $r_1 = r_2 = r$, that is for the monostatic case,

$$\frac{P_R}{P_T} = \left[\sigma G_x G_y \cdot \frac{\lambda^2}{(4\pi)^3 (r)^4} \right] \tag{7.49}$$

Equation (7.49) is called the radar equation. Note that in this case, the power received is proportional to r^{-4}, which indicates that the receiver gets very little power to process. Thus the radar transmitter powers are generally high, say an average power of 500 W and power of the order of 5 kW, to ensure that the receiver gets enough reflected signal power to detect and operate on to get the distance and velocity of the target. In stealth technology, to evade detection, missiles and aircraft are designed to geometrically and materialwise reduce the RCS. The radar cross section (in m^2) of various objects: (pickup truck, 200), (jumbo jet aircraft, 100), (car, 100), (bomber aircraft, 40), (large fighter aircraft, 6), small fighter aircraft, 2), (adult man, 1), (missile, 0.5), (bird, 0.1), (insect, 0.00001), (radar "invisible" aircraft, 0.000001).

7.13 Antennas and Maxwell's equations

7.13.1 Antenna design paramenters

The quality of a well-designed antenna and telecommunication system is determined by the following three quantities:

1. Range (and mobility). The distance over which the telecommunication system should function, and the amount of freedom allowed for the transmitter and the receiver to move.
2. Capacity. The amount of data the system should carry, in number of bits/seconds/Hertz in digital communication systems.
3. Quality. The signal to noise ratio or the bit error rate required of the system.

7.13.2 Maxwell's equations

The study of electromagnetic (EM) fields requires the use of the four Maxwell equations. Ampere's law states

$$\nabla \times \mathbf{H} = \mathbf{J} + \mathrm{j}\omega\varepsilon\mathbf{E} \tag{7.50}$$

which means that a time-varying electric field produces a time-varying magnetic field.

The magnetic field intensity **H** (A/m) is related to the magnetic flux density **B** (Webers/m^2), by the absolute permeability of the medium $\mu = \mu_0\mu_r$, $\mathbf{B} = \mu\mathbf{H}$, where μ_0 is the permeability of free space and μ_r is the relative permeabilty of the medium.

Faraday's law states

$$\nabla \times \mathbf{E} = -\mathrm{j}\omega\mu\mathbf{H} \tag{7.51}$$

which may be expressed as a time-varying magnetic field producing a time-varying electric field. The electric field intensity **E** (V/m) is related to the electric flux density **D** (C/m^2) by $\mathbf{D} = \varepsilon\mathbf{E}$, where $\varepsilon = \varepsilon_0\varepsilon_r$ is the absolute permitivity of the medium, ε_0 is the permittivity of free space and ε_r is the relative permittiviy of the medium.

According to Gauss's law,

$$\nabla \cdot \mathbf{B} = 0 \tag{7.52}$$

which states that magnetic flux lines are continuous. Finally, Coulomb's law may be expressed as

$$\nabla \cdot \mathbf{E} = \frac{\rho}{\varepsilon} \tag{7.53}$$

which means that electric field lines may be continuous, or may start and end on electric charges.

Maxwell's equations relate the two basic physical parameters **E** (electric field intensity) and **B** (magnetic flux density) to the sources **J** (current density) and ρ (electric charge density). In telecommunications, radar and biomedical imaging, the presence and effects of ρ are considered negligible. Thus **J** is the primary source of information about the voice and video signals being transmitted or the object being imaged.

7.13.3 Vector potential functions

Consider the two Maxwell equations that couple the electric field intensity **E** and magnetic flux density **B**. The magnetic flux density itself will be expressed as the curl of the magnetic vector potential **A**. A vector is completely defined only after both its curl and divergence are defined. The vector potential **A** is useful in solving for EM fields generated by an electric current source **J**. Since the magnetic flux **B** is solenoidal, we have

$$\nabla \cdot \mathbf{B} = 0 \tag{7.54}$$

A, $\mathbf{B} = \nabla \times \mathbf{A}$ where **A** is the magnetic vector potential,

Hence **B** can be represented as the curl of another vector because it satisfies the vector identity $\nabla - \nabla_{\times} \equiv 0$. To complete the definition of vector **A**, its divergence will be set to, $\nabla \cdot \mathbf{A} = -\mu\varepsilon\, \partial\phi/\partial t$, where ϕ is the scalar potential.

7.14 Antenna analysis

7.14.1 Introduction

In a telecommunications system, to send electromagnetic energy through air, the launching (at the transmitting station) and the capturing (at the receiver station) of the electromagnetic waves are done by antennas. With antenna structures, we may efficiently launch electromagnetic waves into space in a predetermined manner, making possible wireless transmission of information over very long distances.

With transmitting antennas, on the other hand, the energy radiated must be maximized. Furthermore, an antenna may also be required to optimize the radiation energy in one direction and to suppress it in others, as in the case of earth station to satellite microwave communication. Such antennas are said to be directed because they concentrate the energy transmission in a preferred direction. Without such directionality, the isotropically radiated energy would be nonexistent at the receiving station far away. In other applications, for example radio broadcasting, the antenna must often effectively radiate in all directions (i.e. omnidirectionally), so that the radio receiver antenna may pick up the signal at any location within the radial range of the transmitter.

Thus we see that there are many different needs that an antenna must fulfill. To meet the particular needs of a variety of communications systems, there are various geometrical forms of antennas like the dipole antenna, the helical antenna, the log-periodic antenna and the horn antenna.

The transmitting and receiving field patterns of any antenna are identical. Hence a transmitting antenna may also be used as a receiver antenna.

7.14.2 Types of antennas

7.14.2.1 General classifications

Antennas may be divided into four different classes based on the frequency spectrum in which they are commonly employed. In Table 7.2, we have listed the four major types of antenna that are currently in use. An antenna in general is a transducer which changes one form of energy (e.g. conduction power in a coaxial cable) into another form (radiation power in a free space wave). At the lower end of the frequency spectrum, conduction current elements or wires are used as antennas. At higher frequencies, when the wavelength of the signal is smaller than the antenna length, the wave runs along the antenna wire (traveling wave) before being launched into free space. In order to get the antenna to radiate in a certain prescribed direction and not to radiate in other directions, more than one element of the current element, traveling or apeture antenna are arranged (or arrayed) in such a way that the electromagnetic waves launched in the prescribed direction of each element will add together to give a strong radiated signal in that direction, whereas in the directions in which radiation (or reception) is unwanted, the electromagnetic waves launched by the different elements cancel each other to give zero or negligibly small radiation. At very high frequencies, such as in satellite communications or long range radar, the aperture antennas are used where the antenna is like a window through which traveling waves from a waveguide may be launched or radiated out, or very high-frequency microwave signals traveling in free space may be channeled into receiver electronics through the antenna window or aperture.

Table 7.2: Antenna types.

Antenna type	Frequency range	Wavelength	Antenna size/λ
Current elements	10 KHz–1 GHz	30 m–0.3 m	0.01–1
Travelling wave	1 MHz–10 GHz	300 m–3 cm	1–10
Arrays	5 MHz–50 GHz	60 m–6 mm	1–100
Apertures	0.5–1000 GHz	0.6 m–0.3 mm	1 – 10,000

7.14.2.2 Electromagnetic waves from conductors to free space

Moving electric charges (associated with electric currents) are required to excite the electromagnetic fields carrying voice, text and image information. But the charges are not needed to sustain the fields. For the excitation of electromagnetic fields, we need electric charges flowing in the antenna. But once launched into free space, the electromagnetic fields may exist in electric charge-free space and propagate at the velocity of light through free space without the presence of electric charges. In free space, the time-varying electric fields **E** close upon themselves as circulating displacement currents and may exist without the electric charges to originate from. These closed dispalcement currents produce magnetic fields **H**.

Time-varying currents produce time-varying electric fields. Time-varying electric fields produce time-varying magnetic fields. These time-varying magnetic fields will in turn produce time varying electric fields. This transfer of energy from **E** to **H** and then from **H** to **E**, and so on in space as the energy (or information propagates over space) is represented by the Maxwell equations. In (7.50), although the conduction plus eddy current density **J** is absent in space, the displacement current, $j\omega\varepsilon\mathbf{E}$ (or $\varepsilon\ \partial E/\partial t$) produces time-varying magnetic fields in free space. Then according to (7.51), time-varying magnetic fields produce time-varying electric fields, with **E** and **H** moved along space as each produces the other. Thus we have the production and propagation of electromagnetic waves in free space. The waves radiate away from the transmitting antenna at the velocity of light, c, in vacuum, and at a reduced velocity $c/\sqrt{\varepsilon_r}$ in a dielectric medium of relative permittivity ε_r.

7.14.2.3 Elemental current antennas

Elemental current antennas are generally conducting wires that are arranged such that the signal power radiated is maximized. An inexpensive, but effective, type of wire antenna is the half-wave dipole antenna, of the configuration shown in Fig. 7.6. The input impedance of a half-wave dipole antenna is usually $73 + j42.5\ \Omega$, and becomes $70 + j0\ \Omega$ if the length is slightly reduced, that is tapered. Hence its input impedance almost exactly matches the $70\ \Omega$ characteristic impedance of some coaxial cables. The half-wave dipole antenna, as already indicated, is formed by bending out a wire of length $\lambda/4$ from each end of a two-wire transmission line, where λ is the wavelength of the signal to be transmitted (or received). The total length of the antenna is $\lambda/2$.

An alternative way of forming a half-wave dipole antenna is to use a quarter-wavelength ($\lambda/4$) long wire placed over a perfectly conducting plane (Fig. 7.6(a)). This quarter-wavelength monopole results in a half-wave dipole based upon the theory of images. The theory of images states that when a positive electric charge is placed above a perfectly conducting plane, the plane can be replaced by a negative electric charge placed at the mirror image point of the positive source charge. This means that a positive charge moved vertically up above a perfectly conducting

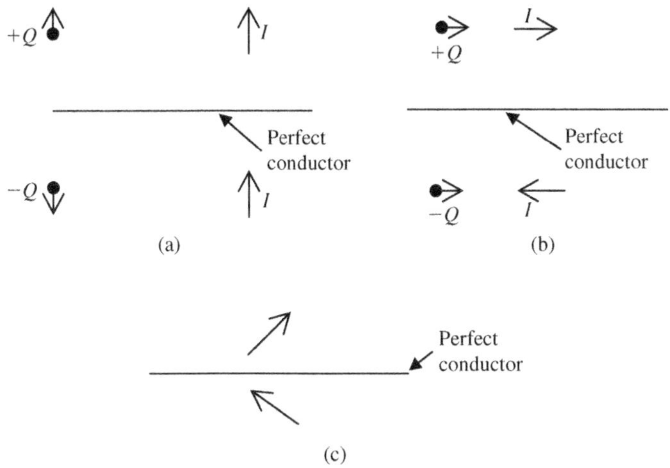

Figure 7.6: Using the concept of mirror images of electric charges to replace perfectly conducting planes.

plane, will have another vertically, downward moving negative charge of the same magnitude at the mirror image point. Hence an electric current moving vertically upward above a perfectly conducting plane will have associated with it a mirror image current that is also moving vertically upward. The input impedance of such a quarter-wavelength monopole (36 Ω) is half that of the half-wavelength dipole (73 Ω). But its directivity is twice that of the half-wave dipole, since its HPBW above ground (39°) is half that of the half-wave dipole (78°). Furthermore, note from Fig. 7.6(b) that a perfectly conducting plane over which a horizontally placed current carrying a element may be replaced by another horizontal current carrying elment at the mirror image point, but carrying current in the opposite direction. Similarly, a perfectly conducting plane over which is placed a slanted current carrying element may be replaced by a mirror image current carrying element where the vertical components of the currents are in the same direction, but the horizontal components of the currents are oppostite to each other as shown in Fig. 7.6(c).

A folded dipole, used in TV reception, is shown in Fig. 7.7. Such folded dipoles are used as receiver antennas for FM broadcast (e.g. at 100 MHz, $\lambda = 1.5$ m). A folded dipole is made by connecting together the top and bottom parts of two dipole wire antennas. The impedance of a half-wave ($L = \lambda/2$) folded dipole is $Z_A = 4Z_{hw}$, so that for $Z_{hw} = 70\,\Omega$, the input resistance of a half-wave dipole, then the impedance of the folded half-wave dipole is $Z_A = 4 \times 70 = 280\,\Omega$, which matches with the intrinsic impedance of a twin-wire transmission line which may be used to connect the center part of one of the two dipoles of the rooftop folded dipole antenna to the TV receiver inside the house.

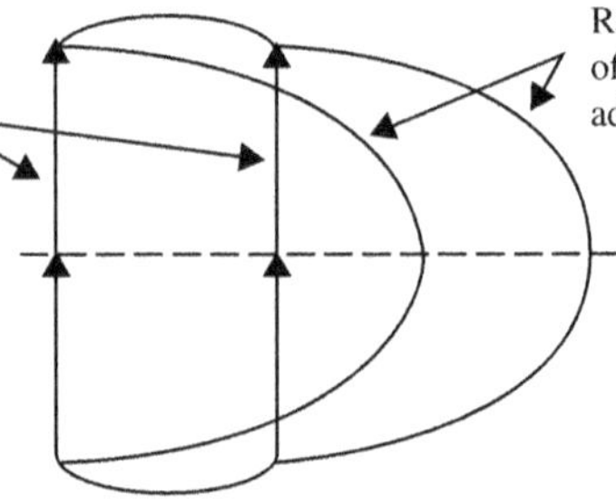

Figure 7.7: Folded Dipole Antenna.

For over-the-horizon transmission of signals, as in the case of over-the-horizon radar, which operates on the principle of the ionospheric reflection of electromagnetic signals radiated by the radar antenna, horizontal dipoles may be used. The ionosphere is at a height of about 300 km above the earth. For the horizontally placed current-carrying wire, the mirror image current flows in the opposite direction to that of the source current placed above the ground. These horizontally held antennas are also used by hobbyists to communicate to someone who is continents away on the earth. The electromagnetic communication signal travels all the way round the globe by repeatedly bouncing off the ionosphere and the earth.

The elemental type of wire antennas are designed to operate at frequencies ranging from 10 kHz to 1 GHz. The size of the antenna (length L) is related to the wavelength by $L/\lambda = 0.01$ to 1.

7.14.3 The basic equations

The wave equation in terms of the vector potential is given by

$$\nabla^2 \mathbf{A} - \mu\varepsilon \frac{\partial^2 \mathbf{A}}{\partial t^2} = -\mu \mathbf{J} \tag{7.55}$$

The wave equation for the scalar potential is given by

$$\nabla^2 \phi - \mu\varepsilon \frac{\partial^2 \phi}{\partial t^2} = -\frac{\rho}{\varepsilon} \tag{7.56}$$

The current density due to electric field produced conduction current and electron drift in magnetic field is given by

$$\mathbf{J} = \sigma(\mathbf{E} + v \times \mathbf{B}) \tag{7.57}$$

which in the absence of a net electron drift in magnetic field reduces to

$$\mathbf{J} = \sigma \mathbf{E} \tag{7.58}$$

which becomes

$$\mathbf{J} = -\sigma \nabla \boldsymbol{\phi} - \sigma \frac{\partial \mathbf{A}}{\partial t} = \mathbf{J_c} + \mathbf{J_e} \tag{7.59}$$

since the electric field is related to the vector and scalar potentials by

$$\mathbf{E} = -\frac{\partial \mathbf{A}}{\partial t} - \nabla \phi \tag{7.60}$$

The conduction current $J_c = -\sigma \nabla \phi$ A/ m^2 only exists where there is a potential drop $\nabla \phi$ across a conductor such as in the case of a wire antenna connected to a signal source. It does not exist outside the antenna and plays only the role of launching the signal into space (transmitting antenna) or being induced in the antenna wire by a free space signal incident upon the wire antenna (receiving antenna). The displacement current term, also in A/m^2, is $\mathbf{J_e} = -\sigma\, \partial \mathbf{A}/\partial t$. It exists at the aperture or opening of an aperture antenna such as a horn antenna or parabolic dish antenna, playing the role of the conduction current in the wire antenna, in launching or receiving signals at the antenna–free space interface. Moreover, in both wire antennas and aperture antennas it is the displacement current which consititutes the electromagnetic wave travelling out from the antenna (transmitting antenna) or incident upon the antenna (receiving antenna).

For a sinusoidal signal at frequency $f = \omega/2\pi$ Hz, the vector potential may be written as

$$\mathbf{A}(x, y, z, t) = R_e(\mathbf{A}(x, y, z)e^{j\omega t}) \tag{7.61}$$

The wave equation becomes

$$\nabla^2 \mathbf{A} + \omega^2 \varepsilon \mu \mathbf{A} = -\mu \mathbf{J} \tag{7.62}$$

where

$$\mathbf{J} = -\sigma \nabla \phi - j\omega\sigma \mathbf{A} \tag{7.63}$$

Separating the current term into conduction and displacement current terms, we get

$$\nabla^2 \mathbf{A} + (\omega^2 \varepsilon \mu - j\omega\sigma\mu)\mathbf{A} = \sigma \mu \nabla \phi = -\mu \mathbf{J_c} \tag{7.64}$$

On the conductor, we have,

$$\omega^2 \varepsilon \mu \ll \omega \sigma \mu \tag{7.65}$$

or

$$\omega \ll \frac{\sigma}{\varepsilon} \tag{7.66}$$

Note that the latter term, $\omega^2\varepsilon\mu\mathbf{A}$, arose from $\varepsilon\mu\,\partial^2\mathbf{A}/\partial t^2$ with a change of sign from $(j\omega)^2$, which in turn came from the displacement current $\varepsilon\,\partial\mathbf{E}/\partial t$. It is this displacement current that enables the propagation of electromagnetic fields in free space. Since outside the conductor, $\sigma = 0$, we have

$$\nabla^2\mathbf{A} + \omega^2\varepsilon\mu\mathbf{A} = \sigma\mu\nabla\phi = -\mu\mathbf{J_c} \tag{7.67}$$

For the scalar potential, the wave equation is

$$\nabla^2\phi + \omega^2\mu\varepsilon\phi = -\frac{\rho}{\varepsilon} \tag{7.68}$$

The wave number k for free space is given by

$$k = \omega\sqrt{\mu\varepsilon} \tag{7.69}$$

which on substitution into (7.67) and (7.68) yields

$$\nabla^2\mathbf{A} + k^2\mathbf{A} = -\mu\mathbf{J_c} \tag{7.70}$$

and

$$\nabla^2\phi + k^2\phi = -\frac{\rho}{\varepsilon} \tag{7.71}$$

Once we solve (7.70) for the vector potential **A**, we may get the magnetic field **H** from the equation,

$$\mathbf{H} = \frac{1}{\mu}\nabla\times\mathbf{A}, \tag{7.72}$$

and subsequently the electric field from the equation,

$$\mathbf{E} = \frac{1}{\sigma + j\omega\varepsilon}\nabla\times\mathbf{H} \tag{7.73}$$

7.14.4 The retarded potentials

$$\phi = \frac{1}{4\pi\varepsilon}\int_3 \frac{\rho}{r}\mathrm{d}R \tag{7.74}$$

When a wave travels in the direction z, the term that defines the wave motion must take the form $e^{-\gamma z}$, where γ is the propagation constant.

$$\phi = \frac{1}{4\pi\varepsilon} \int_3 \frac{\rho e^{j\omega t - \gamma r}}{r} dR \tag{7.75}$$

where the 3 below integral sign signifies a triple integral. This term indicates that the potential $\phi(t)$ at a distance r from the electric charge density ρ source point is due to the source at time instant $(t - (r/v))$, that is $\rho(t - (r/v))$, where v is the velocity at which the wave travels ($v = c$ in free space).

$$A = \frac{\mu}{4\pi} \int_3 \frac{J e^{j\omega t - \gamma r}}{r} dR \tag{7.76}$$

The vector potential at a distance r from the current density source J at time instant t is due to the electric current density at time instant $(t - (r/v))$, that is $J(t - (r/v))$. The potentials A and ϕ are termed the *retarded vector potential* and the *retarded scalar potential*, respectively.

7.15 Elemental current antennas

7.15.1 Infinitesimal antenna (length L is less than 1/50 of the wavelength λ

The wire antenna and the loop antenna are basic, useful forms of elemental wire antennas. The half-wave dipole antenna and the loop antenna are similar to each other. The loop antenna forms a magnetic dipole. A conducting wire is bent into a circular loop. When a high frequency source is connected across a small cut in the loop, the circular loop will radiate out the signal received from the source. The radiation fields at (r, θ, ϕ) and radiation resistance of a loop antenna are given by

$$H_0 = \left(\frac{ka}{2}\right)^2 I \frac{e^{-jkr}}{r} \sin\theta \tag{7.77}$$

$$E_\phi = -\eta H_\theta \tag{7.78}$$

$$R_r = 20\pi^2 \left(\frac{2\pi a}{\lambda}\right)^2 \ \Omega \tag{7.79}$$

where a is the radius of the loop and $k(= 2\pi/\lambda = \omega/c)$ is the wave number.

7.15.2 The complete electromagnetic field equations

In the following analysis, and analytical solution for electric and magnetic waves radiated from antennas, we take the traditional approach of forward solution: given

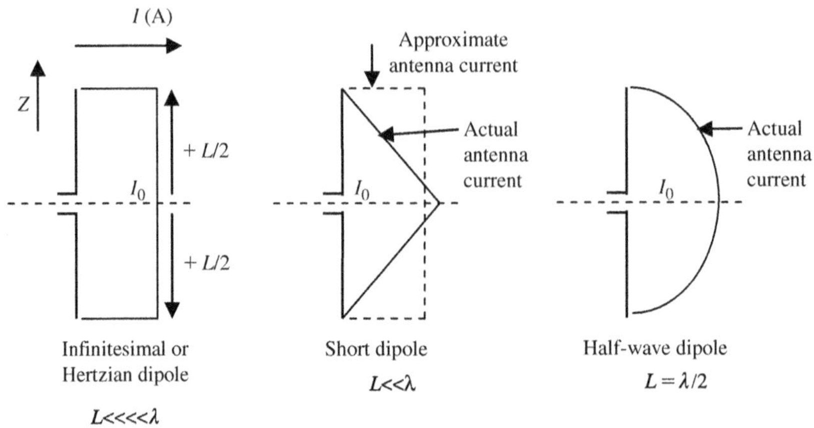

Figure 7.8: Current distribution along the wires of a (a) Hertzian dipole, (b) short dipole, for example, quarter-wave dipole and (c) half-wave dipole.

the antenna geometry and currents as in the case of wire antennas (or electric and magnetic fields appearing at the antenna radiation point, as in the case of aperture antennas such as horn or parabolic reflector antennas) determine the radiated electric and magnetic fields. In the more advanced reverse techniques (as in the case of antenna synthesis and digital beamforming), we specify the electric and magnetic field radiation patterns required, and work backward to get the antenna geometry and the source currents at the antenna.

In the case of the traditional, forward computation, in electric charge free, zero conductivty space, the wave equations for the vector potential and electrostatic potential reduce to

$$\nabla^2 A + k^2 A = 0 \tag{7.80}$$

since there is no conduction current in free space (conductivity is zero). Similarly

$$\nabla^2 \phi + k^2 \phi = 0 \tag{7.81}$$

in electric charge free space. The only electric conduction current and electric charges are found along the antenna.

The wave equation (7.80) has a general solution given by

$$A = \frac{\mu}{4\pi} \int_{3} \frac{J e^{-jkr}}{r} \mathrm{d}R \tag{7.82}$$

In the case of the infinitesimal or Hertzian dipole, the electric current along the antenna wires may be considered constant, that is $I(z) = I_0$ as shown in Fig. 7.8(a). This could be visualized by considering the cosinusoidal current distribution along

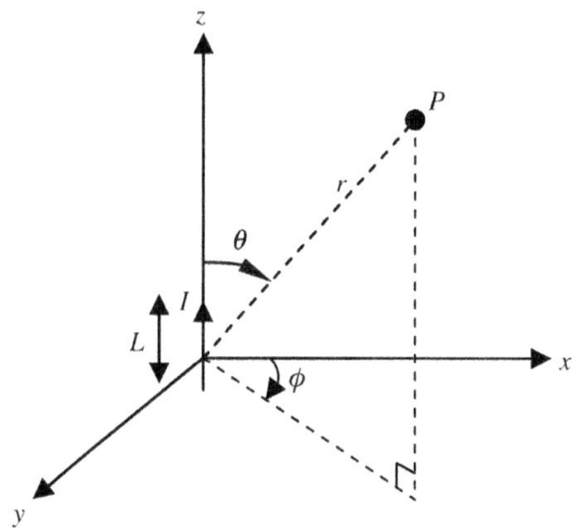

Figure 7.9: An infinitesimal wire antenna positioned at the origin.

a half-wave dipole antenna shown in Fig. 7.8(c). If we now place a very short wire, close to the peak current portion of the current distribution, we can see that the current over the very short wire will be about the peak current over that very short wire, a constant current as shown in Fig. 7.8(a). For constant current distribution along the dipole, the current J (or I) may be brought out of the integration sign in (7.82). Moreover, since the current vector points in the z-direction only (see Fig. 7.9), there will be no vector potential components in the x- and y-directions. The triple integration reduces to a single integration in the z-direction, yielding

$$A_z = \frac{\mu}{4\pi} \int_{-L/2}^{L/2} \frac{I\mathrm{e}^{-\mathrm{j}kr}}{r} \mathrm{d}z = \frac{\mu I \mathrm{e}^{-\mathrm{j}kr}}{4\pi r} \int_{-L/2}^{L/2} \mathrm{d}z = \frac{\mu IL}{4\pi r} e^{-\mathrm{j}kr} \tag{7.83}$$

Now that we have obtained the vector potetial, in order to get the magnetic field, we need to get the curl of the vector potential. It is easier to get the curl of the vector potential in spherical coordinates than in Cartesian coordinates.

Thus we must first convert the vector potential into spherical coordinates. It is not so hard when we remember that the vector potential in Cartesian coordinates is not zero only in the z-direction. The x and y-components of the vector potential are zero. Thus, the spherical coordinate component A_r of the vector potential is (Fig. 7.9)

$$A_\mathrm{r} = A_x \sin\theta \cos\phi - A_y \sin\theta \sin\phi + A_z \cos\theta = A_z \cos\theta = \frac{\mu IL}{4\pi r} \mathrm{e}^{-\mathrm{j}kr} \cos\theta \tag{7.84}$$

Similarly the spherical coordinate component A_θ is given by,

$$\begin{aligned} A_\theta &= A_x \cos\theta \sin\phi + A_y \cos\theta \cos\phi - A_z \sin\theta \\ &= -A_z \sin\theta = -\frac{\mu IL}{4\pi r} \mathrm{e}^{-\mathrm{j}kr} \sin\theta \end{aligned} \tag{7.85}$$

Finally A_ϕ is given by

$$A_\phi = -A_x \cos\phi + A_y \sin\phi = 0 \tag{7.86}$$

Therefore, we get the curl of the vector potential to obtain the magnetic field:

$$\mathbf{H} = \mathbf{H}_\phi = \frac{1}{\mu} \nabla \times \mathbf{A} = \frac{I}{\mu r}\left(\frac{\partial}{\partial r}(rA_\theta) - \frac{\partial A_\mathrm{r}}{\partial \theta}\right) \mathbf{u}_\phi \tag{7.87}$$

with $\widetilde{H}_\mathrm{r} = \widetilde{H}_\theta = 0$

Substituting (7.84) and (7.85) into (7.87), we get

$$\mathbf{H}_\phi = \mathrm{j}\frac{kIL}{4\pi}\left(\left(\frac{1}{r} + \frac{1}{\mathrm{j}kr^2}\right)\sin\theta\right)\mathrm{e}^{-\mathrm{j}kr} \tag{7.88}$$

Since

$$\nabla \times \mathbf{H} = \mathbf{J} + \mathrm{j}\omega\varepsilon\mathbf{E} = \mathrm{j}\omega\varepsilon\mathbf{E} \tag{7.89}$$

we have

$$\mathbf{E} = -\frac{j}{\omega\varepsilon} \nabla \times \mathbf{H} \tag{7.90}$$

which yields

$$E_\mathrm{r} = -\frac{j}{\omega\varepsilon} \frac{1}{r \sin\theta} \frac{\partial}{\partial \theta}(H_\phi \sin\theta) = \eta \frac{IL \cos\theta}{2\pi}\left(\frac{1}{r^2} + \frac{1}{jkr^3}\right)\mathrm{e}^{-jkr} \tag{7.91}$$

and,

$$E_\theta = -\frac{j}{\omega\varepsilon}\left(-\frac{1}{r}\frac{\partial}{\partial r}(rH_\phi)\right) = \mathrm{j}\eta \frac{kIL \sin\theta}{4\pi}\left(\frac{1}{r} + \frac{1}{\mathrm{j}kr^2} - \frac{1}{k^2 r^3}\right)\mathrm{e}^{-\mathrm{j}kr} \tag{7.92}$$

where $\eta = \sqrt{\mu/\varepsilon}$. The electric field has no rotational component:

$$E_\phi = 0 \tag{7.93}$$

Since the free space velocity of electromagnetic waves (of all frequencies) $c = (\mu\varepsilon)^{-1/2}$ we have $(k/(\omega\varepsilon)) = (1/(2\pi f \varepsilon)(2\pi/\lambda) = 1/(c\varepsilon) = (\mu/\varepsilon)^{1/2} = \eta$.

In the far-field region, we may ignore the $(1/r^2)$ and $(1/r^3)$ terms in (7.88), (7.91) and (7.92), yielding the far-field solution for the magnetic field

$$H_\phi = \mathrm{j}\frac{kIL\mathrm{e}^{-\mathrm{j}kr}}{4\pi r}\sin\theta \tag{7.94}$$

For the electric field, we get

$$E_\theta = \mathrm{j}\eta\frac{kIL\mathrm{e}^{-\mathrm{j}kr}}{4\pi r}\sin\theta \tag{7.95}$$

In the far- or radiation-field region, the other components of the electric and magnetic fields are zero:

$$H_\mathrm{r} = H_\phi = E_\mathrm{r} = E_\phi = 0 \tag{7.96}$$

Thus it is seen that for $kr = 2\pi r/\lambda \gg 1$, the fields resemble a uniform plane wave, where the electric and magnetic fields are related by the constant free space impedance.

$$Z = \frac{E_\theta}{H_\phi} = \eta \tag{7.97}$$

Therefore, in the radiation or far-field region, only H_ϕ and E_θ exist.

7.15.3 Radiated power and radiation resistance

From the Poynting vector for power transmitted or radiated, we get the average power radiated by the antenna:

$$P_\mathrm{av} = \frac{1}{2}\mathrm{Re}(\mathbf{E}\times\mathbf{H}^*) = \mathrm{Re}\frac{1}{2}|H_\phi E_\theta| = \frac{1}{2\eta}|E_0|^2 = \frac{\eta}{2}\left(\frac{kIL}{4\pi}\right)^2\frac{\sin^2\theta}{r^2}\ \mathrm{W/m^2} \tag{7.98}$$

And the radiated power per unit solid angle is given by

$$S = \frac{\text{Total power}}{4\pi} = \frac{4\pi r^2 S_\mathrm{av}}{4\pi} = r^2 S_\mathrm{av} = \frac{\eta}{2}\left(\frac{kIL}{4\pi}\right)^2\sin^2\theta\ \ \mathrm{W/sr} \tag{7.99}$$

which is seen to be of the order of $r^2|E_\theta|^2$.

The radiated power P_R moving in the radial direction is obtained by integrating the Poynting power density over a closed sphere of radius r. The integration surface element being $r^2 \sin\theta \, d\theta \, d\phi$. we have

$$P_R = \iint P_{av} \, dS = \int_0^{2\pi}\int_0^{\pi} |P_{av}| r^2 \sin\theta \, d\theta \, d\phi$$

$$= 2\pi \int_0^{\pi} \frac{\eta}{2}\left(\frac{kIL}{4\pi}\right)^2 \frac{\sin^2\theta}{r^2} r^2 \sin\theta \, d\theta$$

$$= 2\pi \frac{\eta}{2}\left(\frac{kIL}{4\pi}\right)^2 \int_0^{\pi} \sin^3\theta \, d\theta = \frac{\eta}{12\pi}(kIL)^2 \tag{7.100}$$

$$= \eta \frac{\pi}{3}\left(\frac{IL}{\lambda}\right)^2 \text{ W}$$

From the definition of radiation resistance R_r, the total radiated power of the antenna is the total power dissipated in the radiation resistance of the antenna.

Hence from $P_R = \frac{1}{2} I^2 R_r$, we have

$$R_r = \frac{2P_R}{I^2} = \eta \frac{2}{3\pi}\left(\frac{\pi L}{\lambda}\right)^2 = 80\left(\frac{\pi L}{\lambda}\right)^2 \ \Omega \tag{7.101}$$

7.15.4 Radiation intensity, directivity and aperture

The maximum power radiated per unit solid angle is

$$S_{max} = \frac{\eta}{2}\left(\frac{kIL}{4\pi}\right)^2 \tag{7.102}$$

Therefore, the directivity of the antenna

$$D = \frac{S_{max}}{\frac{P_R}{4\pi}} = \frac{3}{2} \text{ or } 1.5 \tag{7.103}$$

The maximum effective aperture is given by

$$A_{em} = \frac{\lambda^2}{4\pi} D = \frac{3\lambda^2}{8\pi} \tag{7.104}$$

7.16 General: finite-length linear dipole antenna and radiated signals

7.16.1 Antenna size

In this subsection, we shall work out the radiated electromagnetic field for a wire antenna of length $2L$. The quality factor, or Q factor, of the antenna is defined by

$$Q = \left(\frac{1}{kL}\right)^3 \tag{7.105}$$

The Q factor is ratio of the enrgy stored in the antenna to the energy lost (or radiated) per cycle. The minumum length of the antenna is related to the center frequency f_0 and the bandwidth Δf of the electromagnetic wave (i.e. signal) it is to transmit and receive by the relation

$$L \geq \frac{1}{k}\sqrt[3]{\frac{\Delta f}{f_0}} = \frac{\lambda}{2\pi}\sqrt[3]{\frac{\Delta f}{f_0}} \tag{7.106}$$

The quality factor Q of the antenna determines its bandwidth of operation. To reduce the Q factor, and increase the power radiated by the antenna, either the ground plane must be made of high-conductivity metal or a capacitance that may be added on top of the wire antenna thus effectively increasing the capacitance of the antenna, leading to a reduction in the resonant frequency of the antenna. In a resonant cavity or resonant circuit, where the energy stored must be increased, the Q factor should be made large.

7.16.2 Complete field equations

Consider now a dipole antenna that is $2L$ m long ($z = L$ to $z = -L$), as shown in Fig. 7.10, carrying a current along the wires given by

$$I(z) = \begin{cases} I_\mathrm{m} \sin k(L - z), & 0 \leq z \leq L \\ I_\mathrm{m} \sin k(L + z), & -L \leq z \leq 0 \end{cases} \tag{7.107}$$

where (a) I_m is the peak current, (b) at $z = 0$ the current is $I(0) = I_\mathrm{m} \sin kL$, (c) wave number $k = 2\pi/\lambda$, and (d) at $z = L$ and $z = -L$ the current is zero, $I(\leq) = 0$, as expected at the end points of the wire antenna for conduction current. For a half-wave dipole, $2L = \lambda/2$, $kL = (2\pi/\lambda)(\lambda/4) = \pi/2$ and $I(z) = I_\mathrm{m}$ at $z = 0$.

Consider (Fig. 7.10) the radiation field of a small piece of the wire $\mathrm{d}z'$, which is identical to a Hertzian dipole with the radiation electric field given by (7.95) with $L = \mathrm{d}z$:

$$\mathrm{d}E_\theta = \eta_0 \mathrm{d}H_\phi = \mathrm{j}\eta_0 \frac{kI\,\mathrm{d}z\,\mathrm{e}^{-\mathrm{j}kR}}{4\pi R} \sin\theta\,\mathrm{d}z' \tag{7.108}$$

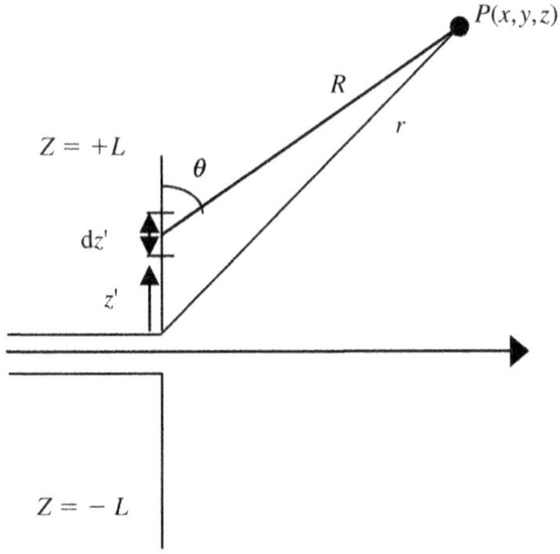

Figure 7.10: The linear antenna.

and

$$\mathrm{d}E_{\mathrm{r}} = \mathrm{d}E_{\phi} = \mathrm{d}H_{\mathrm{r}} = \mathrm{d}H_{\theta} = 0 \tag{7.109}$$

where

$$\begin{aligned} R &= \sqrt{x^2 + y^2 + (z - z')^2} = \sqrt{x^2 + y^2 + z^2 + (z'^2 - 2zz')} \\ &= \sqrt{r^2 + (z'^2 - 2rz' \cos\theta)} \end{aligned} \tag{7.110}$$

Using $z \approx r\cos\theta$ or $R \approx r - z'\cos\theta$ for the far fields, since $z' << r$,

$$\mathrm{d}E_{\theta} = \eta_0\,\mathrm{d}H_{\phi} = \mathrm{j}\eta \frac{Ik\mathrm{e}^{-\mathrm{j}kr}}{4\pi r} \sin\theta\, \mathrm{e}^{\mathrm{j}kz'\cos\theta}\,\mathrm{d}z' \tag{7.111}$$

Therefore, the electric and magnetic fields radiated by the $2L$ long wire antenna are obtained from

$$E_{\theta} = \eta_0 H_{\phi} = \int_{-L}^{+L} \mathrm{d}E_{\theta} = \mathrm{j}\eta \frac{k\mathrm{e}^{-\mathrm{j}kr}}{4\pi r} \sin\theta \int_{-L}^{+L} I\mathrm{e}^{\mathrm{j}kz'\cos\theta}\,\mathrm{d}z' \tag{7.112}$$

For current I, we need to substitute (7.107) into (7.112):

$$E_{\theta} = \eta_0 H_{\phi} = \mathrm{j}\eta \frac{kI_{\mathrm{m}}\mathrm{e}^{-\mathrm{j}kr}}{4\pi r} \sin\theta \left(\int_{0}^{+L} \sin k(L - z')\mathrm{e}^{\mathrm{j}kz'\cos\theta}\,\mathrm{d}z' + \int_{-L}^{0} \sin k(L + z')\mathrm{e}^{\mathrm{j}kz'\cos\theta}\,\mathrm{d}z' \right) \tag{7.113}$$

Using the mathematical relationship

$$\int e^{Ax}\sin(Bx+c)\,dx = \frac{e^{Ax}}{A^2+B^2}[A\sin(Bx+c) - B\cos(Bx+c)] \tag{7.114}$$

in (7.113) we get

$$E_\theta = \eta_0 H_\phi = j\eta\frac{kI_m e^{-jkr}}{2\pi r}\left(\frac{\cos(kL\cos\theta) - \cos kL}{\sin\theta}\right) \tag{7.115}$$

For the $2L$ long dipole, the pattern function or elemental pattern EP from (7.115) is given by

$$\mathrm{EP} = \left(\frac{\cos(kL\cos\theta) - \cos kL}{\sin\theta}\right) \tag{7.116}$$

which for a half-wave dipole antenna, setting $2L = \lambda/2$. $kL = \pi/2$, becomes

$$\mathrm{EP} = \left(\frac{\cos\left(\left(\frac{\pi}{2}\right)\cos\theta\right)}{\sin\theta}\right) \tag{7.117}$$

7.16.3 Radiated power, radiation resistance and directivity

The average power radiated by the $2L$ long antenna may be found from

$$P_{av} = \frac{1}{2}\mathrm{Re}(\mathbf{E}\times\mathbf{H}^*) = \frac{1}{2\eta}|E_\theta|^2 = \frac{I_m^2}{\eta 8\pi^2 r^2}\left(\frac{\cos(kL\cos\theta) - \cos kL}{\sin\theta}\right)^2 \tag{7.118}$$

The power radiated per unit solid angle is given by

$$S(L) = r^2 P_{av} = \eta\frac{I_m^2}{8\pi^2}\left(\frac{\cos(kL\cos\theta) - \cos kL}{\sin\theta}\right)^2 \ \mathrm{W/sr} \tag{7.119}$$

For a half-wave dipole, the power radiated simplifies to

$$P_r = \int_0^{2\pi}\int_0^{\pi} S\left(L = \frac{\lambda}{4}\right)\sin\theta\,d\theta\,d\phi = 30I_m^2(1.22) = 36.6I_m^2 \ \mathrm{W} \tag{7.120}$$

and the radiation resistance of a half-wave dipole antenna is

$$R_r = \frac{2P_r}{I_m^2} = 73.2\,\Omega \tag{7.121}$$

which matches with the characterisitc impedance of a coaxial cable, making the half-wave dipole antenna an attractive antenna with regard to impedance matching, since no additional circuit is needed to match the antenna with the cable connected to it.

The maximum power radiated per unit solid angle is given by

$$S_{\text{max}} = r^2 S_{\text{av}}(\theta = 90°) = 15 I_{\text{m}}^2/\pi \tag{7.122}$$

Hence the directivity of a half-wave dipole antenna is given by

$$D = \frac{S_{\text{max}}}{P_{\text{r}}/4\pi} = \frac{15 I_{\text{m}}^2/\pi}{36.6 I_{\text{m}}^2/4\pi} = 1.64 \tag{7.123}$$

7.17 Antenna radiated signal propagation: an introduction

7.17.1 Antenna waves

How do the electromagnetic waves radiated by antennas travel over free space? Although the space into which the electromagnetic waves are radiated may be free of electric space charges, the properties of the medium such as conductivity, permittivity and permeability will influence the radiated electromagnetic wave amplitude and phase. The electromagnetic waves radiated by antennas, as shown in Fig. 7.11, may be looked at as propagation of electromagnetic wave fronts, electromagnetic rays, electromagnetic signals oscillating in time and space, or as a radiation beam.

Consider the wave fronts of the radiated electromagnetic fields, where the electric and magnetic fields over a wave front are constants. The wave fronts in the far field are almost like a flat plate in three dimensions or a straight line looking at it sideways as shown in Fig. 7.11. Thus the electric field and magnetic field strengths over one wavefront have constant magnitudes E_1 and H_1, respectively.

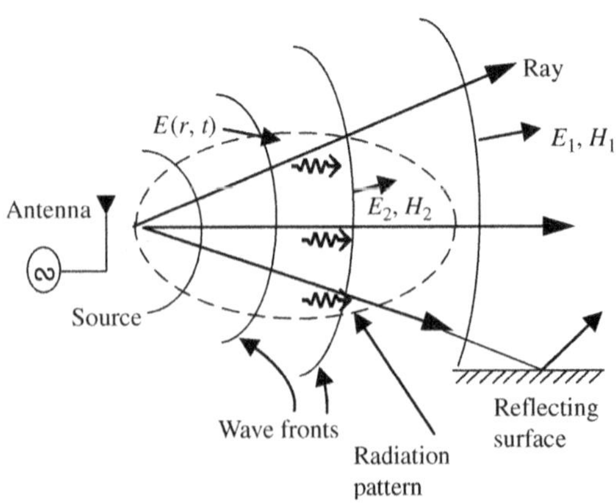

Figure 7.11: Radiated electromagnetic field as moving wave fronts, rays and electromagnetic signals.

In the far-field region of the antenna, note that each wave front may be approximated by a plane over which the electric and magnetic fields are constants. Thus in the far-field or radiation region of an antenna, the radiated electromagnetic fields may be considered as plane waves. The effects of the propagating medium should also be correctly accounted for in the analysis and design of antennas and antennas as part of a communication, remote sensing, security or biomedical system.

7.17.2 The basic equations governing uniform plane waves in a dielectric media

For a plane wave propagating in the z-direction, in a dielectric the electric and magnetic fields are constant on the xy-plane. We have

$$\frac{\partial}{\partial x} = \frac{\partial}{\partial y} \equiv 0; \quad \frac{\partial}{\partial z} \equiv \frac{\mathrm{d}}{\mathrm{d}z} \tag{7.124}$$

The four Maxwell equations for a single-frequency sinusoidal signal simplify to,

$$\nabla \times \mathbf{H} = \mathrm{j}\omega\varepsilon\mathbf{E} \tag{7.125}$$

$$\nabla \times \mathbf{E} = -\mathrm{j}\omega\mathbf{B} \tag{7.126}$$

$$\nabla \cdot \mathbf{D} = 0 \tag{7.127}$$

$$\nabla \cdot \mathbf{B} = 0 \tag{7.128}$$

Let the electric field lie on the zx-plane by orientation of the x-axis, so that

$$\mathbf{E} = u_x E_x + u_z E_z \tag{7.129}$$

Thus (7.126) may be written as

$$-\mathrm{j}(u_x H_x + u_y H_y + u_z H_z) = \begin{vmatrix} u_x & u_y & u_z \\ 0 & 0 & \dfrac{\mathrm{d}}{\mathrm{d}z} \\ E_x & 0 & E_z \end{vmatrix} = u_y \frac{\mathrm{d}}{\mathrm{d}z} E_x \tag{7.130}$$

from which we have

$$\mathbf{H} = u_y H_y \tag{7.131}$$

and

$$H_y = \frac{-1}{\mathrm{j}\omega\mu} \frac{\mathrm{d}}{\mathrm{d}z} E_x \tag{7.132}$$

Differentiating (7.132) with respect to z and using (7.125)

$$\frac{-1}{\mathrm{j}\omega\mu}\frac{\mathrm{d}^2}{\mathrm{d}z^2}E = \frac{\mathrm{d}}{\mathrm{d}z}H_y = -\mathrm{j}\omega\varepsilon E \tag{7.133}$$

which reduces to

$$\frac{\mathrm{d}^2}{\mathrm{d}z^2}E = -\omega^2\varepsilon\mu E^2 = \gamma^2 E \tag{7.134}$$

where the propagation constant

$$\gamma = \sqrt{-\omega^2\varepsilon\mu} = \pm\mathrm{j}\omega\sqrt{\varepsilon\mu} \pm \mathrm{j}\beta \tag{7.135}$$

The phase constant

$$\beta = \omega\sqrt{\varepsilon\mu} \tag{7.136}$$

The wave equation, (7.134), has a solution of the form, writing the electric field as $E(z)$ to show it as a function of distance travelled:

$$E(z) = E^+\mathrm{e}^{-\gamma z} + E^-\mathrm{e}^{\gamma z} \tag{7.137}$$

$$= E^+\mathrm{e}^{-\mathrm{j}\beta z} + E^-\mathrm{e}^{\mathrm{j}\beta z} \tag{7.138}$$

which may be verified by substituting the right-hand side of (7.137) or (7.138) into the left-hand side of the wave equation (7.134) to yield the right-hand side term in (7.134) after differentiating the electric field with respect to z. Substituting (7.137) into (7.132), we get the magnetic field:

$$\begin{aligned} H &= \frac{-1}{\mathrm{j}\omega\mu}\frac{\mathrm{d}}{\mathrm{d}z}E = \frac{-1}{\mathrm{j}\omega\mu}\frac{\mathrm{d}}{\mathrm{d}z}(E^+\mathrm{e}^{-\gamma z} + E^-\mathrm{e}^{\gamma z}) \\ &= \frac{\gamma}{\mathrm{j}\omega\mu}(E^+\mathrm{e}^{-\gamma z} + E^-\mathrm{e}^{\gamma z}) = \frac{E^+}{\eta}\mathrm{e}^{-\gamma z} - \frac{E^-}{\eta}\mathrm{e}^{\gamma z} \end{aligned} \tag{7.139}$$

Both the electric and magnetic fields have a forward wave traveling in the $+z$ direction and a reverse wave traveling in the $-z$ direction (Fig. 7.12). However, although the forward and backward electric field vectors are parallel to each other, in the case of the magnetic field they are directionally opposite to each other as the minus sign indicates. Thus the antenna placed at $z = 0$ radiates in both directions along the z-axis. The forward (E^+, H^+) and backward (E^-, H^-) electric and magnetic fields are related to each other by the *intrinsic (or chracteristic) impedance* of the medium:

$$\eta = \frac{E^+}{H^+} = -\frac{E^-}{H^-} = \frac{\mathrm{j}\omega\mu}{\gamma} = \frac{\mathrm{j}\omega\mu}{j\beta} = \frac{\omega\mu}{\omega\sqrt{\varepsilon\mu}} = \sqrt{\frac{\mu}{\varepsilon}} \tag{7.140}$$

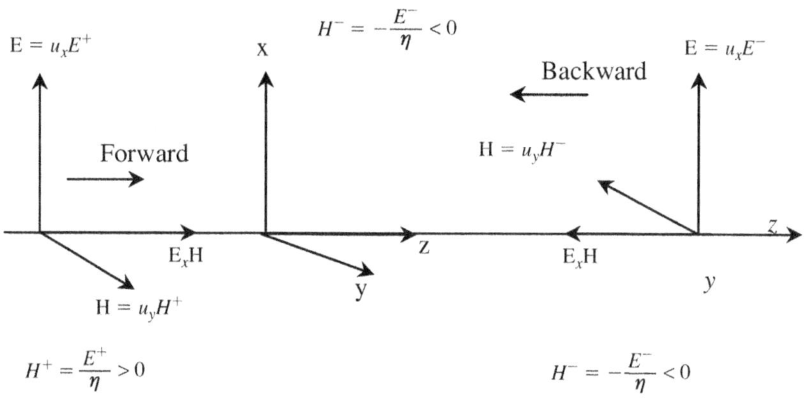

Figure 7.12: Forward and backward waves.

and the magnetic field is

$$H(z) = H^+ \mathrm{e}^{-\gamma z} + H^- \mathrm{e}^{\gamma z} = \frac{E^+}{\eta}\mathrm{e}^{-\gamma z} - \frac{E^-}{\eta}\mathrm{e}^{\gamma z} \tag{7.141}$$

The impedance η is termed an impedance because it is the ratio of the electric field with units V/m to the magnetic field with units A/m, that is, the units of η are V/A = Ω, the same as the impedance of circuit theory. For free space and air, with $\varepsilon_0 = 8.85418 \times 10^{-12}$ F/m and $\mu_0 = 4\pi \times 10^{-7}$ H/m, we have an approximate value 377 Ω.

7.17.3 The generalized wave equation and attenuation

If the antenna has to radiate through a medium that is conducting at the frequency of operation, such as satellite antennas transmitting through heavy rain clouds and thunderclouds, or submarine antennas immersed in seawater, we must include the effect of the conductivity σ of the medium, properly accounting for the signal power lost in the medium through ohmic losses. Hence we return to the more general form of the equaltion of Farday's law:

$$\nabla \times \mathbf{H} = \mathbf{J} + \mathrm{j}\omega D = \sigma \mathbf{E} + \mathrm{j}\omega\varepsilon \mathbf{E} = \mathrm{j}\omega\varepsilon\left(1 - \mathrm{j}\frac{\sigma}{\varepsilon\omega}\right)\mathbf{E} = \mathrm{j}\omega\varepsilon^c \mathbf{E} \tag{7.142}$$

where the complex permittivity is

$$\varepsilon^c = \varepsilon\left(1 - \mathrm{j}\frac{\sigma}{\varepsilon\omega}\right) \tag{7.143}$$

and the propagation constant γ is defined by

$$\gamma^2 = -\omega^2\varepsilon^c\mu = -\omega^2\mu\varepsilon\left(1 - \mathrm{j}\frac{\sigma}{\varepsilon\omega}\right) \tag{7.144}$$

with a real part α termed the attenuation constant and an imaginary part termed the phase constant β. Note that the real part of γ^2 is negative and the imaginary part positive, putting γ^2 in the third quadrant of the argand plane The attenuation constant α causes the signal strength to drop as it travels further away from the transmitting antenna, and the phase constant causes a phase shift of the signal as it travels further away from the antenna:

$$\gamma = \alpha + \mathrm{j}\beta \tag{7.145}$$

is therefore in the first guardant making both α and β positive. The final solutions will be of the identical form to (7.137) and (7.141) although the propagation constant will be of the complex form given in (7.145).

The average power transmitted by the wave is obtained from the Poynting vector,

$$P_{\mathrm{av}} = \frac{1}{2}\mathrm{Re}(\mathbf{E} \times \mathbf{H}^*)\ \mathrm{W/m^2} \tag{7.146}$$

or

$$P_{\mathrm{av}} = \frac{1}{2}H_{\mathrm{m}}\mathrm{E_m}\ \mathrm{W/m^2} \tag{7.147}$$

From (7.144), we may rewrite the propagation constant in the following form:

$$\gamma^2 = \mathrm{j}\omega\mu(\sigma + \mathrm{j}\omega\varepsilon) \tag{7.148}$$

For uniform plane waves in a good insulator, we have $\sigma \ll \omega\varepsilon$, and (7.148) reduces to

$$\gamma = \sqrt{-\omega^2\varepsilon\mu} = \mathrm{j}\omega\sqrt{\varepsilon\mu} \tag{7.149}$$

with

$$\alpha = 0 \tag{7.150}$$

$$\beta = \omega\sqrt{\varepsilon\mu} \tag{7.151}$$

No signal attenuation takes place as in the case of free space.

For uniform plane waves in weakly conducting media, we have $\sigma^2 \ll \omega^2\varepsilon^2$, yielding

$$\gamma = j\omega\sqrt{\varepsilon_\mu}\left(1 - \mathrm{j}\frac{\sigma^j}{\omega\varepsilon}\right)^{\frac{1}{2}} \approx \mathrm{j}\omega\sqrt{\varepsilon\mu}\left(1 - \frac{1}{2}\mathrm{j}\frac{\sigma}{\omega\varepsilon}\right) \tag{7.152}$$

with the attenuation constant

$$\alpha = \frac{1}{2}\sigma\sqrt{\frac{\mu}{\varepsilon}} \tag{7.153}$$

and phase constant

$$\beta = \omega\sqrt{\varepsilon\mu} \tag{7.154}$$

7.17.4 Electromagnetic waves in seawater

Consider the situation where communication is carried out in seawater, as in the case of submarines using electromagnetic waves for communication. Let the forward wave signal radiated be given by

$$E = E_0 \exp(\mathrm{j}\omega t - \gamma z) \tag{7.155}$$

where the propagation constant is

$$\gamma = \sqrt{\mathrm{j}\omega\mu(\sigma + \mathrm{j}\omega\varepsilon)} = \alpha + \mathrm{j}\beta = \frac{1}{\delta} + \mathrm{j}\beta \tag{7.156}$$

where δ $(= 1/\alpha)$ is the skin depth, the distance $(z = \delta)$ inside the seawater at which the electric and magnetic field strength of the signal would have reduced by a factor of $\mathrm{e}^{-z/\delta} = \mathrm{e}^{-1}$.

At high frequency, this simplifies to

$$\alpha = \frac{1}{\delta} = \frac{\sigma}{2}\sqrt{\frac{\mu}{\varepsilon}} \tag{7.157}$$

where typically $\sigma = 4$ and $\varepsilon = 81\varepsilon_0$, giving $\delta = 1.2\,\text{cm}$. At low frequencies, we may ignore the $\mathrm{j}\omega\varepsilon$ term in (7.156), yielding,

$$\gamma = \sqrt{\mathrm{j}\omega\mu\sigma} = \sqrt{\omega\mu\sigma}\cdot\sqrt{\mathrm{j}} = \sqrt{\mathrm{j}\omega\mu\sigma}\left(\frac{1}{\sqrt{2}} + \mathrm{j}\frac{1}{\sqrt{2}}\right) = \frac{1}{\delta} + \mathrm{j}\frac{1}{\delta} \tag{7.158}$$

and the amplitude of the signal decreases with $\exp(-z/\delta)$. At 10 kHz, the attenuation is 87 dB for a distance of 25 m; at 100 Hz it is at 250 m that the signal attenuation is 87 dB. Summarizing the results for uniform, plane waves in submarine communications, we have the following three electromagnetic fields:

(i) For surface-to-submarine communications, where the transmitting antenna is at the surface of the sea and the receiving antenna is at a depth of z m:

$$\mathbf{E} = u_x E_0 \exp[\mathrm{j}(\omega t - \beta z)] \tag{7.159}$$

$$\mathbf{H} = u_y \frac{E_0}{\eta} \exp[\mathrm{j}(\omega t - \beta z)] \tag{7.160}$$

(ii) For submarine to submarine communication underwater, with the transmitting and receiving antennas of length L, and at a distance r m from each other:

$$E_\mathrm{r} = \frac{60\beta IL}{r}\left\{1 - \frac{2}{\mathrm{j}^2} - \mathrm{j}\left(\frac{2}{j} - 1\right)\right\}\cos\phi \tag{7.161}$$

$$E_\phi = \frac{120\beta IL}{r}\left\{\frac{1}{2\mathrm{j}} - \frac{j}{2\mathrm{j}^2}\left[\frac{1}{g}\exp\left(\frac{-(d_\mathrm{r}+d_\mathrm{t})}{\delta}\right)\right]\right\}\sin\phi \tag{7.162}$$

since $1/r \ll 1, E_\mathrm{r}$ is predominant.

(iii) Submarine to surface,

$$\mathbf{E}_a = u_Z \times \frac{60\beta IL}{r}\exp[\mathrm{j}(\omega t - \beta r)] \tag{7.163}$$

$$\times \exp\left[-\mathrm{j}\frac{\mathrm{d}t}{\delta}\left(1 - \frac{\mathrm{j}}{a}\right)\right]\left[\frac{-\sqrt{-\mathrm{j}}}{\sqrt{g}}\mathrm{e}^{-\frac{\mathrm{d}t}{\delta}}\cos\phi\right] \tag{7.164}$$

where I is the antenna current $= 2\sqrt{2P_\mathrm{t}/\omega\mu L}$, L is the Length of antenna for example, $L = 50$ m; for 10 kHz operation frequency, r is the horizontal distance, ϕ is the angle measured in horizontal plane, from a zero along the direction of the submerged horizontal antenna, a is the horizontal distance in units of free space wavelengths/2π, d_t is the depth of transmitting antenna, d_r is the depth of receiving antenna, P_t is the transmitter power, for example 10 kW, σ is the conductivity, $4\,\Omega^{-1}\mathrm{m}^{-1}$ and g is the refraction factor $g = \frac{\sigma}{\omega\varepsilon_0}$ for surface to submarine.

7.17.5 Direct line of sight: space wave

The space wave travelling from the transmitter antenna and the receiver antenna as shown in Fig. 7.13 is made up of two waves. One is the direct line of sight (LOS) wave, considered above in our discussions on plane waves in free space. And the second is the ground reflected wave, which bounces off the earth, thus travelling a longer and slower path to reach the receiver antenna. If the ground may be considered to be a perfect conductor, then the reflected wave may be looked at as the wave radiated by the image current element of the transmitter antenna.

The space wave for telecommunications is limited by the curvature of the earth. When distances between the transmitter and the receiver increase, after a certain distance r, the earth's curvature will block the transmitter antenna from being seen by the receiver antenna. Beyond this distance, space wave communication will fail.

With ground reflected wave, the second type of space wave, the total signal presented to the receiver antenna will be the summation of the direct wave and the

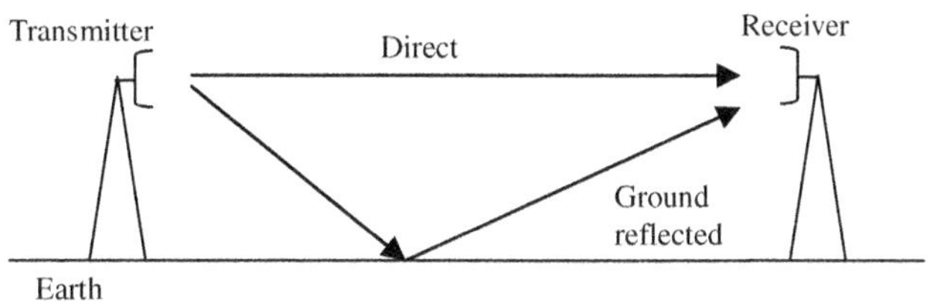

Figure 7.13: Direct and reflected waves adding together at the receiver. The reflected wave will be smaller in magnitude and lagging in phase, although at the same frequency as the direct wave.

ground reflected wave. Since the ground reflected wave travels a longer distance than the direct wave, the phase of the two space wave signals will differ at the receiver, although they are both radiated by the same transmitter antenna. If the phase difference is zero, then both waves will perfectly add. But if the phase difference is 180°(π), then the two waves for sinusoidal radiated signals will cancel each other out, assuming that the attenuation of the ground reflected signal is negligibly small.

Both direct and reflected waves will experience reduction of the electric (and magnetic) field intensity due to (a) the distance traveled (the $1/r$ factor), (b) the finite conductivity of ground and (c) when the free space experiences heavy rains, snow or fog (attenuation). Signal attenuation will occur due to the phase difference between the two signals at the receiver antenna (due to interference) and in the case of the reflected wave, there is additional power loss at the reflection point on the earth's surface (due to absorption). When the telecommunication or radar system operates in free space, with a clear LOS, we may rewrite the Friis equation, (7.44), as

$$P_{\mathrm{r}} = G_{\mathrm{r}}G_{\mathrm{t}}\left(\frac{\lambda}{4\pi r}\right)^2 P_{\mathrm{t}} = G_{\mathrm{r}}\left(\frac{\lambda}{4\pi r}\right)^2 \mathrm{EIRP} \tag{7.165}$$

where r is the distance between the transmitting antenna and the receiving antenna with $r > 2L^2/\lambda$, L being the largest dimension (e.g. length of a dipole antenna or diameter of a dish antenna) of either antenna, $\mathrm{EIRP} = P_{\mathrm{t}}G_{\mathrm{t}}$, the effective isotropic radiated power, P_{t} being the transmitted power, and G_{r} the gain of the receiving antenna, G_{t} is the gain of the transmitting antenna and λ is the signal wavelength.

The radiated power decays as $1/r^2$. If power is lost due to interference or power absorbing (or reflecting) objects between the transmitter and receiver antennas, the received power will be less than that given by Friis equation. From the Friis equation, we note that the path loss is due to the wave spreading out to larger and larger surfaces as it moves away from the transmitter, leading to lower power densities at the far-field receiver, and to lower power being captured by the receiver antenna. The free space path loss

$$L_{\mathrm{p}} = \left(\frac{4\pi r}{\lambda}\right)^2 = \left(\frac{4\pi fr}{c}\right)^2 \tag{7.166}$$

The path loss L_{f} is unitless. In dB,

$$L_{\mathrm{p}}(\mathrm{dB}) = 20\log\left(\frac{4\pi fr}{c}\right)^2 \tag{7.167}$$

or

$$L_{\mathrm{p}} = 20\log\left(\frac{4\pi}{c}\right) + 20\log f + 20\log r \tag{7.168}$$

The path loss given by L_{p} is that for direct LOS propagation, that is for LOS space waves. Dispersion of signals (i.e. electromagnetic waves) may occur due to different travel times of direct and reflected waves.

Diffraction at sharp edges such as building roof, walls and trees gives rise to additional signal paths than just the LOS, direct signal and the ground reflected wave. Analytical or numerical solutions of Maxwell's equations to determine electromagnetic signal degradation due to diffraction effects in densely built cities, demand carefully justified simplifications and intense computation. Instead, empirical models based on measurements are available. The Walfisch–Ikegami model gives the path loss for the direct LOS condition as

$$L = 42.6 + 20\log f_0 + 26\log d \tag{7.169}$$

where d is the distance between the transmitter and receiver and f_0 is the signal frequency. When signal diffractions are present, the path loss is given by

$$L = 32.4 + 20\log f_0 + 20\log d + L_{\mathrm{R}} + L_{\mathrm{B}} \tag{7.170}$$

where the diffraction loss due to rooftops is given by

$$L_{\mathrm{R}} = -16.9 - 10\log \Delta a + 10\log f_0 + 20\log \Delta h + L(0) \tag{7.171}$$

And the diffraction loss due to multiple reflections from buildings, in addition to the single ground reflected wave, is given by

$$L_{\mathrm{B}} = B_0 + B_{\mathrm{a}} + B_{\mathrm{a}}\log d + B_{\mathrm{f}}\log f_0 - 9\log a \tag{7.172}$$

where

Δa is the distance between the mobile antenna and BS,
h_M is the height of the mobile antenna,
$$\Delta h_M = h_R - h_M, \tag{7.173}$$
h_R is the height of the roof,
$$B_0 = -18\log(1 + \Delta hT) \tag{7.174}$$
h_T is the height of the transmitting antenna,
$$\Delta h_T = h_T - h_R \tag{7.175}$$

$$B_a = \begin{cases} 54 - 0.8\Delta h_T & d \geq 0.5\,\text{km} \\ 54 - 1.3\Delta h_T & d < 0.5\,\text{km} \end{cases} \tag{7.176}$$

$$B_d = 18 - 15\left(\frac{\Delta h_T}{h_R}\right) \tag{7.177}$$

$$B_f = \begin{cases} -4 + 0.7\left(\dfrac{f_0}{925} - 1\right) \text{ for suburbs} & (7.178) \\ -4 + 0.7\left(\dfrac{f_0}{925} + 1\right) \text{ for cities} & (7.179) \end{cases}$$

7.17.6 Surface wave

The surface wave is that which travels along the earth's surface.

Since at very high frequencies, the earth is a dielectric medium with finite conductivity, as seen from (7.179) when in the term $(\sigma + j\omega\varepsilon)$ the term $j\omega\varepsilon$ cannot be neglected, a portion of the surface wave energy is absorbed by the earth's surface, as with waves along lossy transmission lines. To minimize the power loss into the earth, the radiated electric field must be vertically polarized (e.g. as with a dipole antenna that is installed perpendicular to the earth's surface).

Horizontally polarized electric field signals E_h dissipate more energy, due to the $J_h = \sigma E_h$ A/m^2 current density induced along the earth surface of conductivity σ. Although the vertically polarized electric field E_v is used for surface wave communications (e.g. at 15 kHz for ship-to-ship and ship–shore communications and maritime navigation), the gradient density of the earth's atmosphere causes E_v to tilt as it propagates, thus increasing power loss due to the horizontal component of the tilted vertically polarized electric field E_v.

7.17.7 Sky wave

The earth's ionosphere is divided into different layers. Each layer will reflect signals, shown by one sky wave in Fig. 7.14. These ionized gas layers above the earth are:

(a) The D-layer, which is approximately 50–100 km above the earth's surface. This layer reflects VLF (3–30 kHz) and LF (3–30 MHz) and MF (0.3 MHz to 3 MHz) electromagnetic waves.

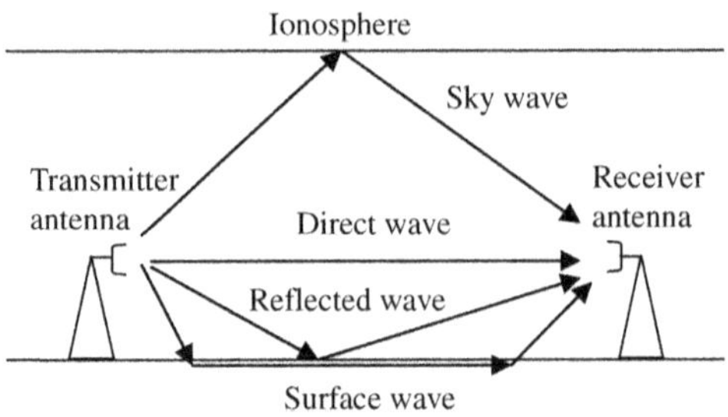

Figure 7.14: Direct wave from transmitter to the receiver; reflected wave from ground; the sky wave which is the reflected wave from the upper ionosphere; the surface wave which travels along the ground and may still enter into the receiver if the receiver is close enough to the ground as in the case of military personnel communication.

(b) The E-layer, which is approximately 100–140 km above the earth's surface. It tends to disappear at night. It is used for MF sky wave communication during daytime.

(c) The F-layer exists from 140–350 km above the earth. It is variable in that at daytime it exists between 140–250 km. In winter time it exists from 250–350 km.

The maximum usable frequency (MUF) of sky waves for communication is given by

$$\text{MUF} = \frac{\text{Critical frequency}}{\cos\theta_c} \tag{7.180}$$

where θ_c is the angle of wave incidence upon the ionosphere measured from the vertical axis and the critical frequency is the maximum frequency of a signal that will be reflected back to the earth by the ionosphere.

7.18 Reflection coefficients for vertically and horizontally polarized electromagnetic signals

In Fig. 7.15(a) is shown a vertically polarized radiating antenna placed along the x- axis at a height h m above a perfectly conducting ground on the xy-plane, carrying also current in the $\mathbf{u}_z$ direction. The ground may be replaced by the image current, shown in Fig. 7.15(a), where the image current is parallel (in the $\mathbf{u}_z$ direction) to the source antenna, at a depth of h m under the ground plane. The vertical reflection coefficient is Γ_V. In Fig. 7.15(b) is shown a horizonatally polarized antenna placed on the zx-plane, at a height h m above the ground, and carrying current in the $\mathbf{u}_x$ direction. Assuming that the ground is a perfect conductor, the ground plane may be replaced by the image of the source antenna placed at a depth h m below the

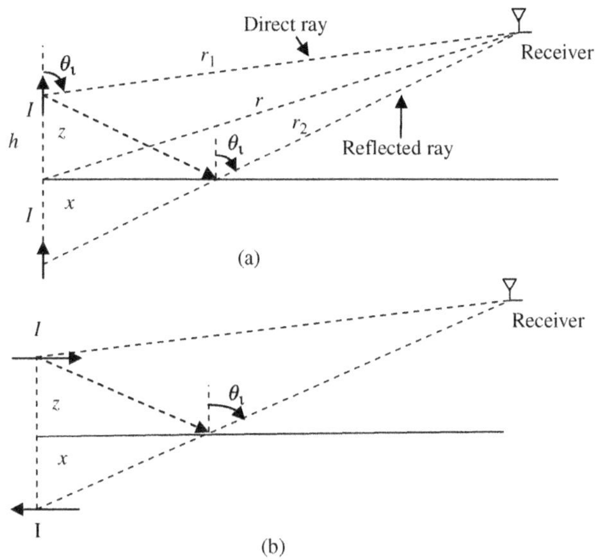

Figure 7.15: (a) Vertically polarized antenna above ground. (b) A horizontally polarized antenna above ground.

ground plane, but with the current flowing in the opposite (i.e. $-\mathbf{u}_x$) direction to that of the source current above the ground. The reflection coefficient for the vertically polarized electric field is Γ_H. Note that $\Gamma_V = \Gamma_H = 1$ for a perfectly conducting ground.

For a virtically polarized dipole antenna, shown in Fig. 7.15(a), the reflected signal (ray) appears to come from the image antenna. The total field received at the receiver antenna is

$$E_\theta = j\omega\mu \sin\theta \left[\frac{IL\,e^{-jkr_1}}{4\pi r_1} + \Gamma_V \frac{IL\,e^{-jkr_2}}{4\pi r_2} \right] \tag{7.181}$$

Using $r_1 = r - h\cos\theta$, $r_2 = r + h\cos\theta$, for the phase, but for the denominator $r_1 = r_2 \approx r$, because for large r the distance difference $r_1 - r_2$ does not significantly affect the magnitude of the signals, we get,

$$E_\theta = \frac{j\omega\mu IL}{4\pi r} e^{-jkr} \sin\theta [e^{jkh\cos\theta} + \Gamma_V e^{-jkh\cos\theta}] \tag{7.182}$$

For a horizontally polarized dipole antenna (see Fig. 7.15(b)),

$$E_\theta = \frac{j\omega\mu IL}{4\pi r} e^{-jkr} \cos\theta [e^{jkh\cos\theta} - \Gamma_H e^{-jkh\cos\theta}] \tag{7.183}$$

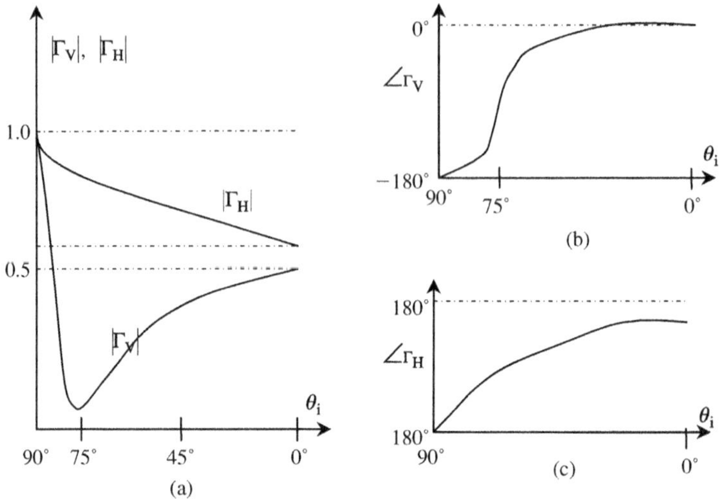

Figure 7.16: (a) The reflection coefficient magnitudes for vertically and horizontally polarized signals for varying signal incident angles θ_i. (b) The reflection coefficient phase angle for vertically polarized signals for varying signal incident angles θ_i. (c) The reflection coefficient phase angles for horizontally polarized signals for varying signal incident angles θ_i.

on the xz-plane and

$$E_\phi = \frac{\mathrm{j}\omega\mu IL}{4\pi r}\,\mathrm{e}^{-\mathrm{j}kr}[\mathrm{e}^{\mathrm{j}kh\cos\theta} + \Gamma_\mathrm{H}\,\mathrm{e}^{-\mathrm{j}kh\cos\theta}] \tag{7.184}$$

on the xy-plane. The amplitudes of the reflection coefficients for vertically and horizontally polarized waves ($|\Gamma_\mathrm{V}|, |\Gamma_\mathrm{H}|$), and their phase angles ($\theta_\mathrm{V}, \theta_\mathrm{H}$) of the vertically and horizontally polarized signals are shown in the sketches Fig. 7.16. For a vertically polarized signal radiated from the antenna in Fig 7.16(a), the reflection coefficient is

$$\Gamma_\mathrm{V} = \frac{-\epsilon_r' \cos\theta_\iota + \sqrt{\varepsilon_r' - \sin^2\theta_\iota}}{\epsilon_r' \cos\theta_\iota + \sqrt{\varepsilon_r' - \sin^2\theta_\iota}} \tag{7.185}$$

and θ_ι is the incident angle, and the complex relative permittivity

$$\epsilon_\mathrm{r}' = \frac{\epsilon'}{\epsilon_0} = \epsilon_\mathrm{r} - \mathrm{j}\frac{\sigma}{\omega\epsilon_0} \tag{7.186}$$

and σ, ϵ_r and ϵ_r are the conductivity, relative permittivity and complex permittivity of the ground, respectively, at frequency f ($\omega = 2\pi f$). At communication frequencies, values for σ and ϵ_r could be about 0.003 S/m and 15 respectively.

The reflection coefficient for a horizontally polarized signal is given by

$$\Gamma_{\mathrm{H}} = \frac{\cos\theta_\iota - \sqrt{\varepsilon_r' - \sin^2\theta_\iota}}{\cos\theta_\iota + \sqrt{\varepsilon_r' - \sin^2\theta_\iota}} \tag{7.187}$$

The reflection coefficient magnitudues and phase angles are sketched in Fig. 7.16. When the signal has to travel through a lossy medium, the attenuation of its signal strength is determined by the attenuation constant given by

$$\alpha = \omega\sqrt{\frac{\mu\epsilon}{2}\left[\sqrt{1+\left(\frac{\sigma}{\omega\varepsilon}\right)^2} - 1\right]} \tag{7.188}$$

and the phase shift that it experiences is defined by the phase constant

$$\beta = \omega\sqrt{\frac{\mu\epsilon}{2}\left[\sqrt{1+(\frac{\sigma}{\omega\varepsilon})^2} + 1\right]} \tag{7.189}$$

The magnitude of the wave impedance is given by

$$|\eta| = \frac{\sqrt{\frac{\mu}{\epsilon}}}{\sqrt[4]{[1+(\frac{\sigma}{\omega\epsilon})^2]}} \tag{7.190}$$

and the impedance phase angle θ_η is determined from

$$\tan 2\theta_\eta = \frac{\sigma}{\omega\epsilon} \quad \text{for } 0 \leq \theta_\eta \leq 45^\circ \tag{7.191}$$

7.19 Antenna measurements

When choosing an antenna for a particular system, for example, to be used in a shore to ship docking navigational system in stormy weather, or for a new radio frequency identification systems (RFID), the two important parameters of the antenna needed are the power gain and the radiation pattern. Other important parameters include antenna impedance and bandwidth.

The gain $G_y$of an antenna$Y$ may be measured using a standard antenna such as a horn antenna X with a known gain G_x. In a test site or in an anechoic chamber, the source antenna output power may be kept constant, and the powers P_x and P_y are the powers received by the antennas X and Y respectively for the same transmission antenna gain and transmitted power. By considering the Friis equation,

$$\frac{P_{\mathrm{R}}}{P_{\mathrm{T}}} = \frac{G_x G_y \lambda^2}{(4\pi r)^2} \tag{7.192}$$

we may determine the gain of the antenna Y under test from

$$G_y = G_x \left(\frac{P_y}{P_x} \right) \tag{7.193}$$

An alternative approach is to use a two-antenna measurement with antenna X transmitting. Then the power received by the antenna Y under test is given by

$$P_{\mathrm{R}} = \frac{P_{\mathrm{T}} G_x G_y \lambda^2}{(4\pi r)^2} \tag{7.194}$$

Thus we may determine G_y if P_{T}, G_{X}, λ and r are known. If G_x is unknown then we could use two identical antennas with unknown gain G_y and make the same measurements to determine G_y from

$$P_{\mathrm{R}} = \frac{P_{\mathrm{T}} G_y G_y \lambda^2}{(4\pi r)^2} \tag{7.195}$$

A third method is Purcell's method, where a large reflecting, conducting sheet is used to reflect back to the antenna under test, the signal that it radiates. In this case it is assumed that the radiation and receiving patterns of the antenna under test do not suffer from mismatch. The effective length of the anechoic chamber should be about twice the distance between the antenna and the reflecting sheet.

Three antennas of unknown antenna gains G_x, G_y and G_z may have their gains determined by using each in turn to be the transmitter to radiate out a power P_{T} and the received powers

$$P_{xy} = \frac{P_{\mathrm{T}} G_x G_y \lambda^2}{(4\pi r)^2} \tag{7.196}$$

$$P_{yz} = \frac{P_{\mathrm{T}} G_y G_z \lambda^2}{(4\pi r)^2} \tag{7.197}$$

$$P_{xz} = \frac{P_{\mathrm{T}} G_x G_z \lambda^2}{(4\pi r)^2} \tag{7.198}$$

Here P_{xy} is the power received by X when Y is transmitting power P_{T} (and vice versa, P_{xy} being the same as P_{yx}). From these we may obtain G_x, G_y and G_z; for instance,

$$G_x = (4\pi) \left(\frac{r}{\lambda} \right) \left[\frac{(P_{\mathrm{xy}} \cdot P_{\mathrm{xz}})}{(P_{\mathrm{T}} P_{\mathrm{yz}})} \right]^{1/2} \tag{7.199}$$

To increase the accuracy of this three element antenna method, to overcome, for instance, inaccuracies due to our lack of knowledge of the path loss or whether the phase center of a horn antenna is at its throat or the aperture, multiple measurements may be made at different separation distances between the transmitting and receiving antennas. These values may now be put together and by extrapolation we may obtain a more accurate value for the gain of each of the three antennas.

The radiation pattern of an antenna may be measured in an open test field so that the far-field radiation field strengths may be directly measured. Open fields are particualrly necessary for antennas operating at frequncies less than 1 GHz. In this case, the measurements may be made at distances where it may be fairly assumed that the wave front of the transmitted radiation field is a plane wave front.

Alternatively, measurements may be made indoors in an anechoic chamber. The walls, floor and ceiling of the room are lined with radio absorbing material (RAM) which would minimize reflections. Therefore, it would be safe to assume that the electromagnetic fields from the source antenna that are picked up by the antenna under test are the direct waves only, and that reflected waves are absent. The antenna under test is mounted in such a way as to move or rotate it (e.g. on a turn table) to measure the received signal strength in different directions with respect to its axis. The RAM is about a few wavelengths thick, made up of carbon-impregnated polyurethane foam and it has a reflection coefficient of about -30 dB. It is shaped in pyramidal form to prevent the build-up of standing waves by providing good impedance matching between free space and the metal sheets that form the walls of the anechoic chamber. Reflections could severely degrade the measured radiation pattern for antennas with low side lobe levels; the reflected signals could create false side lobes in the measured radiation pattern. In general, the required accuracies for amplitude and phase are 0.5 dB and 5°, respectively. To achieve this, the phase error at the edge of the nonplanar wave front should be less than 22.5°. This means that the distance between the source and the antenna under test with an effective aperture diameter L should be $(2/\lambda)L^2$. The amplitude accuracy requirement means that the distance between the antenna under test and the source antenna should be greater than $10L$. The size of the anechoic chamber may be made smaller by adopting one of the following two alternative techniques:

(a) Near-field measurements, by keeping the distance between the source antenna and the antenna under test in the near-field range. Once the raditation pattern for the near field is made, the far-field radiation pattern may be obtained by using the known relationship between the near-field electric field magnitude and phase angle and the far field electric field amplitude and phase angle. In spherical near-field scanning, the antenna under test may be rotated while keeping the probe fixed. In cylindrical near field scanning, the antenna under test may be rotated while the probe is scanned vertically. In planar near-field scanning, the probe is moved over the aperture of the antenna under test.

(b) Compact antenna test range (CATR). The anechoic chamber may be made compact by mounting the source antenna on the roof and allowing its radiated fields to be reflected from a reflector at ground and facing the antenna under test. From the waves received from the source antenna, the reflector is shaped so as to generate a plane wave toward the antenna under test.

7.20 Ionizing and nonionizing electromagnetic fields: health hazards

Electromagnetic fields are divided into ionizing and nonionizing electromagnetic fields, especially in relation to health risks due to electromagnetic fields. Ionizing fields carry sufficient energy to tear apart electrons that are bound to the nucleus of an atom or a molecule. This may cause damage to DNA. At frequencies above 10^{15} Hz (which is in the ultraviolet, or UV band), we have ultraviolet rays, X-rays, and gamma rays, which are ionizing electromagnetic fields. Electromagnetic fields below UV, visible light and infrared regions are called nonionizing fields. Nonionizing fields may cause an electron to jump to a higher energy level, but are not able to free electrons from an atom or molecule. The energy carried by a photon at frequency f is given by $W = hf$, where h is Planck's constant. These may excite electrons and induce high currents (e.g. infrared and microwaves: induce high currents inside the body at microwave frequency, or induce low currents at low frequencies.

The extent of heating and excitiation of nerves depends on the frequency of the signal, the power dissipated by electric and magnetic fields, and the duration over which the biological tissue is exposed to the electromagetic fields. At 1–3000 Hz, the maximum permissible exposure (MPE) level for brain tissues is 1000 A/m of magnetic field, or 0.005 V/m of electric field. In the 10 kHz to 100 MHz range, the MPE are 100–0.1 A/m (for magnetic field) and 100 to 27 V/m (for electric field). At frequencies below 100 kHz, over exposure to electric fields can cause electrostimulation of nerve and muscle cells. At much higher frequncies, 100 MHz to 1 GHz, the MPE levels are given in average power density of 2–100 W/m^2 (on average the MPE is 10 W/m^2 in the 1–100 GHz, range.

The 50 (or 60 Hz) power line magnetic fields should be kept below 720 A/m. The typical values of magneic fields under power lines range from 2 to 6 A/m. For the 1–2 GHz communication signals, the MPE is 1 W/m^2.

7.21 Antenna system for measurements and surveillance

We shall consider here the design of a coil antenna to measure very-low-frequency (VLF: 1–200 kHz) signals. It may be used to check, for instance, the levels of signal strength to ensure health safety of people that live in an electromagnetic environment filled with VLF signals. The stength of the signals to be measured ranges from 0.02–200 μT (or 0.2–2000 mG). Such VLF fields are produced by mountain winds, earthquakes, thunderstorms, TV screens, ground wave military systems, computer sets and medical devices. They are associated with various kinds of health problems including child leukemia, stress, vibration of nerves, tumor and cell structure. Hence measuremnt of these magnetic fields is important. Thus we shall look at a loop antenna that should be dedicated to act as a surveillance receiver antenna. We shall, furthermore, illustrate the use of the loop antenna as a communication transmitter.

When a loop of area A (m^2) is cut by a magnetic flux density B (T), the voltage V induced in an N-turn coil is given by,

$$V = -N\frac{d(BA)}{dt} \text{ V} \tag{7.200}$$

A typical value for the cross-sectional area of a handheld receiver is $5\,cm^2$. If the time harmonic magnetic field cutting the coil is $B = B_m \exp(j\omega t)$, then the induced voltage is,

$$V = -NAB_m(j\omega)\exp(j\omega t) \tag{7.201}$$

so that the maximum magnitude of the voltage induced in the N-turn coil is

$$V_m = \omega NAB_m = 2\pi fNAB_m \tag{7.202}$$

The sensitivity of the sensing antenna is

$$\frac{V_m}{B_m} = \omega NA = 2\pi fNA \text{ V/T} \tag{7.203}$$

Therefore, for a maximum voltage of 5 V, and a coil area of $5\,cm^2$, with a field strength of $200\,\mu T$ at $f = 200$ kHz, the number of turns needed is $N = 40$. Since the voltage magnitude has the frequency also in it, to eliminate the frequency dependance of the measured voltage, making it a function of the magnetic flux only, we should integrate the output of the coil antenna. Since the integration of $V = -NAB_m(j\omega)\exp(j\omega t)$ is $-\left(\frac{1}{j\omega}\right)NAB_m(j\omega)\exp(j\omega t)$, the ω term cancels out, resulting in the output of the integrator to be NAB_m making it a frequency-independent reading of B_m. For the complete system, the antenna should be connected to a 50 Hz (or 60 Hz) high-pass filter to eliminate the 50 Hz (or 60 Hz) noise in the environment close to electric power lines. The output of this filter is integrated to eliminate the frequency term from the magnitude of the measured voltage. Moreover, a second-order high-pass filter with a cut-off frequency of 1 kHz, and a second-order low-pass filter with a cut-off frequency of 200 kHz will yield the signals picked up between the required 1–200 kHz wide band. The output of the combined band pass filter (1–200 kHz band) may be amplified by a power amplifier to display the measured magnetic field strength.

Consider now a single loop of radius a (cross sectional area of $A = \pi a^2\,m^2$ acting as a radiator, carrying a current $I\exp(j\omega t)$ A. The coil is placed on the xy-plane, with its center at the origin (0, 0, 0) of the coordinate system. In order to determine the electric and magnetic fields at an observation point $P(r, \theta, \phi)$, where r is the distance of point $P(r, \theta, \phi)$ from the origin, θ is the angle measured from the z-axis on the zx-plane, and ϕ is the angle measured from the x-axis on the xy-plane;

the vector potential at point $P(r, \theta, \phi)$, is

$$A_\phi = \frac{\mu I}{4\pi} \int_0^{2\pi} \frac{a \sin\phi}{R} \mathrm{e}^{\mathrm{j}(\omega t - kR} \, \mathrm{d}\phi \tag{7.204}$$

where R is the distance between any point on the coil and the observation point $P(r, \theta, \phi)$. For a small loop, such that $a << R$, we may write

$$R = r - a \sin\theta \sin\phi \tag{7.205}$$

for the R appearing in the phase (exponential) term and $R = r$ for the R appearing in the denominator which affects the magnitude. Therefore,

$$A_\phi = \frac{\mu I a}{4\pi r} \mathrm{e}^{\mathrm{j}(\omega t - kr)} \int_0^{2\pi} \mathrm{e}^{jka \sin\theta \sin\phi} \sin\phi \, \mathrm{d}\phi \tag{7.206}$$

which results in

$$A_\phi = \frac{k\mu I a^2 \sin\theta}{4r} \mathrm{e}^{\mathrm{j}(\omega t - kr)} \tag{7.207}$$

Hence, the electric field on the xy-plane is

$$E_\phi = -\frac{\partial A_\phi}{\partial t} = \frac{\omega k \mu I a^2 \sin\theta}{4r} \mathrm{e}^{\mathrm{j}(\omega t - kr)} \ \mathrm{V/m} \tag{7.208}$$

The magnetic field on the zx-plane has two components, which are

$$H_\mathrm{r} = \frac{1}{\mu r \sin\theta} \left(\frac{\partial \sin\theta A_\varnothing}{\partial\theta} \right) = \frac{\mathrm{j}kIa^2 \cos\theta}{2r^2} \mathrm{e}^{\mathrm{j}(\omega t - kr)} \ \mathrm{A/m} \tag{7.209}$$

and

$$H_\theta = \frac{-1}{\mu r} \left(\frac{\partial r A_\varnothing}{\partial r} \right) = -\frac{k^2 I a^2 \sin\theta}{4r} \mathrm{e}^{\mathrm{j}(\omega t - kr)} \ \mathrm{A/m} \tag{7.210}$$

There is no magnetic field on the xy-plane, which means

$$H_\phi = 0 \tag{7.211}$$

The wave impedance is given by

$$\frac{E_\phi}{H_\theta} = -\sqrt{\frac{\mu_o}{\epsilon_o}} \tag{7.212}$$

The power radiated is $P_\mathrm{r} = -E_\phi H_\theta$ W/m^2 which is the Poynting vector in the positive r direction. Note that the small loop (or magnetic dipole) antenna has the same radiation pattern as the Hertzian electric dipole antenna.

For further reading

[1] Balanis, C.A., *Antenna Theory*, New York, USA: John Wiley, 2012.

[2] Bansal, R., Ed., *Engineering Electromagnetics: Applications*, Boca Raton, FL, USA: CRC Press, 2006.

[3] Cheng, D.K., *Field and Wave Electromagnetics*, 2nd Edition, Addison-Wesley, Reading, MA, 1998.

[4] Christopoulos, C., *Principles and Techniques of Electromagnetic Compatibility*, Boca Raton, FL, USA: CRC Press, 1995.

[5] Collin, R.E., *Antennas and Radio Wave Propagation*, New York, USA: McGraw Hill, 1985.

[6] Elliot, R.S., *Electromagnetics*, New York: IEEE Press, 1993.

[7] Hall, M.P.M., Barclay, L.W., Hewitt, M.T., *Propagation of Radiowaves*, London, UK: IEE Press, 1996.

[8] Hall, S.H., Hall, G.W., McCall, J.A., *High Speed Digital System Design: A Handbook of Interconnect Theory and Design Practices*, New York, USA: John Wiley, 2000.

[9] Hansen, R.C., *Phased Array Antennas*, New York, USA: John Wiley, 1998.

[10] Hoole, P.R.P., *Smart Antennas and Signal Processing*, Southampton, UK: WIT Press, 2001.

[11] Hoole, P.R.P., *Electromagnetic Imaging in Science and Medicine*, Southampton, WIT Press, 2000.

[12] Hoole, S.R.H., Hoole, P.R.P., *A Modern Short Course in Engineering Electromagnetics with Computer Applications*, New York: Oxford University Press, 1996.

[13] Inan, U.S. and Inan, A.S., *Electromagnetic Waves*, Addison Wesley, Menlo Park, CA, 1999.

[14] Johnson, R.C. Ed., *Antenna Engineering Handbook*, New York, USA: McGraw Hill, 1993.

[15] Jordon, E.C. and Balmain, K.G., *Electromagnetic Waves and Radiating Systems*, Englewood Cliffs, NJ, USA: Prentice Hall, 1968.

[16] Misra, D.K., *Practical Electromagnetics*, New York, NJ, USA: John Wiley, 2007.

[17] Sampei, S., *Applications of Digital Wireless Technologies and Global Wireless Communications*, Englewood Cliffs, NJ, USA: Prentice Hall, 1997.

[18] Pozar, D.M., and Schaubert, D.H., *Microstrip Antennas: The Theory and Design of Microstrip Antennas and Arrays*, New York, USA: IEEE Press, 1995.

[19] Poljak, D., *Human Exposure to Electromagnetic Fields*, Southampton, UK: WIT Press, 2004.

[20] Ramo, S., Whinnery, J., and Van Duzer, J., *Fields and Waves in Communication Electronics*, New York, USA: John Wiley, 1994.

[21] Sevgi, L., Akklemann, F., and Felsen, L.B., Ground wave propagation modeling: problem matched analytical formulations and direct numerical techniques, *IEEE Antennas and Propagation*, *44*(1), 55–75, 2002.

[22] Skolonik, M.I., *Radar Handbook*, New York, USA: McGraw Hill, 1990.

[23] Wentworth, S.M., *Fundamentals of Electromagnetics with Engineering Applications*, New York, USA: John Wiley, 2005.

[24] Wesolowski, K., *Mobile Communication Systems*, New York, USA: John Wiley, 2002.

[25] Zeimer, R.E. and Tranter, W.H., *Principles of Communications*, New York, USA: John Wiley, 4th Edition, 1995.

8

Antennas II: Electromagnetic radiation and noise

8.1 Introduction

Having considered the fundamentals of electromagnetic field radiation from the most basic and widely used elemental current (wire) antenna, we shall be looking at some advanced antennas that belong to three other types of antennas. In Chaper 7, we also defined and explained the most basic parameters of an antenna, including power gain, radiation patterm, beamwidth and reflection coefficient. Moreover, the two basic system equations common to all four types of antennas, the Friis equation for communication systems and the radar equation for radar systems were also considered. In this chapter, we shall first consider the three other types of antennas first: the array antenna, the traveling wave antenna and the aperture antenna. Second, we shall consider the important subject of electromagnetic interference (EMI) where electromagnetic fields radiated by antennas (or by copper conductors in microelectronic circuits) can interfere, disrupt or even damage other systems and how to accomplish electromagnetic compatibility (EMC) where the EMI problems of a system are brought under control.

Shown in Fig. 8.1 is a low-earth orbit (LEO) satellite antenna A1 and a mobile antenna A2. These two antennas are part of a wireless communication system, and we want to maximize transmission and reception between A1 and A2. However, at antenna A3, which is dedicated to, say a hospital to ambulance communication system, the signals from antennas A1 and A2 are considered noise (or interference). Similarly electromagnetic communication signals radiated by antennas A1 and A2 are also inteference signals and the antenna A4 inside, say the medical diagnosis room of the hospital. Hence to achieve EMC for the hospital system, which seeks to minimize interference, the EMI from antennas A1 and A2 must be suppressed. We shall be looking at three types of antennas, such as the linear array antenna, helical traveling wave antenna and rectangular aperture antenna, which can provide much more focused beams than the first type of antenna, the elemental wire antennas we looked at in Chapter 7 . Focused beam antennas are widely used to avoid interference with neighboring communication systems, as well as to maximize radiation in a given direction. We shall conclude the chapter with a discussion on methods to achieve EMC, other than having focused beam antennas such as the array antenna, in order to maximize reception and minimize EMI.

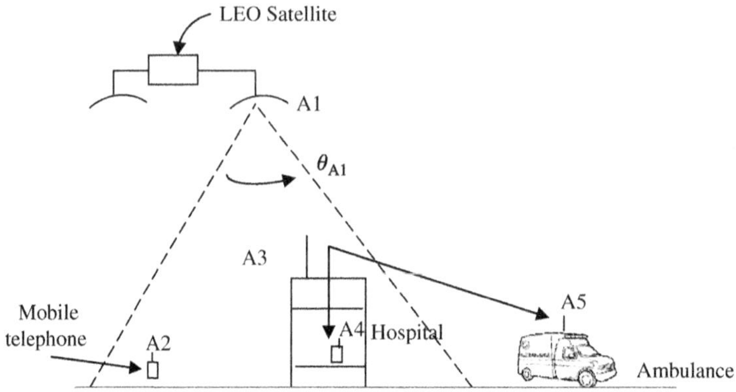

Figure 8.1: The need for more focused radiation and suppression of interference.

With regard to the quality of a signal received (over a communication, radar or biomedical system), the signal-to-noise ratio (SNR) is given by $\frac{S}{N} = \frac{P_r A_t A_r}{r^2 \lambda^2 B k T_s}$, where S is the signal power received, N is the noise power, P_r is the power radiated by the transmitting antenna (in W), A_t is the aperture area of the transmitting antenna (in m^2), A_r is the aperture area of the receiver antenna, r is the distance between the transmitter and the receiver (in m), λ is the wavelength of the carrier signal (in m), B is the bandwidth of the receiver antenna (in Hz), k is the Boltzman's constant and T_s is the temperature of a noise emitting object (in °K). Note that a wide bandwidth, which is attractive in digital communications, leads to poor SNR since interference signals at all the frequencies contained within the wide bandwidth get into the receiver electronics through the receiving antenna. Moreover, as the frequency of operation increases (i.e. as λ decreases), for a given signal power, the noise power (hf^2, where h is Planck's constant) also increases with frequency f.

Antennas find increasing applications in a variety of systems such as from safety and security systems to navigational systems. They are expected to play an increasingly important role in wireless comunication systems, which include the following six types: (a) **Fixed satellite systems.** A link between a satellite antenna A1 in Fig. 8.1, in a satellite to satellite communication link using two satellite antennas pointing toward each other. These are SHF (3–30 GHz) and EHF (30–300 GHz) band communication systems between medium earth orbit (MEO) or geostationary (GEO) satellites and fixed earth stations using large aperture antennas. Array antennas may be used for interplanetary satellites which are further away than the 36,000 km GEO orbit. (b) **Fixed terrestrial systems.** This is a link between two stationary, terrestrial or earth station antennas. These are Very High Frequency on VHF (30–300 MHz) to Extremely High Frequency on EHF systems used to link base stations (BS) of mobile communication systems as well as for fast (i.e. high data rate) networks. (c) **Megacells.** These may be low data rate L (1–2 GHz), S (2–4 GHz) and high data rate Ka (26–40 GHz) band communication systems between LEO satellites and mobile users, providing telephone to Internet services. Each LEO satellite might cover about a 1000 km radius area on the earth's surface. These too are high data rate systems. (d) **Macrocells.** These are VHF and UHF (0.3–3 GHz) communications and broadcast services where a BS placed at a height above the height of the tallest structure in an area provides services to mobile receivers within a radius of 1 km to tens of kilometers. These are high data rate systems. (e) **Microcells.** These are VHF, UHF and EHF mobile communication systems operating within a radius of about 500 m in urban and suburban areas. These are high traffic density systems, that is servicing a large number of users. Although the BS is lower than most of the tall structures around an urban or suburban area where the signals have to go through buildings shadowing, and thus attenuating the signals. (f) **Picocells.** This is a very high density or high data rate inter-room, indoor communication system between fixed and mobile communication units within a room where the communication environment is made up of four walls, furniture and people. Picocell systems are very high data rate systems.

8.2 Array antennas

8.2.1 A basic array antenna: the Yagi antenna

Array antennas are formed by a series of wire or aperture antennas, generally arranged along a single line (linear array) or in a rectangular grid (planar array). In other words, an array antenna is a collection of single antenna elements. One popular type of array antenna is the Yagi antenna used for television signal reception. The advantage of having a series of antenna elements radiating or receiving simultaneously are twofold: (a) increased radiation power and (b) better ability to direct the antenna radiation pattern (beam) in a specific direction. A folded dipole antenna, instead of a half-wave dipole antenna, is used to double the amount of power received; however, the input impedance of the folded dipole is about 388 Ω. Such antennas are used in the frequency range of 5 MHz to 50 GHz, and the typical dimension of the antenna is $L/\lambda = 1-100$, where L is the length of the antenna and λ is the wavelength of the signal. In this simple structure, an array antenna is made by using two half-wave length long wire antennas and connecting them in parallel.

By placing other elements in parallel to this folded dipole, we have the Yagi antenna as shown in Fig. 8.2 which is widely used in television reception. The parasitic elements are not connected to the output cable, but they capture some part of the incident signal from the transmitter and re-radiate it toward the folded dipole receiver which is connected to the output cable which is connected to the television unit inside the house. One extra parasitic element is placed behind the folded dipole to reflect back signals into the folded dipole, thus adding to the total signal strength at the folded dipole. This array of wire antenna elements is pointed toward the television channel transmitter since the antenna radiation pattern is

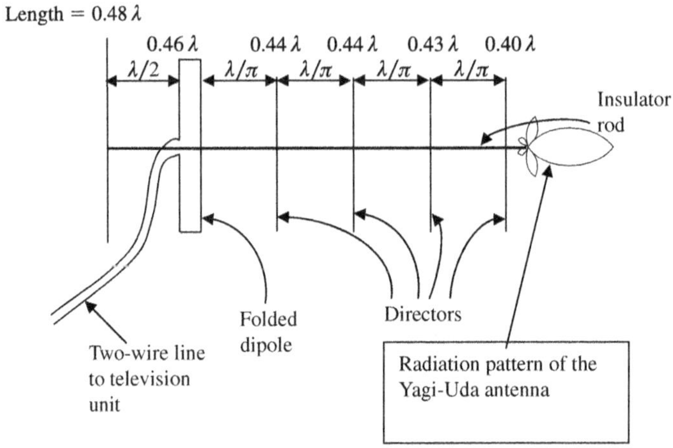

Figure 8.2: Yagi antenna.

maximum along the axis of the array as shown in Fig. 8.2. When installing the antenna, by turning it left and right until signal reception shows maximum received power, we may get the antenna to point directly toward the television transmitter.

An array antenna is an antenna made up of a combination of single antennas arranged in a given pattern, for example five half-wave dipole antennas (called elements) arranged in a straight line with a distance of half-wavelength between adjacent antenna elements. We shall now consider the general theory behind array antennas by first looking at a two-element array antenna and then at a general N-element array antenna. Although we shall develop the theory for a wire antenna array, the final result we shall get for what is called the array factor is the same for whichever antenna element we use. In other words, from antenna element to antenna element, the element radiation pattern may vary according to the particular antenna element that we use (e.g. a half-wave dipole antenna element or a helical antenna element; the array factor is the same for whatever type of antenna element used.

8.2.2 Two-element array antenna

In Fig. 8.3(a) is shown two infinitesimal, Hertzian dipole antennas of length L and each carrying a current I placed along the x-axis with a distance d between them. The phase angle difference between the currents along the two antenna elements is δ. Note that θ is a physical angle and δ is an electrical angle.

From (7.92), for a single infinitesimal, Hertzian dipole of length L, the radiated electric field at a distance r from the dipole antenna is given by

$$\mathbf{E} = \mathbf{u}_\theta \mathrm{j}\eta \frac{kIL}{4\pi r} \mathrm{e}^{-jkr} \sin\theta \tag{8.1}$$

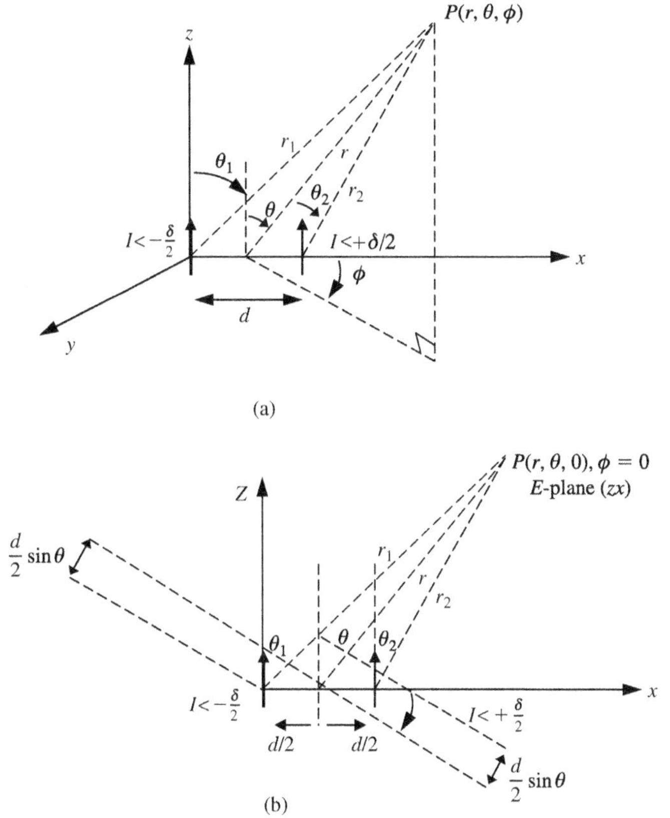

Figure 8.3: A two-element array antenna. Each element is a Hertzian dipole antenna.

Hence the electric fields radiated by the two dipole elements in Fig. 8.3 may be written as

$$\mathbf{E}_1 = \mathbf{u}_\theta \mathrm{j}\eta \frac{kIL}{4\pi r_1} \mathrm{e}^{-\mathrm{j}(kr_1+\delta/2)} \sin\theta_1 \tag{8.2}$$

for the element placed at the origin and

$$\mathbf{E}_2 = \mathbf{u}_\theta \mathrm{j}\eta \frac{kIL}{4\pi r_2} \mathrm{e}^{-\mathrm{j}(kr_2+\delta/2)} \sin\theta_2 \tag{8.3}$$

for the second element placed at a distance d from the origin along the x-axis. Thus the total radiated electric field at point P (i.e. $\mathbf{E}_1 + \mathbf{E}_2$) is given by

$$\mathbf{E} = \mathbf{u}_\theta \mathrm{j}\eta \frac{kIL}{4\pi} \left(\frac{\mathrm{e}^{-\mathrm{j}(kr_1+\delta/2)}}{r_1} \sin\theta_1 + \frac{\mathrm{e}^{-\mathrm{j}(kr_2-\delta/2)}}{r_2} \sin\theta_2 \right) \tag{8.4}$$

The terms r_1 and r_2 appearing in the denominator of (8.4) will not much affect the magnitude of **E**, as seen in Fig. 8.3(b). The difference between r_1, r_2 and r is very small given the fact that the observation point P is in the far-field. Thus, for the denominators, we may set

$$r_1 = r_2 = r \tag{8.5}$$

However for r_1 and r_2 appearing in the exponential terms, which influence the phase angles of the signals radiated by the two elements, even a small difference between r_1 and r_2 can cause a significant difference when the signals are added together to get the total field **E**. A small phase difference between two sinusoidal signals in the time domain can make the two signals add destructively or constructively depending on the phase difference. Hence we need to be more exact in representing r_1 and r_2 by r in the exponential terms.

To use in the phase terms, from Fig. 8.3(b), we seek a more accurate relationship between r_1, r_2 and r:

$$r_1 = r + \frac{d}{2}\sin\theta \tag{8.6}$$

$$r_2 = r - \frac{d}{2}\sin\theta \tag{8.7}$$

Substituting (8.5), (8.6) and (8.7) into (8.4), we get

$$\mathbf{E_p} = \mathbf{u}_\theta \mathrm{j}\eta \frac{kIL\mathrm{e}^{-\mathrm{j}kr}}{4\pi r}\sin\theta(\mathrm{e}^{-\mathrm{j}(kd\sin\theta+\delta)/2} + \mathrm{e}^{\mathrm{j}(kd\sin\theta+\delta)/2}) \tag{8.8}$$

Using Euler's formula $\mathrm{e}^{\mathrm{j}x} = \cos x + \mathrm{j}\sin x$, and $\mathrm{e}^{-\mathrm{j}x} = \cos x - \mathrm{j}\sin x$ (8.8)

$$\mathbf{E_p} = \mathbf{u}_\theta \mathrm{j}\eta \frac{kIL\mathrm{e}^{-\mathrm{j}kr}}{4\pi r}\sin\theta\left[2\cos\left(\frac{kd\sin\theta+\delta}{2}\right)\right] \tag{8.9}$$

For this two-element array antenna, we define the array factor, AF, by

$$\mathrm{AF} = 2\cos\left(\frac{kd\sin\theta+\delta}{2}\right) \tag{8.10}$$

Thus the total electric field at observation point P is

$$\mathbf{E_p} = \mathrm{AF} \times \mathrm{EP} \tag{8.11}$$

where the elemental pattern $\mathrm{EP} = \mathbf{u}_\theta \mathrm{j}\eta \frac{kIL\mathrm{e}^{-\mathrm{j}kr}}{4\pi r}\sin\theta$ for an infinitesimal or Hertzian dipole antenna.

Defining the normalized array factor as $AF_N = AF/N$, where N is the number of elements (for a two-element array, $N = 2$), and the radiation pattern of the array antenna as RP, we have

$$RP = AF_{Nx} \times EP \tag{8.12}$$

where, for the two-element array, $AF_N = AF/2$. Note that we may replace the two antenna elements used by two half-wave dipole antennas, or by two horn antennas or again, by two helical antennas. The resultant radiation pattern is still $RP = AF_N \times EP$, and the array factor will remain the same as that given in (8.10). But the elemental pattern will no longer be $\sin\theta$, but become the radiation pattern of the particular element (e.g. half-wave dipole, horn or helical) used.

For the more general case, for an observation point $P(r, \theta, \phi)$, the array factor is

$$AF = 2\cos\left(\frac{kd\sin\theta\cos\phi + \delta}{2}\right) \tag{8.13}$$

and defining $\psi = kd\sin\theta\cos\phi + \delta$, we have $AF = 2\cos(\psi/2)$ and $AF_N = \cos(\psi/2)$. For the E-plane (zx-plane) pattern, we set $\phi = 0$, and the H-plane (i.e. xy-plane) pattern, we set $\theta = \pi/2$, which should be obvious from Fig. 8.3(a). It is important to note that AF (and AF_N) are symmetrical about the axis of the array (in Fig. 8.3, the x-axis). That is, for instance, if we sketch AF on the E-plane and on the H-plane, for given values of d and δ, we will get identical patterns. However, except for as isotropic radiator, the elemental pattern is not symmetrical about the array axis. Therefore, the final radiation pattern, which is a multiplication of EP and AF_N, will not be symmetrical about the array axis. In the isotropic radiator, the radiation pattern is a sphere and $EP = 1$. In the case of the Infinitesimal, Hertzian dipole antenna, $EP = \sin\theta$ in the E-plane and $EP = 1$ in the H-plane (note that its radiation pattern is not a function of ϕ).

What is important to take note of is this: in the case of the array antenna, for a given signal frequency f (or $k = 2\pi/\lambda = 2\pi f/c$), we can change the radiation pattern of the array antenna by controlling the distance d between the antenna elements, and the electronic phase angle difference δ between the currents in the adjacent individual antenna elements. Now distance d is a physical quantity. We may construct an array antenna with a fixed distance d. Now once installed, the antenna beam may be controlled by electronically controlling the phase angle δ. This is the strength of the array antenna. By merely using a microprocessor which controls the phase angle difference δ between the currents fed to the two-element array antenna as a transmitter, we may change the direction in which the signal is radiated. The array factor (AF or AF_N) and the elemental pattern (EP) will give the radiation pattern of the antenna. Similarly, when the array antenna is acting as a receiver antenna, we may control the electronic angle between the two receiving elements to maximize signal reception in certain directions and to minimize reception in other directions from which interference signals are observed. In other words, the array antenna is a highly versatile antenna in which the antenna can divide the space surrounding the antenna by controlling angle δ. This is called space division multiple access (SDMA). Since δ is altered electrically, the beam direction depending on δ can also be changed rapidly unlike in older radar antenna dish tracking planes which had to be moved mechanically.

We shall later see that in addition to changing the phase angles of the currents, we may use multipliers or dividers to control the magnitudes of the currents as well, in order to get an even more flexible beam steerable antenna by doing these changes in magnitudes and phases through a microprocessor placed at the terminals of the antenna. Hence when we think of "antennas," it is important to remember that a modern antenna is not just the physical structure of the antenna, but also the computer or microprocessor software that is used to make the antenna a more powerful device than it would be just as a physical structure. An antenna is the hardware (the physical structure) plus the software (the software program that controls current magnitudes and phase angles) to give us a flexible SDMA system in telecommunications, biomedical and remote sensing systems.

Consider now the following two cases where two isotropic antennas (i.e. radiating equally in all directions) are placed along the x-axis to form a two-element array antenna:

Case 1.

$$d = \frac{\lambda}{4} \quad \text{or} \quad kd = \frac{\pi}{2}, \quad \delta = -\frac{\pi}{2} \Rightarrow \text{AF} = 2\cos\left(\frac{\pi}{4}(\sin\theta - 1)\right) \tag{8.14}$$

A sketch of AF from (8.14) is as shown in Fig. 8.4(a), which is the endfire array radiation pattern. The elemental pattern (EP) of an isotropic antenna is a circle. The array factor (AF) of the (d, δ) values for Case 1 results in a heart shaped AF, with the peak radiation pointing along the x-axis where the two elements are placed. Since the EP is a circle, the final radiation pattern RP is identical to the AF shape.

Case 2.

$$d = \frac{\lambda}{2} \quad \text{or} \quad kd = \pi, \delta = 0 \Rightarrow \text{AF} = 2\cos\left(\frac{\pi}{2}\sin\theta\right) \tag{8.15}$$

resulting in the broadside array, as shown in Fig. 8.4(b). The EP is a circle. The array factor for the (d, δ) values results in an eight shaped AF with the two peaks pointing along the z-axis, away from the x-axis. As in Case 1, the RP is identical to the AF pattern.

In Fig. 8.5(a) is shown the exact polar plot and the rectangular plot of the normalized array factor AF_N of a two-element antenna. Note the maximum value

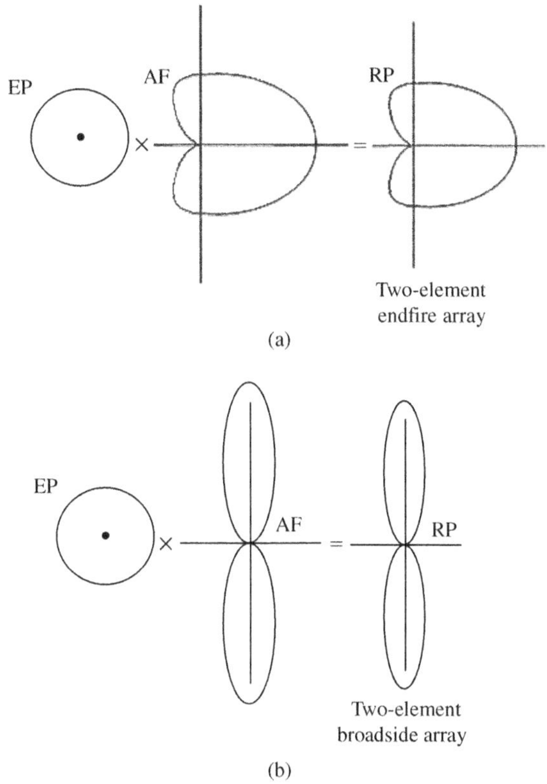

Figure 8.4: End fire and Broadside array antennas: (a) Case 1: Endfire array antenna and (b) Case 2: Broadside array antenna.

is 1.0 and for $\delta = +\pi/2$, the maximum radiation is in the $\theta = \pi$ direction, and not in the $\theta = 0$ direction as shown in Fig. 8.4(a). When analyzing and designing an array antenna, we may use the polar plot or the rectangular plot, or even a three-dimensional plot of the radiation pattern. The rectangular plot is the simplest to draw.

8.2.3 N-element uniform, linear antenna

Consider now the general case of N elements in a Hertzian dipole antenna arranged along the x-axis.

Although we are considering here only the uniform, linear arrangements of antennas, there are other arrangements too that are used. Arranging antenna elements in a square grid is often used with microstrip patch antenna elements. In the uniform, linear antenna as shown in Fig. 8.6, the phase angle difference between each adjacent element is δ, starting with 0 phase angle for the element placed at

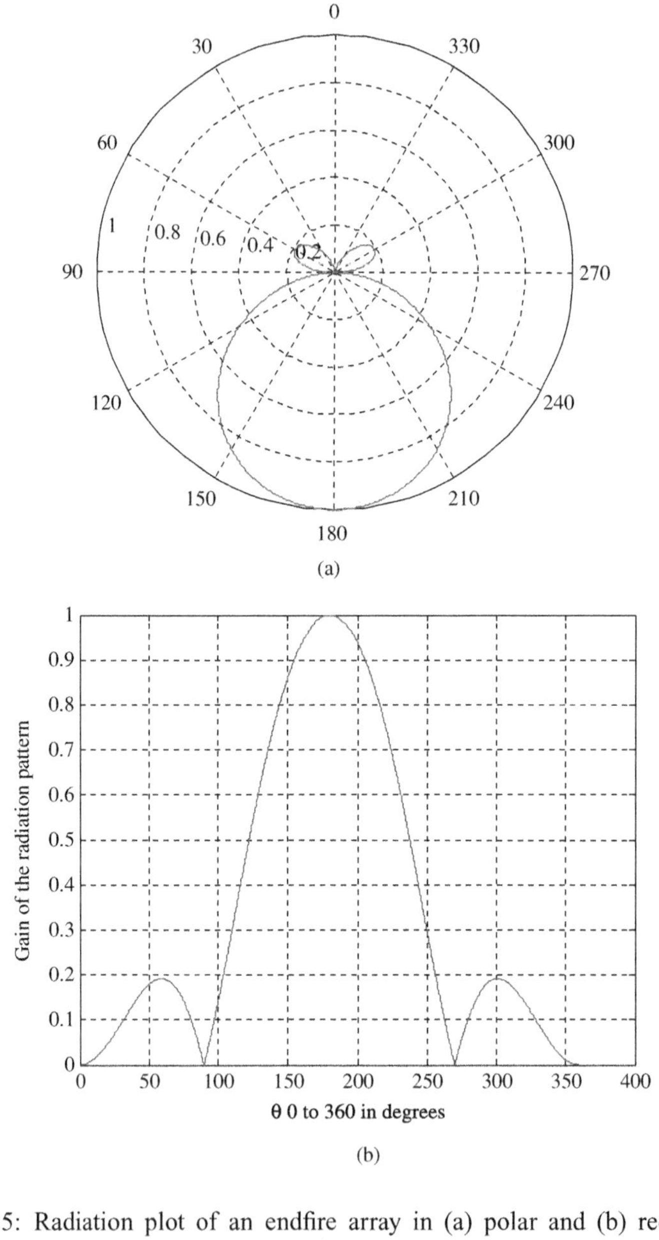

(a) (b)

Figure 8.5: Radiation plot of an endfire array in (a) polar and (b) rectangular form. * Number of elements $N = 2$; Phase excitation, $\delta = 0$; Separation between elements, $d = \lambda/4$

the origin ($x = 0$). With a distant d between adjacent elements the physical length of this array antenna is $(N - 1)\delta$.

The respective currents flowing in each element 1, 2, 3, …, N of the array antenna are $I\angle 0, I\angle\delta, I\angle 2\delta, \ldots, I\angle(N - 1)\delta$. We seek to determine the far-field

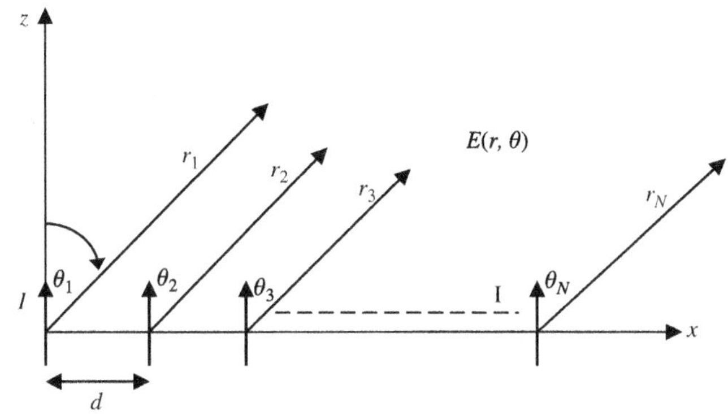

Figure 8.6: N-elment uniform, linear array antenna.

value of the electric field radiated. Therefore, for the angles shown in Fig. 8.6, we may assume that $\theta_1 \approx \theta_2 \approx \theta_3 \approx \cdots \approx \theta_N = \theta$. Moreover, for magnitude computations, we may assume that $r_1 \approx r_2 \approx r_3 \approx \cdots \approx r_N = r$. But for the phase term, a more exact relationship for the distance of each element from the observation point may be taken to be

$$r_1 = r \tag{8.16}$$

$$r_2 \approx r - d \sin\theta \cos\phi \tag{8.17}$$

$$r_3 \approx r - 2d \sin\theta \cos\phi \tag{8.18}$$

$$r_4 \approx r - 3d \sin\theta \cos\phi \tag{8.19}$$

$$\vdots$$

$$r_N \approx r - (N-1)d \sin\theta \cos\phi \tag{8.20}$$

Consider now the electric field in the E-plane, for which $\phi = 0$. The resultant field at an observation point $P(r, \theta, 0)$ is given by

$$E_\theta = E_{\theta 1} + E_{\theta 2} + \cdots + E_{\theta N} \tag{8.21}$$

or

$$E_\theta = \mathrm{j}\frac{kIL \sin\theta \mathrm{e}^{-\mathrm{j}kr}}{4\pi r}\left[1 + \mathrm{e}^{\mathrm{j}(kd\sin\theta+\delta)} + \mathrm{e}^{\mathrm{j}2(kd\sin\theta+\delta)} + \cdots + \mathrm{e}^{\mathrm{j}(N-1)(kd\sin\theta+\delta)}\right] \tag{8.22}$$

$$E_\theta = \mathrm{j}\frac{kIL\sin\theta}{4\pi \mathrm{r}}\mathrm{e}^{-\mathrm{j}kr}\mathrm{AF}(\theta, \delta) \tag{8.23}$$

where the array factor AF combines the terms inside the square barackets of (8.22). The second electric field component

$$E_r(r,\theta) = \mathrm{j}\frac{kIL\cos\theta \mathrm{e}^{-jkr}}{2\pi \mathrm{r}^2}[\mathrm{AF}(\theta,\delta)] \approx 0 \tag{8.24}$$

for far fields, because of the $1/r^2$ term. For simplicity, we denote the array factor by AF, droping the (θ,δ) terms, remembering that AF is a function of (θ,ϕ,δ) in general, or of (θ,δ) on the zx- or E-plane. We have

$$\mathrm{AF}(r,\theta) = 1 + \mathrm{e}^{\mathrm{j}kd\sin\theta+\delta} + \mathrm{e}^{\mathrm{j}2(kd\sin\theta+\delta)} + \cdots + \mathrm{e}^{\mathrm{j}(N-1)(kd\sin\theta+\delta)} \tag{8.25}$$

$$\mathrm{AF}(r,\theta) = \sum_{n-1}^{N} \mathrm{e}^{\mathrm{j}(n-1)(kd\sin\theta+\delta)} = \sum_{n-1}^{N} \mathrm{e}^{\mathrm{j}(n-1)\psi} \tag{8.26}$$

where $\psi = kd\sin\theta + \delta$. In general,

$$\psi = kd\sin\theta\cos\phi + \delta \tag{8.27}$$

If we multiply AF by $\mathrm{e}^{\mathrm{j}\psi}$, we get

$$\mathrm{AF}\cdot\mathrm{e}^{\mathrm{j}\psi} = \mathrm{e}^{\mathrm{j}\psi} + \mathrm{e}^{\mathrm{j}2\psi} + \cdots + \mathrm{e}^{\mathrm{j}N\psi} \tag{8.28}$$

Subtracting (8.25) from (8.28), we get

$$\mathrm{AF}(\mathrm{e}^{\mathrm{j}\psi} - 1) = \left(-1 + \mathrm{e}^{\mathrm{j}N\psi}\right) \tag{8.29}$$

Hence,

$$\mathrm{AF} = \frac{\mathrm{e}^{\mathrm{j}N\psi} - 1}{\mathrm{e}^{\mathrm{j}\psi} - 1}\mathrm{e}^{\mathrm{j}((N-1)\frac{1}{2})\psi} \tag{8.30}$$

$$= \frac{\mathrm{e}^{\mathrm{j}(N/2)\psi} - \mathrm{e}^{-\mathrm{j}(N/2)\psi}}{\mathrm{e}^{\mathrm{j}(N/2)\psi} - \mathrm{e}^{-\mathrm{j}(N/2)\psi}} \tag{8.31}$$

$$= \mathrm{e}^{\mathrm{j}((N-1)/2)\psi}\frac{\sin(N/2))\psi}{\sin(N/2)\psi} \tag{8.32}$$

And the radiated electric field at the observation point $P(r,\theta,0)$ is given by

$$|E(r,\theta)| = |E_\theta|\left|\frac{\sin(N\psi/2)}{\sin(\psi/2)}\right| \tag{8.33}$$

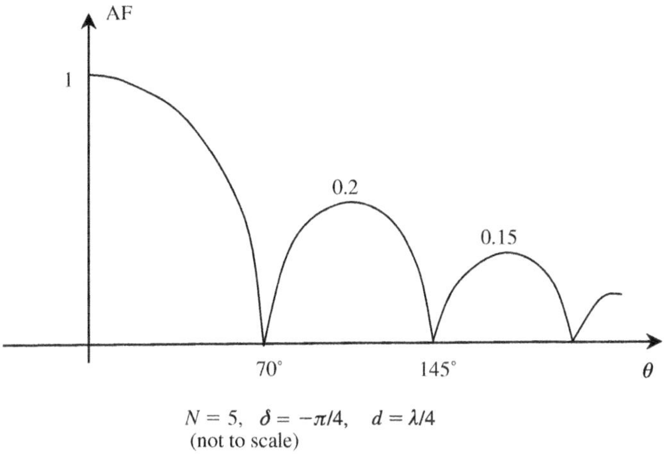

Figure 8.7: The nulls of the array factor.

The electric field E_θ is the electric field radiated by a single element. The radiation pattern (RP) is then the multiplication of the array factor (AF) and the elemental pattern (EP), which is the radiation pattern of the single element (EP $= \sin\theta$ for a Hertzian dipole element). The array factor is given by

$$\mathrm{AF} = \frac{\sin\left(\frac{1}{2}N\psi\right)}{\sin\left(\frac{1}{2}\psi\right)} \tag{8.34}$$

and the normalized array factor is

$$\mathrm{AF}_N = \frac{1}{N}\frac{\sin\left(\frac{1}{2}N\psi\right)}{\sin\left(\frac{1}{2}\psi\right)} \tag{8.35}$$

The array factor for a five-element array antenna, with $\delta = -\pi/4$ and $d = \lambda/4$ is shown in Fig. 8.7.

Consider now some special cases and features of the N-element uniform, linear array antenna.

Case 1. Broadside array antenna. Consider the case when

$$\delta = 0 \quad \text{and} \quad \theta = \frac{\pi}{2}$$

which means that we consider the xy- or the H-plane radiation pattern.

Thus the angle $\psi = kd\sin\theta\cos\phi + \delta$ reduces to $\psi = kd\cos\phi$.

Consider that for a broadside array antenna, the maximum radiation is perpendicular to the axis of the antenna, or along the direction where $\phi = \phi_0 = \pi/2$. Then if the first null is at ϕ_{01}, the first null beamwidth (FNBW, see Fig. 7.3) $\Delta\phi$ is obtained from

$$2\Delta\phi = 2(\phi_{01} - \phi_0) \tag{8.36}$$

Noting that the numerator of the array factor has $N\psi = Nkd\cos\phi = N(2\pi/\lambda)d\cos\phi$ and $\phi_{01} = \phi_0 - \Delta\phi$, we have

$$\cos\phi_{01} = \cos(\phi_0 - \Delta\phi) = \frac{\psi_{01}}{Nkd} \tag{8.37}$$

Thus the FNBW for the broadside array is

$$\text{FNBW}_b = \Delta\phi = \sin^{-1}\left(\frac{\lambda}{Nd}\right) = \frac{\lambda}{Nd}, \quad \text{for } Nd \gg \lambda \tag{8.38}$$

Case 2. Endfire array antenna. Consider the xy- (or H-plane) (i.e. $\theta = \pi/2$, $\sin\theta = 1$). For $\delta = -kd$, and the antenna radiation power being maximum along the x-axis, with $\phi = 0$, the angle $\psi = kd\sin\theta\cos\phi + \delta$ becomes

$$\psi = kd(\cos\phi - 1) \tag{8.39}$$

For the first null angle ϕ_{01}, noting that the peak radiation is along $\phi = 0$ for the endfire array antenna, we get

$$\cos\phi_{01} - 1 = \frac{\psi_{01}}{kd} = -\frac{2\pi}{Nkd} = -\frac{\lambda}{Nd} \tag{8.40}$$

Since the FNBW for the endfire array antenna $\text{FNBW}_e = 2\Delta\phi = 2(\phi_{01} - \phi_0) = 2\phi_{01}$, we have $\Delta\phi = \phi_{01}$. Hence

$$\cos\phi_{01} = \cos(\Delta\phi) = 1 - \frac{(\Delta\phi)^2}{2} \tag{8.41}$$

Thus

$$\frac{(\Delta\phi)^2}{2} = \frac{\lambda}{Nd} \tag{8.42}$$

and we have

$$\text{FNBW}_e = \Delta\phi = \sqrt{\frac{2\lambda}{Nd}} \tag{8.43}$$

Since $Nd > \frac{\lambda}{2}$,

$$\mathrm{FNBW_e} > \mathrm{FNBW_b} \tag{8.44}$$

That is the broadside array antenna has a narrower beam than the endfire array antenna. Moreover we note that as the number of elements N is increased, the beamwidth becomes narrower. The reduction of λ/d also narrows the beamwidth. The width of the main beam of a broadside array is λ/Nd, in comparison with an endfire array, 5 value of $\sqrt{2\lambda/Nd}$. In scanning array antennas in order to obtain a beam as narrow as 3° and thus achieve a radar range of several thousand kilometers, the more focused the beam, the energy can reach a longar distance arrays consisting of several thousand elements (e.g., $N = 3000$) are used.

Consider now the sidelobe location. Sidelobes are found when

$$\sin\left(\frac{N\psi}{2}\right) = 1, \quad \text{for } \frac{N\psi}{2} = \pm(2m+1)\frac{\pi}{2}, \quad m = 1, 2, \ldots \tag{8.45}$$

The first two sidelobes appear when

$$\frac{N\psi}{2} = \pm\frac{3}{2}\pi \quad (m = 1) \tag{8.46}$$

The maxima of minor lobes occur at points where the numerator of (8.35) becomes very small. For $m = 1$, the first sidelobe level is found from

$$\frac{1}{N}\left|\frac{1}{\sin\left(\frac{3\pi}{2N}\right)}\right| = \frac{1}{N}\left|\frac{1}{\frac{3\pi}{2N}}\right| = \frac{2}{3\pi} = 0.212 = 20\log_{10}\left(\frac{1}{0.212}\right) = 13.5\ \mathrm{dB} \tag{8.47}$$

Consider the directivity D of the array antenna (see Section 7.3). For the broadside array antenna,

$$D = 2N\left(\frac{d}{\lambda}\right) = 2\left(\frac{L}{\lambda}\right) \tag{8.48}$$

where $L = Nd$, the length of the array antenna. For the endfire array, directivity

$$D = 4N\left(\frac{d}{\lambda}\right) = 4\left(\frac{L}{\lambda}\right) \tag{8.49}$$

The directivity of the broadside antenna is half that of the endfire array antenna because it simulataneously radiates equally strong beams in both directions $\phi = \pi/2$ and $\phi = -\pi/2$.

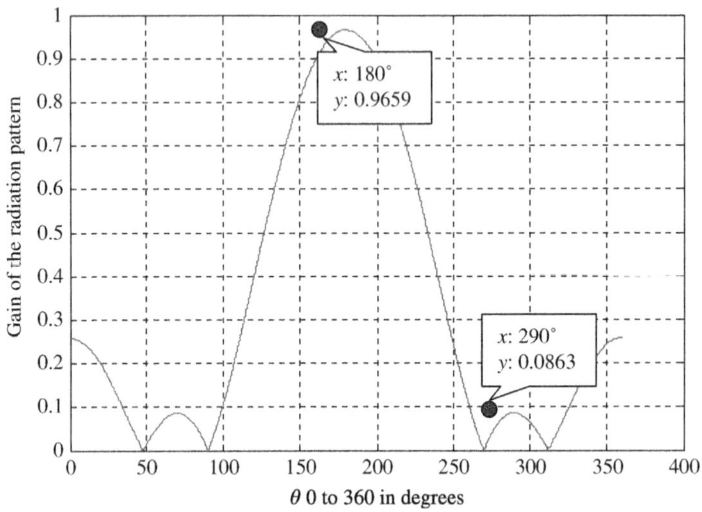

Figure 8.8: Polar and rectangular plot for $N = 2$, $d = \lambda/4$ and $\beta = 2/3\pi$.

To scan between angles $-\theta_s$ and $+\theta_s$, we need to ensure that

$$d < \frac{\lambda}{1 + \sin\theta_s} \tag{8.50}$$

To reduce the grating lobe, that is the angular traversal from beam maximum to null, reduce the visible region $2kd$ in the ψ domain. Grating lobe half-width (i.e. maximum to null) is $2\pi/N$. Thus we seek to design the array antenna such that $2kd < 2\pi - (\pi/N)$ or keep $d < 0.5\lambda(1 - 0.5/N)$.

Thus the ratio of the first sidelobe to the level of the major lobe is obtained from (7.35) to be approximately equal to $2/3\pi$ or $20\log_{10}2/3\pi = -13.46$ dB.

Using the general expression for the array factor, we shall work out the array factor shape for a two-element array on the H-plane (i.e. $\theta = \pi/2$, $\sin\theta = 1$). Consider the case when $N = 2$, $d = \lambda/4$ and $\delta = 2/3\pi$. We shall assume isotropic antenna elements, so that elemental pattern EP = 1 on all planes.

The normalized array factor is given by

$$(\mathrm{AF})_N = \frac{1}{N}\left[\frac{\sin\left(\frac{N}{2}\psi\right)}{\sin\left(\frac{1}{2}\psi\right)}\right] \tag{8.51}$$

where $\psi = kd\cos\theta + \delta$ and $k = 2\pi/\lambda$ when $\phi = 180°$, $\psi = kd\cos\phi + \delta = 0.166667\pi$.

$$(\mathrm{AF})_N = \frac{1}{2}\left[\frac{\sin(\psi)}{\sin\left(\frac{1}{2}\psi\right)}\right] = \frac{1}{2}\left[\frac{0.5}{0.258819}\right] = 0.965926 \cong 1$$

Hence the radiation pattern RP $=$ EP $\times$ $(\mathrm{AF})_N = |-0.965926| \cong 1$.

When $\phi = 290°$, we get $\psi = kd\cos\phi + \delta = 0.837677\pi$. Hence

$$(\mathrm{AF})_N = \frac{1}{2}\left[\frac{\sin(\psi)}{\sin\left(\frac{1}{2}\psi\right)}\right] = \frac{1}{2}\left[\frac{0.488136}{0.967665}\right] = 0.9438 \cong 1$$

The total RP gain is EP $\times$ $(\mathrm{AF})_N = |0.086266| \cong 0.08$.

Thus working out some of the values we may handplot the radiation pattern; or more simply use a computer to plot out the pattern as shown in Fig. 8.8.

To change the electronic angles of each element of an array antenna with respect to the phase of the adjacent element, one of the following techniques may be used:

(a) Phase shift the angles using diode phase shifters, that is by switching onto an antenna element a feedor line element of different length; since the signal takes a longer time to travel over a longer line than a shorter line, the phase angle of an array element that receives a signal through a longer line will be delayed compared to an element that receives a signal through a shorter line. This results in a phase shift between the two adjacent elements.

(b) Phase shift the angles by sending the signal through ferrite phase shifters; since the signal that goes through a given ferrite of permeability μ_x will travel at a speed that is inversely proportional to $(1/\mu_r)^{1/2}$, the phase of the antenna element that receives a signal through a ferritephase shifter will lag behind the element that receives the signal through a copper conductor. In some instances, permittivity ε is also changed by using ferroelectric or plasma phase shifters.

(c) The phase of a signal traveling through a waveguide that has a length L and a cutoff frequency f_c of $2\pi L\sqrt{\mu\varepsilon\left(f^2 - f_c^2\right)}$ at an operating frequency f. The phase of a signal being delivered to (or received from) an element of an array antenna may be changed by changing f_c or f of the waveguide feeding the antenna element.

8.2.4 Additive and multiplicative noise in array antennas

Array antennas help to combat multiplicative and additive noise. These noises may be found in telecommunication, radar, biomedical and remote sensing systems where antennas are used. We shall see how array antennas may be used to maximize the radiation pattern toward a desired direction, or to form radiation nulls toward the directions from which interference signals are trying to get in through the antenna into the electronic amplifier and detector. We shall first give a brief outline of the two kinds of noise commonly found, which pose a major challenge when trying to imporve the quality of antenna performance in mission critical systems such as medical diagnostic systems (e.g. magnetic resonance imaging), radar systems (aircraft landing systems at airports) and collision avoidance antenna based safety systems for vehicles: airborne (aircraft), terrestrial (e.g. cars) or underwater vehicles (e.g. unmanned underwater vehicles):

(a) Additive noise. Additive noise arises from the receiver itself. It includes thermal and shot noise. Moreover interference can come from other equal frequency sources of radiation, such as a mobile telephone in the vicinty of the antenna communicating with another receiver.

(b) Multiplicative noise. Multiplicative noise may arise from the signal being received itself. Consider the following cases: (a) Reflection. The signal being received may impinge on a nearby wall, and thus the receiving antenna is not only getting the direct ray from the transmitter, but also, with a time delay, the reflected ray of the transmitter. (b) Absorption. A part of the signal being received may lose its signal power by having to pass through a wall and thus having some of its signal power being absorbed by the wall material. (c) Scattering. An object partially blocking the transmitted ray may scatter the part impinging on it, preventing a part of the transmitted signal from getting into the receiver antenna. (d) Diffraction. Sharp edges such as roof tops may scatter a part of the signal or may even bend the signal of another unrelated transmitter to make that defracted ray get into our antenna and cause inteference. (e) Refraction. Refraction of transmitted signals is another cause of multiplicative noise.

All these sources of noise are subjects of serious concern and research. For instance, one aspect of research is seeking to characterize the noise using basic electromagnetic theory, in order to develop powerful software which may counteract its effects on the system in which the transmitting and/or receiving antenna is working.

Both additivie and multiplicative nosie give rise to the followig three kinds of losses: (i) **Path loss** caused by trees, buildings, hills and distance. Signal power loss of the order of 150 dB may occur due to path loss. (ii) **Shadowing**. Shadowing occurs over a few hundreds of meters, giving rise to loss of the order of 20 dB. (iii) **Fast fading**. (multiplicative fading). Fast fading varies at every half wavelength distance from the transmitter, giving rise to power loss of the order of 40 dB.

8.2.5 Adaptive or smart array antennas

In this section, we shall use the two-element array to show how suitable values of the current phase angle difference δ between adjacent array antenna elements

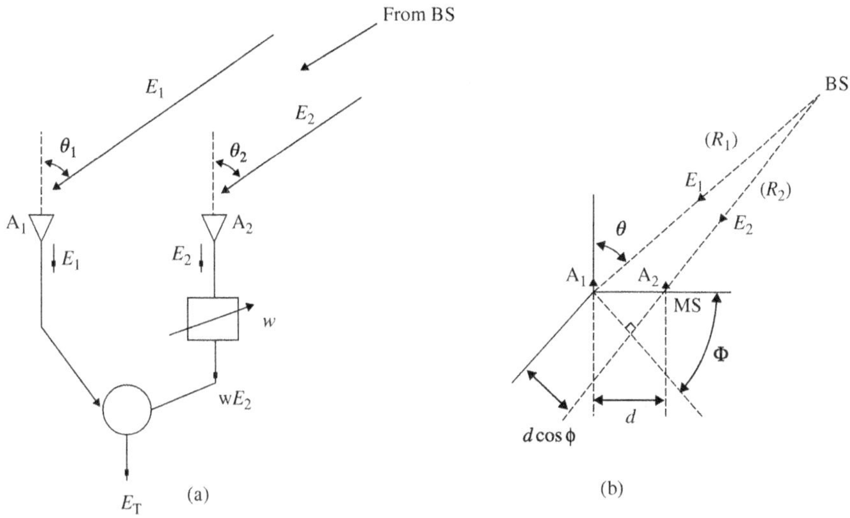

Figure 8.9: (a) A two-element array antenna, with one element having a phase shifter w attached to it. (b) Geometry of the MS with respect to a BS.

may be automatically found by computer software in order to either maximize the array factor toward a desired signal or null the array factor toward the direction of an interference signal. We shall consider first the simplest case where for the two elements, we shall determine the phase angle δ of one of the two elements, letting the phase angle of the second element remain at zero. We also limit our case to where only the angle is being changed, and not the magnitude of the incoming (receiver) or outgoing (transmitter) signal. The discussion applies to a transmitting two-element antenna as well as to a receivng two-element antenna due to the reciprocity of antennas. These antennas are called adaptive antennas in that they automatically adapt the radiation pattern to the position and movement of the antenna. They are also called smart antennas since they may intelligently (using antenna beam forming software) maximize the beam toward the desired signal, as well as null the interference signal.

Consider now the two Hertzian dipole antennas shown in Fig. 8.9, arranged together to form a two-element array antenna, say of a mobile telephone or mobile station (MS). As it moves, it needs to keep in touch with the BS antenna through which the MS communicates to other users, both mobile and fixed. As the MS moves on the horizontal plane, we want to keep the antenna beam focused on the BS. That is, the maximum of the radiation pattern (i.e. the array factor in the case of isotropic radiating elements) should be pointing toward the BS. This is Case 1. In a second case, we shall consider how to null (i.e. zero) the array factor toward another transmitting source which is an interferer, an unwanted signal. This is Case 2.

Figure 8.9 shows the two-element array antenna in receiving mode. An antenna's transmission and radiation patterns are identical at the same operating frequency. At the output of the antenna element is a digital beam steering (beam forming) weight w which must be set either to maximize the MS antenna beam toward the BS antenna or to create a null in the antenna radiation pattern in the direction of an interfering signal (e.g. from another BS or MS). Although both elements are receiving signals from the same BS source, they are represented by E_1 and E_2, where the signal strength may be equal (since the two elements are close to each other), but the phase of the two signals may be considerably different although the distance d between them is small. In Fig. 8.9(b) is shown the first distance difference between the two elements, which is $d \cos \phi$. To get the exact difference between the distances R_1 and R_2, we need to do a further geometrical operation to tilt the $d \cos \phi$ distance to the R_1 line, giving rise to a total difference of $d \sin \theta \cos \phi$. We shall be using this later to represent correctly the phase difference between the two received signals $\mathbf{E}_1$ and $\mathbf{E}_2$.

The total output from the two-element array antenna x with phase shift w, is given by

$$\mathbf{E}_T = \mathbf{E}_1 + w\mathbf{E}_2 \tag{8.52}$$

Consider the BS antenna as a dipole antenna of length L, with a current I flowing through it. If the MS is in the far-field region of the BS, then for magnitude purposes, we may assume that the distance R_1 between the MS and element A_1 of the two-element array antenna and distance R_2 from the element A_2 may be set equal

$$\mathbf{R} = \mathbf{R}_1 \cong \mathbf{R}_2 \tag{8.53}$$

However the short distance of separation between A_1 and A_2 may cause a significant phase (not magnitude) difference between signals E_1 and E_2 which is very important to account for. Setting $R_1 = R$ and referring to Fig 8.9(b), we have

$$R_2 \cong R_1 - d \sin \theta \cos \phi = R - d \sin \theta \cos \phi \tag{8.54}$$

Assuming that each array element is a Hertz dipole antenna, the electric field signal E_1 received at antenna A_1 is given by

$$E_1 = \mathbf{u}_\theta \mathrm{j} \left(\frac{\eta k I L}{4\pi R} \right) \mathrm{e}^{-\mathrm{j}kR} \sin \theta \tag{8.55}$$

where η is the free space intrinsic impedance; L is the length of the BS antenna and I is the induced current flowing along each dipole. The electric field E_2 at the antenna element A_2 is given by

$$E_2 = \mathbf{u}_\theta \mathrm{j}\left(\frac{\eta k I h}{4\pi R}\right)\mathrm{e}^{-\mathrm{j}kR}\sin\theta * (\mathrm{e}^{\mathrm{j}kd\sin\theta\cos\phi}) \tag{8.56}$$

This field picked by the antenna A_2 will be multiplied by the weight w before it arrives at the summation unit. Therefore, the total electric field output that is presented to the low-noise amplifier (LNA), and demodulation and decoder set is given by

$$\mathbf{E}_\mathrm{T} = \mathbf{E}_1 + w\mathbf{E}_2 \tag{8.57}$$

Let $w = w_0\mathrm{e}^{\mathrm{j}\delta} = \mathrm{e}^{\mathrm{j}\delta}$ and $w_0 = 1$. This means that the weighting unit gives a phase shift to δ signal E_2, while leaving the magnitude of E_2 unchanged. Hence

$$\begin{aligned}\mathbf{E}_\mathrm{T} &= \mathbf{E}_1 + \mathbf{E}_2 * w = \mathbf{E}_1 + \mathrm{e}^{\mathrm{j}\delta}\mathbf{E}_2\\ &= \mathbf{u}_\theta \mathrm{j}\left(\frac{\eta k I h}{4\pi R}\right)\mathrm{e}^{-\mathrm{j}kR}\sin\theta\left[1 + \mathrm{e}^{\mathrm{j}(kd\sin\theta\cos\phi+\delta)}\right]\end{aligned} \tag{8.58}$$

Let $\psi = kd\sin\theta\cos\phi + \delta$ and note that

$$1 + \mathrm{e}^{\mathrm{j}\psi} = \mathrm{e}^{\mathrm{j}\psi/2}\left(\mathrm{e}^{-\mathrm{j}\psi/2} + \mathrm{e}^{+\mathrm{j}\psi/2}\right) = \mathrm{e}^{\mathrm{j}\psi/2}\left(2\cos\frac{\psi}{2}\right) \tag{8.59}$$

Therefore,

$$\mathbf{E}_\mathrm{T} = \mathbf{u}_\theta \mathrm{j}\left(\frac{\eta k I h}{4\pi R}\right)\mathrm{e}^{-\mathrm{j}kR}\cdot \mathrm{e}^{\mathrm{j}\psi/2}\sin\theta\left(2\cos\frac{\psi}{2}\right) \tag{8.60}$$

The array factor of the two-element array is given by

$$\mathrm{AF} = 2\cos\frac{\psi}{2} \tag{8.61}$$

which determines the radiation pattern of the MS antenna. Since the AF we obtained is a function of the phase angle δ of the weight, we may change or steer the antenna beam through controlling the array factor (AF) by changing δ.

Let us now consider the two cases mentioned above. These two cases are depicted in Fig. 8.10.

Case 1. We want to maximize the antenna beam toward the BS antenna which radiates the desired signal. See Fig 8.10(a) where the angle $\theta_{2\mathrm{D}}$ is defined. That is, we want to maximize the MS receiver antenna beam, or AF, in the direction $\theta = \theta_{2\mathrm{D}}$ and $\phi = \phi_{2\mathrm{D}}$. If $\theta_{2\mathrm{D}}$ and $\phi_{2\mathrm{D}}$ are not known, we first need to determine $\theta_{2\mathrm{D}}$ and $\phi_{2\mathrm{D}}$ (known as the angles of arrival, AOA) possibly using another computer algorithm to determine the position of the MS with respect to BS or the AOA. Among the well-known software modules to determine AOA are ESPIRIT and MUSIC.

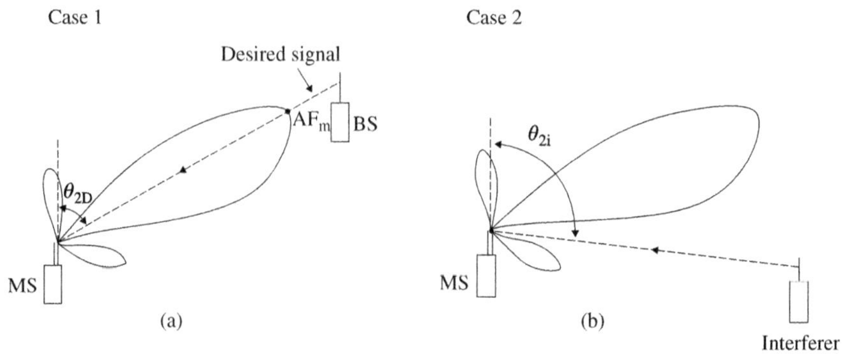

Figure 8.10: Two cases of adaptive beam steering. (a) Case 1. The MS receiver is steered toward the BS to maximize reception. (b) Case 2. The adaptive beam is formed to produce a null in the direction of an interference source.

Other new ways to determine AOA using electromagnetic theory are currently being explored, to reduce the computational time and memory burden.

The maximum value of the array factor

$$\mathrm{AF}_m = 2, \ \text{when} \cos\frac{\psi}{2} = 1, \ \text{or} \ \frac{\psi}{2} = 0 \tag{8.62}$$

When $\psi = 0$, we have

$$kd \sin\theta_{2\mathrm{D}} \cos\phi_{2\mathrm{D}} + \delta = 0 \quad \text{or} \quad \delta = -kd \sin\theta_{2\mathrm{D}} \cos\phi_{2\mathrm{D}} \tag{8.63}$$

which is the δ value we want to determine to set the weight $w = \mathrm{e}^{\mathrm{j}\delta}$ to get AF_m when $\theta_2 = \theta_{2\mathrm{D}}$. Thus the weight w must be set to

$$w = \mathrm{e}^{-\mathrm{j}kd \sin\theta_{2\mathrm{D}} \cos\phi_{2\mathrm{D}}} \tag{8.64}$$

in order to maximize the radiation AF in the desired direction.

Case 2. We want to null the AF in the direction of an interferer such as another mobile station MS_i. Assuming that the position (θ_{2i} and ϕ_{2i}) of the MS_i with respect to the MS is known (Fig. 8.10(b)), we seek $E_\mathrm{T} = 0$ when $\theta_2 = \theta_{2i}$ and $\phi_2 = \phi_{2i}$. Thus to get $\cos\psi/2 = 0$, or $\psi/2 = \pi/2$ or $\psi = \pi \cdot kd \sin\theta_{2i} \cos\phi_{2i} + \delta = \pi$.

Hence $\delta = \pi - kd \sin\theta_{2i} \cos\phi_{2i}$. This means we must set the weight to the following value:

$$w = \mathrm{e}^{-\mathrm{j}(\pi - kd \sin\theta_{2i} \cos\phi_{2i})} \tag{8.65}$$

the weight that must be set in order to get the array factor to become null in the direction of an interfering signal (noise). This is the basic idea behind digital beam steering of an adaptive array antenna or smart antenna. In order to get the maximum point of the transmitting/receiving antenna, the radiation pattern pointing toward the communication receiver/transmitter with which it wants to communicate, the computer software must automatically determine the weight magnitude and phase angle δ, and multiply the outgoing/incoming signals by the complex weight before summing all the outgoing/incoming array antenna signals to get the resultant signal.

8.2.6 A general two-element adaptive array antenna

Consider the general case of a two-element array antenna where the desired radiation pattern is shown in Fig. 8.11(a). To achieve this, the signals received by both elements are modified by weights w_1 and w_2 as shown in Fig. 8.11(b).

The electric fields received at antenna elements 1 and 2 shown in Fig. 8.11 are

$$E_1 = S_1 a_{11} + S_2 a_{21} \tag{8.66}$$

$$E_2 = S_1 a_{12} + S_2 a_{22} \tag{8.67}$$

respectively. The signals from sources S_{s1} and S_{s2} are modified by multiplicative noise factors a_{11} and a_{21} which include the effects of multipath interference, space attenuation, shadowing, fast fading over the space between the transmitters (S_{s1} and S_{s2}) and the receiver antenna element 1, as well as the Doppler phase shifts in the signal due to relative motion between the transmitter and receiver.

The total signal received (for a receiver array antenna) is given by $S = E_1 w_1 + E_2 w_2 = w_1(S_1 a_{11} + S_2 a_{21}) + w_2(S_1 a_{12} + S_2 a_{22})$, which on rearrangement gives

$$S = (w_1 a_{11} + w_2 a_{12})\, S_1 + (w_1 a_{21} + w_2 a_{22})\, S_2 \tag{8.68}$$

If the desired signal output $S = S_1$ the signal transmitted by source S_{s1}, then the multiplication terms associated with S_1 should be 1.0, and that associated with the interference signal S_2 from undesired source S_{s2} should be zero. Thus,

$$(w_1 a_{11} + w_2 a_{12}) = 1 \tag{8.69}$$

$$(w_1 a_{21} + w_2 a_{22}) = 0 \tag{8.70}$$

Solve (8.69) and (8.70) simultaneously to get w_1 and w_2,

$$w_1 = \frac{a_{22}}{a_{11}a_{22} - a_{12}a_{21}} \tag{8.71}$$

$$w_2 = \frac{a_{12}}{a_{12}a_{21} - a_{11}a_{22}} \tag{8.72}$$

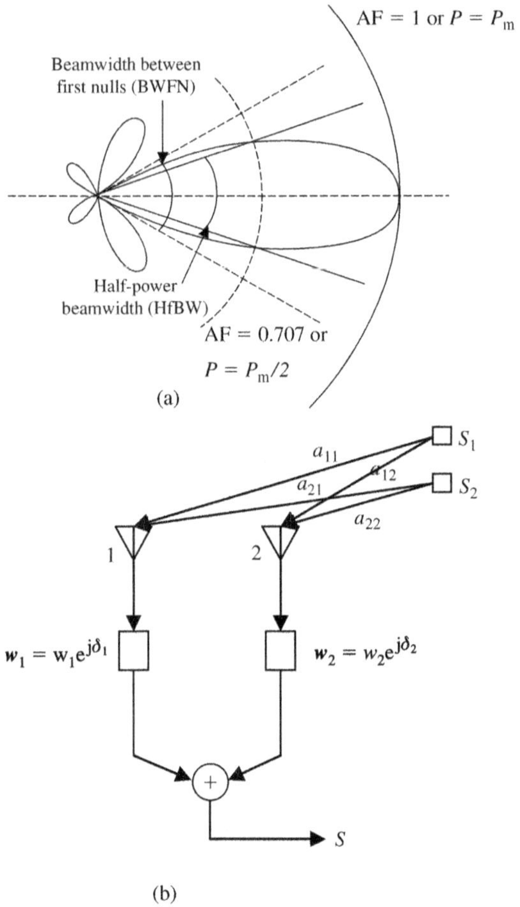

Figure 8.11: (a) Desired antenna radiation pattern. (b) A two-element array antenna designed to maximize reception and minimize interference signals.

w_1 and w_2 derived are linearly combined after being scaled using corresponding weights such that the antenna array pattern is optimized to have maximum possible gain in the direction of the desired signal and nulls in direction of interferers.

8.2.7 The antenna elements of array antennas

It was noted that the final radiation pattern of an array antenna is obtained by multiplying the array factor (AF) by the elemental pattern (EP). The array factor is a function of the number of elements used, the distance between the elements and the elemental current magnitude and its phase angle. Therefore, the AF is independent of the type of antenna element used to make up the array. The EP however will depend on the particualr type of antenna used as the element. It is

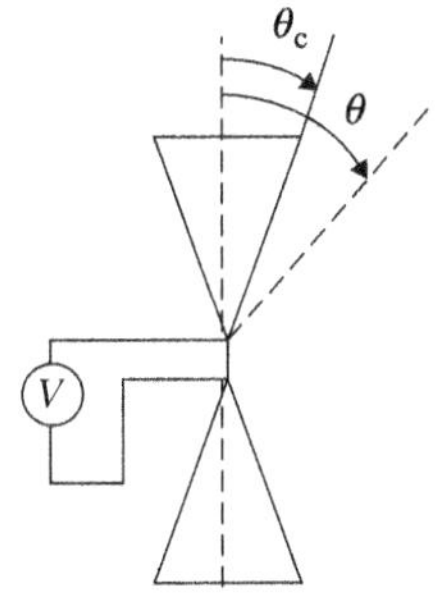

Figure 8.12: Biconical antenna.

possible to mix the antenna types used, such as an electric dipole element and a loop antenna to make up a two-element array antenna, where one element is an electric dipole antenna using two wires and the other element is *a* wire forming a loop, that is a magnetic dipole.

8.2.7.1 Half-wave electric dipole

The current along a half-wave dipole is

$$I(z) = I_0 \sin\left[k\left(\frac{\lambda}{4} - |z|\right)\right] \quad |z| < \frac{\lambda}{4} \tag{8.73}$$

which, on integration, yields an electric field given by

$$E_\theta = \frac{\mathrm{j}2\omega\mu I_0}{4k\pi r}\mathrm{e}^{-jkr}\sin\theta\left[\frac{\cos[(\frac{\pi}{2})\cos\theta]}{\sin^2\theta}\right] \tag{8.74}$$

8.2.7.2 Biconical antenna

The biconical antenna is shown in Fig. 8.12. For an infinitely long Biconical antenna, the radiated magnetic field is given by

$$H_\phi = H_0 \frac{\mathrm{e}^{jkr}}{4\pi r}\frac{1}{\sin\theta} \tag{8.75}$$

The radiation pattern is

$$\mathrm{RP}_{\mathrm{bc}} = \frac{\sin\theta_c}{\sin\theta}, \quad \theta_c < \theta(\pi - \theta_c) \tag{8.76}$$

To obtain the impedance of the antenna, we get the voltage induced on the antenna by the incident field E_θ as

$$V(r) = \int_{\theta_c}^{(\theta-\theta_c)} E_\theta r\, d\theta = \frac{\eta H_0}{2\pi}\mathrm{e}^{jkr}\ln\left(\cot\frac{\theta_c}{2}\right) \tag{8.77}$$

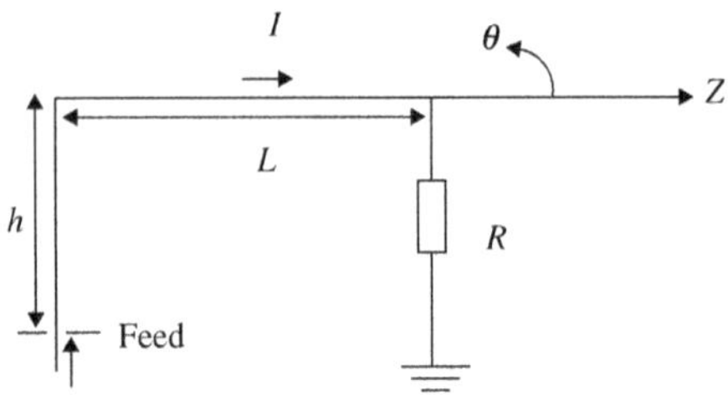

Figure 8.13: Transmission line traveling wave antenna.

The current induced in the antenna is

$$I(r) = \int_{0}^{2\phi} H_{\phi} r \sin\theta \, d\theta = \frac{H_0}{2} e^{-jkr} \tag{8.78}$$

Hence the impedance of the antenna is

$$Z_{bc} = \frac{V(r)}{I(r)} = \frac{\eta}{\pi} \ln\left(\cot \frac{\theta_c}{2} \right) = 120 \ln(\cot \frac{\theta_c}{2})\,\Omega \approx 120 \ln\left(\frac{2}{\theta} \right) \quad \text{for } \theta_c < 20^\circ \tag{8.79}$$

For $\theta_c = 1°, Z_{bc} = 568\,\Omega$, and for $\theta_c = 50°, Z_{bc} = 91\,\Omega$.

8.3 Traveling wave antennas

8.3.1 Basic principle of operation

Traveling wave antennas (TWA) are generally long conductors or helical loops of conductors formed in a zigzag fashion over which an electric current–voltage wave travels close to the velocity of light. As the waves travel along the long conductors, they also radiate signals out into free space, thus forming an antenna. These antennas are used in the frequency range of 1–10 MHz and the size of the antennas (L) is typically of the order L/λ =1–10.

8.3.2 Transmission line traveling wave antenna

The transmission line TWA shown in Fig. 8.13 has a length L much longer than the wavelength of the signal. The electric current along the transmission line TWA has a current of peak amplitude I_0 traveling along the wire in the z-direction.

$$I(z) = I_0 e^{-jkz} \tag{8.80}$$

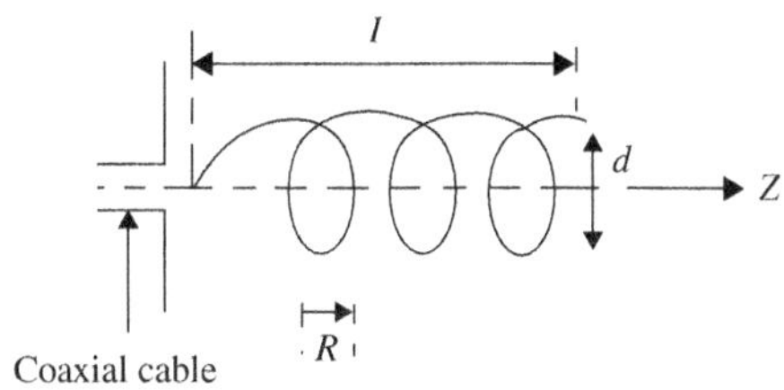

Figure 8.14: Helical antenna.

The radiation pattern is given by

$$RP_{\mathrm{TW}}(\theta) = C \sin\theta \frac{\sin\left[\left(\frac{kL}{2}\right)(1-\cos\theta)\right]}{\left(\frac{kL}{2}\right)(1-\cos\theta)} \tag{8.81}$$

The angle at which maximum radiation occurs is given by

$$\theta_{\mathrm{p}} = \cos^{-1}\left[1 - \frac{0.371}{\left(\frac{L}{\lambda}\right)}\right] \tag{8.82}$$

Typical examples of (L, θ_{p}) values are $(2\lambda, 35°)$, $(4\lambda, 25°)$, $(6\lambda, 20°)$ and $(10\lambda, 17°)$.

As the frequency of the signal is increased such that the length of the line, $L \gg$ wavelength λ, the maximum radiation occurs close to the axis along which the transmission line is placed.

8.3.3 Helical antenna

The helical antenna, as shown in Fig. 8.14, is like a long transmission line antenna with the conductor overall length l shortened by making it a coil. Each loop of the coil is like a loop antenna of radius d. To envision the radiation pattern of the helical antenna, we may picture it as an array of loop antenna elements placed along an axis to give rise to a linear, uniform array antenna carrying the same amplitude of currents, but the phase of each loop lagging behind the phase of the neighboring loop closer to the source end. The phase difference is proportional to the time taken for the signal to travel from one loop to another.

Consider first a helical antenna that mostly radiates in a direction normal to the helix axis, that is in the $\theta = \pi/2$ direction. The radiated electric field of a helical antenna is a sum of

$$\mathbf{E}_{\theta} = \frac{\mathrm{j}\omega\mu Il}{4\pi r} \mathrm{e}^{-\mathrm{j}kr} \sin\theta \, \mathbf{u}_{\theta} \tag{8.83}$$

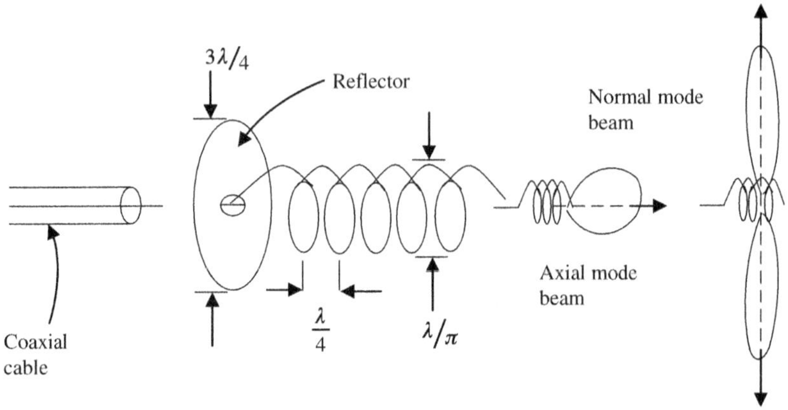

Figure 8.15: Helical (traveling wave) antenna radiating in axial mode (along the helical axis) and normal mode (perpendicular to the helical axis).

and

$$\mathbf{E}_{\phi} = \frac{\eta k^2 \pi d^2 I}{16\pi r} e^{-jkr} \sin\theta \, \mathbf{u}_{\phi} \tag{8.84}$$

where the spacing between adjacent turns is l (e.g. equal to $\lambda/4$, as shown in Fig. 8.15) and d (e.g. equal to λ/π) is the diameter of the helix. Now,

$$\frac{|E_\theta|}{|E_\phi|} = \frac{2/\lambda}{\pi^2 d^2} \tag{8.85}$$

The radiation resistance is $R_r = 640\left(\frac{h}{\lambda}\right)^2 \Omega$, where h is the height above a ground plane (e.g. $h = 0.2\lambda$). The normal mode helical antenna is used in handheld communication units, such as the rubber duct antenna for MS. In this mode, the helical antenna is able to transmit and receive circularly polarized electromagnetic waves (signals).

In the axial mode (see Fig. 8.15), the radiation pattern of the helical antenna is

$$RP_{\mathrm{am}}(\theta) = C\cos\theta \frac{\sin\left(\frac{N\psi}{2}\right)}{N\sin\left(\frac{\psi}{2}\right)} \tag{8.86}$$

where C is a constant depending on the number of turns and the angle:

$$\psi = k\ell\cos\theta + \delta \tag{8.87}$$

Note that in this mode, the helical antenna operates like an array antenna with a number of loop antennas arranged with an inter-element distance ℓ. The angle δ

which is the element to element phase shift is given by

$$\delta = -\left(k\ell + 2\pi + \frac{\pi}{N}\right) \tag{8.88}$$

Typical values for an axial mode helical antenna are: $\delta = 12°, N = 10, \pi d = \lambda$ and $\ell = 0.2\lambda$. The velocity (phase velocity) with which the wave travels along the helix is given by

$$V_{\mathrm{h}} = \frac{c\sqrt{(\pi d)^2 + \ell^2}}{\lambda\left[\frac{\ell}{\lambda} + \frac{2N+1}{2N}\right]} \tag{8.89}$$

and is typically about $0.8c$, where c is the velocity of light. The half-power beamwidth of the helical antenna operating in the axial mode is given by

$$\mathrm{HPBW} = \frac{65°}{\left[\left(\frac{\pi d}{\lambda}\right)\sqrt{N\ell/\lambda}\right]} \tag{8.90}$$

and the gain is

$$G_{\mathrm{h}} = \frac{26000}{\mathrm{HPBW}} = 6.2\left(\frac{\pi d}{\lambda}\right)^2 N\left(\frac{\ell}{\lambda}\right) \tag{8.91}$$

The input impedance

$$R_{\mathrm{h}} = 140\left(\frac{\pi d}{\lambda}\right) \ \Omega \tag{8.92}$$

A typical value for the number of helical turns is $N = 10$, and such a helical antenna operating in the axial mode has half-power beamwidth $\theta_{\mathrm{HPBW}} = 50°$ and power gain $G = 10$. There are other traveling wave antennas such as that shown in Fig. 8.16.

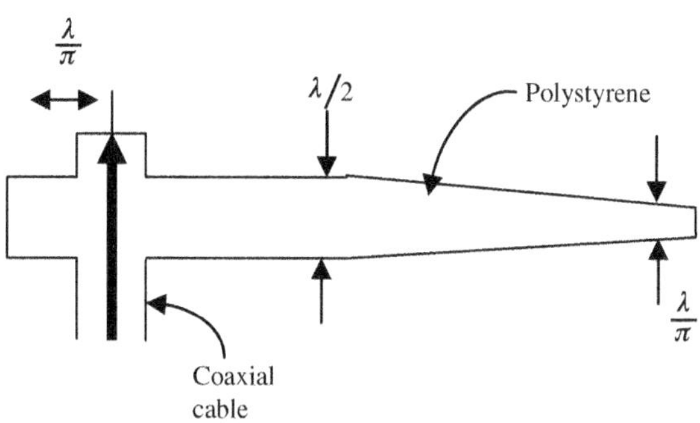

Figure 8.16: Dielectric rod travelling wave antenna made of a dipole radiator inside a coaxial cable terminator, and an insulator radiator that narrows down the radiation pattern.

In Fig. 8.16, the wave launched out from the coaxial cable termination will travel along the high permittivity polystyrene rod, eventually traveling out into space at the λ/π diameter end. The transmitted beam pattern will be directed along the axis of the polystyrene rod. It is similar to the axial mode transmission of a helical antenna.

8.4 Aperture antennas

8.4.1 Basic principles of operation

Aperture antennas are used at very high frequencies and are formed by having apetures or windows through which the electromagnetic signals may be radiated. A widely used type of aperture antenna is the rectangular horn antenna, where electromagnetic signals traveling through a waveguide are radiated out by opening out the waveguide into a horn shape. These may be operated in the reverse direction to receive signals in a highly directive manner. The large parabolic reflector used in satellite communication antennas functions like apertures. Normally a wire antenna (e.g. half-wave dipole antenna) or a horn antenna is placed in front of the parabolic dish (a) to collect the signals focused by the parabolic aperture (in receiving mode) or (b) to project the signals from an oscillator on to the parabolic dish (in transmitting mode). When using a parabolic reflector arrangement, for instance, a horn antenna is placed at the focal point of a parabolic reflector. A low-noise block converter (LNBC) provides low-noise amplification and frequency down conversion (e.g. from 8 GHz down-converted to 800 MHz). The down-converted signal may be carried over a coaxial cable without high ohmic losses. Apertures tend to be highly directive and enable better focusing. In radio astronomy, an array of parabolic reflector type antennas is used to observe signals radiated from a small region of space in the galaxy.

8.4.2 Rectangular aperture

In Fig. 8.17 are shown two rectangular shaped apertures through which the aperture antenna radiates electromagnetic fields. An aperture antenna may be made up of a rectangular waveguide left open at one end, or to be more effective, it is opened up to form a horn shape. The horn end is also a rectangular aperture although its cross sectional area is larger than the aperture area of the waveguide. In Fig. 8.17 is shown the uniform, rectangular aperture where the electric fields are uniformly distributed and the aperture is most effectively used for radiation. In this case, the aperture is considered 100% efficient. However, in real cases, the waveguides which deliver the electromagnetic waves at the aperture for radiation carry nonuniformly distributed electric fields as shown in Fig. 8.17(b).

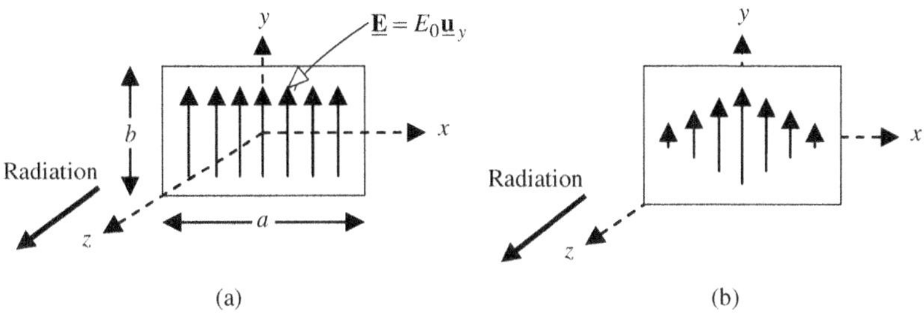

Figure 8.17: Two rectangular apertures having (a) a uniform electric field distribution at the aperture and (b) a nonuniform, cosine electric field distribution at the aperture.

Consider the rectangular aperture of size $a \times b$ with a uniform electric field $\mathbf{E} = \mathbf{E}_0\mathbf{u}_y$ as shown in Fig. 8.17(a). The radiated electric fields in spherical coordinates are given by

$$E_\theta = \frac{jkab\,e^{-jkr}}{2\pi r} E_0 \sin\phi \frac{\sin\left(\frac{\psi_x}{2}\right)}{\left(\frac{\psi_x}{2}\right)} \cdot \frac{\sin\left(\frac{\psi_y}{2}\right)}{\left(\frac{\psi_y}{2}\right)} \tag{8.93}$$

and

$$E_\phi = \frac{jkab\,e^{-jkr}}{2\pi r} E_0 \cos\theta\cos\phi \frac{\sin\left(\frac{\psi_x}{2}\right)}{\left(\frac{\psi_x}{2}\right)} \cdot \frac{\sin\left(\frac{\psi_y}{2}\right)}{\left(\frac{\psi_y}{2}\right)} \tag{8.94}$$

where

$$\psi_x = ka\sin\theta\cos\phi \tag{8.95}$$

$$\psi_y = ka\sin\theta\sin\phi \tag{8.96}$$

On the E-plane (the yz-plane), $\phi = \frac{\pi}{2}$ $\left(\text{and } \frac{3\pi}{2}\right)$, thus

$$\mathbf{E}_\theta = \frac{jkab\,e^{-jkr}}{2\pi r} \frac{\sin\left(\frac{kb\sin\theta}{2}\right)}{\left(\frac{kb\sin\theta}{2}\right)} \mathbf{u}_\theta \tag{8.97}$$

$$\mathbf{E}_\phi = 0 \tag{8.98}$$

On the H-plane (the xz-plane) with $\phi = 0$ (and π), we have

$$\mathbf{E}_\theta = 0 \tag{8.99}$$

$$\mathbf{E}_\phi = \frac{\mathrm{j}kab\,\mathrm{e}^{-\mathrm{j}kr}}{2\pi r} E_0 \cos\theta \frac{\sin\left(\frac{ka\sin\theta}{2}\right)}{\left(\frac{ka\sin\theta}{2}\right)} \tag{8.100}$$

The radiation patterns on these two planes are determined by the values of the wavelength λ $\left(k = \frac{2\pi}{\lambda}\right)$ and the slot size $a \times b$. For a rectangular aperture of large size $(a, b \gg \lambda)$, the $\cos\theta$ term in the equation for $\mathbf{E}_\phi$ becomes $\cos\theta \approx 1$. Hence the radiation patterns on both the E-plane and the H-plane are

$$\mathrm{RP}_E(\theta) = \frac{\sin\left(\frac{kb\sin\theta}{2}\right)}{\left(\frac{kb\sin\theta}{2}\right)} \tag{8.101}$$

and

$$\mathrm{RP}_H(\theta) = \frac{\sin\left(\frac{ka\sin\theta}{2}\right)}{\left(\frac{ka\sin\theta}{2}\right)} \tag{8.102}$$

with very narrow beams. The beam solid angle $\Omega = \frac{\lambda^2}{ab}$ and the directivity $D = 4\pi\frac{ab}{\lambda^2}$ where the maximum effective aperture of the rectangular aperture is (ab).

If the electric field at the aperture is tapered as with a TE_{10} mode from a waveguide, the electric field is, as shown in Fig. 8.17(b),

$$\mathbf{E} = E_0 \cos\left(\frac{\pi x}{2}\right)\mathbf{u}_y \tag{8.103}$$

which is maximum when $x = 0$ (at the center) and zero when $x = \frac{-a}{2}$ or $\frac{+a}{2}$.

The radiation field pattern on the E-plane (yz-plane), with, $\phi = \frac{\pi}{2}$ $\left(\text{or } \frac{3\pi}{2}\right)$ remains as it was for uniform field distribution of $\mathbf{E} = E_0\mathbf{u}_y$. However, the H-plane (zx-plane) pattern becomes

$$\mathrm{RP}_\mathrm{H}(\theta) = \cos\theta \frac{\cos\left(\frac{ka\sin\theta}{2}\right)}{\left[1 - \left(\frac{ka\sin\theta}{\pi}\right)\right]} \tag{8.104}$$

8.4.3 Rectangular horn antenna

In Fig. 8.18 is shown a rectangular waveguide connected to a rectangualr horn antenna. Late last century, these stiff rectangular waveguides were used to bring signals at the UHF and microwave frequencies and above from the communication rooms to outdoor horn antennas. But now with flexible coaxial cables able to perform well up to EHF frequencies, flexible cables are used to connect the communication equipment inside the communication room to the outdoor transmitting and receivng aperture antennas. However, a small segment of rectangular waveguide is used to couple the signals from the coaxial cable to the rectangular aperture antenna.

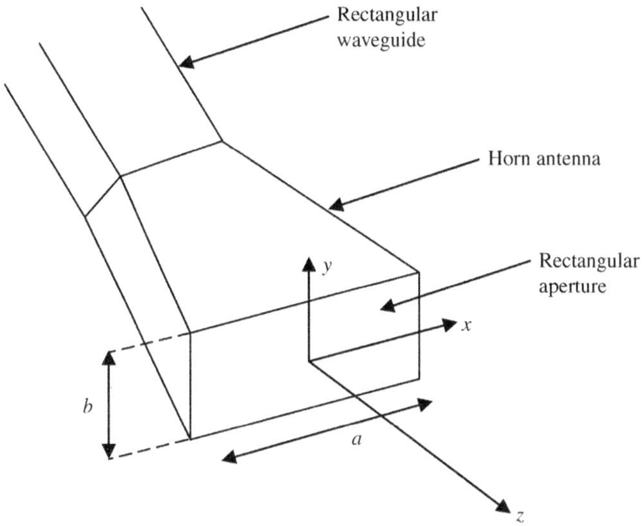

Figure 8.18: A horn antenna connected to a rectangular waveguide.

The horn antenna shown in Fig. 8.18 will radiate according to a three-dimensional radiation pattern, with the beamwidths on the *xy*- and *yz*-planes determined by the aperture height b and width a.

Consider the case when a rectangular waveguide TE_{10} mode electromagnetic fields are delivered to the horn antenna input for the pyramidal horn antenna shown in Fig. 8.18. The radiation pattern on the E (i.e. zx)-plane with $\varphi = 0$ will be

$$E_\varphi(\theta) = \pi^2 \cos\theta \frac{\cos\left(\frac{\pi a}{\lambda}\sin\theta\right)}{\pi^2 - 4\left(\frac{\pi a}{\lambda}\sin\theta\right)} \tag{8.105}$$

and the H (i.e. the yz)-plane pattern will be

$$E_\theta(\theta) = \frac{\sin\left(\frac{\pi b}{\lambda}\sin\theta\right)}{\frac{\pi b}{\lambda}\sin\theta} \tag{8.106}$$

In general, the functions which will define the three-dimensional radiation pattern are sketched in Fig. 8.19(a) and (b), and are

$$\frac{\cos\pi X}{\pi^2 - 4(\pi X)^2} \text{ and } \frac{\sin\pi Y}{\pi Y}$$

where

$$X = \frac{a}{\lambda}\sin\theta \cdot \cos\varphi \tag{8.107}$$

$$Y = \frac{b}{\lambda}\sin\theta \cdot \sin\varphi \tag{8.108}$$

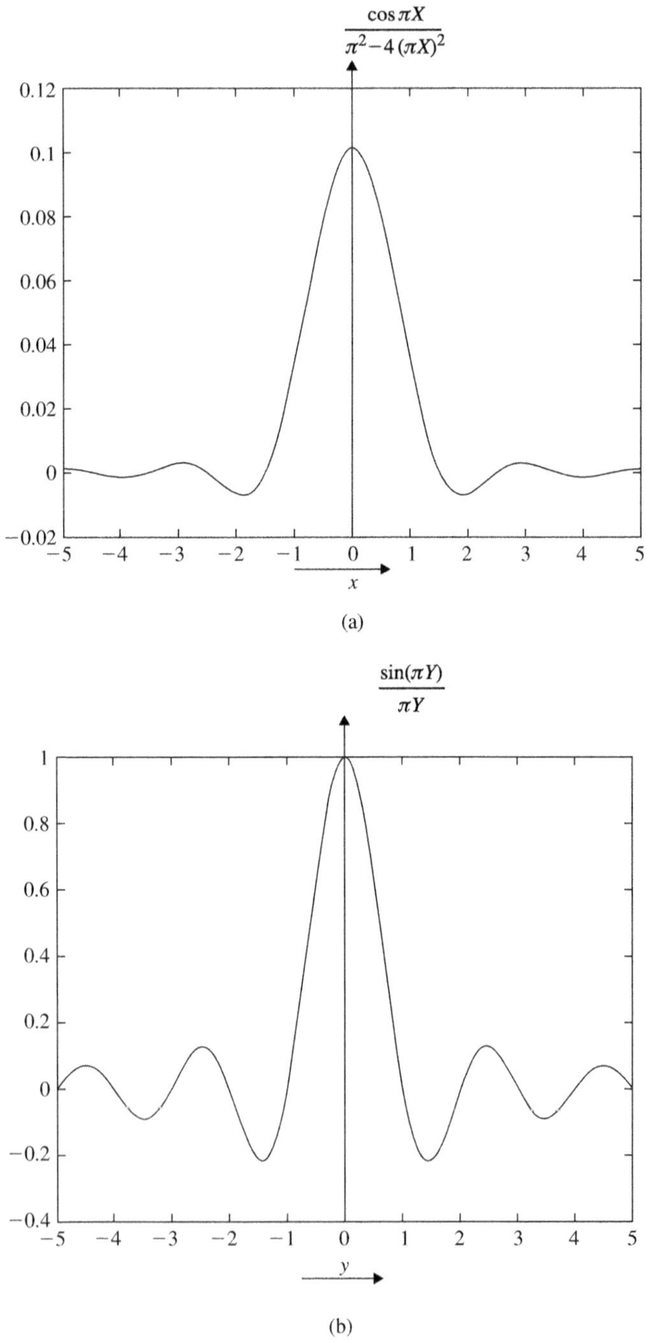

Figure 8.19: The beam pattern of the horn antenna on the plane (a) *zx*- and (b) *yz*-plane.

On the $z > 0$ half-space, as the observation point $P(\theta, \phi)$ varies, the ranges of X and Y are

$$\frac{-a}{\lambda} \leq X \leq \frac{a}{\lambda} \qquad \frac{-b}{\lambda} \leq Y \leq \frac{b}{\lambda}$$

In order to obtain narrow beams we need to make a/λ and b/λ large. On the zx-plane with $\sin(\pi Y)/(\pi Y)$, controlling the radiation pattern, the first null θ_{01} that occurs between the main beam and the firstside lobe is given by

$$\sin\theta_{01} = \frac{\lambda}{b} \tag{8.109}$$

which indicates that the larger the value of b/λ, the smaller that θ_{01} wonld be; that is, the narrower the main beam. Similarly on the yz-plane with the beam $[\cos(\pi X)]/\pi^2 - 4(\pi X)^2$ controlling the beam shape, the first null adjacent to the main beam occurs at

$$\sin\theta_{10} = \frac{3\lambda}{2a} \tag{8.110}$$

which indicates that a narrower beam on the yz-plane is obtained with a larger a/λ. In order to accomplish this with the TE_{10} mode and to prevent higher modes being generated, the horn antenna is placed from its entry point at an angle of about 20°, with a length L large enough to have the TE_{10} mode smoothly radiated out at its $a \times b$ exit. Typical values of a and b are $a < \lambda$ and $b < 0.5\lambda$; a and b should be large enough to produce the desired narrow band.

8.4.4 Circular aperture

In addition to rectangular aperture antennas, we also have circular aperture antennas, as found in the case of a parabolic reflector antenna or a circular horn antenna. The aperture through which radiation takes place is shown in Fig. 8.20(a). The aperture is on the xy-plane and it radiates out on the yz- and the xy-planes.

Consider a circular aperture of the radius a as shown in Fig. 8.20(a) with a uniform electric field $\mathbf{E} = E_0\mathbf{u}_x$ at the aperture ($z = 0$). The radiated electric field observed at point $(r, \theta, \emptyset)$ is given by

$$\mathbf{E}_\theta = \frac{jka^2}{2\pi} E_0 \cos\phi \cdot \mathrm{RP}_p(\theta)\mathrm{u}_\theta \tag{8.111}$$

and

$$\mathbf{E}_\emptyset = \frac{jka^2}{2\pi} E_0 \cos\theta \sin\phi \cdot \mathrm{RP}_\mathrm{p}(\theta)\mathrm{u}_\phi \tag{8.112}$$

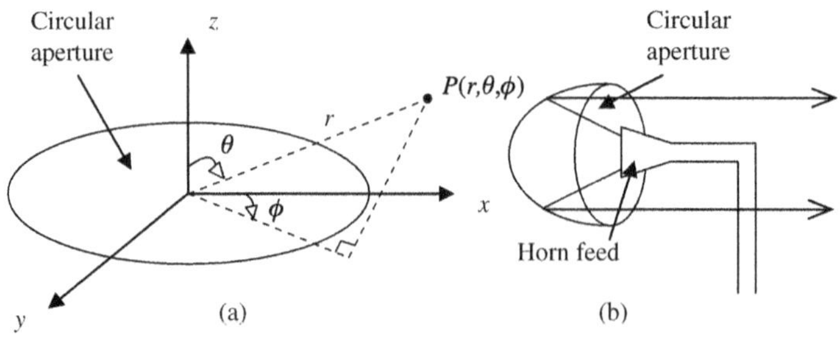

Figure 8.20: Circular apperture antenna. (a) Radiation takes place through the circular aperture. (b) The cassegrain spheroidal antenna with a circular aperture.

where

$$\mathrm{RP_p}\,(\theta) = \frac{J_1(ka\sin\theta)}{(ka\sin\theta)} \tag{8.113}$$

J_1 being a Bessel function of the first kind and first order. A popular circular aperture antenna is the spherical reflector antenna as shown in Fig. 8.20(b), which is widely used in microwave, satellite communications. A primary radiating source is a horn antenna as shown in Fig. 8.20(b) which radiates electromagnetic signals on to the parabolic reflector. The reflector in turn reradiates the signal through the circular aperture, producing a much narrower beam than possible with a horn antenna.

Considering the circular aperture as containing an array of electric field lines radiating out of the aperture, the radiaition pattern is given by the array factor $\mathrm{AF}_N = \frac{\sin\frac{N\psi}{2}}{N\psi} = \left|\frac{\sin\left(\frac{\pi L\sin\theta}{\lambda}\right)}{\left(\frac{\pi L\sin\theta}{\lambda}\right)}\right|$ and the beamwidths in spherical coordinates will be $\phi_\mathrm{B} = \frac{1.1284}{N_\mathrm{h}d_\mathrm{h}}$ and $\theta_\mathrm{B} = \frac{1.1284}{N_\mathrm{v}d_\mathrm{v}}$. The 3 dB bandwidth is $\phi_{3\,\mathrm{dB}} = 0.886(\frac{\lambda}{\mathcal{A}})$. The effective aperture area is $A_\mathrm{e} = \tau_\mathrm{e}\frac{\pi}{4}d^2 = \frac{\pi}{4}\left(\sqrt{\tau_\mathrm{e}}d\right)^2$ where we have defined by the effective diameter as $d_\mathrm{e} = \sqrt{\tau_\mathrm{e}}d$. The 3 dB bandwidth is therefore given by $\phi_{3\,\mathrm{dB}} = 0.886\frac{\lambda}{d_\mathrm{e}} = \frac{0.886}{\sqrt{\tau_\mathrm{e}}d}$ rad.

8.4.5 Lens aperture antenna

In Fig. 8.21 is shown a circular aperture antenna obtained by placing a dielectric lens in front of a dipole antenna. The lens defracts the radiated electromagnetic rays as shown and produces a narrow plane wave beam.

The reflector placed behind the half-wave dipole antenna prevents radiation in the reverse direction and focuses all the beams from the dipole antenna on to the dielectric lens.

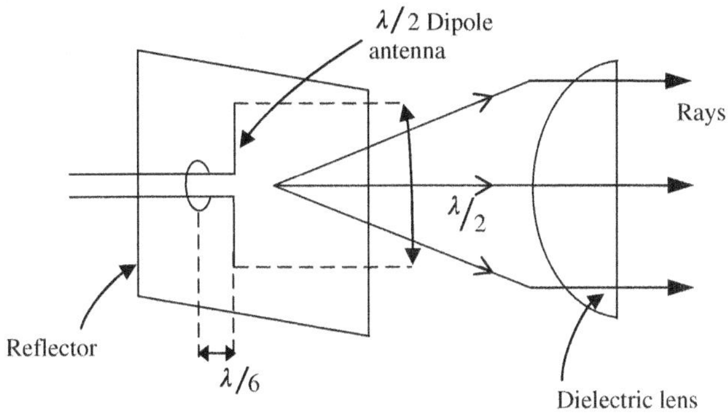

Figure 8.21: Dielectric lens aperture antenna.

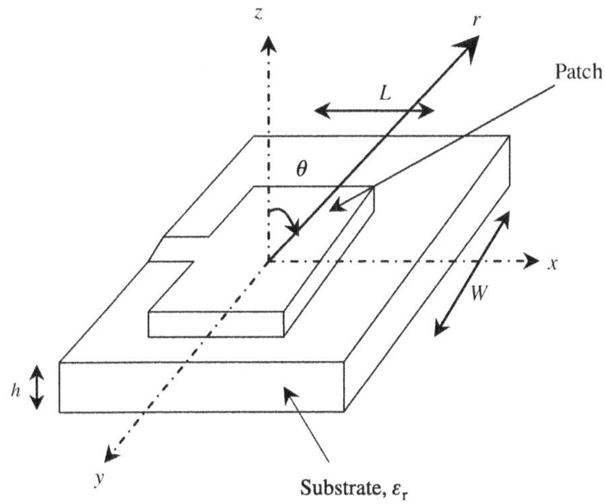

Figure 8.22: The patch antenna.

8.4.6 Patch antenna

Consider the patch antenna shown in Fig. 8.22. The rectangular patch, of size $L \times W$, is formed on top of the substrate shown, and the gap between the patch and the ground plane electrode under the substrate is h. The patch is connected to the source through a thin copper conductor as shown and may be fabricated together with electronic circuits. The radiation of electromagnetic fields takes place through the aperture formed by the gap between the patch and the ground plane.

For a half-wave patch shown in Fig. 8.22, $L = 0.49\lambda_{\text{d}} = 0.49\frac{\lambda}{\varepsilon_r}$, where the wavelength λ_{d} of the signal coming in from the source by travelling between the incoming copper conductor and the ground plane (by way of reflection as between two parallel plates of a waveguide) is less than the free space wavelength λ ($\lambda_{\text{d}} = \lambda/\varepsilon_{\text{r}}$, where ε_{r}

is the relative permittivity of the substrate). With the maximum radiation occurring on zx-plane for the antenna placed on the xy-plane, the broad radiation pattern (i.e. relatively large beamwidth) is given by

$$E_\theta = E_0 \cos\phi \mathrm{PF}(\theta,\phi) \tag{8.114}$$

and

$$E_\phi = -E_0 \cos\theta \sin\phi \, \mathrm{PF}(\theta,\phi) \tag{8.115}$$

where the patch factor $\mathrm{PF}(\theta,\phi)$ is given by

$$\mathrm{PF}(\theta,\phi) = \frac{\sin\left[\frac{kW}{2}\sin\theta\sin\phi\right]}{\frac{kW}{2}\sin\theta\sin\phi}\cos\left[\frac{kW}{2}\sin\theta\cos\phi\right] \tag{8.116}$$

from which we may obtain both the E-plane (i.e. $\phi = 0°$) and H-plane (i.e. $\theta = 90°$) patterns. The impedance (resistive impedance) of the antenna is given by

$$Z_\mathrm{n} = 90\frac{\varepsilon_\mathrm{r}^2}{\varepsilon_\mathrm{r}-1}\left(\frac{L}{W}\right)^2 \Omega \tag{8.117}$$

which has typical values ranging from 100 to 400 Ω. The bandwidth of the antenna (in Hz) is

$$\mathrm{BW} = \begin{cases} 3.77\dfrac{\varepsilon_\mathrm{r}-1}{\varepsilon_\mathrm{r}^2}\dfrac{W}{L}\dfrac{h}{\lambda}, & \text{for } h \ll \lambda \\ 3.77\dfrac{\varepsilon_\mathrm{r}-1}{\varepsilon_\mathrm{r}^2}\dfrac{h}{\lambda}, & \text{for } W \ll L \end{cases} \tag{8.118}$$

For $f = 4\,\mathrm{GHz}$, $\varepsilon_\mathrm{r} = 25$ and $h = 1.2\,\mathrm{mm}$, we have a patch size $\left(L = 0.5\lambda/\sqrt{\varepsilon_\mathrm{r}} = W\right)$ of the order of 3 × 3 cm.

In Fig. 8.23 is shown patch antenna elements forming an array antenna. Array antennas consisting of several hundreds of patch elements which are easily fabricated on a substrate. They are light and offer little wind resistance when placed on moving vehicles (e.g. on the outer structure of an aircraft, above the pilot's cabin, for aircraft to satellite communication).

8.5 EMI and EMC

8.5.1 Defining EMI and EMC

Electrical and electronic circuits, devices and systems are submerged in a space filled with electromagnetic fields. These electromagnetic fields may be produced

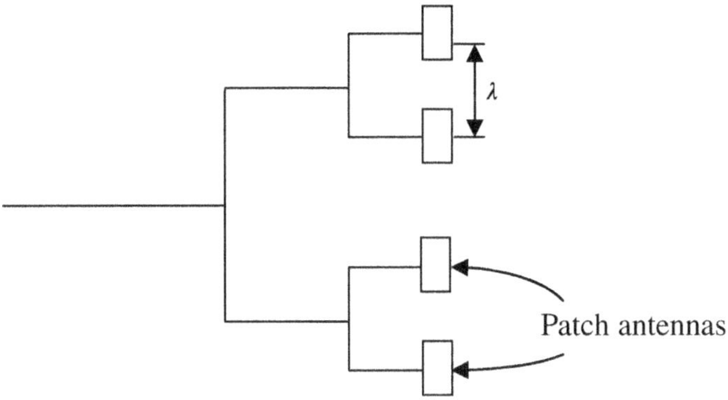

Figure 8.23: Patch antennas forming an array antenna.

by other devices and systems or could be self-produced. An example of the former is a digital circuit board which is part of an electronic control system which is surrounded by electromagnetic fields produced by an aircraft to satellite communication system. An example of the latter case is the self-produced fields of the control circuit board amplifier currents producing electromagnetic fields which may be coupled to a neighboring device such as a filter circuit. Now in the above two cases, the electromagnetic fields of the communication system are termed noise from the perspective of the control system. Similarly, the amplifier produced electromagnetic fields in the space are noise to the filter circuit. Such noise may affect and alter the performance of the control system in the first case and degrade the performance in the second case. When noise alters or degrades the performance of a device or system, it is called interference.

EMI is a major source of concern in many electrical and electronic circuits and systems. EMI refers to the undesirable effects of noise on electrical and electronic components (e.g. resistor, capacitor and inductor), devices and systems. Consider the aircraft control system for instance. In fighter military aircraft, the center of gravity of the aircraft is moved to its tail end, instead of being designed to be at the center, in order to allow for rapid upward and downward movements during a fight. Hence, the control system that holds the aircraft horizontally poised is a mission-critical, real-time system. EMI produced by a nearby lightning flash can cause that small voltage pulse in the control circuit which could send the aircraft spinning out of control.

Therefore, it is important that electronic systems be designed to be insusceptible to electromagnetic noise (EMN) as well as not causing EMI problems to other devices or systems. Electromagnetic susceptibility (EMS) is defined as the undesirable response of devices and systems to EMI. The device or system is said to have achieved EMC (electromagnetic compatibility) when it is neither a source of EMI nor susceptible to or disturbed by EMI due to other sources of EMI. Thus

EMC defines the ability of the devices and systems to function as desired inspite of EMI. Electrostatic discharges (ESD) which may be caused by EMI, as well as generate EMI, are the process by which electric charges are discharged on to electronic circuits and MEMS. In most cases, ESD leads to electric breakdown of insulators.

The intentional sources of EMI are the communication and control devices that radiate electromagnetic fields as part of the operational requirements. The transmitting communication antenna, for instance, is designed to maximize radiation of the electromagnetic fields which will carry text, voice or video information. Examples of intentional EMI sources are:

I. Analog and digital radio and television broadcast stations.

II. Wireless mobile communication systems, such as cellular telephone MS and BS.

III. Wireless networking devices.

IV. Jammers such as high-power density antenna radiators pointed toward enemy, or business rival transmitting antennas to cause signal interference and possible damage to circuits.

V. Unlicensed and unauthorized private communication stations and equipment.

Unintentional sources of EMI are devices and systems, such as amplifier circuits an electric motors the universal motors, which are designed to keep radiation power to a minimum. These include:

I. Wired communication and computer networks.

II. Variable speed, high-energy control systems.

III. Overhead power lines.

IV. Lighting systems.

V. Cell phones, radar and television or radio transmitters.

VI. Medical equipment such as electrosurgical units' defibrillators.

VII. Cathode ray tube (CRT) and plasma displays, electronic devices such as computers, instruments and television.

VIII. Aging and malfunctioning electric services.

IX. ESD, including from humans

X. Fire detection and alarm systems.

An isotropic radiator which transmits out signals at P_T W will radiate a power intensity of $P_T/4\pi r^2$ W/m^2 at a distance r from the transmitter. The electric field it radiates is roughly given by $E = (5.5/r)\ (P_T)^{1/2}$ V/m. Similarly, a half-wave dipole antenna radiates a power intensity of $P_T/4\pi r^2$ and an electric field of $(7.01/r)\,(P_T)^{1/2}$ V/m. A cell phone or a handheld transmitter with transmitted power of about 0.5–10 W radiates an electric field of about $(A/r)\ (P_T)^{1/2}$ V/m, where A varies from 0.45–3.35 and has a typical value of 1.6. Digital equipment for instance used in a domestic environment should be tested to ensure that the radiated electric field at distance $r = 3$ m is less than 40 dBμV/m in the frequency range of (1–8.8 MHz), 43.5 dBμV/m (8.8–216 MHz), 46 dBμV/m (216–960 MHz) and 54 dBμV/m (above 960 MHz). For commerical, business or industrial equipment, the equivalent values at different frequencies are 48.5, 54, 56.6 and 60 dBμV/m.

With high-frequency transmitters, it should be remembered that these radiation fields behave differently from near fields. In the far field, the electric field to magnetic field ratio E/H is equal to the free space impedance 120π (or 377) Ω. For an electric dipole antenna, $E/H < 120\pi$ for high current and low voltage, and $E/H > 120\pi$ for low voltage and high current. Moreover, since E is proportional to $1/r^3$ and H is proportional to $1/r^2$ in the near field, the wave impedance $Z = E/H$ drops with increasing distance. However, in a magnetic dipole such as a loop antenna, in the near field, E is proportional to $1/r^2$ and the H field is proportional to $1/r^3$ in the near field, so the wave impedance Z increases with r.

EMI also affects signals and circuits. A sinusoidal signal may be distorted by EMI; such harmonic distortion occurs when power electronic circuits, for instance, may change the sinusoidal wave pattern. Moreover, a digital data stream with many bits of 1s and 0s may be corruptuted by EMI signals being superimposed on to the digital data stream.

8.5.2 Signals and EMI

In Fig. 8.24(a) is shown the typical waveshape of curents which flow along a lightning channel and through any object that becomes attached to the lightning channel, such as an aircraft or a windmill blade (e.g. an 80 m long blade). The lightning current wavefront has a convex shape as shown. In this, the wavefront part of the current has a very large rise rate, for example 90 kA/μs, since the changes occur at microsecond timescales. The current reaches its peak value of around 20 kA in around 1 μs. As the lightning current discharges, the electrically charged lightning channel on the aircraft body it is attached to, will experience a voltage collapse such as that shown in Fig. 8.24(b). Moreover, the lightning current radiates electric and magnetic fields, which in turn may interact with sensitive navigation, communication and aircraft control circuits and equipment, resulting in EMI problems which need to be resolved solved for aircraft safety and to ensure longer lifetime of equipment.

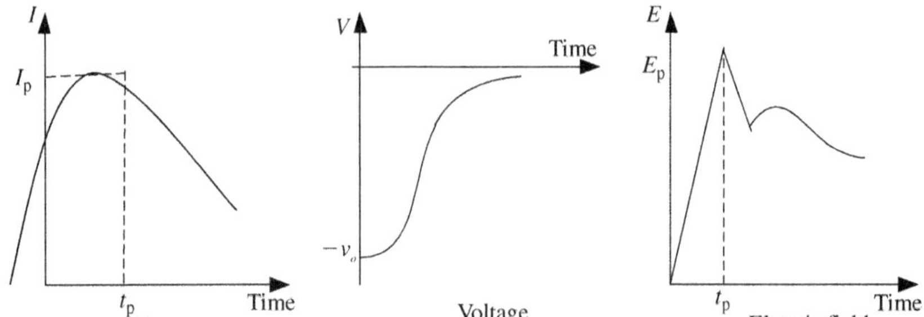

Figure 8.24: Natural lightning generated (a) currents, (b) voltages and (c) radiated electric field.

Digital electronics and communication systems are major sources of EMI since digital pulses, where each digital pulse is made up of a large number of cosine waves at different frequencies, occupy a broad spectrum.

Consider a digital pulse with a rise time t_r and a fall time $t_f = t_0 - t_m$ as shown in Fig. 8.25(a). Moreover, let the rise time (t_r) and fall time (t_f) of the pulse be equal: $t_r = t_f = t_0 = t_m$ and the pulse be repeated every T_0 s. Then the pulse may be represented by a linear superposition of its Fourier components given by

$$V(t) = a_0 = \sum_{n=1}^{\infty} (a_n \cos(n\omega_0 t)) \tag{8.119}$$

where the Fourier coefficients are

$$a_0 = \frac{V_0}{T_0}(t_m + t_0) \tag{8.120}$$

$$u_n = \frac{2V_0}{\pi n^2 \omega_0 (t_0 - t_m)} [\cos(n\omega_0 t_m) - \cos(n\omega_0 t_m)] \tag{8.121}$$

and $\omega_0 = \frac{2\pi}{T_0}$ or $f_0 = \frac{1}{T_0}$.

Typical digital pulses with $f_0 = 1$ MHz and $t_r = 1$ ns have significant frequency components up to about the 20th harmonic ($n = 20$). For higher clock frequencies (e.g. 4 GHz), the significant frequency components are much higher. This means that digital systems are sources of EMI over a large frequency spectrum.

Moreover, digital systems are increasingly more prone to being affected by EMI since more and more they being made into low-voltage (e.g. 1.5 V and less), low-power devices. This means whereas a typical bit "1" voltage is presently 6 V

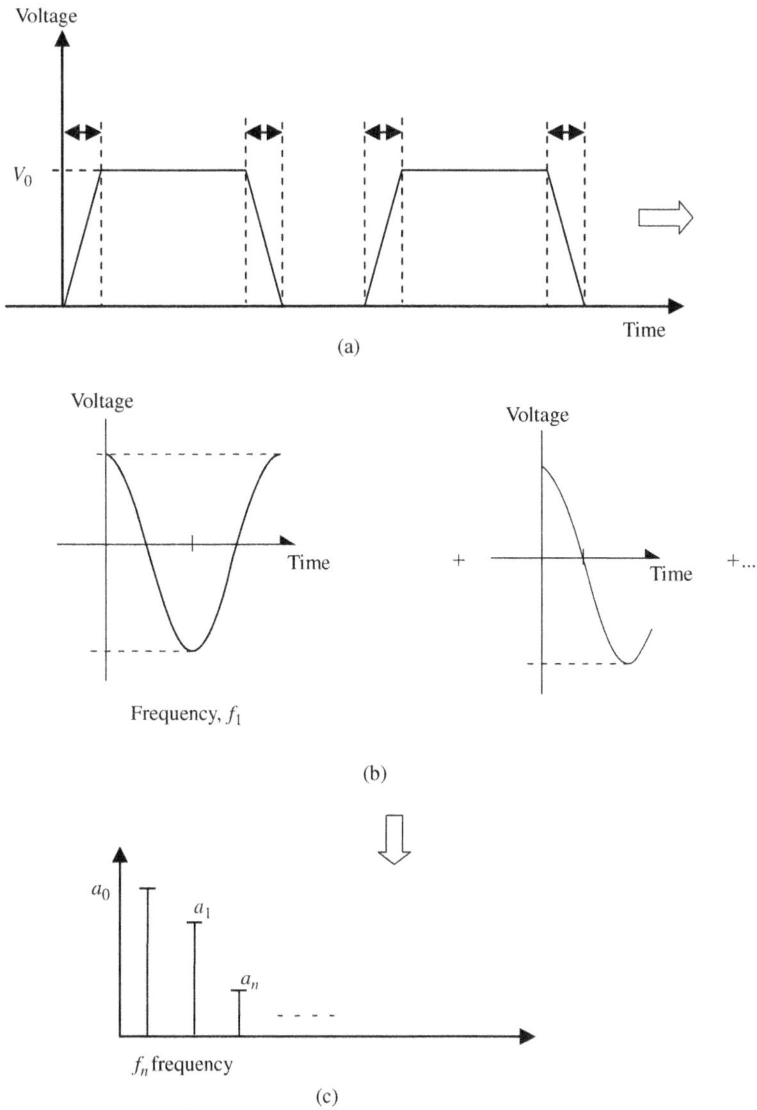

Figure 8.25: (a) Digital pulses in time domain. (b) The several sinusoidal signals that make up a single digital pulse. (c) The frequency spectrum of a single digital pulse.

(i.e. $V_0 = 6\,\text{V}$), future systems are expected to operate at 1.5 V (i.e $V_0 = 1.5\,\text{V}$). Hence, say an EMI which may cause a noise voltage of 0.5 V may not appear significant in a 6 V digital system, but it will be a significant interferer at $V_0 = 1.5\,\text{V}$.

The spectral bandwidth of a digital system is roughly given by $\text{BW} = \frac{1}{2t_r}$. This means that for a rise time of 10 ns, BW = 50 MHz; and for $t_r = 1$ ns,

BW = 500 MHz. Hence EMI-caused signal degradation occurs over a wide spectrum of frequencies as the digital pulse is driven closer to the ideal zero rise time ($t_r = 0$).

8.5.3 EMI control

There are several general ways of controlling EMI, although total eradication of EMI may not be possible:

1. **Mitigation.** First, study the EMS or sensitivity of the device or system. A spectrum analyzer and a broadband antenna for instance may be used to assess the frequency spectrum of the device. Second, study the electromagnetic environment, for instance, of the room in which a particular device is going to be used, using a spectrum analyzer. Now place the instrument or device where there will be minimal EMI at the frequencies and where the device is most active. Such an approach should be adapted when planning to build or refurbish a room for a specific, specialized electrical operation (e.g. an eye clinic with highly specialized electronic or laser devices.
2. **Detection.** If a device or system is malfunctioning, first ensure that the failure is not due to internal system defect or breakdown. If not, then an investigation should be carried out to determine whether EMI was the culprit and if so, determine the source of the EMI. Unused sources inside the room, such as the fast-rising short duration pulses, would also be EMI sources.
3. **Correction.** The susceptibility to EMI may be due to, for instance, the conductive coating on a piece of equipment that may have been worn out, thus making the equipment unshielded from EMI. Moreover, it is important to ascertain that EMI is from outside the room before shielding the room is thought of as a corrective action.
4. **Prevention.** It is important both to keep to the guidelines to prevent EMI and to do regular maintenance to minimize EMI sources. In commercial aircraft, for instance, mobile telephones are required to be turned off because they are possible EMI sources for the aircraft navigational system.

8.5.4 Grounding

Grounding, shielding and filtering are very important methods of reducing EMI and establishing EMC.

When grounding a circuit, we seek to ensure that circuits and systems do not produce unwanted voltage drops along the common return path of a circuit ground.

In Fig. 8.26(a) is shown the grounding of two points of an electronic circuit. In such grounding of a circuit or device, we should note that no guarantee can be given

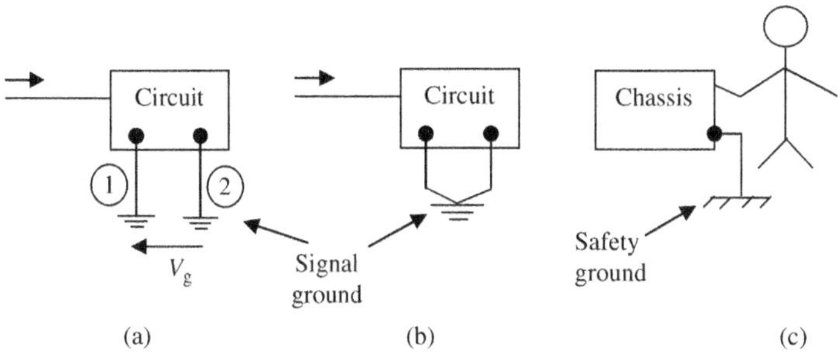

Figure 8.26: (a) Wrong signal grounding, giving rise to volt drop V_g across the grounding points during a voltage surge. (b) Correct signal grounding to keep all ground points at the same voltage to prevent circulation current during voltage surges. (c) Chassis grounding for safety.

that the electric potentials at the grounding points (1) and (2) will be equal and zero. There may be defects of grounding of the two points, defects such as one of the two points using a long lead to ground or the connection of one of the leads to ground not being a short; long grounding wires will cause a potential (i.e. a voltage) drop across the wire, such that the ground and the electronic circuit point which requires to be grounded may be at two different potentials. Moreover, defective grounding at one of the two ground points will result in a voltage drop V_g between two ground points. The potential drop V_g will cause currents to circulate inside the circuit which may cause EMI problems. Thus signal grounding should seek to do single point grounding as shown in Fig. 8.26(b). Moreover, bond wires or straps should be less than $\lambda/20$ long, where λ is the wavelength of the circuit operating frequency. Keep the lengths of the wires short. Moreover, anyone working on electronic circuits should ensure that the hands that will touch the circuit are well grounded though wearing ground straps.

Another kind of grounding is safety grounding as illustrated in Fig. 8.26(c). In earthing the chassis of an electronic device, any person that may touch the equipment chassis that has become electrically charged will be protected by the grounding conductor which discharges the chassis. In safety grounding practice where the equipment chassis is grounded so that any person touching a faulty equipment where there has been a short circuit to the chasis will be protected by the low impedance grounding path diverting all electric charges and currents down to the earth, thereby keeping the electric current through the human body at zero. It is very important that the impedance of the safely grounding lead should be short and that it is properly connected to both chassis and the ground with zero impedance at soldering or connecting points. If the human who touches the chassis is wet, then his surface body impedance is relatively small which will compete with the low impedance of the ground lead or conductor to conduct discharging currents.

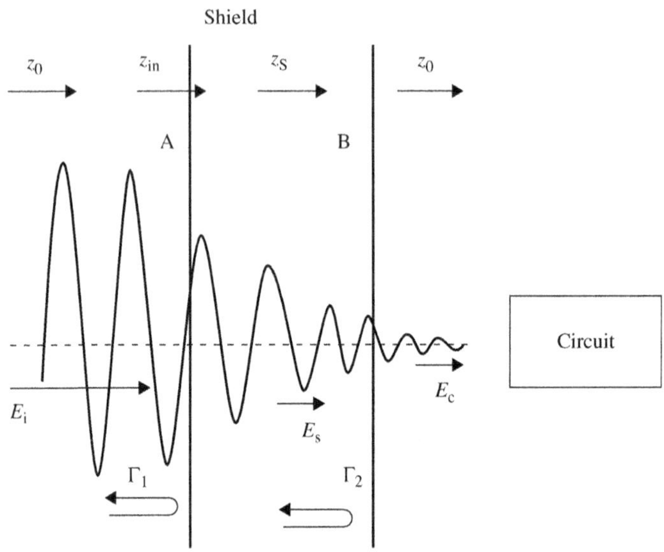

Figure 8.27: A shield enables the EMI signal E_i to be attenuated so that the EMI signal power reaching the shielded circuit is very small. If the attenuation constant of the shield is α, it will attenuate the electromagnetic field by $\exp(-\alpha z_s)$, where z_s is the distance the signal has travelled inside the shield. The power of the EMI signal will be attenuated by $\exp(-2\alpha z_s)$.

8.5.5 Shielding

Where necessary, use shielding which is a way of preventing externally produced electromagnetic fields (e.g. switching voltage pulse on a power line) from inducing currents and voltages (EMI) on circuit board micro-strip lines or the longer conductors that connect, say, the source and the circuit. Telecommunication lines should always be shielded. The shield may be a metal screen placed over the circuit and the incoming conductors, with the shield connected to a zero reference ground point of the circuit being protected. In Fig. 8.27 is shown an EMI causing signal being significantly eliminated before it gets to a circuit.

The electric field incident on the shield is rapidly attenuated to a smaller signal strength of E_S. The electric field E_C which finally impinges on the circuit is much smaller than the original EMI signal E_i.

The shielding effectiveness is given by

$$S_e = 20 \log\left(\frac{E_C}{E_i}\right) \tag{8.122}$$

Inside the shield metal, the electric field drops to e^{-1} after it has moved a distance $\delta = 1/\sqrt{\pi f \mu \sigma}$, known as the skin depth. The shield may be made of copper or aluminum plates. The thickness and quality of the conductors used for shielding should be adequate to prevent the desired interference signals from going out or coming in. For instance, when either very thin or poor quality aluminum conductors (or wire mesh) are used to shield a room where integrated chips are being tested, the test room signals could leak out and interfere with wireless, mobile communication (or mobile computing) signals of a communication service provider.

In order to work out some approximate relationships for input impedance z_{in}, the shield impedance z_s and the reflection coefficients at the two surfaces A and B of Fig. 8.27, we may use the following mathematical equations. The input impedance looking in the direction of wave propagation direction at surface A is given by

$$z_{\text{in}} = z_1 \left[\frac{z_0 + z_S \tanh(\gamma d)}{z_s + z_0 \tanh(\gamma d)} \right] \tag{8.123}$$

where the propagation constant and shield impedance are given by

$$\gamma = \sqrt{j\omega(\sigma + j\omega\varepsilon)}, \tag{8.124}$$

and

$$z_S = \sqrt{\frac{j\omega\mu}{\sigma + j\omega\varepsilon}} \tag{8.125}$$

respectively. The reflection coefficients are

$$\Gamma_1 = \frac{z_S - z_0}{z_S + z_0} \quad \text{at surface A} \tag{8.126}$$

$$\Gamma_2 = \frac{z_0 - z_S}{z_0 + z_S} \quad \text{at surface B} \tag{8.127}$$

8.5.6 Filters

Filters are used at the circuit level to combat interference. Filters may reflect the interference so that the interference signal is turned back from proceeding any further into the circuit. In addition to reflecting, the filter may attenuate the signal. Ferrite chokes are used to reduce EMI levels inside a highly EMI sensitive piece of equipment. A ferrite choke is placed between the power line and the equipment, as shown in Fig. 8.28(a).

When the source and the load have low impedances, then the T-filter shown in Fig. 8.28(b) may be used. When the source and load have high impedances, then the

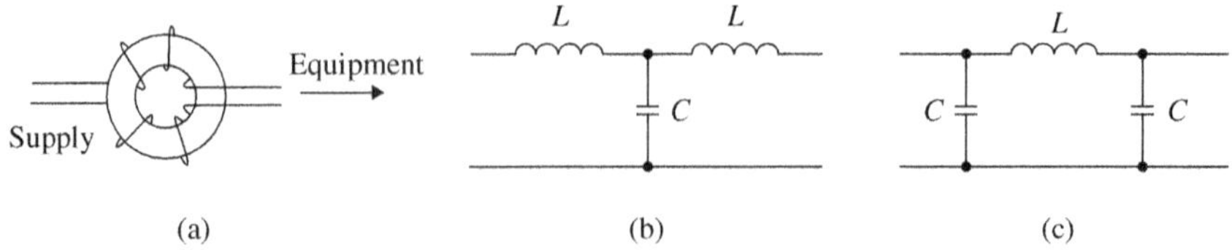

Figure 8.28: Filters for EMC.

π-filter shown in Fig. 8.28(c) may be used. Since the noise power is only reflected back, it still exists in the circuit and could be attenuated using resistors such as a short length of coaxial cable with a lossy ferrite material. The effectiveness of a filter is measured by calculating the insertion loss, defined as

$$I_{\mathrm{L}} = 10 \ln\left(\frac{P_{\mathrm{L}}}{P_{\mathrm{Lf}}}\right) \tag{8.128}$$

where P_{L} is the power delivered to the load without the filter and P_{Lf} is the power delivered to the load with the filter.

8.5.7 Doppler frequency

Consider a transmitter (T) radiating time-harmonic signals (sinusoids) at a frequency j. The receiver (R) will receive that signal at the same frequency if both transmitter and receiver are stationary with respect to each other, as shown in Fig. 8.29(a). The waveforms picked up by the receiver will be spherical electromagnetic waves as shown in Fig. 8.29(a). Both transmitter and receiver are stationary in this case.

The radiated electromagnetic waveforms move at the velocity of light (c) toward the receiver R, at frequency f; $f = 1/T$, where T is the period of the wave. However, if the transmitter moves toward the receiver, the received signal frequency will be greater than the transmitted signal frequency f. As shown in Fig. 8.29(b), the received signal in the direction of T movement will have a shorter time period T_1; hence $f_1 = 1/T_1 > f = 1/T$. Similarly, the wave traveling out in the opposite direction to that of the transmitter will have a time period that will be longer than T; thus the frequency of the signal will be $f_2 = 1/T_2$, which is lower than f. The change in the radiated signal frequency is called the Doppler effect. And the new frequency is called the Doppler frequency, and the change in frequency (e.g. $f_1 - f$) is called the Doppler shift.

Consider now the situation shown in Fig. 8.29(b) where the transmitter moves at a velocity v and at an angle θ with respect to the receiver. The signals radiated by the transmitter, at instant $t = 0$, will reach the receiver at $t_0 = \frac{r_0}{c}$; at the next time

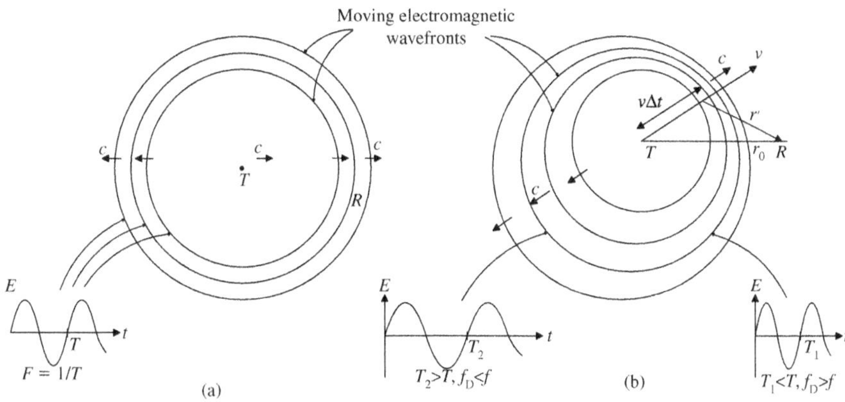

Figure 8.29: Doppler frequency shifts (a) when the transmitter and receiver are stationary with respect to each other. (b) When the transmitter and receiver move closer to each other at a velocity v.

instant Δt, the signal from the transmitter will reach the receiver at

$$t_1 = t + \Delta t = \frac{r}{c} + \Delta t = \frac{1}{c}\left[r_0^2 - 2r_0(v\Delta t)\cos\theta + (v\Delta t)^2\right]^{1/2} + \Delta t \quad (8.129)$$

which, on assuming $r_0^2 \gg (v\Delta t)^2$, becomes

$$t_1 \cong \frac{r_0}{c}\left[1 - \frac{v\Delta t}{r_0}\cos\theta\right] + \Delta t \quad (8.130)$$

Thus the time elapse

$$\Delta t' = t_1 - t_0 = \Delta t - \frac{v\Delta t}{c}\cos\theta = \Delta t\left(1 - \frac{v}{c}\cos\theta\right) \neq \Delta t \quad (8.131)$$

Let $\Delta t' = T = 1/f$, then the frequency of the signal when it reaches the receiver is given by

$$f_{\mathrm{D}} = \frac{1}{\Delta t} = \frac{1}{\Delta t}\left(\frac{1}{(1 - (\frac{v}{c})\cos\theta)}\right) = \frac{f}{(1 - (\frac{v}{c})\cos\theta)} \cong f\left(1 + \frac{v}{c}\cos\theta\right) \quad (8.132)$$

assuming that $(v/c)^2 \ll 1$.

We may make the following remarks on (8.132).

(i) Maximum frequency shift occurs when $\theta = 0$ and $f_{\mathrm{D}} = f\left(1 + \left(\frac{v}{C}\right)\right)$.

(ii) When the transmitter is moving toward the receiver, v is positive and $f_{\mathrm{D}} > f$.

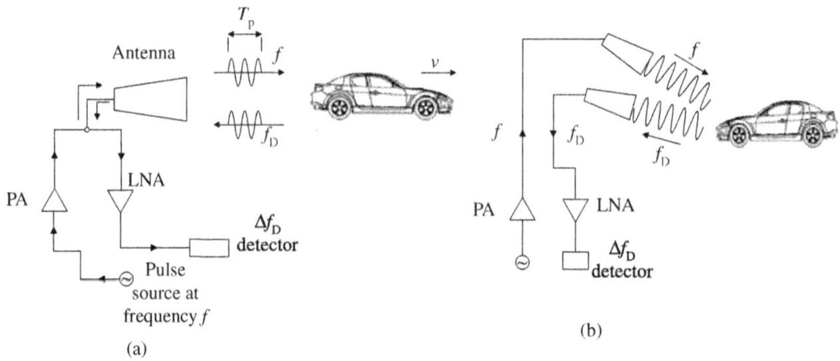

Figure 8.30: Doppler frequency shift and pulsed radar: (a) monostatic radar and (b) bistatic radar.

(iii) When T is moving away from the receiver, $f_D < f$.

(iv) The Doppler frequency shift $\Delta f_D = f_D - f$, that is $\Delta f_D = \frac{v}{c}\cos\theta$.

(v) When $v = 0$, $f_D = f$ and $\Delta f_D = 0$.

In telecommunication systems, Δf_D is undesirable and needs to be compensated for. In older systems, a measurement of Δf_D will yield the velocity of the target under observation. In a pulsed wave radar (PWR) system, shown in Fig. 8.30(a), the vehicle is moving away from the radar. In Fig. 8.30(b), the vehicle is moving toward the radar. In monostatic radar (Fig. 8.30(a)), the same antenna is used for transmitting and receiving the radar signal. Hence the timing should be such that when the received pulse is captured, the transmitted pulse should be switched out to prevent interference between the transmitted and received pulses.

In the bistatic radar shown in Fig. 8.30(b), the transmission antenna and the receiving antenna are two separate antennas and may be placed at two separate locations as well. Therefore, the waveform used by the transmitter may be a continuous wave or a pulsed wave.

In the PWR, a microwave source frequency f produces an output signal for a fixed time period T_p. A power amplifier (PA) is used to amplify the signal before it is transmitted. The moving vehicle will reflect the signal back toward the radar antenna. The antenna will receive the reflected signal at f_D, where the frequency shift $f_D - f$ is a function of the target velocity, direction and relative angle θ with respect to the direct line of sight of the antenna with respect to the direction of the target's movement. The reflected signal strength (at frequency f_D) will be very weak, so that an LNA is used to amplify it before the received signal is processed to obtain parameters such as the velocity of the target (car) from (8.132).

For further reading

[1] Balanis, C.A., *Antenna Theory*, USA: John Wiley, New York, 2012
[2] Bansal, R. Ed., *Engineering Electromagnetics: Applications*, CRC Press, Boca Raton, FL 2006.
[3] Cheng, D.K., *Field and Wave Electromagnetics*, 2nd Edition, Addison-Wesley, Reading, MA 1998.
[4] Christopoulos, C., *Principles and Techniques of Electromagnetic Compatibility*, USA: CRC Press, Boca Raton, FL 1995.
[5] Collin, R.E., *Antennas and Radio Wave Propagation*, USA: McGraw Hill, New York, 1985.
[6] Elliot, R.S., *Electromagnetics*, IEEE Press, New York, 1993.
[7] Hall, M.P.M., Barclay, L.W., and Hewitt, M.T., *Propagation of Radiowaves*, London, UK: IEE Press, 1996.
[8] Hall, S.H., Hall, G.W., and McCall, J.A., *High Speed Digital System Design: A Handbook of Interconnect Theory and Design Practices*, USA: John Wiley, New York, 2000.
[9] Hansen, R.C., *Phased Array Antennas*, USA: John Wiley, New York, 1998.
[10] Hoole, P.R.P., *Smart Antennas and Signal Processing*, UK: WIT Press, Southampton, 2001.
[11] Inan, U.S. and Inan, A.S., *Electromagnetic Waves*, USA: Addison Wesley, Menlo Park, CA, 1999.
[12] Johnson, R.C. Ed., *Antenna Engineering Handbook*, USA: McGraw Hill, New York, 1993.
[13] Jordon, E.C. and Balmain, K.G., *Electromagnetic Waves and Radiating Systems*, USA: Prentice Hall, Englewood Cliffs, NJ, 1968.
[14] Misra, D.K., *Practical Electromagnetics*, New Jersey, USA: John Wiley, New York, 2007.
[15] Sampei, S., *Applications of Digital Wireless Technologies and Global Wireless Communications*, USA: Prentice Hall, Englewood Cliffs, NJ, 1997.
[16] Pozar, D.M., and Schaubert, D.H., *Microstrip Antennas: The Theory and Design of Microstrip Antennas and Arrays*, USA: IEEE Press, New York, 1995.
[17] Ramo, S., Whinnery, J., and Van Duzer, J., *Fields and Waves in Communication Electronics*, USA: John Wiley, New York, 1994.
[18] Sevgi, L., Aklemann, F., and Felsen, L.B., Ground Wave Propagation Modeling: Problem Matched Analytical Formulations and Direct Numerical Techniques, *IEEE Antennas and Propagation*, *44*(1), 55–75, 2002.
[19] Skolonik, M.I., *Radar Handbook*, New York, USA: McGraw Hill, New York, 1990.
[20] Wentworth, S.M., *Fundamentals of Electromagnetics with Engineering Applications*, USA: John Wiley, New York, 2005.
[21] Wesolowski, K., Mobile Communication Systems, USA: John Wiley, New York, 2002.
[22] Zeimer, R.E. and Tranter, W.H., *Principles of Communications*, USA: John Wiley, 4th Edition, New York, 1995.

9

Influence of radical pair reactions on biological rhythms

Chathurika D. Abeyrathne, Malka N. Halgamuge and Peter M. Farrell

The University of Melbourne

9.1 Introduction

The concept of magnetoreception in animals is based on detecting weak magnetic fields by means of the magnetic field-dependent radical pair mechanism that occurs in the photoreceptors in eyes. Blue light sensitive cryptochromes engaged in this radical pair reaction are involved in synchronizing biological rhythms. This raises the possibility that weak magnetic fields may influence, or be made to influence, biological rhythms through the radical pair mechanism. This chapter analyzes the effect of magnetic fields on radical pair reactions, and discusses the likelihood of altering biological rhythms through these reactions. Our results show that, at zero and very weak magnetic fields (magnetic field, $B << A$, where A is the hyperfine coupling strength), the effect of hyperfine interaction is significant, thus increasing the magnetic field-dependent rate constant k_{isc}. This rate constant k_{isc} decreases when B is somewhat higher than A (i.e., in medium magnetic fields) due to the Zeeman effect. At higher magnetic fields ($B >> A$) k_{isc} again starts to increase due to the Δg mechanism. The Δg mechanism provides an additional source of spin conversion if two radicals have slightly different g-factors (see Section 9.4.3). In addition, our results predict that the perturbation of chemical reaction rates in circadian rhythms changes the phase and concentration of these rhythms. The key words that sum up what this is about are biological effect, biological rhythm, magnetoreception, magnetic field-dependent rate constant (k_{isc}), radical pair mechanism and weak magnetic fields.

A magnetic field is produced by moving electrical charges, whereas an electric field is produced by stationary charges. The main characteristics of electromagnetic fields are their frequency and their strength. Electromagnetic fields of different strengths and frequencies influence biological systems in different ways. The electromagnetic spectrum is shown in Fig. 9.1. The high frequency waves (with lower wavelengths) have enough energy to ionize an atom or molecule. The low frequency

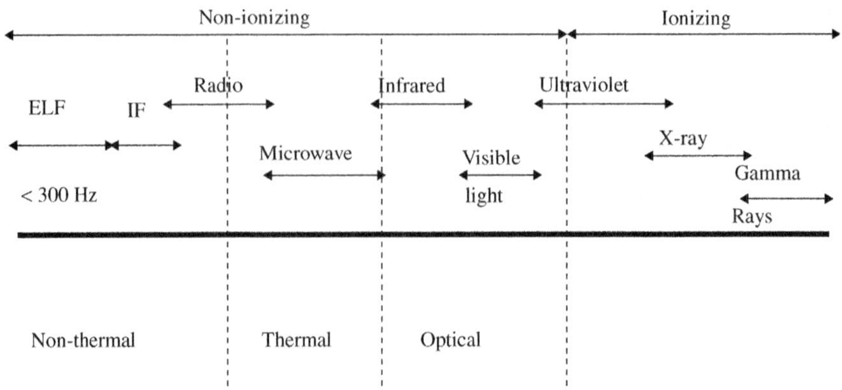

Figure 9.1: The electromagnetic spectrum: At higher frequencies (lower wavelengths), electromagnetic radiation carries sufficient energy to ionize an atom or molecule, whereas low frequencies cause non-ionizing radiation. Frequencies up to 300 Hz are extremely low frequencies (ELF), 300 Hz to 10 MHz are intermediate frequencies (IF), 10–300 MHz are radio frequencies (RF) and 300 MHz to 300 GHz are microwaves.

waves cause non-ionizing radiation and produce thermal and non-thermal effects. Visible light in the non-ionizing frequency range influences radical reactions by exciting electrons. Electricity and all appliances using electricity, that is, 50–60 Hz electric power, are in the extremely low frequency (ELF) range (<300 Hz). These fields that are in the ELF range, where the radiation is below the levels that cause heating, are thought to be harmless. However, for many years researchers have been concerned about the biological effects of these nonthermal fields. Evidence is still inconclusive and there is a controversy over the possible biological effects of exposure to weak, low frequency electromagnetic fields.

Recently, there has been a surge of interest in, and research on, magnetoreception and biological rhythm synchronization in animals [1, 2]. Magnetoreception occurs in photoreceptors which contain blue light sensitive cryptochromes (there are two types of cryptochromes called cryptochrome 1 (CRY 1) and cryptochrome 2 (CRY 2)). Recent evidence shows a radical pair mechanism in cryptochromes, which can be influenced by weak magnetic fields [3]. As an example, seedlings grown in a magnetic field of 500 μT have shown higher levels of cryptochrome [4], which has been shown to be involved in biological rhythm synchronization [5]. This suggests that the magnetic field affects biological rhythms through cryptochrome [6], and this is supported by experimental observations [1, 5]. The sensitivity of Drosophila's circadian rhythm to the magnetic fields is observed to depend on the activation of the cryptochrome and on the applied field strength [1]. In another experiment, flies exposed to static magnetic fields were shown to have slower clock rhythms, and this effect was present in blue light, which activates cryptochromes [5].

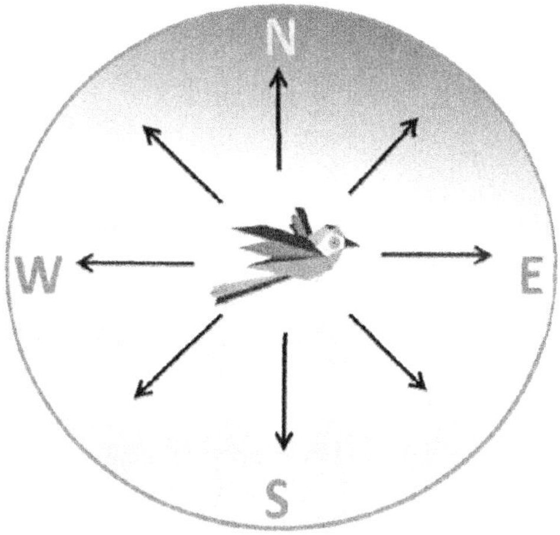

Figure 9.2: Magnetoreception in birds: Birds use information about the earth's magnetic field to discover their migration path. The magnetic vector and intensity of the magnetic field give them their direction and position.

In this chapter, the influence of magnetic fields on biological systems is explained by reviewing and analyzing the possibility of influencing biological rhythms through radical pair reactions. Magnetoreception, which is the ability to sense the magnetic field, in animals and biological rhythms is discussed in Sections 9.2 and 9.3. Section 9.4 describes the radical pair mechanism. In Section 9.5, the influence of magnetic fields on the chemical reaction rates and biological rhythms is analyzed. Conclusions are presented in Section 9.6.

9.2 Magnetoreception in animals

Birds, animals and insects, such as tortoises, fishes, rats, chickens, pigeons, monkeys, flies and termites detect the earth's magnetic field [7, 8]. Birds can use the earth's magnetic field information to discover their migration path (Fig. 9.2). Information about direction and position are given by the magnetic field vector and intensity of the magnetic field, respectively [2]. Direction finding involves the radical pair mechanism in photoreceptors, and positional detection involves ion rich particles, such as magnetite. Photoreceptors are located in the eyes and convert light into electrical signals (Fig. 9.3). Light-activated flavin adenine dinucleotide (FAD) formed by cryptochrome proteins in the photoreceptors undergoes the radical pair process [2]. Cryptochromes are blue-green photoreceptor proteins. Thus, magnetoreception in animals depends on the intensity and wavelength of light [9]. For example, birds obtain magnetic information from the entire short-wavelength part (high energy) of the visible light. They show a strong preference for a specific direction if the light is of higher energy (blue, green) and preferred direction

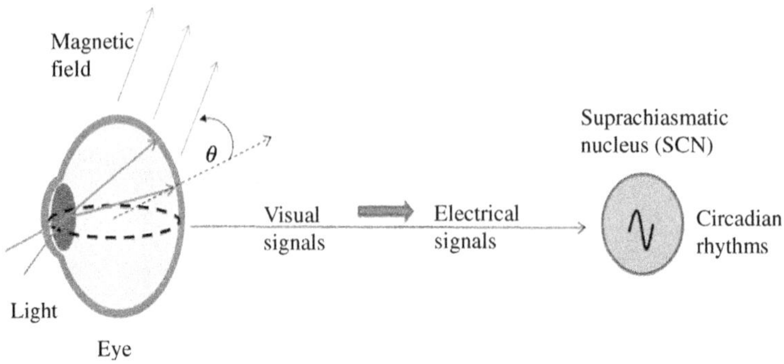

Figure 9.3: The light signal detected by the photoreceptors in the eyes is converted into electrical signals. These electrical signals are then transmitted to different areas in the brain, including the thalamus and hypothalamus.

becomes random if the energy of light is low (red, yellow). Magnetoreception is further limited to a narrow range of magnetic field intensities [9]. Some experiments demonstrate a pigeon's ability to detect differences in magnetic field intensity [10]. The magnetic orientation behavior can be disrupted by a very small oscillating magnetic field. Migratory birds were found to be disoriented when exposed to weak fields of 0.1–10 MHz, at a 24° or 48° angle [2, 11].

9.3 Biological rhythms

Biological rhythms refer to the timing and duration of biological processes and occur in animals, plants and microbial organisms. There are several important biological rhythms, such as (i) circadian rhythms, roughly in 24-h cycles; (ii) infradian rhythms, cycles longer than a day like reproduction cycles; (iii) ultradian rhythms, cycles shorter than 24 h, such as the cycle of growth hormone production and (iv) gene oscillations where some genes are expressed more during certain hours of the day [12, 13]. These rhythms include the sleep–wake cycle, body temperature, blood pressure, hormone levels, number of immune cells in the blood, cellular regeneration, migration and leaf movements [14]. The individual cells of microbes regulate their circadian rhythms, whereas in mammals and birds the primary circadian "clock" is located in the suprachiasmatic nucleus (SCN) of the hypothalamus. The SCN receives information about light through the photoreceptors in the eyes, interprets it, and passes it to the other areas as shown in Fig. 9.4 [2]. The peripheral oscillators, which are more or less independent circadian rhythms, are found in the lung, liver, pancreas, spleen, thymus and the skin [14].

The clock regulations involve several genes and proteins [14, 15]. Cryptochrome is involved in the biological rhythm synchronization of plants and animals [16]. These are regulated through positive and negative feedback loops, as shown in

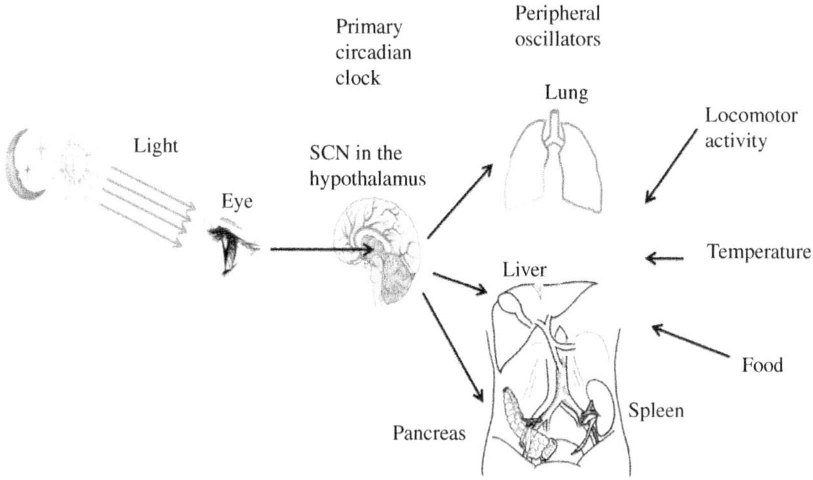

Figure 9.4: The suprachiasmatic nucleus (SCN) of the hypothalamus receives information about light through the photoreceptors in the eyes. These light signals are converted to electrical signals and passed to other areas of the brain. Signals from the SCN entrain peripheral oscillators, which are found in the lung, liver, pancreas, spleen, thymus and the skin. In addition, other signals such as temperature, food and locomotor activity shift these peripheral oscillators independently of the SCN.

Fig. 9.5 [14, 15]. Figure 9.5 shows the production and degradation of a gene and protein where protein affects gene production by a negative feedback [17].

This circadian rhythm model can be defined as

$$\frac{d[G]}{dt} = \frac{r_G}{1 + ([P]/k)^n} - q_G[G] \tag{9.1}$$

$$\frac{d[P]}{dt} = r_P[G](t - \tau)^m - q_P[P] \tag{9.2}$$

where $[G]$ and $[P]$ are the concentration of the genes and proteins, r_G and q_G are the gene production and degradation rate constants, r_P and q_P are the protein production and degradation rate constants, τ is the total duration of protein production from the gene, n is the Hill coefficient (which determines the degree of cooperativeness of the protein binding to a receptor), k is a scaling constant and m is the nonlinearity in the protein production [17]. The process of shifting the rhythm depends on factors, such as light intensity, wavelength (or color) of light, temperature and certain chemicals [18, 19].

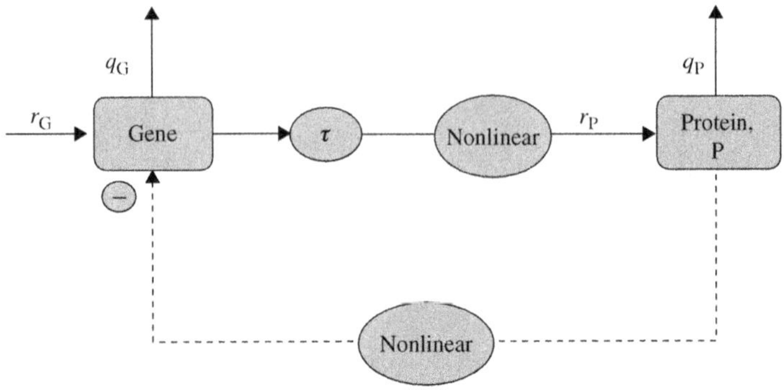

Figure 9.5: A model for the circadian rhythm [17]. This shows the production and degradation of a gene and protein where the protein affects the gene production by a negative feedback. Here, the gene production and degradation rate constants are r_G, q_G, whereas the protein production and degradation rate constants are r_P, q_P, and delay τ is the total duration of protein production from the gene.

9.4 Radical pair mechanism

A radical is an atom or a molecule with an unpaired electron. The electron nuclear spin motion of the radical pair is explained by the hyperfine interaction, exchange interaction and Zeeman effect. This is given by the Hamiltonian:

$$H = \sum a_{1k} S_1 I_k + \sum a_{2l} S_2 I_l - J\left(\frac{1}{2} + 2S_1 S_2\right) + \mu B(g_1 S_1 + g_2 S_2) \tag{9.3}$$

where S_1 and S_2 are the electron spins, a_{1k} and a_{2l} are the hyperfine interaction constants, I_k and I_l are the nuclear spins, J is the exchange interaction, μ is the Bohr constant, B is the magnetic field, and g_1 and g_2 are the g-factors of two radicals [20]. The first and second terms in (9.3) represent the hyperfine interaction between the unpaired electron spins and the nuclear spins. The third term is the exchange interaction between the unpaired electron spins. The Zeeman effect, which accounts for the interaction between the electron spin and the magnetic field, is given by the last term.

9.4.1 Hyperfine interaction

The hyperfine interaction is the intra-radical coupling between the magnetic moment of an atomic nucleus and an unpaired electron [21]. As shown in Fig. 9.6, the electron–nuclear coupling depends on the orientation of the molecule in an external

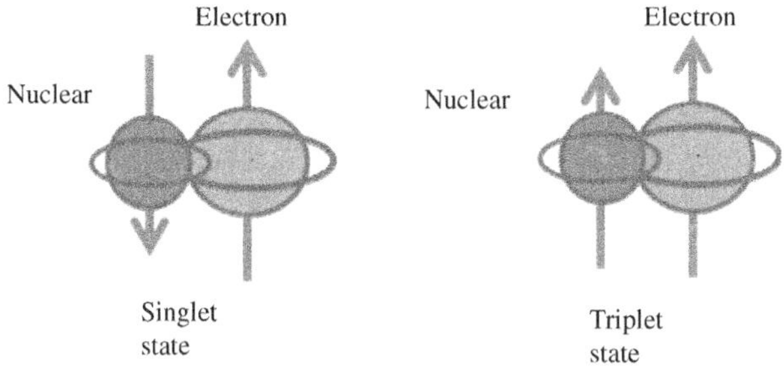

Figure 9.6: The hyperfine interaction: The spin state is described as a singlet state when the orientation of the spins is antiparallel. Triplet states result when the spins are aligned parallel.

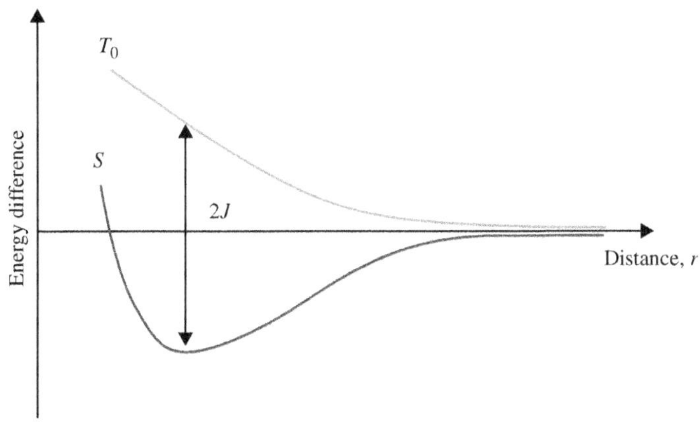

Figure 9.7: The energy difference ($2J$) between singlet (S) and triplet (T_0) states decreases with the increasing pair separation, r.

magnetic field [22]. The spin state is described as triplet states T: T_{-1}, T_0, T_+, when the orientation of the spins is parallel [23]. A singlet state (S) results when the spins are aligned antiparallel. The hyperfine interaction allows it to be modified by an external magnetic field and drives the interconversion of the singlet and triplet states of the radical pair [24].

9.4.2 Exchange interaction

The interaction between the unpaired electron spins of the two radicals is called the exchange interaction, J [25]. The energy difference between the singlet and triplet states is $2J$ [26]. The increasing pair separation reduces the exchange interaction and at a certain distance J becomes negligible (Fig. 9.7).

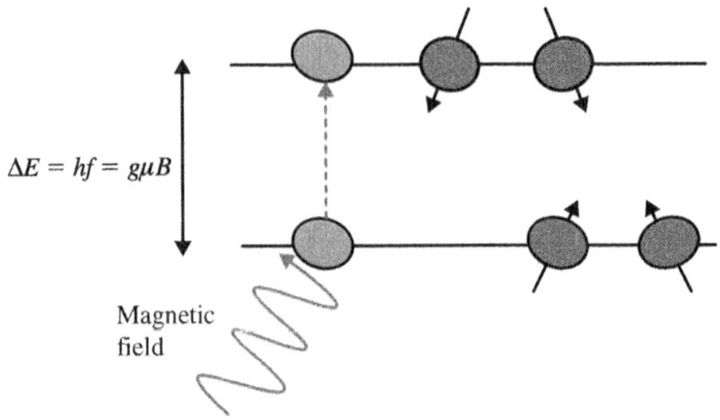

Figure 9.8: The Zeeman effect: The electrons of atoms and the molecules can jump from one energy level to another, either by absorbing energy or by radiating energy. The energy difference is given by $\Delta E = hf = g\mu B$, where h is Planck's constant, f is the frequency of radiation, g is the g-factor, μ is the Bohr constant and B is the magnetic field.

9.4.3 The Zeeman effect

The electrons of atoms and molecules exist in different energy levels and they can jump from one energy level to another, either by absorbing energy or by radiating energy. The size of this jump, that is, the difference between the energy levels (ΔE), determines the frequency of the radiation. These frequencies are called spectral lines, and the splitting of a spectral line into several components in the presence of a static magnetic field is called the Zeeman effect (Fig. 9.8). The distance between the Zeeman sublevels is proportional to the magnetic field, as given by the last term in (9.3) [27]. An electron is at its highest energy level when its magnetic moment is antiparallel to the magnetic field. When the magnetic moment of an electron is parallel to the magnetic field it is at its lowest energy state.

9.4.4 Singlet-triplet interaction

A light-activated molecule or atom transfers an electron to another molecule or atom, resulting in a radical pair. This radical pair exists in either a singlet (S) or triplet (T) state. The hyperfine interaction determines the frequency of the oscillation between the S and T states [23]. These singlet and triplet pairs react to give distinct products, reducing the number of radical pairs. The Zeeman effect influences the frequency of this process and the products formed from S and T states [23]. The influence of the exchange interaction (J) is stronger at low magnetic field strengths. This interaction lifts the T states away from the S states, reducing singlet–triplet interconversion in the contact region [20]. However, with increasing radical pair separation, J becomes negligible and singlet–triplet interconversion becomes

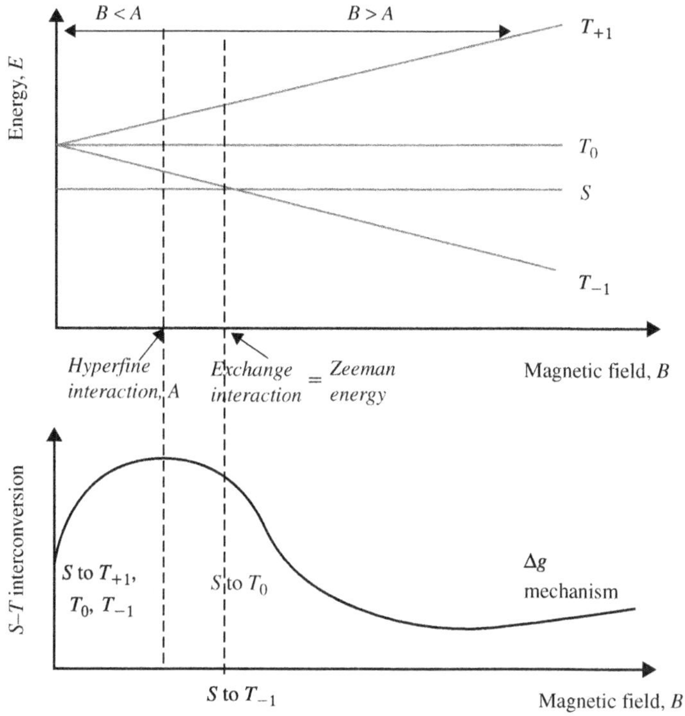

Figure 9.9: Energy levels of singlet (S) and triplet (T) states. The singlet–triplet interconversion is enhanced if the exchange interaction is negligible and when the magnetic field strength is smaller than the hyperfine coupling strength ($B << A$). At medium magnetic fields ($B \geq A$) the Zeeman interaction shifts the energy of triplet states, and only S to T_0 interconversion is possible. When the electron exchange energy matches the Zeeman energy, S to T_{-1} conversion is facilitated.

feasible. Weak magnetic fields, where the magnetic field strength is smaller than the hyperfine coupling strength ($B << A$), enhance singlet–triplet interconversion when the exchange interaction is zero ($J = 0$) [28]. This increases the number of T states that can be converted into S states (Fig. 9.9). The Zeeman interaction shifts the energy of the T states away from the S state energy if the field strength is significant compared to the hyperfine interaction ($B \geq A$). This reduces the triplet state radicals so that only the T_0 state is available for singlet–triplet interconversion [25]. Even at zero magnetic fields, singlet–triplet interconversion is not possible if J is large. However, hyperfine interaction becomes feasible and S to T_{-1} interconversion can be facilitated when the electron exchange energy matches the Zeeman energy [26]. Independent of hyperfine interaction, an additional S to T_0 spin interconversion is induced if the two radicals have different g-factors. This mechanism is significant for quite large magnetic fields, since the difference in g-factors is quite small [25].

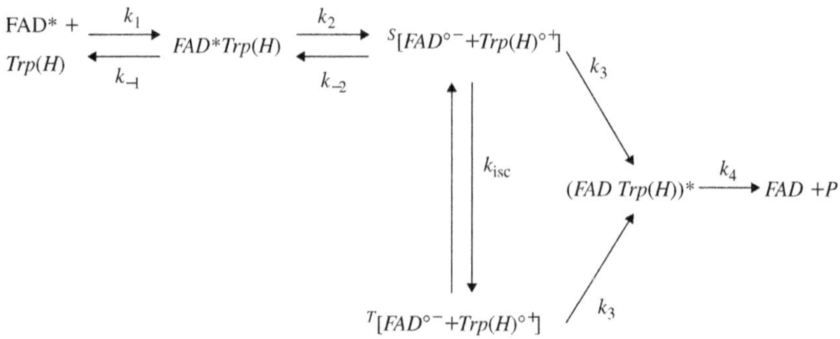

Figure 9.10: Magnetic field-dependent radical pair reaction. Light-absorbed *FAD** enzyme and *Trp(H)* acid react at the forward rate constant k_1 and form the *FAD*Trp(H)* enzyme–acid complex. The singlet enzyme–acid complex, $^S[FAD^{\circ-} + Trp(H)^{+}]$, is produced at the forward rate constant k_2 and singlet–triplet interconversion occurs due to the magnetic field-dependent rate constant, k_{isc}. The triplet enzyme–substrate complex is $^T[FAD^{\circ-} + Trp(H)^{+}]$. Singlet and triplet states form the enzyme–substrate complex *(FAD Trp(H))** through the forward reaction rate constant k_3. The product *P* and *FAD* release are done at the rate constant k_4.

9.4.5 Radical pair reaction

Figure 9.10 shows a magnetic field-dependent radical pair reaction scheme for light-activated FAD formed by cryptochrome proteins in the photoreceptors [29, 30]. Light-absorbed *FAD* and amino acid tryptophan (*Trp*) subjected to the radical pair formation is described as *FAD** and *Trp(H)*. This *FAD** and *Trp(H)* form an enzyme–acid complex, *FAD* Trp(H)*, through forward reaction at a rate of k_1. The backward reaction from *FAD* Trp(H)* to *FAD** and *Trp(H)* is provided by k_{-1}. The forward reaction rate, k_2 produces the singlet enzyme–acid complex, $^S[FAD^{\circ-} + Trp(H)^{\circ+}]$. The reaction from a singlet state to *FAD*Trp(H)* is formed at a backward reaction rate k_{-2}. Singlet–triplet interconversion occurs due to the magnetic field-dependent rate constant k_{isc} where $^T[FAD^{\circ-} + Trp(H)^{\circ+}]$ is the triplet enzyme–acid complex. An enzyme–acid complex, *(FAD Trp(H))** is formed from the singlet and triplet states at a forward reaction rate constant k_3. The products, *P* and enzyme release are done at the rate constant k_4. The chemical rates involved in this reaction scheme are in the microsecond to nanosecond time range [30]. The magnetic field effect is observed when $k_4 >> k_2$ [31, 32].

The resulting differential equations for the enzyme kinetics are

$$\frac{d[E]}{dt} = k_{-1}[ES] - k_1[E][S] + k_4[ES] \tag{9.4}$$

$$\frac{d[S]}{dt} = k_{-1}[ES] - k_1[E][S] \tag{9.5}$$

$$\frac{d[ES]}{dt} = k_1[E][S] - (k_{-1} + k_2)[ES] + k_{-2}{}^S[E^{\circ\circ}S] \tag{9.6}$$

$$\frac{d^S[E^{\circ\circ}S]}{dt} = k_2[ES] - (k_{-2} + k_3 + k_{isc})^S[E^{\circ\circ}S] + k_{isc}{}^T[E^{\circ\circ}S] \tag{9.7}$$

$$\frac{d^T[E^{\circ\circ}S]}{dt} = k_{isc}{}^S[E^{\circ\circ}S] - (k_{isc} + k_3)^T[E^{\circ\circ}S] \tag{9.8}$$

$$\frac{d[ES^*]}{dt} = k_3({}^S[E^{\circ\circ}S] + {}^T[E^{\circ\circ}S]) - k_4[ES^*] \tag{9.9}$$

$$\frac{d[P]}{dt} = k_4[ES^*] \tag{9.10}$$

where $[E] = FAD^*$, $[S] = Trp(H)$, $[ES] = FAD^*Trp(H)$, ${}^S[E^{\circ\circ}S] = {}^S[FAD^- + Trp(H)^{\circ +}]$, ${}^T[E^{\circ\circ}S] = {}^T[FAD^{\circ -} + Trp(H)^+]$ and $[ES^*] = (FAD\ Trp(H))^*$. The reaction rate is

$$\nu^F = \frac{d[P]}{dt} = k_2[ES](1 - {}^SP_g) \tag{9.11}$$

where ${}^SP_g$ is the recombination probability and $(1 - {}^SP_g)$ is the probability that the radical pair does not recombine [31]. Assume these reactions are at a quasi-steady state, that is, the concentration of the enzyme–substrate complexes changes more slowly than the product ($[P]$) and substrate ($[S]$). This gives, $d[ES]/dt = d^S[E^{\circ\circ}S]/dt = d^T[E^{\circ\circ}S]/dt = d[ES^*]/dt = 0$. Also, it is assumed that the total concentration does not change over time. Thus, the total concentration is

$$[E]_{tot} = [E] + [ES] + {}^S[E^{\circ\circ}S] + {}^T[E^{\circ\circ}S] + [ES^*] \tag{9.12}$$

Equating (9.6)–(9.9) to zero and using (9.12) we can obtain ν^F from [30]

$$\nu^F = \frac{V^F_{max}[S]}{[S] + K^F_M} \tag{9.13}$$

where

$$V_{\max}^{F} = \frac{k_4[E]_{\text{tot}}}{1+\left(\frac{k_4}{k_2}\right)\left\{1+\frac{1}{k_3}\left(k_2+\frac{k_{-2}}{f(k_{\text{isc}})}\right)\right\}},$$

$$K_M^F = \frac{\frac{k_4}{k_2}\{K_1 + K_{n1}k_{-2}/(f(k_{\text{isc}})k_3)\}}{1+\left(\frac{k_4}{k_2}\right)\left\{1+\frac{1}{k_3}\left(k_2+\frac{k_{-2}}{f(k_{\text{isc}})}\right)\right\}}, \quad f(k_{\text{isc}}) = \frac{(2k_{\text{isc}}+k_3)}{(k_{\text{isc}}+k_3)},$$

$$K_1 = \frac{(k_{-1}+k_2)}{k_1} \quad \text{and} \quad K_{n1} = \frac{k_{-1}}{k_1}$$

Equating (9.8) and (9.10), the recombination probability becomes

$$^{S}P_g = \frac{k_{-2}}{(k_{-2}+k_3 f(k_{\text{isc}}))} \tag{9.14}$$

Salikhov *et al.* [3] have obtained this probability using the exponential model for zero and high magnetic fields ($B >> A$). The recombination probability for zero magnetic field is

$$^{S}P_g(0) = \frac{k_{-2}}{(k_{-2}+k_3)}\left\{\frac{1+3C_1(a/4)^2}{1+3C_2(a/4)^2}\right\} \tag{9.15}$$

where

$$C_1 = \frac{2k_3+k_{-2}}{k_3[(k_3+k_{-2}/2)^2+(2J+a/2)^2]} \quad \text{and} \quad C_2 = \frac{(2k_3+k_{-2})}{(k_3+k_{-2})}C_1$$

This limit for the high magnetic field ($B >> A$) is [31]

$$^{S}P_g(B) = \frac{k_{-2}}{(k_{-2}+k_3)}\left\{\frac{1+(C_1'+C_2')\Omega_s + C_1'C_2'\Omega_d^2}{1+2C_2'\Omega_s+(C_2^1)^2\Omega_d^2}\right\} \tag{9.16}$$

where

$$\Omega_{s,d} = (\Delta\omega)^2 \pm (a/4)^2, \quad C_1' = \frac{2k_3+k_{-2}}{k_3[(k_3+k_{-2}/2)^2+(2J)^2]} \quad \text{and}$$

$$C_2' = \frac{(2k_3+k_{-2})}{(k_3+k_{-2})}C_1'$$

Equating (9.14)–(9.16), the magnetic field-dependent rate constants k_{isc} for zero ($B = 0$) and high field ($B >> A$: hyperfine constant in Tesla) are obtained [31]:

$$k_{\text{isc}}(0) = 3\frac{(k_{-2}+2k_3)}{((k_{-2}/2+k_3)^2+(2J+a/2)^2)}(a/4)^2 \tag{9.17}$$

$$k_{\text{isc}}(B) = \frac{(k_{-2} + 2k_3)}{((k_{-2}/2 + k_3)^2 + (2J)^2)} \left\{ \frac{(\Delta\omega)^2 + (a/4)^2 + \gamma((\Delta\omega)^2 - (a/4)^2)^2}{1 + \gamma((\Delta\omega)^2 + (a/4)^2)} \right\} \tag{9.18}$$

where

$$\gamma = \frac{(k_{-2} + 2k_3)^2}{k_3(k_{-2} + k_3)[(k_{-2}/2 + k_3)^2 + (2J)^2]} \quad \text{and} \quad \Delta\omega = \frac{\Delta g \mu B}{2\hbar}$$

Here, B is the magnetic field, μ is the Bohr constant, $\hbar$ is the reduced Planck constant, Δg is the difference between g-factors of two radicals and a is the hyperfine interaction constant (rad/s). $\Delta\omega$ becomes negligible for zero and very weak magnetic fields ($B << A$). Hence, only the hyperfine interaction drives singlet–triplet crossing between the S and T_{-1}, T_0, T_+ states when J is negligible. The high field limit is obtained for $B >>> A$ where only S to T_0 interconversion exists. If the g-values are different, intersystem crossing between S and T_0 states is provided additionally to the hyperfine interaction where $\Delta\omega \approx \Delta g \mu B/(2\hbar) + \sum a_{1k} m_{1k} - \sum a_{2k} m_{2k}$ and m is the spin states of the nucleus on radicals [26].

9.5 Results and discussion

The energy of magnetic interactions per particle involved in the radical-pair process is lower than the average thermal energy per particle under environmental conditions. This is because the living cells are electrically noisy, due to the random movements of ions and charged molecules. On the other hand, the thermal relaxations are in the nanosecond range and thus the question of whether the radical pair is altered by the thermal noise is still controversial. However, the effect of magnetic fields on radical pair reactions is assumed to be possible, since the spin of electrons is weakly coupled to the thermal bath [33]. Further, it is argued that the effects of magnetic fields are frequency independent up to a few megahertz, because the lifetime of the radical pair is in the microsecond range. Nevertheless, there is experimental evidence for the influences of static and 50, 60 Hz magnetic fields on radicals [34, 35]. In this chapter, only the possibility of low ($B << A$), medium ($B \geq A$) and high ($B >> A$) magnetic fields' effect on radical pairs is discussed, neglecting the effect of thermal relaxation time, which is however a relevant area for further research.

The variation of the magnetic field-dependent rate constant k_{isc} with the magnetic field indicates the amount of radicals available under a certain magnetic field. The ratio of $k_{\text{isc}}(B)/k_{\text{isc}}(0)$ can be obtained from (9.17) and (9.18). To consider AC magnetic fields' effect, B in $\Delta\omega$ is replaced by $B = B_{\text{DC}} + B_{\text{AC}} \cos \omega_{\text{AC}} t$ where B_{DC} and B_{AC} are the strength of DC and AC magnetic fields and ω_{AC} is the frequency of the AC magnetic field. Figures 9.11–9.13 and 9.15–9.16 are obtained for zero exchange interaction ($J = 0$). Therefore, the singlet to triplet interconversion due to

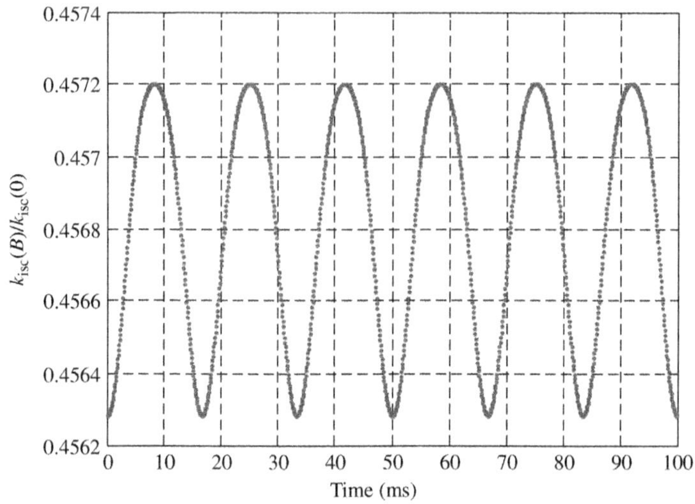

Figure 9.11: The ratio of $k_{isc}(B)/k_{isc}(0)$ with time. Here, $k_{isc}(B)$ and $k_{isc}(0)$ are the magnetic field rate constants for a high magnetic field and a zero magnetic field, where $k_{-2} = 1.6 \times 10^6\,\mathrm{s}^{-1}$, $k_3 = 3.3 \times 10^6\,\mathrm{s}^{-1}$, $J = 0$, $\Delta g = 0.05$, $B_{dc} = 50\,\mu\mathrm{T}$, $\omega_{ac} = 376.9911$ rad/s (60 Hz), $B_{ac} = 30\,\mu\mathrm{T}$ and $A = 25\,\mu\mathrm{T}$.

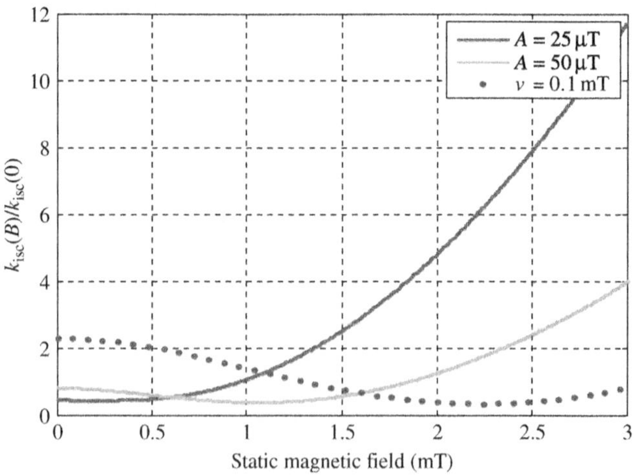

Figure 9.12: The ratio of $k_{isc}(B)/k_{isc}(0)$ with static magnetic field. Here, $k_{isc}(B)$ and $k_{isc}(0)$ are the magnetic field rate constants for a high magnetic field and a zero magnetic field, where $k_{-2} = 1.6 \times 10^6\,\mathrm{s}^{-1}$, $k_3 = 3.3 \times 10^6\,\mathrm{s}^{-1}$, $J = 0$, $\Delta g = 0.05$, $\omega_{ac} = 376.9911$ rad/s (60 Hz), $B_{ac} = 30\,\mu\mathrm{T}$, $t = 100\,\mu\mathrm{s}$, and $A = 25\,\mu\mathrm{T}$, $50\,\mu\mathrm{T}$ and 0.1 mT.

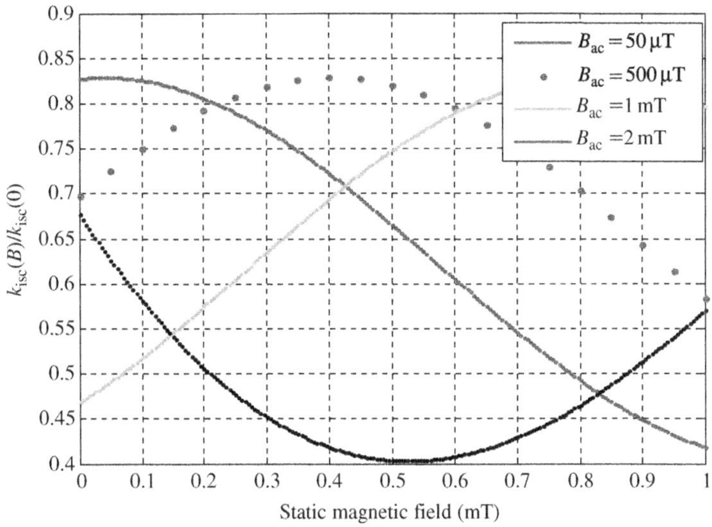

Figure 9.13: The ratio of $k_{isc}(B)/k_{isc}(0)$ with static magnetic field. Here, $k_{isc}(B)$ and $k_{isc}(0)$ are the magnetic field rate constants for a high magnetic field and a zero magnetic field, where $k_{-2} = 1.6 \times 10^6\ s^{-1}$, $k_3 = 3.3 \times 10^6\ s^{-1}$, $J = 0$, $\Delta g = 0.05$, $\omega_{ac} = 376.9911$ rad/s (60 Hz), $A = 50\ \mu T$, $t = 10$ ms and $B_{ac} = 50\ \mu T$, 500 μT, 1 mT and 2 mT.

hyperfine interaction is viable. The variation of $k_{isc}(B)/k_{isc}(0)$ with time is different for different magnetic field combinations. The positive and negative half cycles of $k_{isc}(B)/k_{isc}(0)$ are different when $B_{dc} = 50\ \mu T$ and $B_{ac} = 30\ \mu T$, as shown in Fig. 9.11. Further, we observed that the positive and negative half cycles are similar for $B_{dc} = 150\ \mu T$ and $B_{ac} = 30\ \mu T$, whereas for $B_{dc} = 250\ \mu T$ and $B_{ac} = 30\ \mu T$ the second positive cycle is less than the first positive cycle. This implies the importance of the magnetic field strength and reaction duration with regard to radical pair reactions.

In Fig. 9.12, the magnetic field effect on the magnetic field-dependent rate constant k_{isc} is depicted as a function of the DC magnetic field strength at different values of the hyperfine interaction constant. The effect of AC magnetic field strengths on $k_{isc}(B)/k_{isc}(0)$ is shown in Fig. 9.13, when $\omega_{ac} = 376.9911$ rad/s (60 Hz), $A = 50\ \mu T$ and $t = 10$ ms. The ratio of $k_{isc}(B)/k_{isc}(0)$ increases with an increasing hyperfine interaction constant for $A >> B$ (Fig. 9.12). This ratio starts to decrease with increasing magnetic field strength, although above a certain magnetic field strength the ratio again increases. At medium magnetic field strengths, when the magnetic field strength is comparably higher than the hyperfine interaction constant ($B \geq A$), the Zeeman effect shifts the energy of the T states away from the S state energy, reducing $k_{isc}(B)/k_{isc}(0)$. At these magnetic field strengths the Δg

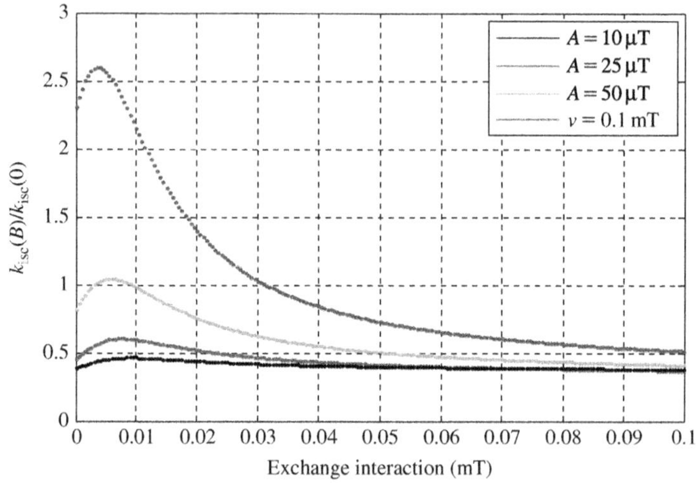

Figure 9.14: The ratio of $k_{isc}(B)/k_{isc}(0)$ with exchange interaction. Here, $k_{isc}(B)$ and $k_{isc}(0)$ are the magnetic field rate constants for a high magnetic field and a zero magnetic field, where $k_{-2} = 1.6 \times 10^6\,s^{-1}$, $k_3 = 3.3 \times 10^6\,s^{-1}$, $\Delta g = 0.05$, $B_{dc} = 50\,\mu T$, $\omega_{ac} = 376.9911$ rad/s (60 Hz), $B_{ac} = 30\,\mu T$, $t = 100\,\mu s$ and $A = 10\,\mu T$, $25\,\mu T$, $50\,\mu T$ and 0.1 mT.

mechanism is not feasible. Above a certain magnetic field strength $k_{isc}(B)/k_{isc}(0)$ increases when the magnetic field strength is reasonably large enough ($B >> A$), to have a realistic effect of the Δg mechanism.

The variation of $k_{isc}(B)/k_{isc}(0)$ with exchange interaction and hyperfine interaction constant is shown in Fig. 9.14. This ratio decreases with increasing exchange interaction and increases with increasing hyperfine constant. This is because when the hyperfine interaction constant is comparably higher than the exchange interaction, singlet–triplet conversion is possible, and when the exchange interaction is large it reduces the singlet–triplet interconversion. Figure 9.15 illustrates the influence of the difference in g-factors ($\Delta g = g_2 - g_1$) of two radicals on the magnetic field-dependent reaction rate k_{isc}. Generally, the difference in g-factors between two radicals is very small, and the effect is viable only in strong magnetic fields. The ratio of $k_{isc}(B)/k_{isc}(0)$ varies with increasing angular frequency of the AC magnetic field, ω_{ac}, as shown in Fig. 9.16. This is due to the variation of effective magnetic field strength B in singlet to triplet interconversion frequency. These results indicate that the magnetic field-dependent rate constant k_{isc} depends on the strength and frequency of the external magnetic field, the exchange interaction (J), the hyperfine interaction constant (A) and the difference in g-factors (Δg) of the radicals.

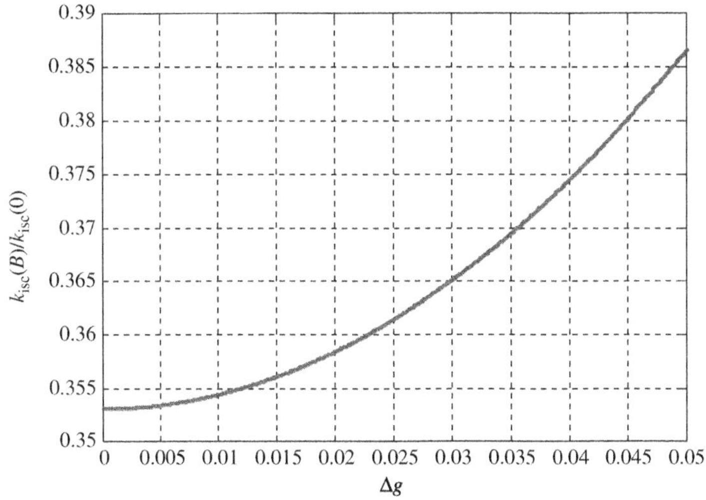

Figure 9.15: The ratio of $k_{isc}(B)/k_{isc}(0)$ with $\Delta g = g_2 - g_1$. Here, $k_{isc}(B)$ and $k_{isc}(0)$ are the magnetic field rate constants for a high magnetic field and a zero magnetic field, where $k_{-2} = 1.6 \times 10^6\,\text{s}^{-1}$, $k_3 = 3.3 \times 10^6\,\text{s}^{-1}$, $J = 0$, $B_{dc} = 50\,\mu\text{T}$, $\omega_{ac} = 376.9911$ rad/s (60 Hz), $B_{ac} = 30\,\mu\text{T}$, $t = 100\,\mu\text{s}$ and $A = 10\,\mu\text{T}$.

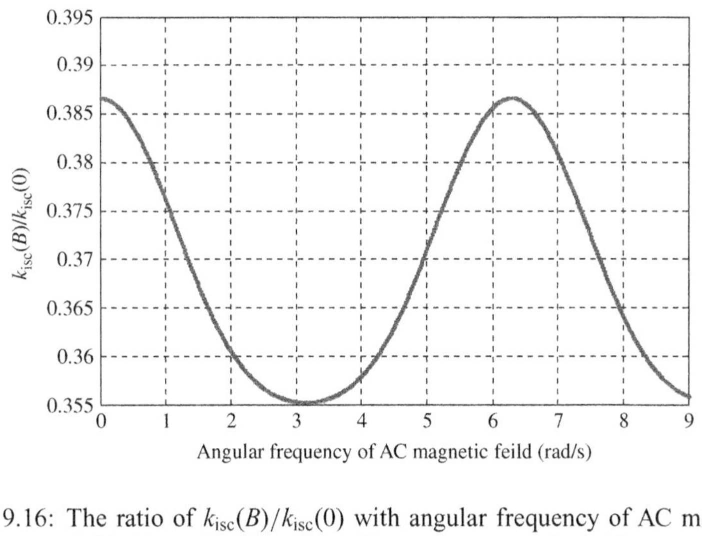

Figure 9.16: The ratio of $k_{isc}(B)/k_{isc}(0)$ with angular frequency of AC magnetic field. Here, $k_{isc}(B)$ and $k_{isc}(0)$ are the magnetic field rate constants for a high magnetic field and a zero magnetic field, where $k_{-2} = 1.6 \times 10^6\,\text{s}^{-1}$, $k_3 = 3.3 \times 10^6\,\text{s}^{-1}$, $J = 0$, $\Delta g = 0.05$, $B_{dc} = 50\,\mu\text{T}$, $B_{ac} = 30\,\mu\text{T}$, $t = 1$ s and $A = 10\,\mu\text{T}$.

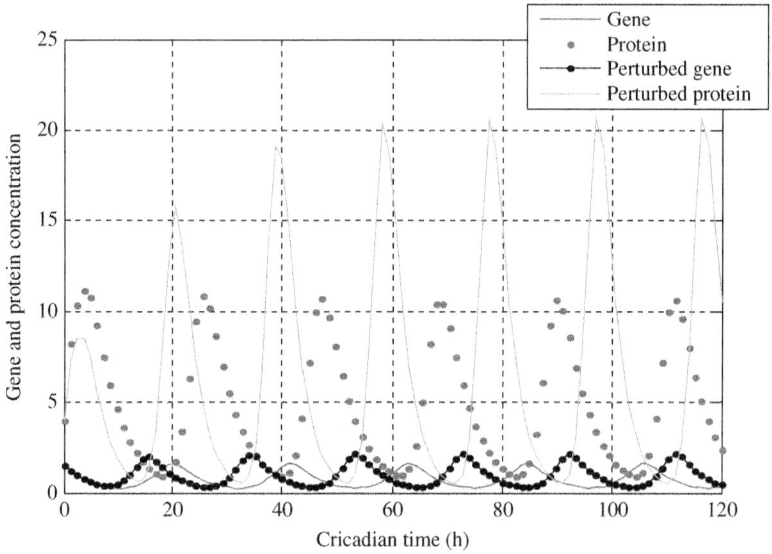

Figure 9.17: Influence of chemical reaction rates on the circadian rhythm. Here, $r_G = 1\,h^{-1}, r_P = 1\,h^{-1}, q_G = 0.21\,h^{-1}, q_P = 0.21\,h^{-1}, n = 2, m = 3$ and $k = 1$. The perturbed chemical reaction rate, $q_P = 0.35\,h^{-1}$ and due to this the concentration and phase of gene and protein oscillations change.

The light signals detected by the photoreceptors in the eyes are converted into electrical signals and transmitted to different areas of the brain, including the thalamus and hypothalamus. The SCN in the hypothalamus detects these signals and drives the circadian rhythms. The cryptochrome protein is involved in this circadian rhythm synchronization; *FAD*, which is formed by cryptochromes, gets activated by blue light and undergoes radical pair reactions in the photoreceptors, as explained in Fig. 9.10. External magnetic fields affect these reactions [29], which in turn influence the length of time that cryptochrome stays in its activated state. The products formed by cryptochrome can be changed when the magnetic field alters chemical reaction rates in the photoreceptors, which modifies the signals that go to the SCN because the activation of cryptochrome may affect the light sensitivity of retinal neurons. The changes to the cryptochrome in the photoreceptors can alter reaction rates of genes and proteins involved in circadian rhythms [6]. The sensitivity of the circadian rhythms to variation of gene and protein production and degradation rates has been investigated by determining how the period of oscillations alters when each parameter in turn is varied. The phase and concentration of protein and gene oscillations are changed when the protein degradation rate in (9.3) is perturbed, as shown in Fig. 9.17. After some time the perturbations of rate constants lead the rhythm to a new steady state (a new oscillation pattern). However, an initial difference in the phase of oscillations cannot usually disappear with time [36].

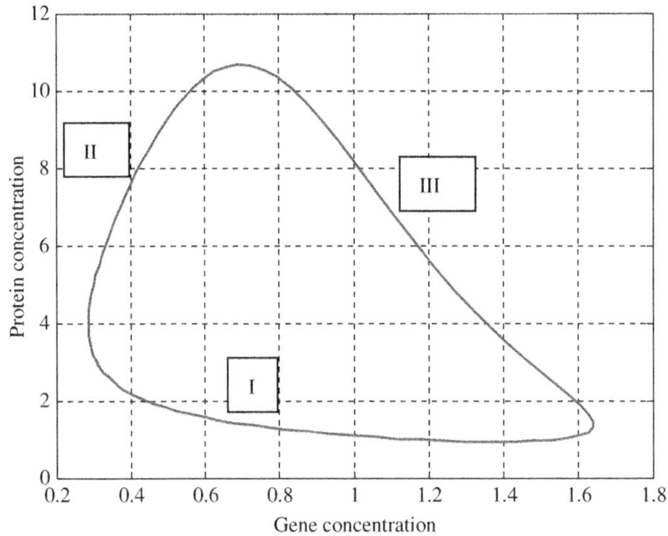

Figure 9.18: Limit cycle of the circadian rhythm for 24 h. Here, $r_G = 1\,h^{-1}$, $r_P = 1\,h^{-1}$, $q_G = 0.21\,h^{-1}$, $q_P = 0.21\,h^{-1}$, $n = 2$, $m = 3$, and $k = 1$. The triangular form of the limit cycle indicates three different stages of the circadian rhythm. In stage I, $[G]$ disappears, while $[P]$ remains small (conversion of $[G]$ to $[P]$). In stage II, $[G]$ remains small while $[P]$ is accumulated. In stage III it verifies the conversion of $[P]$ to $[G]$.

Figure 9.18 shows the limit cycle contour with the gene concentration $[G]$ plotted versus the protein concentration $[P]$ which is obtained from (1). A limit cycle represents a closed and steady oscillation which either repels or attracts neighboring states of the system. The triangular form of the limit cycle of $[P]$ and $[G]$ indicates three different stages of the circadian rhythm [37]. In stage I, it determines the conversion of $[G]$ to $[P]$. During this transformation $[G]$ disappears, while $[P]$ remains small. In stage II, $[G]$ remains small while $[P]$ is accumulated where $[G]$ disappears, while $[P]$ remains small. In stage III, $[P]$ is converted to $[G]$. This demonstrates that in different portions of the limit cycle different chemical processes determine the motion of the system [37]. Figures 9.19–9.22 show the limit cycles of the circadian rhythm when reaction rates q_G, q_P, r_G and r_P are changed. As shown in Fig. 9.19, when the protein degradation rate increases, the transformation of $[G]$ to $[P]$ increases (Fig. 9.5), in turn raising $[P]$ and $[G]$ concentrations. An increase in the gene degradation reduces $[P]$ and $[G]$ concentrations (Fig. 9.20) because of the decreasing gene concentration engaged in the circadian rhythm. The concentration changes due to different protein production rates r_P are small, as illustrated in Fig. 9.21. Different gene production rates affect protein and gene concentrations more than different protein production rates (Fig. 9.22). As demonstrated in Figs. 9.19–9.22, limit cycles for different rate constants do not follow the same path.

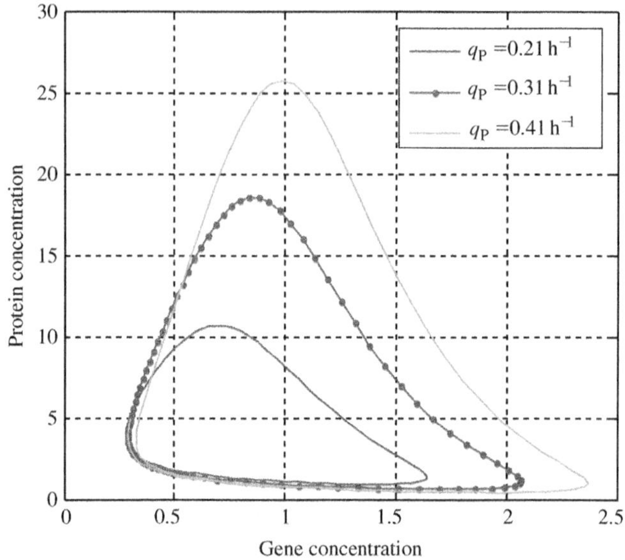

Figure 9.19: Limit cycle: Influence of protein degradation rate q_P on the circadian rhythm. Here, $r_G = 1\,h^{-1}$, $r_P = 1\,h^{-1}$, $q_G = 0.21\,h^{-1}$, $q_P = 0.21$, $0.31, 0.41\,h^{-1}$, $n = 2$, $m = 3$ and $k = 1$.

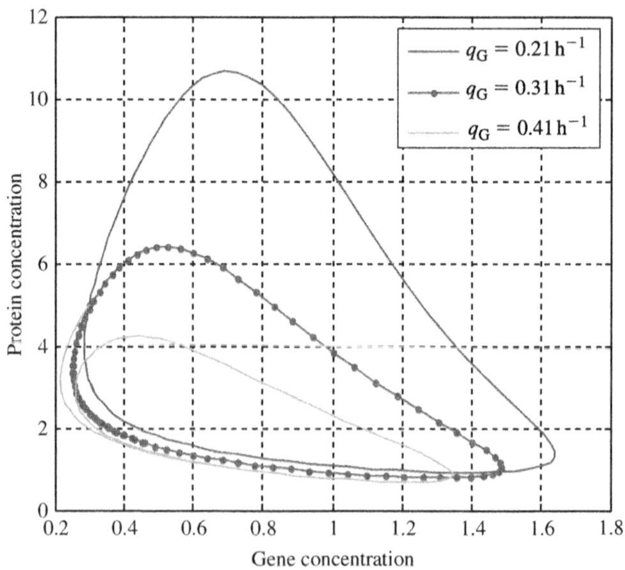

Figure 9.20: Limit cycle: Influence of gene degradation rate q_G on the circadian rhythm. Here, $r_G = 1\,h^{-1}$, $r_P = 1\,h^{-1}$, $q_G = 0.21, 0.31, 0.41\,h^{-1}$, $q_P = 0.21\,h^{-1}$, $n = 2$, $m = 3$ and $k = 1$.

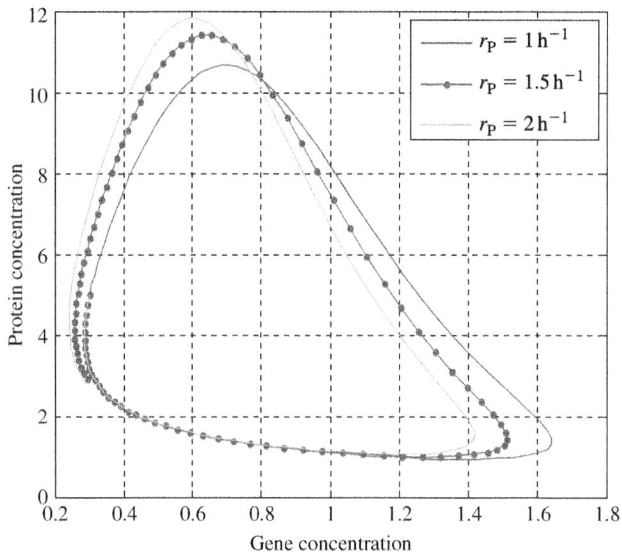

Figure 9.21: Limit cycle: Influence of protein production rate r_P on the circadian rhythm. Here, $r_G = 1\,h^{-1}$, $r_P = 1,\ 1.5,\ 2\,h^{-1}$, $q_G = 0.21\,h^{-1}$, $q_P = 0.21\,h^{-1}$, $n = 2$, $m = 3$ and $k = 1$.

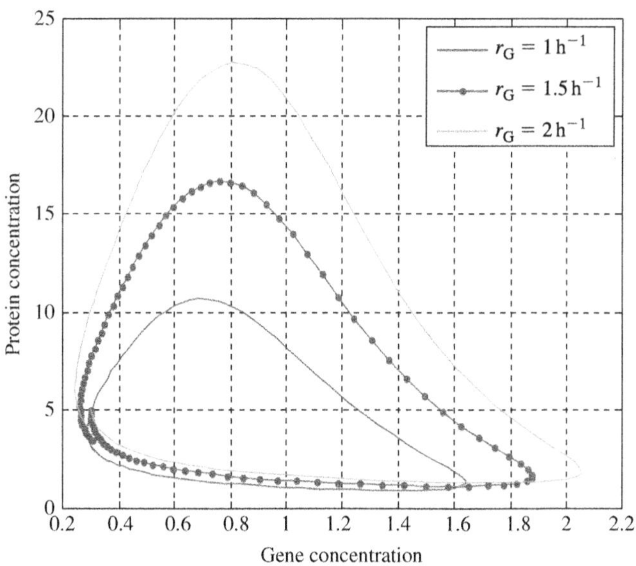

Figure 9.22: Limit cycle: Influence of gene production rate r_G on the circadian rhythm. Here, $r_G = 1,\ 1.5,\ 2\,h^{-1}$, $r_P = 1\,h^{-1}$, $q_G = 0.21\,h^{-1}$, $q_P = 0.21\,h^{-1}$, $n = 2$, $m = 3$ and $k = 1$.

These cycles are shifted, confirming the phase advances or delays due to different rate constants as illustrated in Fig. 9.17.

The magnetic fields can affect radical pair reactions at the photoreceptors. This changes the number of radicals due to alteration of S to T interconversion (magnetic field-dependent rate constant). An increase in free radicals can affect various biological processes, such as release of calcium, cell growth and death and gene expression [38]. Further, there is a probability that the changes in chemical reaction rates in the photoreceptor, caused by the magnetic fields, influence the biological rhythms' reaction rates. Biological systems use oscillations to convey and propagate information efficiently within an extended system [39]. Hence, any disruption to rhythms in the longer term is believed to have significant adverse health consequences. The disruption of the circadian rhythm inhibits melatonin production, which may increase the risk of developing cancer [40, 41]. For example, night shift work can have a negative effect on health due to the disturbances of the normal circadian rhythms. Thus, it is postulated that when the radical pair process is influenced the magnetic fields such that the biological rhythms are altered, biological effects are caused [29]. However, it is not yet known how changes in the cryptochromes in photoreceptors change the cryptochromes of biological rhythms. Exactly how this information interacts, affecting biological systems, remains to be discovered.

9.6 Conclusion

The question of whether the radical pair mechanism is altered by thermal noise is still unanswered, since the relaxation time of the thermal noise is in the nanosecond range. In this analysis, we assumed that the radical pair process can occur because the electron spins are weakly bound to the thermal bath. The effect of magnetic fields on the chemical reaction rates of the radical pair mechanism is observed. The magnetic field-dependent rate constant k_{isc} is increased, thus increasing radicals in very weak magnetic fields (magnetic field, $B << A$, hyperfine coupling strength) when the exchange interaction is zero. When the magnetic field strength is comparable with the hyperfine interaction constant ($B \geq A$) the rate constant k_{isc} is reduced, thus reducing the amount of radicals due to the Zeeman effect. Our results show that in a high magnetic field ($B >> A$), k_{isc} is increased thus increasing singlet–triplet interconversion when the g-factors of radicals are slightly different. The variation of the magnetic field-dependent rate constant depends on the strength and frequency of magnetic fields, exchange interaction, hyperfine interaction constant and the difference in g-factors of the radicals. Our results predict that the perturbation of chemical reaction rates in circadian rhythms changes the phase and concentration of these rhythms. We suggest that biological rhythms can be influenced due to radical pair reactions that occur in the living organism. Thus, the biological effects can arise due directly to magnetic fields when the biological rhythms are influenced by the radical pair mechanism.

For further reading

[1] Gegear, R.J., Casselman, A., Waddell, S., Reppert, S.M., Cryptochrome mediates light-dependent magnetosensitivity in Drosophila, *Nature, 454*, 1014–1018, 2009.

[2] Wiltschko, W., Wiltschko, R., Magnetoreception in birds: two receptors for two different tasks, *Journal of Ornithology, 148*, S61–S76, 2007.

[3] Salikhov, K.M., Molin, Y.N., Sagdeev, R.Z., Buchachenko, A.L., *Spin Polarization and Magnetic Effects in Radical Reactions*, Hungary: Elsevier Science Publishers, 1984.

[4] Ahmad, M., Galland, P., Ritz, T., Wiltschko, R., Wiltschko, W., Magnetic intensity affects cryptochrome-dependent responses in *Arabidopsis thaliana, Planta, 225*, 615–624, 2007.

[5] Yoshii, T., Ahmad, M., Helfrich-Forster, C., Cryptochrome mediates light-dependent magnetosensitivity of Drosophila's circadian clock, *PLoS Biology, 7*(4), 0813–0819, 2009.

[6] Zaporozhan, V., Ponomarenko, A., Mechanisms of geomagnetic field influence on gene expression using influenza as a model systems: basics of physical epidemiology, *International Journal of Environmental Research and Public Health, 7*, 938–965, 2010.

[7] Fitzgerald, T.M., Taylor, P.D., Migratory orientation of juvenile yellow-rumped warblers (*Dendroica coronata*) following stopover: sources of variation and the importance of geographic origins, *Behavioral Ecology and Sociobiology, 62*, 1499–1508, 2009.

[8] Wiltschko, W., Wiltschko, R., Magnetic compass of European robins, *Science, 176*, 62–64, 2009.

[9] Rieper, E., Gauger, E., Morton, J.J.L., Benjamin, S.C., Vedral, V., Quantum coherence and entanglement in the avian compass, *arXiv:0906.3725*, 2009.

[10] Anea, C.B., Zhang, M., Stepp, D.W., Simkins, G.B., Reed, G., Fulton, D.J., et al., Vascular disease in mice with a dysfunctional circadian clock, *Journal of the American Heart Association, 119*, 1510–1517, 2009.

[11] Ritz, T., Thalau, P., Phillips, J.B., Wiltschko, R., Wiltschko, W., Resonance effects indicate a radical-pair mechanism for avian magnetic compass, *Nature, 429*, 177–180, 2004.

[12] Aguzzi, J., Puig, P., Company, J.B., Hydrodynamic, non-photic modulation of biorhythms in the Norway lobster *Nephrops norvegicus* (L.), *Deep-Sea Research I, 56*, 366–373, 2009.

[13] Liu, S., Cai, Y., Sothern, R.B., Guan, Y., Chan, P., Chronobiological analysis of circadian patterns in transcription of seven key clock genes in six peripheral tissues in mice, *Chronobiology International, 24*(5), 793–820, 2007.

[14] Weber, F., Remodeling the clock: coactivators and signal transduction in the circadian clockworks, *Naturwissenschaften, 96*, 321–337, 2009.

[15] Yamato, M., Ishida, N., Iwatani, H., Yamato, M., Rakugi, H., Ito, T., Kid-1 participates in regulating ERK phosphorylation as a part of the circadian clock output in rat kidney, *Journal of Receptors and Signal Transduction, 29*(2), 94–99, 2009.

[16] Maeda, K., Henbest, K.B., Cintolesi, F., Kuprov, I., Rodgers, C.T., Liddell, P.A. et al., Chemical compass model of avian magnetoreception, *Nature, 453*, 387–390, 2009.

[17] Scheper, T.O., Klinkenberg, D., Pennartz, C., Pelt, J.V., A mathematical model for the intracellular circadian rhythm generator, *The Journal of Neuroscience, 19*(1), 40–47, 199.

[18] Elvitigala, T., Stckel, J., Ghosh, B.K., Pakrasi, H.B., Effect of continuous light on diurnal rhythms in *Cyanothece* sp. ATCC 51142, *BMC Genomics, 10*, 226, 2009.

[19] Shimada, H., Numazawa, K., Sasaki, T., Kato, N., Ebisawa, T., Introduction of tau mutation into cultured rat1-R12 cells by gene targeting, using recombinant adeno-associated virus vector, *Cellular and Molecular Neurobiology, 29*, 699–705, 2009.

[20] Werner, H.J., Schulten, Z., Schulten, K., Theory of the magnetic field modulated geminate recombination of radical ion pairs in polar solvents: application to the pyrene-N,N-dimethylaniline system, *The Journal of Chemical Physics, 67*(2), 646–663, 1977.

[21] Schulten, K., Biological effects of static and extremely low frequency magnetic fields, *BGA Schriften, 86*(3), 133–140, 1986.

[22] Miura, T., Maeda, K., Arai, T., The spin mixing process of a radical pair in low magnetic field observed by transient absorption detected nanosecond pulsed magnetic field effect, *The Journal Physical Chemistry A, 110*, 4151–4156, 2006.

[23] Timmel, C.R., Henbest, K.B., *A Study of Spin Chemistry in Weak Magnetic Fields*, The Royal Society, vol. 362, pp. 2573–2589, 2004.

[24] Plenio, M.B., Huelga, S.F., Dephasing-assisted transport: quantum networks and biomolecules, *New Journal of Physics, 10*, 2009.

[25] Engstrom, S., "Magnetic Field Effects on Free Radical Reactions in Biology," Chap. 6 in F.S. Barnes and B. Greenebaum (eds.,) *Bioengineering and Biophysical Aspects of Electromagnetic Fields*, Roca Baton, FL: CRC Press, 2006.

[26] Gould, I.R., Turro, N.J., Zimmt, M.B., Magnetic field and magnetic isotope effects on the products of organic reactions, *Advances in Physical Organic Chemistry*, eds. Gold, V., Bethell, D. London: Academic Press Inc. Ltd, 20, pp. 1–51, 1984.

[27] Wang, K., Ritz, T., Zeeman resonances for radical-pair reactions in weak static magnetic fields, *Molecular Physics, 104*, 1649–1658, 2006.

[28] Ritz, T., Adem, S., Schulten, K., A model for photoreceptor-based magnetoreception in birds, *Biophysical Journal, 78*, 707–718, 2000.

[29] Abeyrathne, C.D., Halgamuge, M.N., Farrell, P.M., Effect of magnetic field on the biological clock through the radical pair mechanism, *International Conference on Bioscience, Biotechnology, and Biochemistry, Rome*, vol. 64, pp. 18–23, 2010.

[30] Solovyov, I.A., Greiner, W., Theoretical analysis of an iron mineral-based magnetoreceptor model in birds, *Biophysical Journal, 93*, 1493–1509, 2007.

[31] Eichwald, C., Walleczek, J., Model for magnetic field effects on radical pair recombination in enzyme kinetics, *Biophysical Journal, 71*, 623–631, 1996.

[32] Eichwald, C., Walleczek, J., Magnetic field perturbations as a tool for controlling enzyme-regulated and oscillatory biochemical reactions, *Biophysical Chemistry, 74*, 209–224, 1999.

[33] Adair, R.K., Effects of very weak magnetic fields on radical pair reformation, *Bioelectromagnetics, 20*, 255–263, 1999.

[34] Zmyslony, M., Rajkowska, E., Mamrot, P., Politanski, P., Jajte, J., The effect of weak 50 Hz magnetic fields on the number of free oxygen radicals in rat lymphocytes *in vitro*, *Bioelectromagnetics, 25*, 607–612, 2004.

[35] Regoli, F., Gorbi, S., Machella, N., Tedesco, S., Benedetti, M., Bocchetti, R. et al., Pro-oxidant effects of extremely low frequency electromagnetic fields in the land snail *Helix aspersa*, *Free Radical Biology & Medicine, 39*, 1620–1628, 2005.

[36] Kholodenko, B.N., Demin, O.V., Westerhoff, H.V., Control analysis of periodic phenomena in biological systems, *The Journal of Physical Chemistry B, 101*, 2070–2081, 1997.

[37] Busse, H., Hess, B., Information transmission in a diffusion-coupled oscillatory chemical system, *Nature, 244*(5413), 203–205, 1973.

[38] MacArthur, J.D., Cell phones and the brain, *The Townsend Letter for Doctors and Patients*, pp. 1–13, 2002.
[39] Feijo, J.A., Sainhas, J., Holdaway-Clarke, T., Cordeiro, M.S., Kunkel, J.G., Hepler, P.K., Cellular oscillations and the regulation of growth: the pollen tube paradigm, *BioEssays, 23*, 86–94, 2001.
[40] Mostafaie, N., llay, E.K., Sauerzapf, E., Bonner, E., Kriwanek, S., Cross, H.S., et al., Correlated down regulation of estrogen receptor beta and the circadian clock gene per1 in human colorectal cancer, *Molecular Carcinogenesis, 48*, 642–647, 2009.
[41] Velissaris, D., Karamouzos, V., Polychronopoulos, P., Karanikolas, M., Chronotypology and melatonin alterations in minimal hepatic encephalopathy, *Journal of Circadian Rhythms, 7*, 6, 2009.

10

Radio frequency electromagnetic fields from mobile phones

Malka N. Halgamuge

The University of Melbourne, Australia

10.1 Introduction

Today, technology is cheaper and more widely available than ever before. Hence, most people are exposed to microwaves, either passively or by choice, most of the time. Electromagnetic radiation in the microwave range is used for mobile phone communication. There are more than 4.3 billion mobile phone users in the world [1] and more than 17.2 million (~86% of the population) in Australia in 2006. Usage is still increasing among all age groups. With the advance of communication technology in the radio frequency (RF) region, studying the biological effects of electromagnetic fields (EMFs) has become a significant area of research. Magnetic fields (MFs) act only on moving ions whereas electric fields (EFs) act on static as well as moving ions.

This research area has generated conflicting results and thus uncertainty regarding possible adverse health effects [2, 3]. The International Committee of Non-Ionizing Radiation Protection (ICNIRP) for limited exposure to low frequency EMFs and microwaves, aims to protect against nerve stimulation and body heating, respectively. About 30 years ago, the question arose whether weak, low frequency EMFs constituted a major health hazard. This question has still not been answered satisfactorily, particularly in the case of long exposure. Many epidemiological studies on residential and occupational exposure to extremely weak low frequency (ELF) MFs show varying, but in general positive, associations with different cancer forms, such as leukemia, brain tumors and breast cancer. A positive association was found especially for leukemia in children with residential exposure from power lines. Risks exceeded about 50% for exposure levels averaging over 0.4 μT [4]. The National Institute of Environmental Health Sciences (NIEHS) of the US classifies EMFs as a "possibly" carcinogenic source (that might transform normal cells into cancer cells). Other investigations focused on: pregnancy outcomes in women working with visual display units, neurologic and psychiatric characteristics (Alzheimer's disease, multiple sclerosis, amyotrophic lateral sclerosis (ALS), suicide and depression) and electrical hypersensitivity, but J.C. Lin reported these results as non-conclusive [5]. A basic problem with all epidemiological studies so far has been the lack of any concept of dose or an exposure metric established from laboratory studies. In order to develop a concept of dose, an accepted interaction model for EMF is urgently needed.

Mobile phones are low powered RF transmitters that can be used to communicate through wireless media (peak powers in the range of 0.1–2 W). Information is transmitted from the mobile phone antenna to the base station antenna through high frequency (between 450 and 2700 MHz) EMFs. Radiation intensity is very high close to the antenna and reduces rapidly with distance. Every time a phone call is made or received the mobile phone user unavoidably absorbs radiation. This radiation is very strong while the phone is transmitting, although the phone emits less radiation when we have good signal strength (i.e., a good connection). Digital phones can radiate 1000 times more power when used in weak signal areas, e.g., rural areas where radio base stations are far away from the user. Statistical results have shown cell phone users in rural areas were 50% more likely to suffer bad health effects than those living in urban areas. The specific absorption rate (SAR) is the amount of electromagnetic radiation absorbed into the body. It varies with the mobile phone brand and model. SAR values for particular mobile phones can be found directly on phone makers' websites, as well as on third-party web sites [6]. The best-known brands on the market have a low SAR, even though low SARs may not be absolutely safe. Mobile phone radiation affects the temporal lobe (behind the ear), and therefore could affect memory function in developing brains. Nonetheless, there are still no conclusive results and therefore uncertainty regarding the possible adverse health effects that prevail. A new, third generation of mobile communication is becoming increasingly important, but the impact of this radiation is largely unknown. A comprehensive epidemiological study of the occurrence and distribution of a disease resulting from this technology would take a long time. This is a common problem where technological advances are very rapid. Epidemiological studies using real people will not be able to answer this question until after 10–15 years of exposure. Therefore, a laboratory study of possibly dangerous biological effects is useful.

10.2 Biological effects

10.2.1 EMF exposure limits

The biological effects of EMFs can vary for different reasons: the position of the biological system with respect to the field; availability of EFs or MFs; amplitude and gradient of the field; frequency spectrum and exposure duration [7]; biological system (plants, animals and chemicals) and also the direction of the EMF within the system. Measurements also vary due to: exposure conditions of the control system; instruments used to measure the fields; background EMFs; coil configurations used

Table 10.1: The Exposure Limit Values.

Frequency Range	Features	Limiting Factor
Up to 10 MHz	Non-thermal-induced body current leads to nerve excitation (by inducing electric currents in the body)	(**J**) Current density (A/m^2)
300 MHz to 10 GHz	Microwaves – thermal – body heating (by setting water dipoles into rotation – high penetration depth)	(SAR) Specific absorption rate (W/kg)
10–300 GHz	Skin burn due to body heating – very low penetration depth	(S) Power density (W/m^2)

to generate MFs and environmental conditions (different seasons, temperature, light, humidity, etc.).

The electromagnetic spectrum that covers the waves which may cause adverse effects in biological tissues is in the range of 3 kHz to 300 GHz [8]. The recommendations of the ICNIRP for exposure limit values for different frequencies are given in Table 10.1. The physical quantities or limiting factors, such as current density (**J**), specific energy absorption rate (SAR) and power density (S), depend on the frequency range of the field. EFs and MFs are measurable quantities external to the body, whereas current density and the SAR are internal quantities. Hence, we absorb a negligible amount of energy, so we have no measurable temperature rise due to exposure to low frequency EFs and MFs. Exposure to above 100 kHz, however, implies a significant absorption of energy and temperature increases.

EF **Es** arise from electric charges (measured in V/m). External EFs induce charges on the body surface. This changing surface charge at the body boundary produces a current density inside the body volume. At low frequencies, the body behaves like a good conductor. In addition, EFs are generally highly filtered from the cell interior. MFs give rise to the motion of electric charges or current. An MF can be identified in two ways: magnetic flux density, **B** (Tesla (T) or Gauss (G)) or MF strength, **H**, Ampere per meter (A/m) where $\mathbf{B} = \mu\mathbf{H}$ and μ is the magnetic permeability constant. According to Ohm's law, the internal EF and current density are linked as $\mathbf{J} = \sigma\mathbf{E}$ where σ is the electrical conductivity of the medium [8]. MFs are strongest close to the device and weaken with distance. Alternating current (AC) induces MFs, and the magnitudes of the induced field and current density are proportional to the loop radius of the body. The strongest electric currents are induced when the loop dimensions are greatest. However, the body is not electrically homogeneous. MFs penetrate biological bodies without significant attenuation. This attenuation effect depends on the frequency and electrical properties of the tissue. In Section 10.4, we investigate this by using a finite-difference time domain (FDTD) simulation tool.

MFs are present only when an appliance is turned on, whereas EFs can exist without any current flow [9]. The acceptable exposure level to ELF MFs is still controversial in terms of MF strength. Practically, MFs are more difficult to shield, whereas EFs can be shielded successfully. They penetrate deeply into the cells and tissues of the exposed person. In MFs, induced fields are the same as external fields, whereas EFs inside biological tissues are diminished by several orders of magnitude from their values in air external to the tissues [9]. The characteristics of EMFs are a function of distance from the radiating antenna. These distances are measured as a function of the wavelength. The near field effect comes from the source being very close to the antenna, where distance between the transmitter and the observed point, R, is small compared to the wavelength, λ, (e.g., $R << \lambda/2\pi$). The far field area is outside of the near field area. The far field effect, or effect of the base station, is where the distance between the transmitter and the observed point, R, is large compared to the wavelength (e.g., $R >> \lambda/2\pi$).

The possible biological effects from electromagnetic radiation from mobile communication systems are considered to depend on the mean power level and the low frequency amplitude modulation of the power. The main features from third-generation systems using CDMA (Code Division Multiple Access) are a new set of frequencies in the power spectrum, such as 1500 Hz in fast power control and 3.84 MHz associated with the chip rate [10]. Still, years after introduction, virtually nothing is known about the health effects of 3G and 4G mobile phones and base stations working at 2100 MHz. In a Dutch study, however, a significant decrease in well-being was reported for subjects exposed to low-level (0.07 mW/kg) 3G UMTS (Universal Mobile Telecommunications System) whereas no decrease was found during exposure to GSM (Global System for Mobile Communications) fields. A number of cognitive tasks were also influenced more by UMTS fields than by GSM 900 and GSM 1800. The reason for the higher sensitivity to UMTS radiation is unknown, but it can be speculated that if the interaction mechanisms present frequency resonances for the frequency of amplitude modulation, the more broadband-like modulation in UMTS might be more likely to include possible resonance frequencies (albeit with low power density) than GSM radiation.

10.2.2 Specific absorption rate

The SAR is the amount of the electromagnetic radiation absorbed per unit mass of the human body in Watts per kilogram (W/kg). The maximum SAR for branded mobile phones is set by the Federal Communication Commission (FCC). In the USA, the FCC has set an SAR limit of 1.6 W/kg, averaged over a volume of 1 g of tissue, for the head. In Europe, when averaged over a volume of 10 g of tissue, the limit is 2 W/kg. SAR values are greatly dependent on the size of the averaging volume. Assessments between different measurements cannot be completed while we lack information on the averaging volume used [6].

Heat is dissipated through the body surface by evaporation or radiated heat transfer; hence, power density in W/m^2 of body surface area may imply the animal's electromagnetic power dissipation [11]. The SAR for a body is obtained by calculating the SAR at each point in the body and averaging over the whole body or the part of the body being considered. Hence, FDTD is used as a numerical method to carry out the SAR calculations [12]. SAR is equal to $\sigma_e E^2/\rho$ where ρ is mass density of the object in kg/m^3. According to ICNIRP guidelines [8], SAR is used to measure field exposure between 100 kHz and 10 GHz. SAR can be noticeable as heat since it provides a measure of internal field that could affect the biological system, hence, SARs are a key parameter in dosimetry. Generally, wetter materials (muscle, blood) lose more than the drier material (bone, nails) and therefore absorb more energy from EMFs [11].

Some cities try to minimize their citizens' exposure as a precaution. For example, a new rule in San Francisco states that every mobile phone company should put up signs stating the SAR of their devices. This is the first city in the world to have radiation warnings about mobile phones.

10.2.3 Blood–brain barrier effect

The brain is divided into lobes, each with an individual function (Fig. 10.1). In the cerebral cortex, (i) the temporal lobe is located around the inner ears and below the lateral sulcus [13] on both cerebral hemispheres of the mammalian brain. The temporal lobe plays a significant portion in the development of long-term memory, and the left temporal lobe expression, unclear naming, comprehension, verbal memory, and other language functions, (ii) The frontal lobe functions in learning,

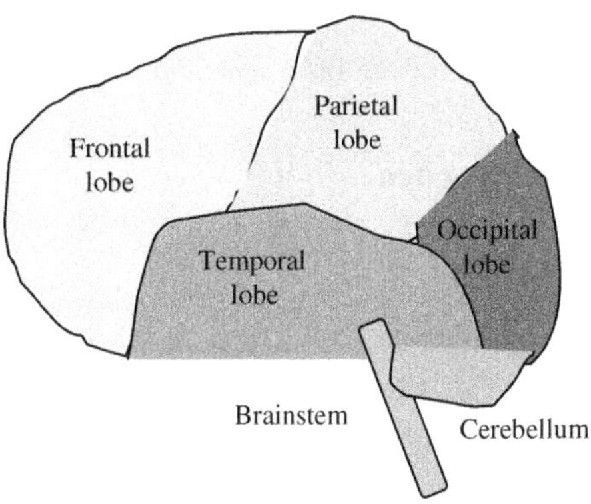

Figure 10.1: Four lobes in the brain: frontal lobe, temporal lobe, parietal lobe and occipital lobe.

behavior, and movement, (iii) the occipital lobe has mainly visual functions and (iv) the parietal lobe receives and correlates sensory information, and serves in problem solving and abstract thinking [14].

The blood–brain barrier (BBB) protects the brain from harmful substances in the blood by separating blood and cerebrospinal fluid (CSF) in the central nervous system (CNS). The BBB is an extremely complex system and is even watertight [15]. However, toxic substances, for example, the albumin (protein in the blood) do leak through the BBB and thus, it is important to quantify and study the toxicological effects of this leakage.

A research group in Sweden has experimentally studied the effects of RF EMFs on the BBB using a series of nearly 2,000 animals. They discovered that albumin leakage through the BBB was considerably higher in exposed rats than in non-exposed animals [15, 16]. An *in vitro* proteomic study [17, 18] supported these findings by showing the effect of low intensity GSM (900 MHz) EMFs on the BBB of the human endothelial cell line. Another group has discovered [19] indications of albumin leakage in rats exposed for 2 h to GSM (900 MHz) by non-thermal SAR values of 0.12, 0.5 and 2.0 W/kg. Breaking the BBB does not indicate the presence of disease; however, it does indicate that something unnatural has happened. In 2007, Salford et al. showed that an SAR of 1 mW/kg gave rise to more pronounced albumin leakage than high SARs [20]. The reasons for these heightened effects of the lowest examined SAR values are not yet discovered.

10.2.4 Extremely low frequency magnetic field from mobile phone battery

The battery current supplying power to a mobile phone follows the amplitude modulation of the RF signal. In the GSM, mobile phones provide current pulses with a repetition rate of 217 Hz and each pulse lasts 0.5 ms. UMTS-3G regulates the signal strength in milliseconds (ms) and battery current varies in the kHz range. The main motivation for this pilot experiment was the leakage of human albumin through the BBB due to the RF from the phone and ELF from the phone battery.

10.3 Experiments in Sweden

10.3.1 Experiment setup

Twenty inbred Fischer 344 rats, sixteen males and four females were taken from the animal house. Each rat's weight was approximately 260–470 g and the rats' ages were 12–26 weeks, equivalent to human teenagers (frequent users of mobile phones). The rats' weight was separated equally between groups. During the experiment rats were exposed to GSM-RF 915 MHz for 2 h at 0.4 mW/kg SAR level. In this pilot experiment, we took effects from RF from the phone and ELF from the phone battery. Therefore, we considered external fan mounted (fan group: RF + ELF) and

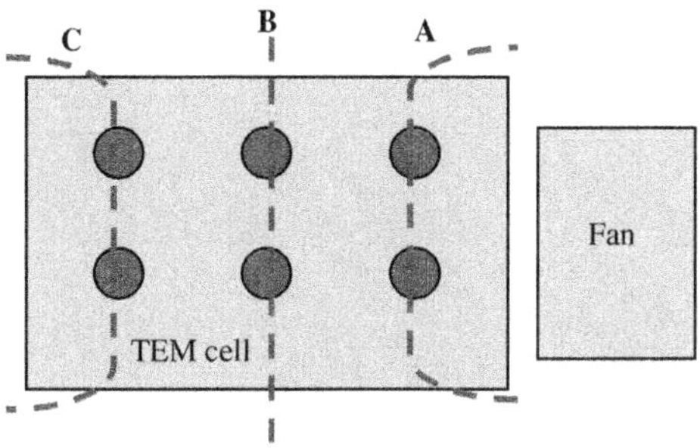

Figure 10.2: Measured field strength inside the TEM cell due to 50 Hz fan exposure: A – closest to the fan is 1.7–1.9 μT, B – middle of compartment is 0.5–0.8 μT, C – Furthest from fan is 0.1–0.3 μT.

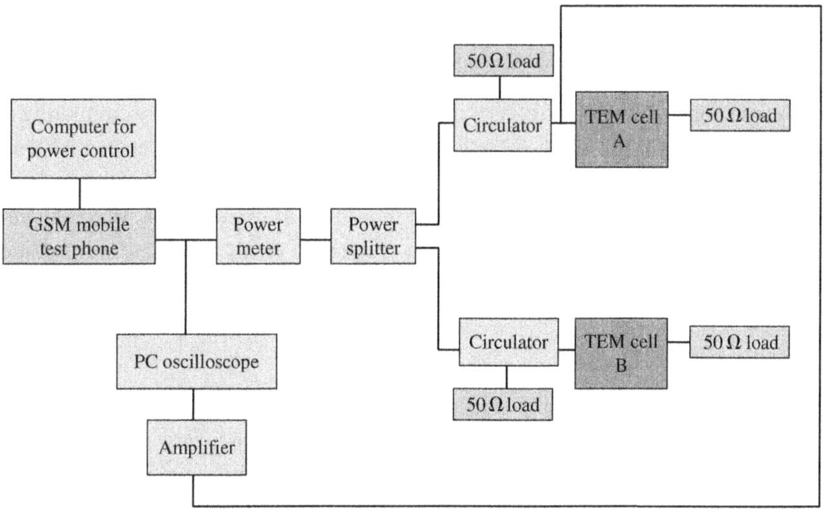

Figure 10.3: Block diagram for RF exposure system. For sham exposures the same TEM cells were used, however in this case, they were not connected to the GSM mobile test phone and consequently no RF EMFs were directed into the TEM cells.

no external fan mounted (no fan group: RF only). Rates were randomly divided between groups: eight males and two females in the external fan mounted group and eight males and two females in the no-external fan mounted group. Two rats were exposed to EMF in an arrangement shown in Fig. 10.2. The block diagram of the RF exposure system is shown in Fig. 10.3. Transverse electromagnetic or TEM cells with rats were placed on a ventilation table during the exposures. A genuine, programmable, GSM phone was used in this experiment whose power output was connected by a coaxial cable to the TEM cell (Fig. 10.3). Voice modulation was not used in this experiment.

Figure 10.4: Slices with 0.5 μm are stained.

The TEM cell, surrounded in a wooden box 15 × 15 × 15 cm, was placed under constant lighting conditions and in a temperature controlled by circulating air through 18 holes in the side wall and top of the wooden box. TEM cells are known to create uniform EMFs for standard measurements. The field strength inside the TEM cell due to 50 Hz fan exposure was measured at three points. Two rats were exposed at a time to the communication frequency fields. The rats were placed in plastic trays with 12 × 12 × 7 cm to avoid association with the central plate and outer conductor. Paper at the bottom of the tray absorbed urine and collected faces.

During the experiment, the orientation of the rats with respect to the propagation direction of GSM radiation was considered. Therefore, rats were kept in the TEM cells and could not move or change orientation. Rats were not anesthetized during the 2-h exposure, after which the rats' weight was measured. Chloral hydrate anesthesia was given to both control and expose rats 1 h after the GSM exposure to avoid stress and blood pressure rising during the perfusion–fixation procedure, the purpose of which is to preserve the tissue for examination. The rats' brains were perfused with saline (buffer) for 3–4 min. It is important to either clean vessels of blood, or fix the blood in vessels, thus allowing no albumin to leak to the brain from blood. Aorta descendants were clamped to stop blood flowing out only from the brain. The blood was removed first and then the brain fixed to avoid blood flow. We also collected other organs, such as skin, liver, spleen, kidney, eyes, testes and bone marrow. Extracted organs were ultra frozen at −85° to −68°C. First, we extracted the head and eyes and placed them in formalin. Brains were fixed in 4% formaldehyde for 5–6 min and immersed in 4% formaldehyde for more than 24 h, after which brains were dehydrated and embedded in paraffin. Brains were cut into 5 μm thick slices across whole coronal sections of the brain 3, 7 and 11 mm from the tip of the frontal pole (Fig. 10.4). Albumin and fibrinogen were verified immunohistochemically.

10.3.2 Experiment results

Albumin leakage and dark neurons were analyzed. The number of albumin spots had increased to some extent in rats exposed to both RF and ELF. Significance was tested using a two-sided Mann–Whitney test and a value of $p=0.06$ was found. The team at Lund University has spent 16 years investigating a different risk with a total series of about 2000 animals. Most experiments during these years were acute 2 h exposures and the SARs varied between 0.3 and 200 mW/kg (real mobile phone radiation) up to 1 W/kg, with an RF generator and several different modulations. The highest frequency of occurrence tended to be for lower SAR values. Dark neurons were not analyzed at that time and modulation was found not to be very important. The long-term effects (55 weeks) were not proven [21]. Here, albumin leakage during the earlier stages of the whole-year exposure period was observed to be comparable to the albumin leakage over a short period. Neurons that would normally not lead to senility in people (mostly in their 60s) may now do so in people in their 30s. The Sweden group has revealed albumin leakage at powers as low as 0.5 mW, a level that exists as far as 1.8 m away from a mobile phone's antenna.

It is shown that radiation from mobile phones can allow harmful proteins, such as albumin and toxins, through the BBB into the brain tissue in rats. A lab-based experiment was made using rats to investigate the biological effects of non-thermal RF EMFs of GSM mobile phones. Rats used in this study were aged between 12 and 26 weeks and their brains are regarded as being in the same stage of development as those of teenagers, with developing brains and thinner and smaller skulls. The main objective was to ascertain the leakage of the rats' albumin through the BBB due to RF from the phone and ELF from the phone battery and neuron death. We found that albumin leakage increased somewhat when rats were exposed to RF from the phone and ELF from the phone battery. This was the pilot experiment; the results of the extended experiment are detailed in [22].

10.4 Finite-difference time domain for SAR calculation

10.4.1 Background

Numerical computer simulations are frequently used to compute the EFs and MFs in the human body. The main benefit of these methods is their capability to work with complex geometries [12]. Owing to its flexibility and computational efficiency, the FDTD method is a/the most commonly used numerical method in bioelectromagnetic dosimetry [12] to calculate SARs and induced currents that engage whole-body or partial-body exposures. These currents can be spatially uniform or non-uniform (far field or near field) to sinusoidally varying EMFs [23]. However, to calculate SAR values in biological material it is important to have a clear understanding of the effect of model parameters, such as permittivity values (dielectric value) and conductivity, both the predicted localized and whole-body SAR values [24].

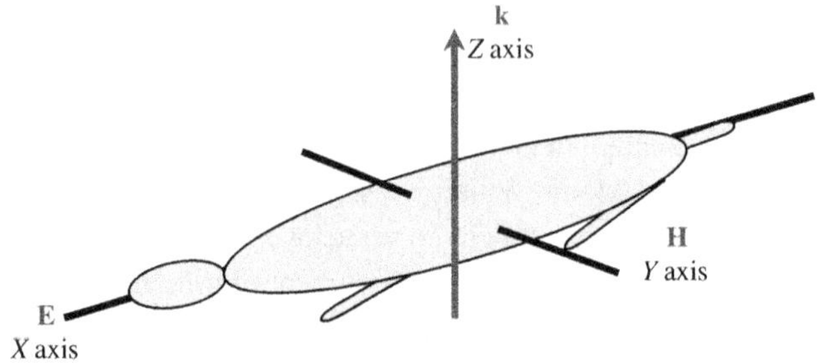

Figure 10.5: +EHK polarization: Direction of incidence of the radiation on the rat with radio frequency wave.

10.4.2 Polarization effect on SAR calculation

The field strength inside the body influences the orientation of the **E** and **H** fields with respect to the body. The polarization of the incident field is the above orientation. The orientation of the object is identified as the incident field vector **E–H** traveling in the direction **K** parallel to the long axis of the body [11], defined by the incident electric field measured in V/m EF (**E**); MF measured in A/m (**H**) and direction of propagation (k) parallel to the long axis of the body. For example, if **E** is parallel to the long axis we call this E polarization. Therefore, all possible exposures from the front and from the side of the tissue are considered. Although, the program suggests that the input power of the radiation can be varied, this was not implemented. The input power was always 1 mW/cm^2. This provides a whole-body SAR in the rat of about 0.3 W/kg, depending on the orientation of the rat in the radiation field. There is a high power absorption at higher frequencies for same field exposure. High frequency signals attenuate more than low frequency signals as they spread through tissues [26].

EHK polarization (E $||\, x$, H $||\, y$, K $||\, z$) shows the direction of incidence of the radiation on the human body or object as in Fig. 10.5. When incident **E**-field is more parallel to the object than perpendicular to it, the SAR is higher. Hence, the average SAR is higher for **E** polarizations. The SAR is higher when the cross-section of the body perpendicular to the incident **H**-field is higher than when the section is smaller. Therefore, the SAR increases when the body is longer and thinner and decreases as it becomes shorter and fatter for **E** polarization [11].

10.4.3 FDTD simulation results

We used the FDTD tool to calculate SAR values. For this calculation, we varied the frequency between 60 and 3000 Hz. In our simulation, we considered all different polarizations (Table 10.2). In the simulation, whole-body total power was measured in W/cm^2 and the whole-body average SAR in W/kg. We considered a rat model with 3,99,588 cells and a total mass of 0.3343 kg. We changed the exposed frequency and the polarization and observed the different SAR values for different organs.

Table 10.2: Maximum (Max), Minimum (Min) and Average (Avg) SAR Values for All Tissues. SAR indicates the amount of electromagnetic energy absorbed by biological tissues. (The SAR values for given frequencies of waves that adversely impact biological tissues are given in bold for noting.)

	GSM Frequency						UMTS Frequency		
	900 MHz			1800 MHz			2100 MHz		
Polarization	Max	Min	Avg	Max	Min	Avg	Max	Min	Avg
EHK	**7.947**	1.419E–03	5.195E–01	1.118	1.441E–03	4.4519E–01	**14.413**	1.374E–03	4.349E–01
EKH	**7.657**	4.359E–04	4.755E–01	1.334	4.640E–04	4.2235E–01	**17.910**	4.191E–04	3.673E–01
HEK	5.452	9.959E–04	2.942E–01	**5.207**	1.757E–03	3.3137E–01	6.077	4.967E–04	3.380E–01
HKE	4.859	7.378E–04	2.819E–01	3.347	1.305E–03	2.2942E–01	4.112	2.658E–04	2.494E–01
KEH	4.662	5.219E–04	3.308E–01	**6.904**	3.019E–04	3.2398E–01	8.802	1.531E–04	3.168E–01
KHE	4.698	4.019E–04	3.521E–01	2.958	1.581E–03	2.2275E–01	3.755	9.326E–04	2.517E–01

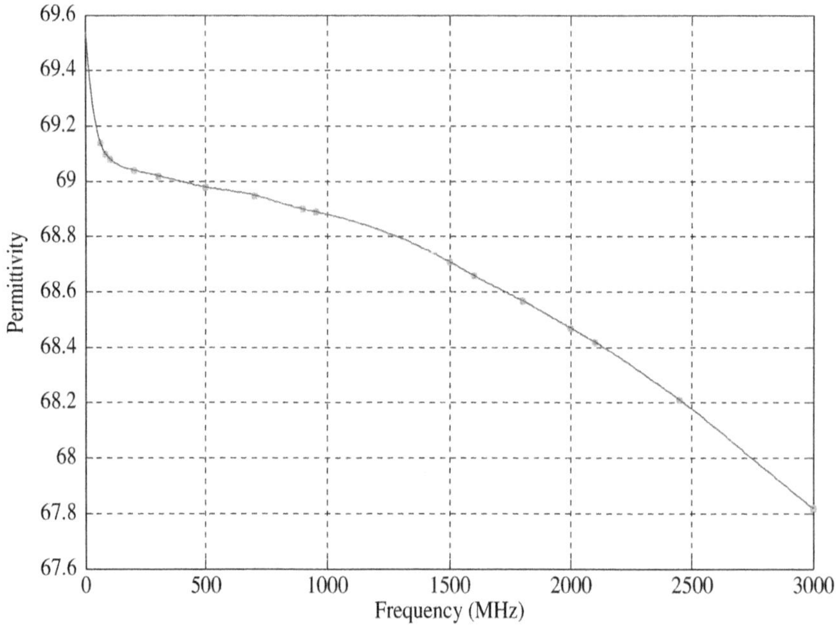

Figure 10.6: Changes in permittivity values in body fluid at different frequencies.

The values in Table 10.2 have been calculated from the FDTD simulation tool. We used both GSM (900 and 1800 MHz) and UMTS (2100 MHz) frequency bands. Here, we observe the changes due to different polarization by maintaining the same frequency. Our results show that the **E** polarization gives higher SAR values than other polarizations when using GSM 900 MHz and UMTS 2100 MHz bands. Hence, the SAR increases when the body is longer and thinner with **E** polarization.

Figure 10.6 shows changes in permittivity values in body fluid at different frequencies and Fig. 10.7 in all organs at different frequencies. Usually, permittivity decreases with frequency. Some body parts, such as nails, blood vessels and body fat, do not seem to be varying much with the frequencies. Other organs, however, such as the heart, liver and white matter, do vary with the frequency, especially with frequencies <100 MHz. Based on Fig. 10.8, internal SAR values are not uniform, hence different organs' SAR values are different. Figure 10.9 confirms this, and also shows high resonance values in all organs around 500 MHz frequencies. It also shows that the lungs and heart tend to contribute more SAR values than the other organs.

Figure 10.10 shows average, minimum and maximum SAR values for all tissues. Here, we observed high resonance values at frequencies around 500 MHz as well as 2450 MHz when we consider maximum SAR values. This resonance diminishes with average and minimum SAR values. However, the lung, intestine and liver showed high SAR values at 500 MHz, as shown in Figs. 10.11–10.13.

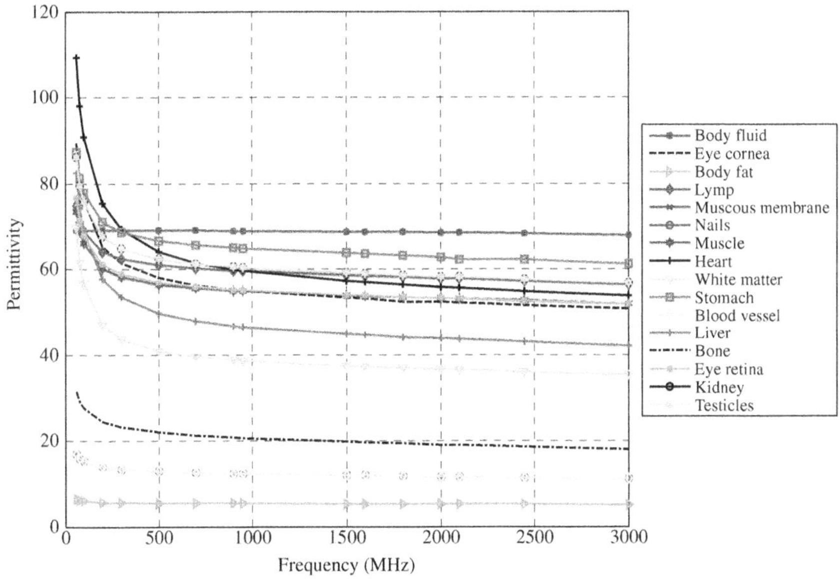

Figure 10.7: Changes in permittivity values in different organs at different frequencies.

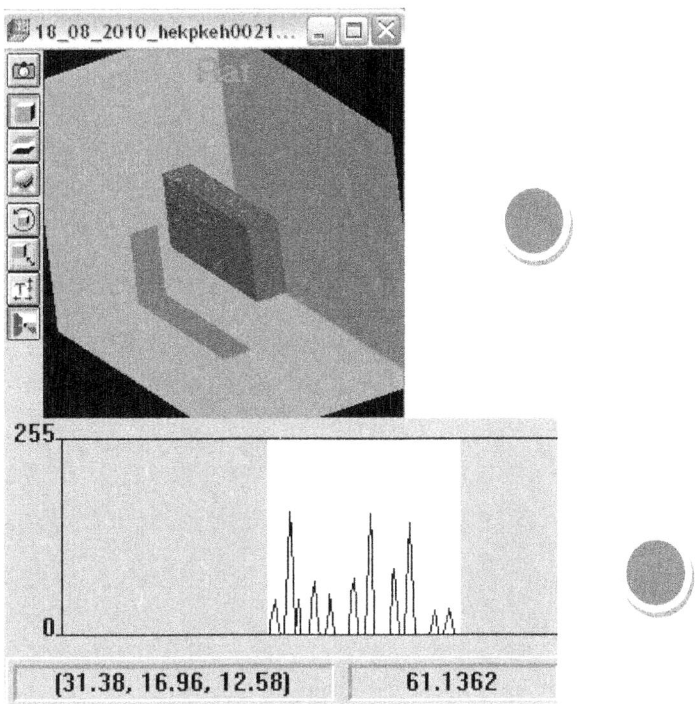

Figure 10.8: Changes in SAR values: Internal SAR values are not uniform; hence, different organs have different values.

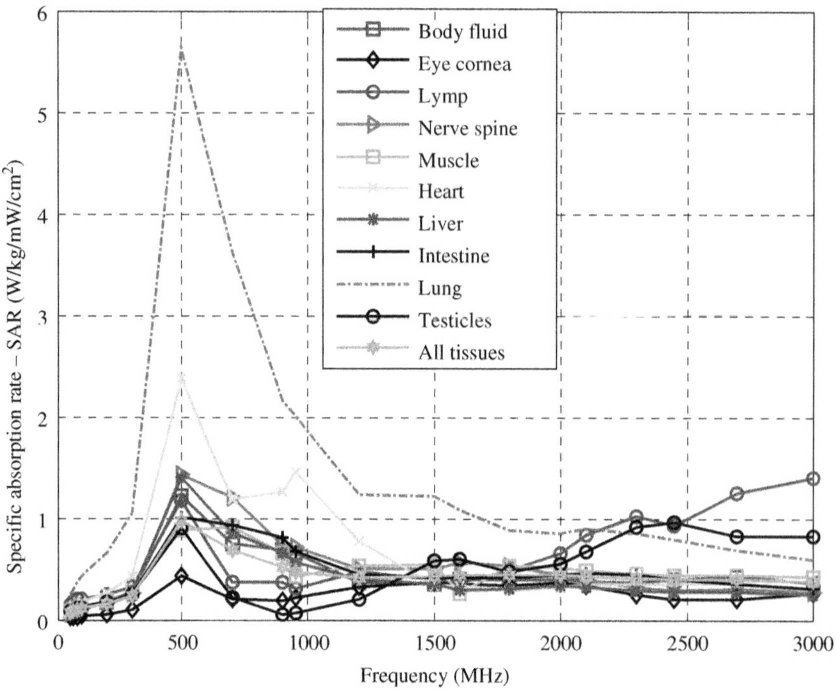

Figure 10.9: Changes in average-specific absorption rate values in different organs at different frequencies.

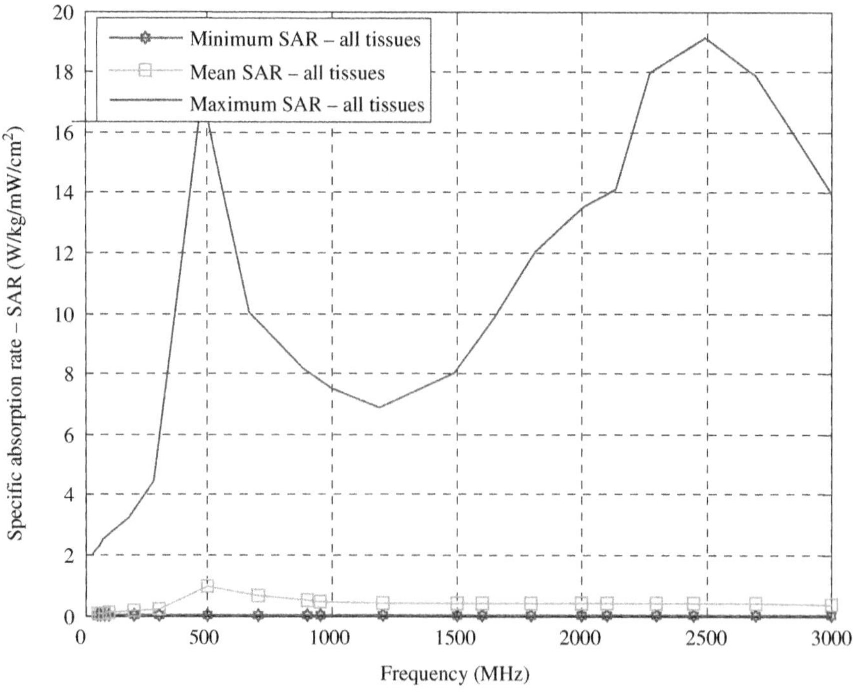

Figure 10.10: Changes in average-, minimum- and maximum-specific absorption rate values in all tissues at different frequencies.

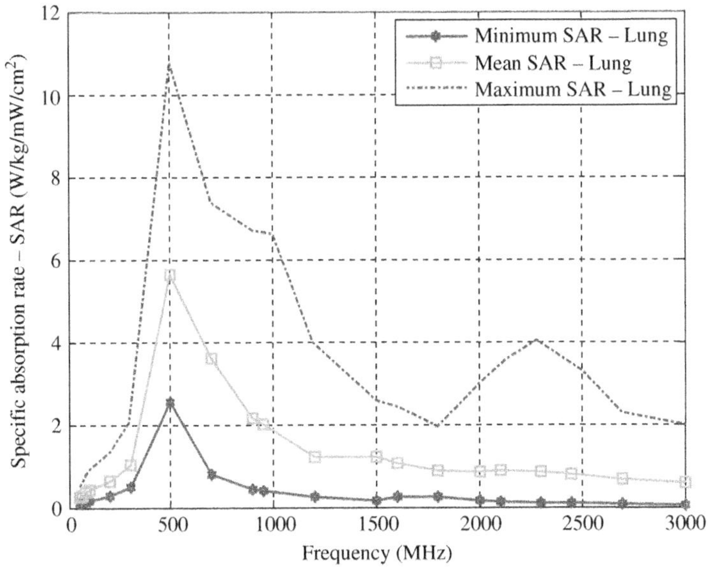

Figure 10.11: Changes in average-, minimum- and maximum-specific absorption rate values in lung at different frequencies.

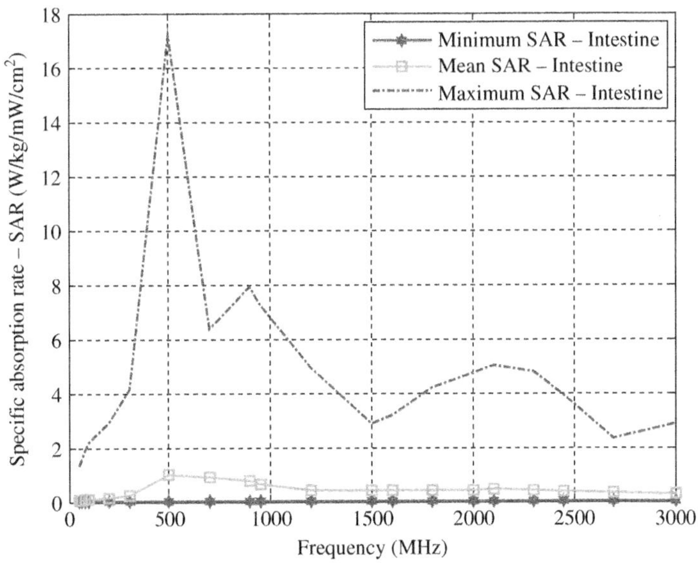

Figure 10.12: Changes in average-, minimum- and maximum-specific absorption rate values in intestine at different frequencies.

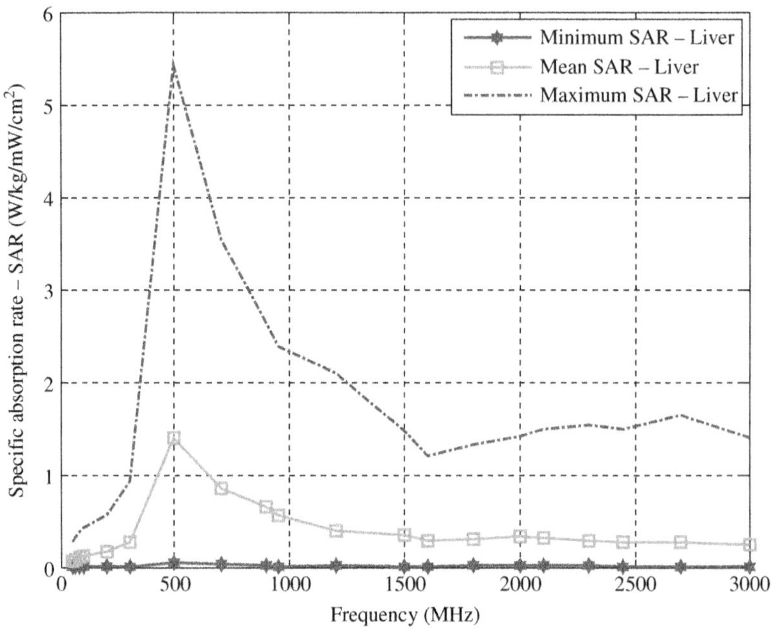

Figure 10.13: Changes in average-, minimum- and maximum-specific absorption rate values in liver at different frequencies.

The average, minimum and maximum SAR values follow the same pattern in the heart (Fig. 10.14), though it is different in muscle (Fig. 10.15). Again, higher SAR values were observed in the 500 MHz range with nerve spins (Fig. 10.16), and around 3000 MHz with lymph nodes (Fig. 10.17). The resonance or peak SAR for the eye's cornea occurs not only at 500 MHz but also at 1500 and 2000 MHz (Fig. 10.18). Finally, we observed high SARs in body fluid at the 500–800 MHz frequency range (see Fig. 10.19). Generally, for E polarization we observed resonance around 500 MHz for the considered rat model.

10.5 Discussion

10.5.1 EMF effects on children

Children's growing skulls are thinner than adults' and less resistant to radiation, and their nervous systems are still developing. Moreover, they will be exposed to more radiation during their lives. A rat's brain has similarities to the human brain, particularly when we compare the BBB and neurons. There is good reason to be alarmed that mobile phones could have the same effect on humans. It cannot be ignored that this may promote the development of neuro degenerative diseases after many years of intensive use. However, experiments conducted by a research

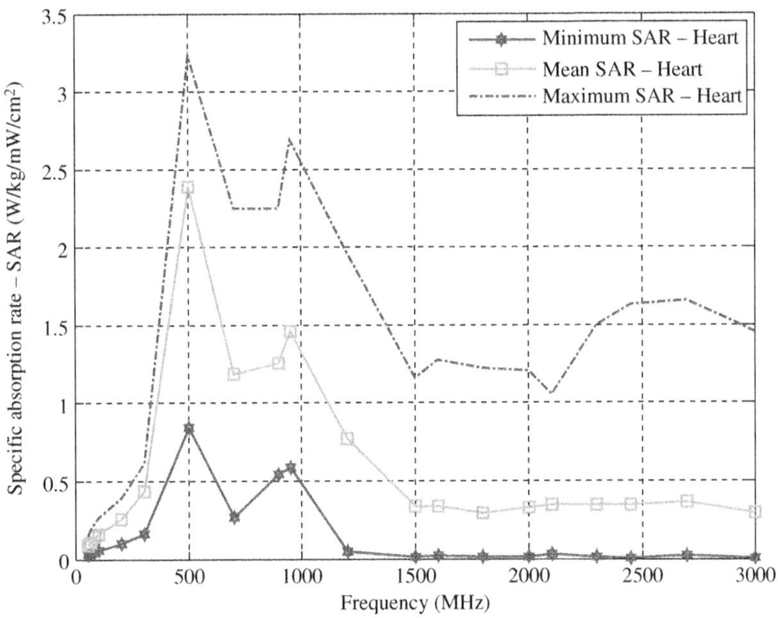

Figure 10.14: Changes in average-, minimum- and maximum-specific absorption rate values in heart at different frequencies.

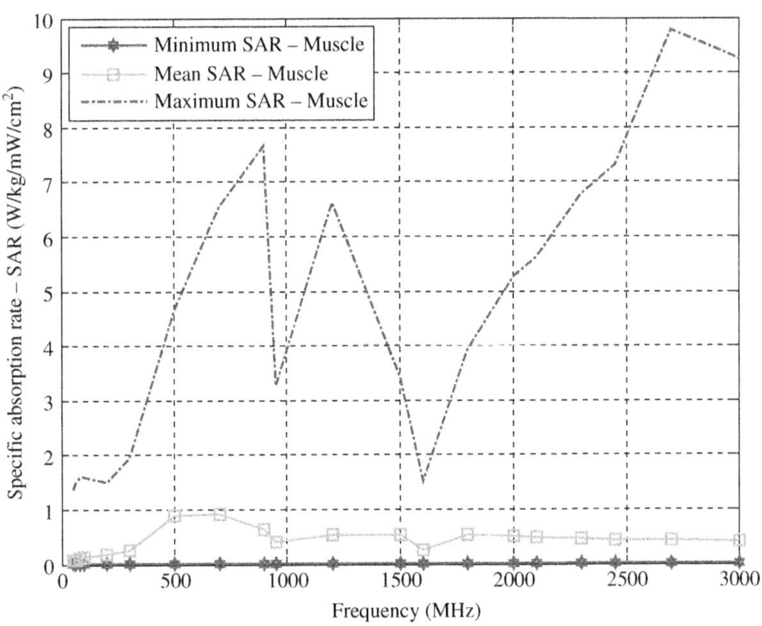

Figure 10.15: Changes in average-, minimum- and maximum-specific absorption rate values in muscle at different frequencies.

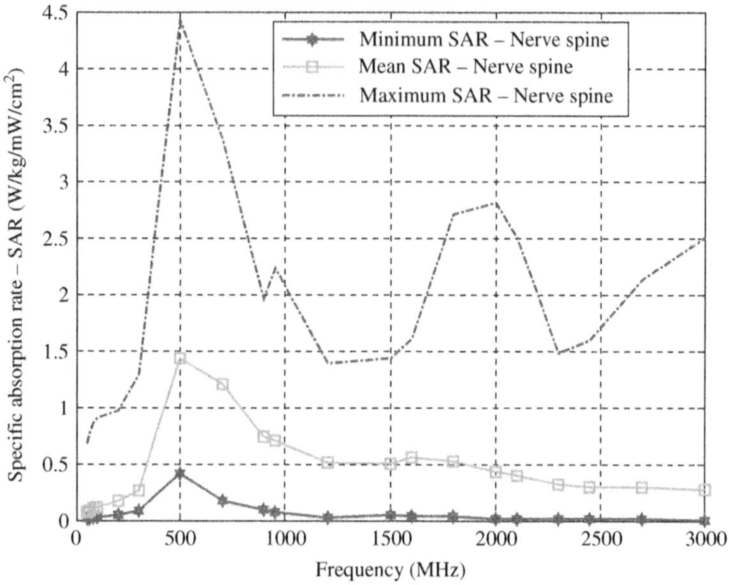

Figure 10.16: Changes in average-, minimum- and maximum-specific absorption rate values in nerve spine at different frequencies.

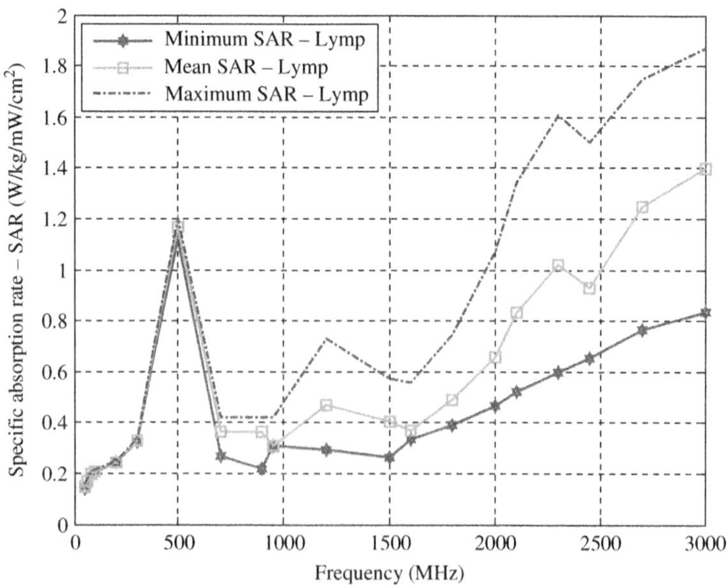

Figure 10.17: Changes in average-, minimum- and maximum-specific absorption rate values in lymph node at different frequencies.

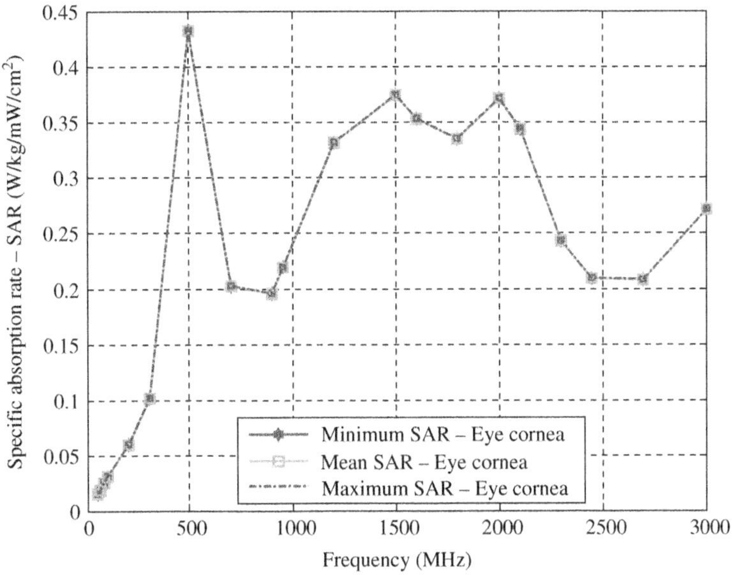

Figure 10.18: Changes in average-, minimum- and maximum-specific absorption rate values in eye cornea at different frequencies.

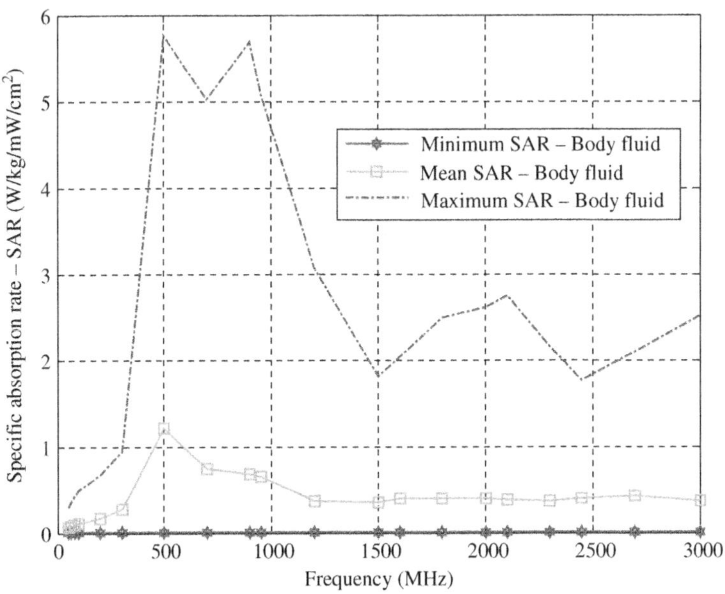

Figure 10.19: Changes in average-, minimum- and maximum-specific absorption rate values in body fluid at different frequencies.

group in Sweden, investigating the long-term effects (55 weeks) provided a different result; observable albumin leakage during the earlier stages of the whole-year exposure period but not considerable leakage after 55 weeks. There is a possibility that neurons would repair themselves as time goes on, so perhaps the effects are reversible. Most of the Swedish group's findings support that non-thermal RF fields affect living organisms. Some other studies agree with them, whereas other studies find no effects.

10.5.2 Possible precautions

The World Health Organization (WHO) report released on May 17, 2010 declares little or no risk of brain tumours for most long-term users of cell phones. Different studies on the long-term health effects of using mobile phones have found conflicting results. In reality, mobiles have saved far more lives than they could probably ever cost. Nonetheless at this stage of research, it is better to take precautions. For example: (i) avoid long conversations on mobile phones and send text messages (SMS) instead. (ii) Children under the age of 16 should be cautious and those under the age of 9 should not use mobile phones. (iii) Use a speaker on your mobile phone if possible. (iv) Use hands free kits or Bluetooth to avoid absorbing radiation in head tissue. (v) Try to avoid using mobile phones in areas where the signal is weak. This means mobile phone companies must work harder to establish their connections. (vi) Avoid using mobile phones in enclosed metal spaces. For example, vehicles and elevators may require the device to work harder in establishing a connection. The metal also acts as a Faraday cage, trapping the radiation and reflecting it back upon the occupants. (vii) It is not advisable to talk on a mobile phone while data is being uploaded or downloaded. (viii) Wait for the call to connect before placing the phone next to the ear. More radiation is present during connection time. (ix) Keep your mobile phone out of your pocket and off your belt when on or in use, because body tissues in the lower body areas absorb radiation more quickly than the head.

According to a new rule in San Francisco, every mobile phone company must place signs stating the SAR of their devices. This is the first city in the world to have radiation warnings about mobile phones.

10.6 Conclusions

The FDTD method was used to determine the permittivity and SAR in different organs at different frequencies from electromagnetic sources with different polarizations. Our results show the orientation of the field plays an essential role in energy absorption from EMFs. For E polarization, we observed a resonance around 500 MHz for the rat model. Another major observation is the higher albumin leakage when rats are exposed to both RF and ELF EMFs. Different studies have found conflicting results about the health effects of using mobile phones in the long term. At this stage of research, it is better to take precautions. Further theoretical and experimental research must be completed in order to evaluate these issues. All these studies should be looked at with the interests of the funding agency or company in mind, so unbiased research in this area is extremely important.

Acknowledgments

I acknowledge Leif Salford without whom I would have never even thought of RF EMFs, and Jacob Eberhardt for invaluable discussion and support during my stay in Sweden and still today. Many people have contributed to the rat experiment in Sweden: Lars Malmgren; Bertil Perrson; Arne Brun; Henrietta Nittby; Mehri Kaviani; and also Susanne Stromblad and Catarina Blennow at the Rausing laboratory with technical assistance. This work was supported by the Solander Fellowship from the Solander Foundation and the University of Melbourne, Australia.

For further reading

[1] GSM Market Data Summary, GSM Association, http://www.gsmworld.com/newsroom/market data/market_data_summary.htm. Retrieved 2010-01-30, 2009.

[2] Halgamuge, M.N., Persson, B.R.R., Salford, L.G., Mendis, P., Eberhardt, J.L., Comparison between two models for interactions between electric and magnetic fields and proteins in cell membranes, *Environmental Engineering Science, 26*(10), 1473–1480, 2009.

[3] Halgamuge, M.N., Abeyrathne, C.D., Mendis, P., Analysis of biological effect of AC–DC electromagnetic fields using the Lorenz model, *Biomedical Engineering and Information Systems: Technologies, Tools and Applications*, *Medical Information Science Reference*, Chapter 2, ISBN: 978-1-61692-004-3, pp. 31–53, 2010.

[4] NRC, Possible health effects of exposure to residential electric and magnetic fields, National Academy Press, Washington, Tech. Rep., 1996.

[5] Lin, J.C., Studies on microwaves in medicine and biology: from snails to humans, *Bioelectromagnetics*, vol. 25, Wiley, 2004, pp. 146–159.

[6] SAR Values, http://www.sarvalues.com/eu-complete.html.

[7] Halgamuge, M.N., Abeyrathne, C.D., Behavior of charged particles in a biological cell exposed to AC–DC electromagnetic fields, *Environmental Engineering Science, 28*(1), 2010.

[8] ICNIRP Guidelines, *Guidelines for Limiting Exposure to Time Varying Electric, Magnetic and Electromagnetic Fields (up to 300 GHz)*, International Commission on Non-ionizing Radiation Protection, 1998.

[9] Habash, R, Biological effects of electromagnetic fields, *Handbook of Engineering Electromagnetic Fields*, Boca Raton, FL: CRC Press, 2004.

[10] Andersen, J.B., Mogensen, P., Pedersen, G.F., *Review of Radio Science*, Wiley-Interscience, pp. 935–941, July 2002.

[11] Durney, C.H., Massoudi, H., Iskander, M.F., Radiofrequency Radiation Dosimetry Handbook, *Technical Report*, Brooks Air Force Base, 1986.

[12] Furse, C., Christensen, D.A., Durney, C.H., *Basic Introduction to Bioelectromagnetics*, 2nd ed., Florida: CRC Press, 2009.

[13] Chi, J.G., Dooling, E.C., Gilles, F.H., Gyral development of the human brain, *Annals of Neurology*, *1*(1), 86–93, 1977.

[14] Bertoti, D., Davis, F.A., Functional Neurorehabilitation Through the Life Span, *Philadelphia*, USA: F.A. Davis Company, 2004.

[15] Salford L.G., Nittby, H., Brun, A., Grafstrom, G., Malmgren, L., Sommarin, M., Eberhardt, J., Widegren, B., Persson, B.R.R., The mammalian brain in the electromagnetic fields designed by man with special reference to blood–brain barrier function, neuronal damage and possible physical mechanisms, *Progress of Theoretical Physics Supplement*, Japan: Physical Society of Japan, 173, pp. 283–309, 2008.

[16] Salford, L.G., Brun, A.E., Eberhardt, J.L., Malmgren, L., Persson, B.R.R., Nerve cell damage in mammalian brain after exposure to microwaves from GSM mobile phones, *Environmental Health Perspectives (USA: National Institute of Environmental Health Sciences) 111*(7), 881–883, 2003.

[17] Leszczynski, D., Joenvaara, S., Reivinen, and J., Kuokka, R., "Non-thermal activation of the hsp27/p38MAPK stress pathway by mobile phone radiation in human endothelial cells: Molecular mechanism. For cancer- and blood-brain barrier-related effects," *Differentiation. 70*(2–3), 120–129, 2002.

[18] Nylund, R. & Leszczynski, D., "Proteomics Analysis of human endothelial cell line EA.hy926 after exposure to GSM 900 radiation," *Proteomics 4*(5), 1359–1365, 2004.

[19] Toře, F., Dulou, P.E., Haro, E., Veyret, B., Aubineau, P., *Proc. of the 5th International congress of the EBEA*, Helsinki, Finland, p. 43, 2001; *Proc. of the 24th Annual Meeting of the BEMS*, p. 61, 2002.

[20] Salford, L.G., Nittby, H., Brue, A., Grafstrom, G., Eberhardt, J.L., Malmgren, L., Persson, B.R.R., Non-thermal effects of EMF upon the mammalian brain: the Lund experience, *Environmentalist, 27*, 493–500, 2007.

[21] Nittby, H., Grafström, G., et al., Cognitive impairment in rats after long-term exposure to GSM-900 mobile phone radiation, *Bioelectromagnetics, 29*, 219–232, 2008a.

[22] Nittby, H., Grafström G., Eberhardt, J.L., et al., Radiofrequency and extremely low-frequency electromagnetic field effects on the blood–brain barrier, *Electromagnetic Biology and Medicine, 27*(2), 103–126, 2008b.

[23] Gandhi, O.P., Gao, B.Q., Chen, J., A frequency-dependent finite difference time-domain formulation for induced current calculations in human beings, *Bioelectromagnetics, 13*(6), 543–555, 1992.

[24] Gajsek, P., Hurt, W.D., Ziriax, J.M., Parametric dependence of SAR on permittivity values in a man model, *IEEE Transactions on Biomedical Engineering, 48*(10), 1169–1177, 2001.

[25] Ida, N., *Engineering Electromagnetics*, 2nd ed, Belgium: Springer-Verlag, New York LLC, 2004.

[26] Bernardi, P., Cavagnaro, M., Pisa, S., Piuzzi, E., Specific absorption rate and temperature elevation in a subject exposed in the far field of radio-frequency sources operating in the 10–900 MHz range, *IEEE Transactions on Biomedical Engineering, 50*, 295–304, 2003.

[27] Durney, C.H., Christensen, D.A., *Basic Introduction to Bioelectromagnetics*, Boca Raton: CRC Press, *Tech. Rep.*, 2000.

[28] Polk, C., Postow, E., CRC Handbook of Biological Effects of Electromagnetic Fields, Boca Raton: CRC Press, *Tech. Rep.*, 1986.

Appendix A Electric field intensity

A.1 The dipole antenna

Consider a cellular mobile network with M base stations having an antenna located on a tower radiating electromagnetic signals to a mobile station (mobile phone). The area enclosed within the boundary is called a cell. We consider the EF distribution for the GSM band frequency spectrum.

Length of the current element Δl carrying a current I_0 generates fields. The wave number, β, is constant for a given wave ($\beta = 2\pi/\lambda$), where λ is the wavelength of the signal. The distance between the transmitter antenna and the position of observation or mobile user is R. The angle θ can be calculated from the tangent value which is the height difference between the transmitter and receiver antennas divided by R (Fig. 10.20).

The EMFs and MFs radiated by an antenna can be illustrated as follows. The MF and EF intensities of the electric dipole as in [25] are given by

$$H_\phi = \frac{I_0 \Delta l \beta^2 \, \mathrm{e}^{-\mathrm{j}\beta R} \sin\theta}{4\pi} \left(\frac{1}{\mathrm{j}\beta R} + \frac{1}{(\mathrm{j}\beta R)^2} \right) \tag{10.1}$$

$$E_R = \frac{\eta I_0 \Delta l \beta^2 \, \mathrm{e}^{-\mathrm{j}\beta R} \cos\theta}{4\pi} \left(\frac{1}{(\mathrm{j}\beta R)^2} + \frac{1}{(\mathrm{j}\beta R)^3} \right) \tag{10.2}$$

$$E_\theta = \frac{\eta I_0 \Delta l \beta^2 \, \mathrm{e}^{-\mathrm{j}\beta R} \sin\theta}{4\pi} \left(\frac{1}{\mathrm{j}\beta R} + \frac{1}{(\mathrm{j}\beta R)^2} + \frac{1}{(\mathrm{j}\beta R)^3} \right) \tag{10.3}$$

Here, the R and θ components of the MF are zero since the MF intensity must be on the plane perpendicular to current. The electric (E-field) and magnetic (B-field)

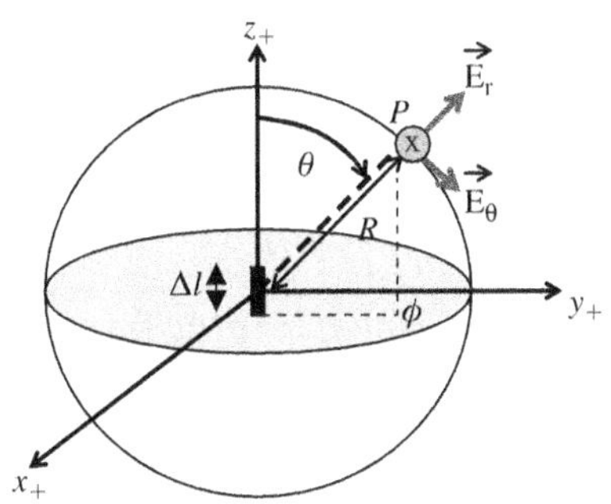

Figure 10.20: Electric field of the dipole antenna in spherical coordinates. Here $0 \le \theta \le \pi$ and $0 \le \phi \le 2\pi$.

components are perpendicular to each other. Therefore, $E_\theta = 0$ since $H_\phi = 0$ and $E_R = 0$. Note that the unit of EF is V/m and MF is A/m.

A.2 Near field effect

A near field is considered as near the source where the distance between transmitter and observed point, R, is smaller than the wavelength [25]. Therefore, this can be written as $R << \lambda/2\pi$, where λ is the wavelength. To be in the near field we must consider R as much smaller than one wavelength from the transmitting antenna. Otherwise, it is considered as the far field. Now, we can get $\beta R << 1$, therefore, the term $e^{-j\beta R}$ tends to 1 [25]. The MF and EF intensities in the near field can be rewritten from (10.1), (10.2) and (10.3) as

$$H_\phi \approx \frac{I_0 \Delta l \sin\theta}{4\pi R^2} \tag{10.4}$$

$$E_R \approx \frac{I_0 \Delta l \beta^2 \cos\theta}{2\pi j\omega\varepsilon R^3} \tag{10.5}$$

$$E_\theta \approx \frac{I_0 \Delta l \sin\theta}{4\pi j\omega\varepsilon R^3} \tag{10.6}$$

where ω is the radian frequency of the signal and ε is the permittivity of free space.

A.3 Far field effect

The far field area is considered as outside of the near field area. In the far field the distance between the transmitter and the observed point, R, is larger than the wavelength [2]. Therefore distance $R >> \lambda/2\pi$. To be in the far field we must be a few wavelengths away from the transmitting antenna, otherwise it is considered as near field. The terms containing $1/R^2$ and $1/R^3$ in (10.1) and (10.3) can be ignored given that they are much smaller in comparison to a term with $1/R$. We can rewrite the magnetic and electric field intensities in the far field from (10.1) and (10.3) as

$$H_\phi = \frac{j\beta I_0 \Delta l\, e^{-j\beta R} \sin\theta}{4\pi R} \tag{10.7}$$

$$E_\theta = \frac{j\beta\eta I_0 \Delta l \beta^2\, e^{-j\beta R} \sin\theta}{4\pi R} \tag{10.8}$$

where η is the wave or insintric impedance. For free space $\eta = E_\theta/H_\phi = 377 \approx 120\pi\ \Omega$. According to this equation, the electric field intensity depends only on frequency and the distance R times the current I_0 and the loop area. This is not a realistic model since there is a difference between calculated and measured electric field strengths. However, this is a very important analytical tool when designing cellular networks. The reason for this difference is the effects of multipath propagation and co-channel cells which cannot be neglected.

A.4 Biological effects

Now we apply the previously defined electric field intensity model to biological tissues. The relationship between internal and external electric fields is given by [27]

$$E_{\text{in}} \cong \frac{\sigma_{\text{air}} + i\omega\varepsilon_0}{\sigma_{\text{br}}(\omega) + i\omega\varepsilon_{\text{br}}(\omega)} E_{\text{ext}} \tag{10.9}$$

where σ_{air} and σ_{br} are the conductively of air and the brain, ε_0 and ε_{br} are the permittivity of air and the brain, and E_{ext} is the external electric field. As in [27] SAR is given by

$$\text{SAR} = \frac{\sigma_{\text{br}}(\omega) + \omega\varepsilon_{\text{br}}(\omega)}{\rho} E_{\text{in}}^2 \tag{10.10}$$

where ρ is the mass density.

The cell is the smallest unit of living organisms also called the building block of life. Human organisms are multi-cellular with approximately 10^{14} cells each with a typical size of 10 μm. Cell membrane voltage changes due to inducing internal field E_{in} can be given by, as in [28]

$$\Delta V(\theta) \cong 1.5 E_{\text{in}} R_{\text{cell}} \cos(\alpha) \tag{10.11}$$

where $R_{\text{cell}} \approx 10\,\mu\text{m}$ and α is the angle between the field and the normal to sphere of the cell body. Exposure to electromagnetic fields is a research area that has generated contradictory results and therefore vagueness about possible adverse biological effects [7].

Nomenclature

Physical constants

Permittivity of free space, $\varepsilon_0 = 8.854 \times 10^{-12}$ F/m
Permeability of free space, $\mu_0 = 4\pi \times 10^{-7}$ H/m
Boltzmann's constant, $k = 1.38 \times 10^{-23}$ J/K
Planck's constant, $h = 6.63 \times 10^{-34}$ J s
Electric charge of an electron, $e = -1.602 \times 10^{-19}$ C
Mass of an electron, $m_e = 9.11 \times 10^{-31}$ kg
Mass of a proton, $m_p = 1.67 \times 10^{-27}$ kg
Speed of electromagnetic waves and light in free space, $c = 3 \times 10^8$ m/s
Impedance of free space, z_0 or $\eta_0 = 376.7\ \Omega$

Conductivity, σ, at room temperature (20°C, 68°F) in S/m

Conductors:

Silver (6.17×10^7)
Copper (5.8×10^7)
Brass (2.56×10^7)
Tungsten (1.83×10^7)
Iron (1.03×10^7)
Mercury (1.0×10^6)
Graphite (3.0×10^4)
Seawater (4.0)
Intrinsic germanium (2.2)
Ferrite (1.0×10^{-2})

Intrinsic semiconductors:

Intrinsic silicon (0.44×10^{-3})

Insulators:

Distilled water (1.0×10^{-4})
Glass (1.0×10^{-12})
Mica (6.17×10^{-15})
Fused quartz (6.17×10^7)

Biomaterials (in S/m):

Skin (0.61 at 230 MHz; 3.72 at 5.8 GHz)
Breast tissue (0.03 at 230 MHz; 0.42 at 5.8 GHz)
Muscle (0.75 at 230 MHz; 4.5 at 5.8 GHz)
Bone (0.08 at 230 MHz; 1.15 at 5.8 GHz)
Nerves (0.4 at 230 MHz; 2.94 at 5.8 GHz)
Heart (0.86 at 230 MHz; 5.86 at 5.8 GHz)
Lung (0.48 at 230 MHz; 3.45 at 5.8 GHz)
Liver (0.58 at 230 MHz; 4.64 at 5.8 GHz)

Relative permittivity, ε_r, at low frequencies

Air, 1.0006
Bakelite, 4.8
Glass, 6.0
Lucite, 3.2
Nylon, 3.6
Plexiglas, 3.45
Polystyrene, 2.5
Quartz, 3.8
Dry soil, 3.0
Teflon, 2.1
Water, 80

Biomaterials:

Skin (53 at 230 MHz; 35.1 at 5.8 GHz)
Breast tissue (5.57 at 230 MHz; 0.42 at 5.8 GHz)
Muscle (0.75 at 230 MHz; 4.5 at 5.8 GHz)
Bone (13.7 at 230 MHz; 9.67 at 5.8 GHz)
Nerves (38.5 at 230 MHz; 27.2 at 5.8 GHz)
Heart (72.6 at 230 MHz; 49 at 5.8 GHz)
Lung (41.8 at 230 MHz; 31.1 at 5.8 GHz)
Liver (55.8 at 230 MHz; 38.1 at 5.8 GHz)

Relative permeability

Diamagnetic materials:

Bismuth, 0.9999834
Silver, 0.99998
Copper, 0.999991
Vacuum, 1.0 (nonmagnetic)

Ferromagnetic materials:

Cobalt, 250
Nickel, 600

Mild steel, 2000
Iron, 5000

Frequency bands: general

Very low frequency (VLF), 3–30 kHz
Low frequency (long waves) (LF), 30–300 kHz
Medium frequency (medium wave), MF 0.3–3 MHz
High frequency (short waves) (HF) 3–30 MHz
Very high frequency (VHF), 30–300 MHz
Ultrahigh frequency (UHF), 0.3–3 GHz
Super high frequency (centimeter wave) (SHF), 3–30 GHz
Extra high frequency (millimeter wave) (EHF), 30–300 GHz

Frequency bands: microwaves

L-band, 1–2 GHz
S-band, 3–4 GHz
C-band, 4–8 GHz
X-band, 8–12 GHz
Ku-band, 12–18 GHz
K-band, 18–26 GHz
Ka-band, 26–40 GHz
V-band, 40–75 GHz
W-band, 75–111 GHz

Basic mathematical relations

$e^{j\theta} = \cos\theta + j\sin\theta$
$e^{-j\theta} = \cos\theta - j\sin\theta$
$e^{j\theta} + e^{-j\theta} = 2\cos\theta$
$e^{j\theta} - e^{-j\theta} = j2\sin\theta$
$\cos(-\theta) = \cos\theta$
$\sin(-\theta) = -\sin\theta$
$\cos(A+B) = \cos A\cos B - \sin A\sin B$
$\cos(A-B) = \cos A\cos B + \sin A\sin B$
$\sin(A+B) = \sin A\cos B + \cos A\sin B$
$\sin(A-B) = \sin A\cos B - \cos A\sin B$
$x\cos A - y\sin B = (x^2+y^2)^{1/2}\cos(A+\tan^{-1}(y/x))$
$x\cos A + y\sin B = (x^2+y^2)^{1/2}\cos(A-\tan^{-1}(y/x))$
$\sin(2A) = 2\sin A\cos B$
$\cos(2A) = \cos^2 A - \sin^2 A = 1 - 2\sin^2 A = 2\cos^2 A - 1$
$2\sin^2 A = 1 - \cos(2A)$
$2\cos^2 A = 1 + \cos(2A)$

$$\mathbf{A}\cdot\mathbf{B} = |\mathbf{A}||\mathbf{B}|\cos\theta$$

$$\mathbf{A}\times\mathbf{B} = |\mathbf{A}||\mathbf{B}|\sin\theta\,\mathbf{u}_n$$

$$\text{grad}\, f(n_1,n_2,n_3) = \nabla f(n_1,n_2,n_3) = \left(\frac{1}{h_1}\frac{\partial f}{\partial n_1}\mathbf{u}_{n1} + \frac{1}{h_2}\frac{\partial f}{\partial n_2}\mathbf{u}_{n2} + \frac{1}{h_3}\frac{\partial f}{\partial n_3}\mathbf{u}_{n3}\right)$$

$$\nabla\cdot\mathbf{F}(x,y,z) = \frac{\partial F_x}{\partial x} + \frac{\partial F_y}{\partial y} + \frac{\partial F_z}{\partial z}$$

$$\nabla\cdot\mathbf{F}(r,\phi,z) = \frac{1}{r}\frac{\partial(rF_r)}{\partial r} + \frac{1}{r}\frac{\partial F_\phi}{\partial \phi} + \frac{\partial F_z}{\partial z}$$

$$\nabla\cdot\mathbf{F}(r,\theta,\phi) = \frac{1}{r^2}\frac{\partial(r^2F_r)}{\partial r} + \frac{1}{r\sin\theta}\frac{\partial(F_\theta\sin\theta)}{\partial\theta} + \frac{1}{r\sin\theta}\frac{\partial F_\phi}{\partial\phi}$$

$$\oint_s \mathbf{F}\cdot \mathrm{d}s = \int_v \nabla\cdot\mathbf{F}\,\mathrm{d}v$$

$$\nabla\times\mathbf{F}(x,y,z) = \left\{\frac{\partial F_z}{\partial y} - \frac{\partial F_y}{\partial z}\right\}\mathbf{u}_x + \left\{\frac{\partial F_x}{\partial z} - \frac{\partial F_z}{\partial x}\right\}\mathbf{u}_y + \left\{\frac{\partial F_y}{\partial x} - \frac{\partial F_x}{\partial y}\right\}\mathbf{u}_z$$

$$\nabla\times\mathbf{F}(r,\phi,z) = \left\{\frac{1}{r}\frac{\partial F_z}{\partial\phi} - \frac{\partial F_\phi}{\partial z}\right\}\mathbf{u}_r + \left\{\frac{\partial F_r}{\partial z} - \frac{\partial F_z}{\partial r}\right\}\mathbf{u}_\phi + \frac{1}{r}\left\{\frac{\partial(rF_\phi)}{\partial r} - \frac{\partial F_r}{\partial\phi}\right\}\mathbf{u}_z$$

$$\nabla\times\mathbf{F}(r,\theta,\phi) = \frac{1}{r\sin\theta}\left\{\frac{\partial(F_\phi\sin\theta)}{\partial\theta} - \frac{\partial F_\theta}{\partial\phi}\right\}\mathbf{u}_r + \frac{1}{r}\left\{\frac{1}{\sin\theta}\frac{\partial F_r}{\partial\phi} - \frac{\partial(rF_\phi)}{\partial r}\right\}\mathbf{u}_\theta + \frac{1}{r}\left\{\frac{\partial(rF_\theta)}{\partial r} - \frac{\partial F_r}{\partial\theta}\right\}\mathbf{u}_\phi$$

$$\oint_c \mathbf{F}\cdot\mathrm{d}l == \int_s (\nabla\times\mathbf{F})\cdot\mathrm{d}s$$

$$\nabla\times\nabla\phi = 0$$

$$\nabla\cdot\nabla\times\mathbf{A} = 0$$

$$\nabla(\phi\psi) = \phi(\nabla\psi) + \psi(\nabla\phi)$$

$$\nabla\cdot(\phi\mathbf{A}) = \mathbf{A}\cdot(\nabla\phi) + \phi(\nabla\cdot\mathbf{A})$$

$$\nabla\times(\phi\mathbf{A}) = (\nabla\phi)\times\mathbf{A} + \phi(\nabla\times\mathbf{A})$$

$$\mathbf{A}\cdot(\mathbf{B}\times\mathbf{C}) = \mathbf{B}\cdot(\mathbf{C}\times\mathbf{A}) = \mathbf{C}\cdot(\mathbf{A}\times\mathbf{B})$$

$$\mathbf{A}\times\mathbf{B}\times\mathbf{C} = \mathbf{B}(\mathbf{A}\cdot\mathbf{C}) - \mathbf{C}(\mathbf{A}\cdot\mathbf{B})$$

Symbols used in text

A, Vector magnetic potential (Webers/m)
$\mathbf{A}_e$, Antenna aperture area (m^2)
B, Magnetic flux density (T)
C, Capacitance (F)
D, Electric flux density (C/m^2)
E, Electric field intensity (V/m)
f, Frequency (Hz)
F, Force (N)

G, Conductance (S)
G, Antenna gain
H, Magnetic field intensity (A/m)
I, Current (A)
J, Current density (A/m^2)
l, Length (m)
$\boldsymbol{L}$, Inductance (H)
m, Magnetic dipole moment (A m^2)
M, Mutual inductance (H)
M, Magnetic polarization vector (A/m)
N, Number of turns
P, Power (W)
Q, q, Electric charge (C)
r, Distance (m)
R, Resistance (Ω)
R, Distance vector (m)
s, Surface area (m^2)
S, Standing wave ratio
t, time (s)
$\boldsymbol{T}$, Temperature (K)
T, Period (s)
T, Torque (N m)
v, Volume (m^3)
v, Velocity (m/s)
V, Potential difference (V)
W, Work, energy (J)
y, Normalized admittance
Y, Admittance (S)
z, Normalized impedance
Z, Impedance (Ω)
Z_0, Characteristic impedance of transmission line (Ω)
Z_0, η_0, Characteristic impedance of free space (Ω)
$\mathbf{u_x}$, $\mathbf{u_y}$, $\mathbf{u_z}$, Unit vectors in Cartesian coordinates
$\mathbf{u_r}$, $\mathbf{u_\phi}$, $\mathbf{u_z}$, Unit vectors in cylindrical coordinates
$\mathbf{u_r}$, $\mathbf{u_\theta}$, $\mathbf{u_\phi}$, Unit vectors in spherical coordinates
α, Angle (rad)
α, Attenuation constant (Neper/m)
β, Angle (rad)
β, Phase constant (rad/m)
γ, Propagation constant (1/m)
δ, Angle (rad)
δ, Skin depth (m)
Γ, Reflection coefficient

θ, Angle (rad)
τ, Transmission coefficient
ε, Permittivity (F/m)
ε_r, Relative permittivity
μ, Permeability (H/m)
μ_r, Relative permeability
ρ_l, Line electric charge density (C/m)
ρ_s, Surface electric charge density (C/m^2)
ρ_v, Volume electric charge density (C/m^3)
σ, Conductivity (S/m)
σ, Radar cross section (m^2)
ϕ, Angle (rad)
ω, Radian frequency (rad/s)
Ω, Beam solid angle (sr)

List of prefixes (multipliers)

tetra (10^{12})
giga (10^{9})
mega (10^{6})
kilo (10^{3})
hecto (10^{2})
deka (10)
deci (10^{-1})
centi (10^{-2})
milli (10^{-3})
micro (10^{-6})
nano (10^{-9})
pico (10^{-12})
femto (10^{-15})
atto (10^{-18})

CPSIA information can be obtained at www.ICGtesting.com
Printed in the USA
BVOW06*1336270715
410058BV00003B/3/P

9 781845 647988